Hadron Physics

Cover image

The cover image shows the manner in which QCD vacuum fluctuations are expelled from the interior region of a baryon, like the proton. The positions of the three quarks composing the proton are illustrated by the coloured spheres. The surface plot illustrates the reduction of the vacuum action density in a plane passing through the centres of the quarks. The vector field illustrates the gradient of this reduction. The positions where the vacuum action is maximally expelled from the interior of the proton are illustrated, exposing the presence of flux tubes. The transition to flux-tube formation occurs when the distance of the quarks from the centre of the triangle is greater than 0.5 fm. The diameter of the flux tubes remains approximately constant as the quarks move to large separations. As it costs energy to expel the vacuum field fluctuations, a linear confinement potential is felt between quarks in baryons, as well as mesons.

Cover image courtesy of Derek B Leinweber, CSSM, University of Adelaide, Australia (http://www.physics.adelaide.edu.au/theory/staff/leinweber/VisualQCD/Nobel/).

Scottish Graduate Series

Hadron Physics

I J D MacGregor
University of Glasgow, Scotland, UK

R Kaiser
University of Glasgow, Scotland, UK

CRC Press
Taylor & Francis Group
Boca Raton London New York

CRC Press is an imprint of the
Taylor & Francis Group, an **informa** business

A CHAPMAN & HALL BOOK

Copublished by Scottish Universities Summer School in Physics.

CRC Press
Taylor & Francis Group
6000 Broken Sound Parkway NW, Suite 300
Boca Raton, FL 33487-2742

First issued in paperback 2019

© 2006 by Taylor & Francis Group, LLC
CRC Press is an imprint of Taylor & Francis Group, an Informa business

No claim to original U.S. Government works

ISBN-13: 978-1-58488-705-8 (hbk)
ISBN-13: 978-0-367-39062-4 (pbk)

Library of Congress Cataloging-in-Publication Data

Catalog record is available from the Library of Congress

**Visit the Taylor & Francis Web site at
http://www.taylorandfrancis.com**

**and the CRC Press Web site at
http://www.crcpress.com**

SUSSP proceedings

/continued

SUSSP proceedings (continued)

Lecturers

Gunnar Bali	University of Glasgow, UK
Stanislav Belostotski	St Petersburg Nuclear Physics Institute, Russia
Stanley J Brodsky	Stanford Linear Accelerator Center, USA
Matthias Burkardt	New Mexico State University, USA
Volker D Burkert	Jefferson Laboratory, USA
Frank E Close	University of Oxford, UK
Christine T H Davies	University of Glasgow, UK
Kees de Jager	Jefferson Laboratory, USA
Nicole d'Hose	DAPNIA/SPhN, France
Meinulf Göckeler	Universität Regensburg, Germany
Roger Horsley	University of Edinburgh, UK
Paul Hoyer	University of Helsinki, Finland
Boris L Ioffe	ITEP, Moscow, Russia
Volker Metag	Universität Giessen, Germany
Bernard Metsch	Universität Bonn, Germany
Wolf-Dieter Nowak	DESY, Germany
Angels Ramos	Universitat de Barcelona, Spain
Günther Rosner	University of Glasgow, UK
Dirk Ryckbosch	Ghent University, Belgium
Björn Seitz	Universität Giessen, Germany
Wolfram Weise	Technische Universität München, Germany

Organising committee

Professor G Rosner	University of Glasgow	*Co-Director*
Professor S Belostotski	St Petersburg NPI	*Co-Director*
Professor D Branford	University of Edinburgh	*Secretary*
Dr D G Ireland	University of Glasgow	*Treasurer*
Dr I J D MacGregor	University of Glasgow	*Co-Editor*
Dr R Kaiser	University of Glasgow	*Co-Editor*
Dr J D Kellie	University of Glasgow	*Steward*
Dr D Protopopescu	University of Glasgow	*Steward*
Professor C T H Davies	University of Glasgow	*Steward*
Ms C Wilson	University of Glasgow	*Administration*

International advisory committee

Professor S J Brodsky	SLAC, USA
Dr E de Sanctis	INFN Frascati, Italy
Professor K Goeke	Ruhr-Universität Bochum, Germany
Professor P Hoyer	University of Helsinki, Finland
Dr R Kaiser	University of Glasgow
Dr J-M Laget	CEA Saclay, France & Jefferson Lab, USA
Professor A Martin	University of Durham, UK
Professor U Meissner	Universität Bonn, Germany
Professor P Mulders	Vrije Universiteit Amsterdam, The Netherlands
Professor V Metag	Universität Giessen, Germany
Professor G Rosner	University of Glasgow
Professor D Ryckbosch	Ghent University, Belgium
Professor G Schierholz	DESY- Zeuthen, Germany
Professor W Weise	Technische Universität München, Germany

Directors' preface

Hadron Physics, as the physics of the strong interaction whose fundamental theory is QCD, is coming into its own as a scientific subject. There are exciting new results in hadron spectroscopy and new perspectives in hadron structure. Taken together with the increasing precision of lattice QCD calculations and plans for new experimental facilities these developments paint a picture of a vibrant and growing sub-field of physics. At the time of the Hadron Physics summer school we did not yet know that the 2004 Nobel Prize would be awarded to Gross, Politzer and Wilczek for their contributions to QCD. But this fact nicely underlines our optimism and enthusiasm for Hadron Physics.

This graduate text contains the invited Hadron Physics lectures presented at the 58th Scottish Universities Summer School in Physics (SUSSP58) held in St Andrews from 22nd – 28th August 2004, augmented by selected contributions to the associated topical workshop held from 29th August – 1st September 2004. The concept of the school was to give students the opportunity to learn both theoretical and experimental approaches in Hadron Physics from some of the leading experts in the world. Fifty-seven students attended 28 lectures over 7 days. The students also had the opportunity to present their own research in short talks or through posters. The following 3-day workshop was organised by 3 projects/networks within the EU FP 6 Integrated Infrastructure Initiative (I3) HadronPhysics and included 14 invited and 24 contributed talks.

We believe that this collection of lectures from the school and invited talks from the workshop provides a comprehensive overview of Hadron Physics and should prove useful for both graduate students and senior scientists.

Summer schools are remembered not only for the lectures, but also for the dinners, excursions, discussions and for the people that one got to know. There was plenty of all of this at this summer school: Outings to the Fife fishing ports of Crail and Anstruther and to Falkland palace, a traditional Burns supper and more informal activities like bonfires and swimming on the West Beach. The Burns supper included a healthy dose of Scottish culture provided by Douglas MacGregor's recital of Robert Burns' legendary "Address to the Haggis". The workshop programme included a free Whisky tasting and quiz on one of the evenings.

We would like to thank the friendly and helpful staff of John Burnett Hall who provided an ideal environment for both school and workshop. The support from the other staff of St. Andrews, conference organisers, janitors and technicians was also exceptional.

The summer school was supported by NATO as an Advanced Study Institute, the UK Engineering and Physical Sciences Research Council (EPSRC), the UK Particle Physics and Astronomy Research Council (PPARC), the Institute of Physics (Nuclear and Particle Physics Division and Nuclear Physics Group), the Royal Society of Edinburgh, the US National Science Foundation, and the EU FP6 I3 HadronPhysics. The associated workshop was supported by the EU FP6 I3 HadronPhysics, directly and via the Networks "Computational Lattice Hadron Physics", "Structure and Dynamics of Hadrons" and the Joint Research Activity "Generalised Parton Distributions".

We would like to thank all lecturers and participants for their contributions and for their enthusiasm for hadron physics, which made this a memorable school and workshop.

Günther Rosner and Stanislav Belostotski

Co-Directors, July 2005

Editors' note

Hadron Physics is an extremely active research area straddling the boundary between the traditional disciplines of nuclear and particle physics. Scientifically it is of vital importance in extrapolating our knowledge of quark-gluon physics at the sub-nucleon level to provide a wider perspective of strongly interacting hadrons which make up the vast bulk of known matter in the Universe. The vitality of the field is evident in the very large numbers of young physicists who participated.

The summer school and associated workshop have given us a unique opportunity to produce a graduate text which maps out our contemporary knowledge of this topic. The book contains the lectures given at SUSSP58 together with selected invited talks presented at the topical workshop. The summer school lectures are detailed and pedagogical in their approach and are intended to provide a sound and thorough foundation for research students and young postdoctoral researchers working in Hadron Physics. There is a balance between theoretical and experimental physics which reflects the symbiotic relationship that exists between both approaches. The associated workshop contributions give an up-to-date account of on-going research activity, which nicely complements the more general summer school lectures.

To provide a fuller picture of activity in the subject we also include short summaries of selected student talks given at the summer school, and talks in three specific areas of current research highlighted in the associated workshop: Computational lattice hadron physics; The structure and dynamics of hadrons; and Generalised parton distributions. The standard of these presentations was excellent. Unfortunately there is not sufficient space to provide summaries from twenty-seven other students who presented their work in posters at the summer school.

It is the SUSSP editorial policy that only the presenters of each talk can be authors of the written versions of the presentations. In cases where the talks describe work carried out in the context of a larger group or collaboration, the contribution of this larger group is recognised via the acknowledgments section at the end of each paper. English spellings are used throughout.

We conclude this note by wishing you a useful and enjoyable read.

I J Douglas MacGregor and Ralf Kaiser

Co-Editors, July 2005

Contents

Lectures and invited talks

Summer school student talks

The structure and dynamics of hadrons

Computational lattice hadron physics

Generalised parton distributions

Nucleon electromagnetic form factors

Kees de Jager

Jefferson Lab, Newport News,
VA 23606, USA

1 Introduction

Although nucleons account for nearly all the visible mass in the universe, they have a complicated structure that is still incompletely understood. The first indication that nucleons have an internal structure, was the measurement of the proton magnetic moment (Frisch and Stern 1933, Estermann and Stern 1933) which revealed a large deviation from the value expected for a point-like Dirac particle. The investigation of the spatial structure of the nucleon, resulting in the first quantitative measurement of the proton charge radius, was initiated by the HEPL (Stanford) experiments in the 1950s, for which Hofstadter was awarded the 1961 Nobel prise. The first indication of a non-zero neutron charge distribution was obtained by scattering thermal neutrons off atomic electrons. The recent revival of its experimental study through the operational implementation of novel instrumentation has instigated a strong theoretical interest.

Nucleon electro-magnetic form factors (EMFFs) are optimally studied through the exchange of a virtual photon, in elastic electron-nucleon scattering. The momentum transferred to the nucleon by the virtual photon can be selected to probe different scales of the nucleon, from integral properties such as the charge radius to scaling properties of its internal constituents. Polarisation instrumentation, polarised beams and targets, and the measurement of the polarisation of the recoiling nucleon have been essential in the accurate separation of the charge and magnetic form factors and in studies of the elusive neutron charge form factor.

2   Theory of electron scattering and form factor measurements

The nucleon EMFFs are of fundamental importance for the understanding of the nucleon's internal structure. Under Lorentz invariance, spatial symmetries and charge conservation, the most general form of the electromagnetic current inside a nucleon can be written as:

$$J_{EM}^\mu = F_1(Q^2)\gamma^\mu + \frac{\kappa}{2M_N}F_2(Q^2)i\sigma^{\mu\nu}q_\nu , \tag{1}$$

where F_1 denotes the helicity non-flip Dirac form factor, F_2 the helicity flip Pauli form factor, $Q^2 = -q^2$, and κ the nucleon anomalous magnetic moment. The remaining variables are defined in Figure 1. The second term, usually referred to as the Foldy contribution, carries the information about the nucleon anomalous magnetic moment and thus is of relativistic origin. It is useful to introduce the isospin form-factor components, corresponding to the isoscalar (s) and isovector (v) response of the nucleon,

$$F_i^s = \frac{1}{2}(F_i^p + F_i^n); \quad F_i^v = \frac{1}{2}(F_i^p - F_i^n); \quad (i = 1, 2) . \tag{2}$$

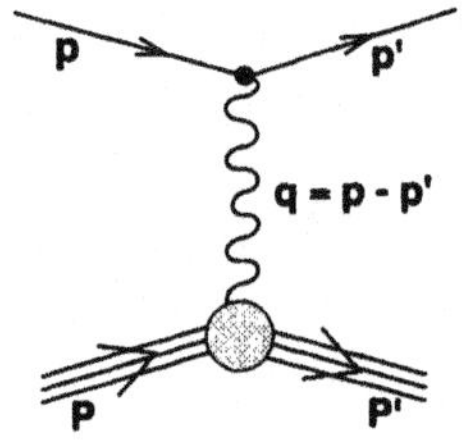

Figure 1. *The Feynman diagram for the scattering of an electron with four-momentum $p = (E_e, \vec{p})$ through an angle θ_e off a nucleon with mass M_N and four-momentum P. In this diagram a single virtual photon with four-momentum $q = p - p' = (\omega, \vec{q})$ is exchanged. The four-momenta of the scattered electron and nucleon are $p' = (E'_e, \vec{p'})$ and P', respectively.*

The form factors can be continued analytically into the complex plane and can be related in different regions through a dispersion relation of the form

$$F(t) = \frac{1}{\pi}\int_{t_0}^\infty \frac{\text{Im}\,F(t')}{t' - t}dt' , \tag{3}$$

with $t = -Q^2$, $t_0 = 9(4)M_\pi^2$ for the isoscalar (isovector) case and M_π the pion mass. In the isovector case the minimum t-value is determined by the threshold for the $e^+e^- \to \pi^+\pi^-$ reaction, in the isovector case the lightest hadronic state involves three pions. In the positive Q^2-region, called spacelike, form factors can be measured through electron scattering, in the negative Q^2-region, called timelike, form factors can only be measured through the creation or annihilation of a $N\bar{N}$-pair.

In the plane wave born approximation the cross section for elastic electron-nucleon scattering can be expressed in the Rosenbluth (1950) formula as:

$$\frac{d\sigma}{d\Omega} = \sigma_M[(F_1^2 + \kappa^2\tau F_2^2) + 2\tau(F_1 + \kappa F_2)^2 \tan^2(\frac{\theta_e}{2})] , \tag{4}$$

where $\tau = Q^2/(4M_N^2)$ and $\sigma_M = (\frac{\alpha_{QED}\cos\theta_e/2}{2E_e\sin^2\theta_e/2})^2\frac{E_e'}{E_e}$ is the Mott cross section for scattering off a point-like particle, with α_{QED} denoting the fine-structure constant. Again, the remaining variables are defined in Figure 1. F_1 and F_2 are now clearly identified as the Dirac and Pauli form factors. Hofstadter determined the values of F_1 and F_2 by measuring the cross section at different scattering angles, but the same value of Q^2 and drawing intersecting ellipses. Hand, Miller and Wilson (1963) expressed Equation 4 in an alternate form

$$\frac{d\sigma}{d\Omega} = \sigma_M[\frac{(G_E^p)^2 + \tau(G_M^p)^2}{1+\tau} + 2\tau(G_M^p)^2\tan^2(\frac{\theta_e}{2})]$$
$$= \frac{\sigma_M}{\epsilon}[\tau(G_M^p)^2 + \epsilon(G_E^p)^2](\frac{1}{1+\tau}) , \qquad (5)$$

with $\epsilon = 1/[1 + 2(1+\tau)\tan^2(\frac{\theta_e}{2})]$ the linear polarisation of the virtual photon and

$$G_E(Q^2) = F_1(Q^2) - \tau\kappa F_2(Q^2) ; \qquad G_E^p(0) = 1 ; \qquad G_E^n(0) = 0 ;$$
$$G_M(Q^2) = F_1(Q^2) + \kappa F_2(Q^2) ; \qquad G_M^{p,n}(0) = \mu_{p,n} , \qquad (6)$$

with $\mu_{p,n}$ denoting the magnetic moment of the proton and neutron, respectively. This equation illustrates that the electric and magnetic Sachs form factors G_E^p and G_M^p can be separated in a straightforward way by performing cross-section measurements at fixed Q^2 as a function of ϵ, over a range of (θ_e, E_e) combinations. This technique has become somewhat erroneously known as the Rosenbluth separation technique.

In the Breit frame, which for elastic scattering is equivalent to the electron-nucleon centre-of-mass frame, the Sachs form factors can be identified with the Fourier transform of the nucleon charge and magnetisation density distributions. In this frame the incoming electron has momentum $\vec{p} = +\vec{q}/2$ and hits a nucleon which has equal but opposite momentum $\vec{P} = -\vec{q}/2$. The exchanged photon carries momentum $\vec{q}$ but no energy. In the Breit frame the electromagnetic current of the proton simplifies into the following expression

$$J_{EM}^\mu = e\{G_E + (\vec{\sigma} \times \vec{q})G_M\} . \qquad (7)$$

Through the mid-1990s practically all available proton EMFF data had been collected using the Rosenbluth separation technique. This experimental procedure requires an accurate knowledge of the electron energy and the total luminosity. In addition, because the G_M^p contribution to the elastic cross section is weighted with Q^2, data on G_E^p suffer from increasing systematic uncertainties with increasing Q^2-values. The then available world data set (Bosted *et al.* 1995) was compared to the so-called dipole parameterisation G_D, which corresponds to two poles with opposite sign close to each other in the time-like region. In coordinate space G_D corresponds to exponentially decreasing radial charge and magnetisation densities, albeit with a non-physical discontinuity at the origin:

$$G_D = \left(\frac{\Lambda^2}{\Lambda^2 + Q^2}\right)^2 \qquad \text{with } \Lambda = 0.84 \text{ GeV and } Q \text{ in GeV.} \qquad (8)$$

For G_E^p, G_M^p/μ_p and G_M^n/μ_n the available data agreed to within 20% with the dipole parameterisation. Both the G_E^p and the G_M^p/μ_p data could be fitted adequately with an identical parameterisation. However, the limitation of the Rosenbluth separation was

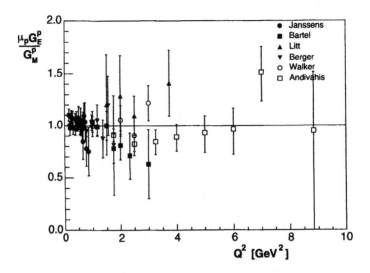

Figure 2. *The ratio $\mu_p G^p_E / G^p_M$ from Rosenbluth separation. Data are from Janssens et al. (1966), Bartel et al. (1966), Litt et al. (1970), Berger et al. (1971), Walker et al. (1994) and Andivahis et al. (1994). The errors shown in all figures are the quadratic sum of the statistical and systematic contributions.*

evident from the fact that different data sets for $\mu_p G^p_E / G^p_M$ scattered by up to 50% at Q^2-values larger than 1 GeV^2 (Figure 2). Although no fundamental reason has been found for the success of the dipole parameterisation, it is still used as a base line for comparison of data because it removes the largest variation with Q^2 and enables small differences to be seen.

3   Instrumentation for form factor measurements

More than 40 years ago Akhiezer *et al.* (1958) (followed ∼20 years later by Arnold *et al.* (1981)) showed that the accuracy of nucleon charge form-factor measurements could be increased significantly by scattering polarised electrons off a polarised target (or equivalently by measuring the polarisation of the recoiling proton). However, it took several decades before technology had sufficiently advanced to make the first of such measurements feasible and only in the past few years have a large number of new data with significantly improved accuracy become available. The next few sections introduce the various techniques. The figure of merit for different polarisation techniques is defined as the product of the luminosity, the square of the degree of polarisation, or analysing power, and the efficiency. For G^p_E measurements the highest figure of merit at Q^2-values larger than a few GeV^2 is obtained with a focal plane polarimeter. Here, the Jacobian focusing of the recoiling proton kinematics allows one to couple a standard magnetic spectrometer for the proton detection to a large-acceptance non-magnetic detector for the detection of the scattered electron. For studies of G^n_E one needs to use a magnetic spectrometer to detect the scattered electron in order to cleanly identify the reaction channel.

As a consequence, the figure of merit of a polarised $\overrightarrow{^3\text{He}}$ target is comparable to that of a neutron polarimeter.

3.1 Polarised beam

Various techniques are available to produce polarised electron beams, but photo-emission from GaAs has until now proven to be optimal (Aulenbacher 2002). A thin layer of GaAs is illuminated by a circularly polarised laser beam of high intensity, which preferentially excites electrons of one helicity state to the conductance band through optical pumping. The helicity sign of the laser beam can be flipped at a rate of tens of Hertz by changing the high voltage on a Pockels cell. The polarised electrons that diffuse to the photocathode surface are then extracted by a 50-100 kV potential. An ultra-high vacuum environment is required to minimise surface degradation of the GaAs crystal by backstreaming ions. Initially, the use of bulk GaAs limited the maximum polarisation to 50% because of the degeneracy of the $P_{3/2}$ sublevels. This degeneracy is removed by introducing a strain in a thin layer of GaAs deposited onto a thicker layer with a slightly different lattice spacing. Although such strained GaAs cathodes have a significantly lower quantum efficiency than bulk GaAs cathodes, this has been compensated by the development of high-intensity diode or Ti-sapphire lasers. Polarised electron beams are now reliably available with a polarisation close to 80% at currents of $\geq 100\ \mu A$.

The polarised electrons extracted from the GaAs surface are first pre-accelerated and longitudinally bunched and then injected into an accelerator. Typically, the polarisation vector of the electrons is manipulated in a Wien filter, a system of crossed magnetic and electric fields and magnetic quadrupole lenses, so that the electrons are fully longitudinally polarised at the target. If the beam is injected into a storage ring for use with an internal target, a Siberian snake (Derbenev and Kondratenko 1973) is needed to compensate for the precession of the polarisation.

Three processes are used to measure the beam polarisation: Mott (Steigerwald 2001) scattering, Møller (Hauger *et al.* 2001) scattering or Compton (Baylac *et al.* 2001) scattering. Any of these results in a polarimeter with an accuracy approaching 1%. In a Mott polarimeter the beam helicity asymmetry is measured in scattering polarised electrons off atomic nuclei. This technique is limited to electron energies below ~ 20 MeV and multiple scattering effects have to be estimated by taking measurements at different target foil thicknesses. In a Møller polarimeter polarised electrons are scattered off polarised atomic electrons in a magnetised iron foil. In this technique the major uncertainties are in the corrections for atomic screening and in the foil magnetisation, unless the polarising field is strong enough to fully saturate the magnetisation. A potentially superior alternative (Chudakov and Luppov 2004) has been proposed in which the electrons are scattered off a sample of atomic hydrogen, polarised to a very high degree in an atomic beam, and trapped in a superconducting solenoid. Finally, in a Compton polarimeter the beam helicity asymmetry is measured in scattering polarised electrons off an intense beam of circularly polarised light, produced by trapping a laser beam in a high-finesse Fabry-Perot cavity. The electron beam in a storage ring is sufficiently intense that a laser beam can be directly scattered off the electron beam without the use of an amplifying cavity. Only the last two methods, the atomic hydrogen Møller and the Compton polarimeter, have no effect on the quality of the electron beam and thus can be used continuously during an experiment.

3.2   Polarised targets

Two different techniques are used, depending on the intensity of the electron beam, to produce polarised targets for protons. In storage rings where the circulating beam can have an intensity of 100 mA or more, but the material interfering with the beam has to be minimised, gaseous targets are used, whereas solid targets can be used as external targets. Because free neutrons are not available in sufficient quantity, effective targets, such as deuterium or ^{3}He, are necessary, and the techniques used to polarise deuterons are similar to those used for protons. For ^{3}He, gaseous targets are used both as internal and external targets.

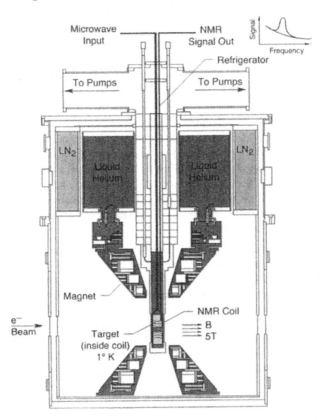

Figure 3. *Layout of a polarised hydrogen/deuterium target using the dynamic nuclear polarisation technique.*

Solid polarised targets that can withstand electron beams with an intensity of up to 100 nA all use the dynamic nuclear polarisation technique (Crabb *et al.* 1995, Averett *et al.* 1999). A hydrogenous compound, such as NH_3 or LiD, is doped, *eg* by radiation damage, with a small concentration of free radicals. Because the magnetic substate occupation in the radicals follows the Boltzmann distribution, the free electrons are polarised to more than 99% in a ~ 5 T magnetic field at a temperature ~ 1 K (see Figure 3). A radiofrequency (RF) field is then applied to induce transitions to states with a preferred orientation of the nuclear spin. Because the relaxation time of the electrons is much shorter than that of the nuclei, polarised nuclei are accumulated. This technique has been successful in numerous deep-inelastic lepton scattering and nucleon form-factor experiments; it has provided polarised hydrogen or deuterium targets with an average polarisation of $\sim 80\%$ or $\sim 30\%$, respectively.

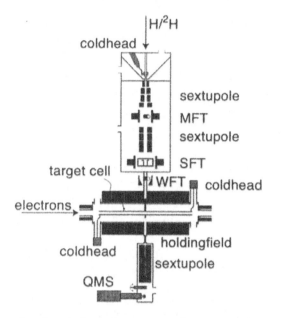

Figure 4. *Layout of an Atomic Beam Source target for polarised hydrogen/deuterium.*

Internal hydrogen/deuterium targets (Steffens and Haeberli 2003) are polarised by the atomic beam source (ABS) technique, which relies on Stern-Gerlach separation and RF transitions (see Figure 4). First, a beam of atoms is produced in an RF dissociator through a nozzle cooled with liquid nitrogen. Then, atoms with different electron spin directions are separated through a series of permanent (or superconducting) sextupole magnets and transitions between different hyperfine states are induced by a variety of RF units (MFT/ SFT/ WFT). The result is a highly polarised beam with a flux up to 10^{17} atoms/s. This beam is then fed into an open-ended storage cell, which is cooled and coated to minimise recombination of the atoms bouncing off the cell walls. The circulating electron beam, passing through the long axis of the storage cell, encounters only the flowing atoms. The polarisation vector is oriented with a set of coils, producing a field of ~ 0.3 T in order to minimise depolarisation by the RF structure of the circulating electron beam. The diameter of the storage cell is determined by the halo of the electron beam. A target thickness of 2×10^{14} nuclei/cm^2 has been obtained at a vector polarisation of more than 80%.

Polarised hydrogen or deuterium atoms can also be produced by spin-exchange collisions between such atoms and a small admixture of alkali atoms that have been polarised by optical pumping. The nucleus is then polarised in spin-temperature equilibrium. Although the nuclear polarisation obtained in such a laser driven source (LDS) is smaller than through the ABS technique, the flux can be more than 10^{18} atoms/s. Moreover, an LDS offers a more compact design than an ABS. A figure of merit comparable to that of the ABS at the HERMES experiment has recently been achieved by the MIT group (Clasie *et al.* 2003).

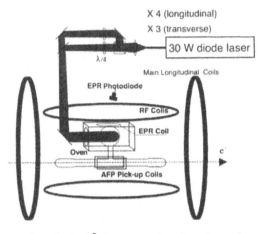

Figure 5. *Layout of a polarised ^{3}He target using the spin-exchange technique.*

Polarised ^{3}He is attractive as an effective polarised neutron target because its ground state is dominated by a spatially symmetric s-state in which the proton spins cancel, so that the spin of the ^{3}He nucleus is mainly determined by that of the neutron. Corrections for the (small) d-state component and for charge-exchange contributions from the protons can be calculated accurately at Q^2-values smaller than 0.5 GeV^2 (Golak *et al.* 2001) and larger than ~ 2 GeV^2 (Sargsian 2001). Direct optical pumping of ^{3}He atoms is not possible because of the energy difference between the ground state and the first excited state. Instead ^{3}He is polarised, either by first exciting the atoms to a metastable 2^3S_1 state and optically pumping that state, which then transfers its polarisation to the ground state by metastability-exchange collisions, or by optically pumping a small admixture of rubidium atoms, which then transfer their polarisation to the ^{3}He atoms through spin-exchange collisions (see Figure 5). In internal targets only the metastability-exchange technique has been used because of the possible detrimental effects of the rubidium admixture on the storage ring environment. With beam on target, polarisation values of up to 46% at target thicknesses of 1×10^{15} nuclei/cm^2 have been obtained. For external targets the spin-exchange technique (Alcorn *et al.* 2004) has been used to optically pump a glass target cell filled with 10 atm of ^{3}He with a 0.1% rubidium admixture. After the spin-exchange collisions the polarised ^{3}He diffuses into a 25 cm long cell which the electron beam traverses. Polarisations in excess of 40% have been reached with beam on target. A pair of 5 mT Helmholz coils is used to orient the polarisation vector, and care must be taken to minimise depolarising magnetic field gradients. Alternatively, the metastability technique (Surkau *et al.* 1997) has been used to polarise ^{3}He under atmospheric pressure which is then compressed to a density of more than 6 atm.

3.3 Recoil polarimeters

Focal-plane polarimeters have long been used at proton scattering facilities to measure the polarisation of the scattered proton. In such an instrument (Alcorn *et al.* 2004) the azimuthal angular distribution is measured of protons scattered in the focal plane of a magnetic spectrometer by an analyser, which often consists of carbon. From this angular distribution the two polarisation components transverse to the proton momentum can be derived

$$f(\theta, \phi) = f_0(\theta)[1 + P_n^{pol} A_y(\theta) \cos\phi + P_t^{pol} A_y(\theta) \sin\phi] . \qquad (9)$$

To extract the longitudinal polarisation component the nucleon's spin is precessed with a dipole magnet. The analyser is preceded by two detectors, most often wire or straw chambers, to measure the track of the incident proton; it is followed by two more detectors to track the scattered particle. The thickness of the analyser is adjusted to the proton momentum, limiting multiple scattering while optimising the figure of merit. In order to determine the two polarisation components in the scattering plane at the target, care must be taken to accurately calculate, on an event-by-event basis, the precession of the proton spin in the magnetic field of the spectrometer.

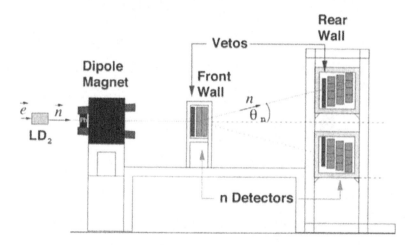

Figure 6. *Layout of a G_E^n measurement with a recoil polarimeter, showing the liquid deuterium target, the spin-precessing dipole magnet and the front and rear plastic scinillators.*

Neutron polarimeters follow the same basic principle. Here, plastic scintillator material is used as an active analyser, preceded by a veto counter to discard charged particles. This eliminates the need for the front detectors. Sets of scintillator detectors are used to measure an up-down asymmetry in the scattered neutrons, which is sensitive to a polarisation component in the scattering plane, perpendicular to the neutron momentum. In modern neutron polarimeters (Ostrick *et al.* 1999) the analyser is preceded by a dipole magnet, with which the neutron spin can be precessed (see Figure 6).

4 Experimental results

4.1 Proton electric form factor

In elastic electron-proton scattering a longitudinally polarised electron will transfer its polarisation to the recoil proton. In the one-photon exchange approximation the proton can attain only polarisation components in the scattering plane, parallel (P_l) and transverse (P_t) to its momentum. This can immediately be seen from the expression of the proton current in the Breit frame which separates into components proportional to G_E and G_M (see Equation 7). The ratio of the charge and magnetic form factors is directly proportional to the ratio of these polarisation components:

$$\frac{G_E^p}{G_M^p} = -\frac{P_t}{P_l}\frac{E_e + E_e'}{2M}\tan(\frac{\theta_e}{2}) \ . \tag{10}$$

The polarisation-transfer technique was used for the first time by Milbrath *et al.* (1998) at the MIT-Bates facility. The proton form factor ratio was measured at Q^2-values of 0.38 and 0.50 GeV^2 by scattering a 580 MeV electron beam polarised to $\sim$ 30%. A follow-up measurement was performed at the MAMI facility (Pospischil *et al.* 2001) at a Q^2-value of 0.4 GeV^2.

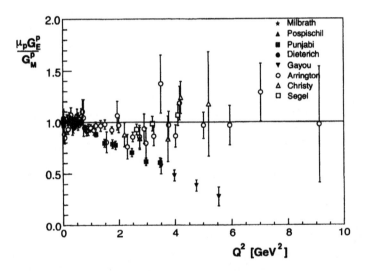

Figure 7. *The ratio $\mu_p G_E^p / G_M^p$ from polarisation transfer (Milbrath et al. 1998, Pospischil et al. 2001, Punjabi et al. 2003, Jones et al. 2000, Dieterich et al. 2001 and Gayou et al. 2002), compared to recent Rosenbluth data (Qattan et al. 2004 and Christy et al. 2004) and the reanalysis by Arrington (2003) of older SLAC data.*

The greatest impact of the polarisation-transfer technique was made by the two recent experiments (Punjabi *et al.* 2003, Gayou *et al.* 2002) in Hall A at Jefferson Lab, which measured the ratio G_E^p/G_M^p in a Q^2-range from 0.5 to 5.6 GeV^2. Elastic *ep* events were selected by detecting electrons and protons in coincidence in the two identical high-resolution spectrometers. At the four highest Q^2-values a lead-glass calorimeter was used to detect the scattered electrons in order to match the proton angular acceptance. The polarisation of the recoiling proton was determined with a focal-plane polarimeter in the hadron spectrometer, consisting of two pairs of straw chambers with a carbon or polyethylene analyser in between. The data were analysed in bins of each of the target coordinates. No dependence on any of these variables was observed (Punjabi *et al.* 2003). Figure 7 shows the results for the ratio $\mu_p G_E^p / G_M^p$. The most striking feature of the data is the sharp, practically linear decline as Q^2 increases:

$$\mu_p \frac{G_E^p(Q^2)}{G_M^p(Q^2)} = 1 - 0.13(Q^2 - 0.29) \quad \text{with} \ Q^2 \text{ in GeV}^2. \tag{11}$$

Since it is known that G_M^p closely follows the dipole parameterisation, it follows that G_E^p falls more rapidly with Q^2 than G_D. This significant fall-off of the form-factor ratio is in clear disagreement with the results from the Rosenbluth extraction. Arrington (2003) has performed a careful reanalysis of earlier Rosenbluth data. He selected only experiments in which an adequate ϵ-range was covered with the same detector. The results (Figure 7) do not show the large scatter seen in Figure 2. Recently, Christy *et al.* (2004) analysed an extensive data set on elastic electron-proton scattering collected in Hall C at Jefferson Lab as part of experiment E99-119. The results are evidently in good agreement with Arrington's reanalysis. Qattan *et al.* (2004) performed a high-precision Rosenbluth extraction in Hall A at Jefferson Lab, designed specifically to significantly reduce the systematic errors compared to earlier Rosenbluth measurements. The main improvement came from detecting the recoiling protons instead of the scattered electrons, so that the proton momentum and the cross section remain practically constant when one varies ϵ at a constant Q^2-value. In addition, possible dependences on the beam current are minimised. Special care was taken in surveying the angular setting of the identical spectrometer pair. One of the spectrometers was used as a luminosity monitor during an ϵ scan. Preliminary results of this experiment, covering Q^2-values from 2.6 to 4.1 GeV^2, are in excellent agreement with previous Rosenbluth results. This basically rules out the possibility that the disagreement between Rosenbluth and polarisation-transfer measurements of the ratio G_E^p / G_M^p is due to an underestimate of ϵ-dependent uncertainties in the Rosenbluth measurements.

4.2 Two-photon exchange

In order to resolve the discrepancy between the results for G_E^p / G_M^p from the two experimental techniques, an ϵ-dependent modification of the cross section is necessary. In two-(or more-)photon exchanges (TPE) the nucleon undergoes a first virtual photon exchange which can lead to an intermediate excited state and then a second one or more, finally ending back in its ground state (Figure 8). The TPE contributions to elastic electron scattering have been investigated both experimentally and theoretically for the past fifty years. In the early days such contributions were called dispersive effects (Offermann *et al.* 1991). Lately, they have been relocated to radiative corrections in the so-called box diagram. Almost all analyses with the Rosenbluth technique have used radiative corrections originally derived by Mo and Tsai (1969), revisited by Maximon and Tjon (2000), that only include the infrared divergent parts of the box diagram (in which one of the two exchanged photons is soft). Thus, terms in which both photons are hard (and which depend on the hadronic structure) have been ignored.

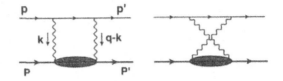

Figure 8. *The Feynman diagrams depicting two-photon exchanges.*

The most stringent tests of TPE on the nucleon have been carried out by measuring the ratio of electron and positron elastic scattering off a proton. Corrections due to TPE will have a different sign in these two reactions. Unfortunately, this (e^+e^-) data set is quite limited (Arrington 2004a), only extending (with poor statistics) up to a Q^2-value of ~ 5 GeV^2, whereas at Q^2-values larger than ~ 2 GeV^2 basically all data have been measured at ϵ-values larger than ~ 0.85. Other tests, also inconclusive, searched for non-linearities in the ϵ-dependence or measured the transverse (out-of-plane) polarisation component of the recoiling proton, of which a non-zero value would be a direct measure of the imaginary part of the TPE amplitude.

Several studies have provided estimates of the size of the ϵ-dependent corrections necessary to resolve the discrepancy. Because the fall-off of the form-factor ratio is linear with Q^2, and the Rosenbluth formula also shows a linear dependence of the form-factor ratio (squared) with Q^2 through the τ-term, a Q^2-independent correction linear in ϵ would cancel the disagreement. An additional constraint that any ϵ-dependent modification must satisfy, is the (e^+e^-) data set. Guichon and Vanderhaeghen (2003) introduced a general form of a TPE contribution from the so-called box diagram in radiative corrections into the amplitude for elastic electron-proton scattering. This resulted in the following modification of the Rosenbluth expression:

$$d\sigma \propto \tau + \epsilon \frac{\tilde{G}_E^{\,2}}{\tilde{G}_M^{\,2}} + 2\epsilon(\tau + \frac{\tilde{G}_E}{\tilde{G}_M})Y_{2\gamma} , \qquad (12)$$

where $Y_{2\gamma} = \mathrm{Re}\,\frac{\nu\tilde{F}_3}{M^2 G_M}$ and $\tilde{G}_M$, $\tilde{F}_2$ and $\tilde{F}_3$ are equal to G_M, F_2 and 0, respectively, in the Born approximation. $Y_{2\gamma}$ and the "two-photon" form factors $\tilde{G}_E$ and $\tilde{G}_M$ were fitted (Guichon and Vanderhaeghen 2003) to the Rosenbluth and polarisation transfer data sets. This resulted in a value of ~ 0.03 for $Y_{2\gamma}$ with very little ϵ- or Q^2-dependence.

Arrington (2004b) performed a fit to the complete data set, investigating two different modifications to the cross section with a Q^2-independent linear ϵ-dependence of 6 % over the full ϵ-range. Both modifications have the same ϵ-dependence, but one does not modify the cross section at small values of ϵ, whereas the other leaves the cross section unchanged at large values of ϵ. He found that the second gave a much better description of the complete data set. Moreover, it was in good agreement with the data set for the ratio of electron-proton and positron-proton elastic scattering.

Blunden *et al.* (2003) carried out the first calculation of the elastic contribution from TPE effects, albeit with a simple monopole Q^2-dependence of the hadronic form factors:

$G(Q^2) = \Lambda^2/(Q^2 + \Lambda^2)$. They obtained a practically Q^2-independent correction factor with a linear ϵ-dependence that vanishes at forward angles ($\epsilon = 1$). However, the size of the correction only resolves about half of the discrepancy. A later calculation (Melnitchouk 2003) which used a more realistic form factor behaviour, resolved up to 80% of the discrepancy.

A different approach was used by Chen *et al.* (2004), who related the elastic electron-nucleon scattering to the scattering off a parton in a nucleon through generalised parton distributions. TPE effects in the lepton-quark scattering process are calculated in the hard-scattering amplitudes. The handbag formalism of the generalised parton distributions is extended in an unfactorised framework in which the x-dependence is retained in the scattering amplitude. Finally, a valence model is used for the generalised parton distributions. The results for the TPE contribution nearly reconcile the Rosenbluth and the polarisation-transfer data and retain agreement with positron-scattering data.

Hence, it is becoming more and more likely that TPE processes have to be taken into account in the analysis of Rosenbluth data and that they will affect polarisation-transfer data only at the few percent level. Of course, further effort is needed to investigate the model-dependence of the TPE calculations. Experimental confirmation of TPE effects will be difficult, but certainly should be continued. The most direct test would be a measurement of the positron-proton and electron-proton scattering cross-section ratio at small ϵ-values and Q^2-values above 2 GeV^2. Positron beams available at storage rings are too low in either energy or intensity, but a measurement in the CLAS detector at Jefferson Lab, a more promising venue, has been proposed (Brooks *et al.* 2004). A measurement of the beam or target single-spin asymmetry normal to the scattering plane, which directly accesses the imaginary part of the box diagrams, would provide a sensitive test of TPE calculations. Also, real and virtual Compton scattering data can provide additional constraints on calculations of TPE effects in elastic scattering. Rosenbluth analyses have so far been restricted to simple PWBA, Coulomb distortion effects should certainly be included too. Additional efforts should be extended to studies of TPE effects in other longitudinal-transverse separations, such as proton knock-out and deep-inelastic scattering (DIS) experiments.

4.3 Proton magnetic form factor

An extensive data set (Borkowski *et al.* 1975) with a good accuracy is available up to a Q^2-value of more than 30 GeV^2 from unpolarised cross-section measurements (Figure 9). Because G_M^p dominates in a Rosenbluth extraction at larger Q^2-values, the G_M^p data have only a minor sensitivity to the discrepancy between the Rosenbluth extraction and the polarisation-transfer technique. Brash *et al.* (2002) have shown that the G_M^p data must be renormalised upwards by $\sim 2\%$ if one assumes the polarisation-transfer data to be correct.

4.4 Neutron magnetic form factor

Early data on G_M^n were extracted from inclusive quasi-elastic scattering off the deuteron. However, modeling of the deuteron wavefunction, required to subtract the contribution from the proton, resulted in sizable systematic uncertainties. A significant break-through

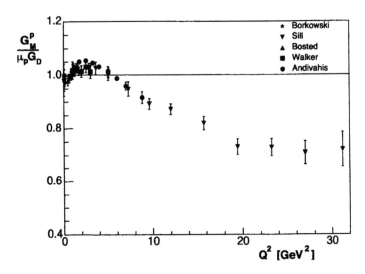

Figure 9. *The proton magnetic form factor G_M^p, in units of $\mu_p G_D$, as a function of Q^2. Data are from Borkowski et al. (1975), Bosted et al. (1990), Sill et al. (1993), Walker et al. (1994) and Andivahis et al. (1994).*

was made by measuring the ratio of quasi-elastic neutron and proton knock-out from a deuterium target. This method has little sensitivity to nuclear binding effects and to fluctuations in the luminosity and detector acceptance. The basic set-up used in all such measurements is very similar: the electron is detected in a magnetic spectrometer with coincident neutron/proton detection in a large scintillator array. The main technical difficulty in such a ratio measurement is the absolute determination of the neutron detection efficiency. Such measurements have been pioneered for Q^2-values smaller than 1 GeV^2 at Mainz (Anklin *et al.* 1994, Anklin *et al.* 1998, Kubon *et al.* 2002) and Bonn (Bruins *et al.* 1995). The Mainz G_M^n data are 8%-10% lower than those from Bonn, at variance with the quoted uncertainty of $\sim$2%. This discrepancy would require a 16%-20% error in the detector efficiency.

A study of G_M^n at Q^2-values up to 5 GeV^2 has recently been completed in Hall B by measuring the neutron/proton quasi-elastic cross-section ratio using the CLAS detector (Brooks and Vineyard 1994). A hydrogen target was in the beam simultaneously with the deuterium target. This made it possible to measure the neutron detection efficiency by tagging neutrons in exclusive reactions on the hydrogen target. Preliminary results indicate that G_M^n is within 10% of G_D over the full Q^2-range of the experiment (0.5-4.8 GeV^2).

Inclusive quasi-elastic scattering of polarised electrons off a polarised ^{3}He target offers an alternative method to determine G_M^n through a measurement of the beam asymmetry (Donnelly and Raskin 1986):

$$A = -\frac{(\cos\theta^* v_{T'} R_{T'} + 2\sin\theta^* \cos\theta^* v_{TL'} R_{TL'})}{v_L R_L + v_T R_T} , \qquad (13)$$

where θ^* and ϕ^* are the polar and azimuthal target spin angles with respect to $\vec{q}$, R_i

denote various nucleon response functions, and v_i the corresponding kinematic factors. By orienting the target polarisation parallel to $\vec{q}$, one measures $R_{T'}$, which in quasi-elastic kinematics is dominantly sensitive to $(G_M^n)^2$. For the extraction of G_M^n corrections for the nuclear medium (Golak *et al.* 2001) are necessary to take into account effects of final-state interactions and meson-exchange currents. The first such measurement was carried out at Bates (Gao *et al.* 1994). Recently, this technique was used to measure G_M^n in Hall A at Jefferson Lab in a Q^2-range from 0.1 to 0.6 GeV^2(Xu *et al.* 2000). This experiment provided an independent, accurate measurement of G_M^n at Q^2-values of 0.1 and 0.2 GeV^2, in excellent agreement with the Mainz data. At the higher Q^2-values G_M^n could be extracted (Xu *et al.* 2003) in plane wave impulse approximation, since final-state interaction effects are expected to decrease with increasing Q^2.

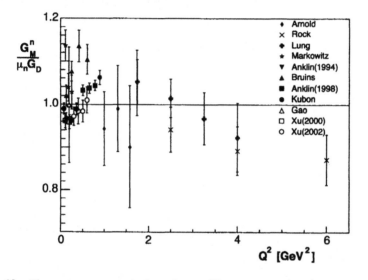

Figure 10. *The neutron magnetic form factor G_M^n, in units of $\mu_n G_D$, as a function of $\vec{Q^2}$. Results from ^{3}He are indicated by open symbols. Data are from Arnold et al. (1988), Rock et al. (1992), Lung et al. (1993), Markowitz et al. (1993), Anklin et al. (1994), Bruins et al. (1995), Anklin et al. (1998), Kubon et al. (2002), Gao et al. (1994), Xu et al. (2000) and Xu et al. (2003).*

Figure 10 shows the results of all completed G_M^n experiments.

4.5   Neutron electric form factor

Analogously to G_M^n, early G_E^n experiments used deuterium (quasi-)elastic scattering to extract the longitudinal deuteron response function. Due to the smallness of G_E^n, the use of different nucleon-nucleon potentials resulted in a 100% spread in the resulting G_E^n values (Platchkov *et al.* 1990). In the past decade a series of double-polarisation measurements of neutron knock-out from a polarised ^{2}H or ^{3}He target have provided accurate data on G_E^n. The ratio of the beam-target asymmetry with the target polarisation

perpendicular and parallel to the momentum transfer is directly proportional to the ratio
of the electric and magnetic form factors,

$$\frac{G_E^n}{G_M^n} = -\frac{P_x}{P_z}\frac{E_e + E_e'}{2M}\tan(\frac{\theta_e}{2}),\tag{14}$$

where P_x and P_z denote the polarisation component perpendicular and parallel to $\vec{q}$.

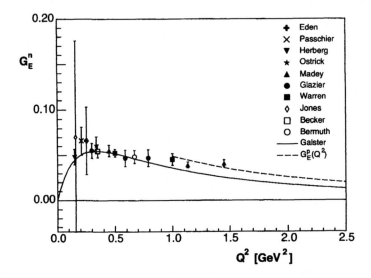

Figure 11. *The neutron electric form factor G_E^n as a function of Q^2. Results from*
$\overset{\rightarrow}{^3\text{He}}$ *are indicated by open symbols. Data are from Eden et al. (1994), Passchier et al.
(1999), Herberg et al. (1999), Ostrick et al. (1999), Madey et al. (2003), Glazier et al.
(2005), Warren et al. (2004), Zhu et al. (2001), Jones et al. (1991), Becker et al. (1999),
Golak et al. (2001), Bermuth et al. (2003) and Rohe et al. (1999). The full curve shows
the Galster et al. (1971) parameterisation; the dashed curve represents the Q^2-behaviour
of G_E^p.*

A similar result is obtained with an unpolarised deuteron target when one measures
the polarisation of the knocked-out neutron as a function of the angle over which the
neutron spin is precessed with a dipole magnet:

$$\frac{G_E^n}{G_M^n} = -\tan(\delta)\sqrt{\frac{\tau(1+\epsilon)}{2\epsilon}};\tag{15}$$

here, δ denotes the precession angle where the measured asymmetry is zero.

Again, the first such measurements were carried out at Bates, both with a polarised
$\overset{\rightarrow}{^3\text{He}}$ target and with a neutron polarimeter. Figure 11 shows results obtained through
all three reactions $^2\text{H}(\vec{e}, e'n)$, $^2\text{H}(\vec{e}, e'\vec{n})$ and $^3\text{He}(\vec{e}, e'n)$. At low Q^2-values corrections
for nuclear medium and rescattering effects can be sizeable: 65% for ^2H at 0.15 GeV^2

and 50% for ^{3}He at 0.35 GeV^2. These corrections are expected to decrease significantly with increasing Q^2, although no reliable calculations are presently available for ^{3}He above 0.5 GeV^2. There is excellent agreement between the results from the different techniques. Moreover, medium effects have clearly become negligible at ~ 0.7 GeV^2, even for ^{3}He. The latest data from Hall C at Jefferson Lab, using either a polarimeter or a polarised target (Madey *et al.* 2003, Warren *et al.* 2004, Zhu *et al.* 2001), extend up to $Q^2 \approx 1.5$ GeV^2 with an overall accuracy of $\sim 10\%$, in mutual agreement. From ~ 1 GeV^2 onwards G_E^n appears to exhibit a Q^2-behaviour similar to that of G_E^p. Schiavilla and Sick (2001) have extracted G_E^n from available data on the deuteron quadrupole form factor $F_{C2}(Q^2)$ with a much smaller sensitivity to the nucleon-nucleon potential than from inclusive (quasi-)elastic scattering. The 30-years-old Galster parameterisation (Galster *et al.* 1971) continues to provide a fortuitously good description of the data.

4.6 Timelike form factors

In the timelike region EMFF measurements have been made at electron-positron storage rings or by studying the inverse reaction (only for the proton form factors), antiproton annihilation on a hydrogen target. The rather limited data set on timelike form factors is shown in Figure 12. The quality of the data does not allow a separation of the charge and magnetic form factors; G_M has been extracted from the data using the G_E-values calculated by Iachello and Wan (2004). Clearly G_D which gives a very good description of the spacelike magnetic form factors, does not describe the data in the timelike region, at least from threshold down to -6 GeV^2. Iachello and Wan (2004), Hammer *et al.* (1996) and Dubnicka *et al.* (2003) have carried out an analytic continuation of their VMD calculations (see section 5). Iachello's model provides a consistent description of the magnetic form factors in the timelike region. An extension of the data set in the timelike region and of theoretical efforts to obtain a consistent description of all EMFFs in both the space- and timelike regions is highly desirable.

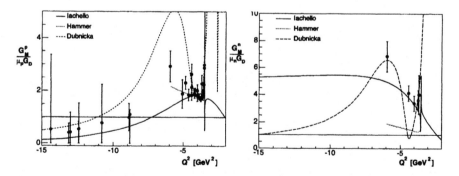

Figure 12. *The magnetic form factors (divided by G_D) in the time-like region as a function of Q^2, compared to the calculations by Iachello and Wan (2004), Hammer et al. (1996) and Dubnicka et al. (2003). See Iachello and Wan (2004) for references to the experimental data.*

4.7   Experimental review and outlook

In recent years highly accurate data on the nucleon EMFFs have become available from
various facilities around the world, made possible by the development of high luminosity
and novel polarisation techniques. These have established some general trends in the
Q^2-behaviour of the four EMFFs. The two magnetic form factors G_M^p and G_M^n are
close to identical, following G_D to within 10% at least up to 5 GeV^2, with a shallow
minimum at $\sim$ 0.25 GeV^2 and crossing G_D at $\sim$ 0.7 GeV^2. G_E^p/G_M^p drops linearly
with Q^2 and G_E^n appears to drop from $\sim$ 1 GeV^2 onwards at the same rate as G_E^p.
Measurements that extend to higher Q^2-values and offer improved accuracy at lower
Q^2-values, will become available in the near future. In Hall C at Jefferson Lab, Perdrisat
et al. (2001) will extend the measurements of G_E^p/G_M^p to 9 GeV^2 with a new polarimeter
and large-acceptance lead-glass calorimeter. Wojtsekhowski *et al.* (2002) will measure
G_E^n in Hall A at Q^2-values of 2.4 and 3.4 GeV^2 using the ^{3}He$(\vec{e}, e'n)$ reaction with a 100
msr electron spectrometer. The Bates Large Acceptance Spectrometer Toroid facility
(BLAST, http://www.mitbates.mit.edu) at MIT with a polarised hydrogen and deuteron
target internal to a storage ring will provide highly accurate data on G_E^p and G_E^n in a Q^2-
range from 0.1 to 0.8 GeV^2. Gao *et al.* (2001) have shown that the proton charge radius
can be measured with unprecedented precision by measuring the ratio of asymmetries
in the two sectors of the BLAST detector. Thus, within a couple of years G_E^n data with
an accuracy of 10% or better will be available up to a Q^2-value of 3.4 GeV^2. Once the
upgrade to 12 GeV (Cardman *et al.* 2004) has been implemented at Jefferson Lab, it will
be possible to extend the data set on G_E^p and G_M^n to 14 GeV^2 and on G_E^n to 5 GeV^2.

The charge and magnetisation rms radii are related to the form-factor slope at $Q^2 = 0$:

$$< r_E^2 >= \int \rho(r) r^4 dr = -6 \frac{dG(Q^2)}{dQ^2} \bigg|_{Q^2=0}$$

$$< r_M^2 >= \int \mu(r) r^4 dr = -\frac{6}{\mu} \frac{dG(Q^2)}{dQ^2} \bigg|_{Q^2=0} , \qquad (16)$$

with $\rho(r)$ ($\mu(r)$) denoting the radial charge (magnetisation) distribution. Table 1 lists the
results. For an accurate extraction of the radius Sick (2003) has shown that it is necessary
to take into account Coulomb distortion effects and higher moments of the radial distribution. His result for the proton charge radius is in excellent agreement with the most
recent three-loop QED calculation (Melnikov and van Ritbergen 2000) of the hydrogen
Lamb shift. Within error bars the rms radii for the proton charge and magnetisation
distribution and for the neutron magnetisation distribution are equal. The value for the
neutron charge radius was obtained (Kopecky *et al.* 1997) by measuring the transmission
of low-energy neutrons through liquid ^{208}Pb and ^{209}Bi. The Foldy term $\frac{3}{2}\frac{\kappa}{M_n^2} = -0.126$
fm^2 is close to the value of the neutron charge radius. Isgur (1999) showed that the Foldy
term is canceled by a first-order relativistic correction, which implies that the measured
value of the neutron charge radius is indeed dominated by its internal structure.

In the Breit frame the nucleon form factors can be written as Fourier transforms of
their charge and magnetisation distributions. However, if the wavelength of the probe is
larger than the Compton wavelength of the nucleon, *ie* if $|Q| \geq M_N$, the form factors are

Observable	value ± error	Reference
$< (r_E^p)^2 >^{1/2}$	0.895 ± 0.018 fm	(Sick 2003)
$< (r_M^p)^2 >^{1/2}$	0.855 ± 0.035 fm	(Sick 2003)
$< (r_E^n)^2 >$	-0.119 ± 0.003 fm^2	(Kopecky *et al.* 1997)
$< (r_M^n)^2 >^{1/2}$	0.87 ± 0.01 fm	(Kubon *et al.* 2002)

Table 1. *Values for the nucleon charge and magnetisation radii*

not solely determined by the internal structure of the nucleon. Then, they also contain dynamical effects due to relativistic boosts and consequently the physical interpretation of the form factors becomes complicated. Recently, Kelly (2002) has extracted spatial nucleon densities from the available form factor data. He selected a model for the Lorentz contraction of the Breit frame in which the asymptotic behaviour of the form factors conformed to perturbative quantum chromo-dynamics (pQCD) scaling at large Q^2-values and expanded the densities in a complete set of radial basis functions, with constraints at large radii. The neutron and proton magnetisation densities are found to be quite similar, narrower than the proton charge density. He reports a neutron charge density with a positive core surrounded by a negative surface charge, peaking at just below 1 fm, which he attributes to a negative pion cloud. Alternatively, he extracts the radial distributions of the u and d quarks which both show a secondary lobe which he interprets as an indication of an orbital angular momentum (OAM) component in the quark distributions. Friedrich and Walcher (2003) observe a bump/dip at $Q \approx 0.5$ GeV with a width of ~ 0.2 GeV, a feature common to all EMFFs. A fit to all four EMFFs was performed, assuming a dipole behaviour for the form factors of the constituent quarks and an $l = 1$ harmonic oscillator behaviour for that of the pion cloud. They then transformed their results to coordinate space, neglecting the Lorentz boost, where they find that the pion cloud peaks at a radius of ~ 1.3 fm, slightly larger than Kelly did, close to the Compton wavelength of the pion. Hammer *et al.* (2004) argue from general principles that the pion cloud should peak much more inside the nucleon, at ~ 0.3 fm. However, they assign the full $N\bar{N}2\pi$ continuum to the pion cloud which includes different contributions than just the one-pion loop that Kelly (2002) (and Friedrich and Walcher (2003)) assign to the pion cloud. The structure at ~ 0.5 GeV, common to all EMFFs, is at such a small Q^2-value that its transformation to coordinate space should be straightforward.

5 Model calculations

The recent production of very accurate EMFF data, especially the surprising G_E^p data from polarisation transfer, has prompted the theoretical community to intensify their investigation of nucleon structure. Space limitations compel us to focus on only a few highlights. The interested reader is encouraged to read the original publications; the review by Thomas and Weise (2001) is an excellent introduction.

The u-, d- and s-quarks are the main building blocks of the nucleon in the kinematic domain relevant to this review. Its basic structure involves the three lightest vector mesons (ρ, ω and ϕ) which have the same quantum numbers as the photon. Consequently,

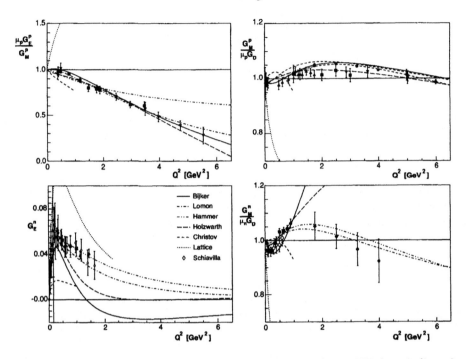

Figure 13. *Comparison of various calculations with available EMFF data, indicated by the same symbols as in Figures 7, 9, 10 and 11. For G_E^p only polarisation-transfer data are shown. Not shown are the data for G_E^n from Jones et al. (1991) and Eden et al. (1994) and the data for G_M^n from Rock et al. (1992), Arnold et al. (1988), Markowitz et al. (1993), Gao et al. (1994) and Bruins et al. (1995). For G_E^n the results of Schiavilla and Sick (2001) have been added. The calculations shown are from Bijker and Iachello (2004), Lomon (2001, 2002), Hammer and Meissner (2004), Holzwarth (1996, 2004), Christov et al. (1995), Kim H C et al. (1996) and Ashley (2004). Where applicable, the calculations have been normalised to the calculated values of $\mu_{p,n}$.*

one should expect these vector mesons to play an important role in the interaction of the photon with a nucleon. The first EMFF models were based on this principle, called vector meson dominance (VMD), in which one assumes that the virtual photon, after becoming a quark-antiquark pair, couples to the nucleon as a vector meson. The EMFFs can then be expressed in terms of coupling strengths between the virtual photon and the vector meson and between the vector meson and the nucleon, summing over all possible vector mesons. In the scattering amplitude a bare-nucleon form factor is multiplied by the amplitude of the photon interaction with the vector meson. With this model Iachello *et al.* (1973) predicted a linear drop of the proton form factor ratio, similar to that measured by polarisation transfer, more than 20 years before the data became available. Gari and Krümpelmann (1985, 1992) extended the VMD model to conform with pQCD scaling at large Q^2-values. The VMD picture is not complete, as becomes obvious from the fact that the Pauli isovector form factor F_2^V is much larger than the isoscalar one F_2^S. An improved description requires the inclusion of the isovector $\pi\pi$ channel through dispersion relations (Höhler *et al.* 1976, Mergell *et al.* 1996). By adding more parameters, such as the width of the ρ-meson and the masses of heavier vector mesons (Lomon 2001, 2002), the VMD models succeeded in describing new EMFF data as they became available, but with little predictive power. Figure 13 confirms that Lomon's calculations provide an excellent description of all EMFF data. Bijker and Iachello (2004) have extended the original calculations by also including a meson-cloud contribution in F_2, but still taking only two isoscalar and one isovector poles into account. The intrinsic structure of the nucleon is estimated to have an rms radius of ~ 0.34 fm. These new calculations are in good agreement with the proton form-factor data, but do rather poorly for the neutron. The most recent dispersion-theoretical analysis (Hammer *et al.* 2004), using four isoscalar and three isovector mesons, results in an excellent description of G_M^p and G_M^n, but only reasonably describes G_E^p and G_E^n. Subsequent studies (Fuchs *et al.* 2004) have further developed this combined approach to include chiral perturbation theory. However, such models can only be used at small Q^2-values, $\leq 0.4\ \text{GeV}^2$.

Many recent theoretical studies of the EMFFs have applied various forms of a relativistic constituent quark model (RCQM). Nucleons are assumed to be composed of three constituent quarks, which are quasi-particles where all degrees of freedom associated with the gluons and $q\bar{q}$ pairs are parameterised by an effective mass. Because the momentum transfer can be several times the nucleon mass, the constituent quarks require a relativistic quantum-mechanical treatment. Three possibilities exist for such a treatment: the instant form, where the interaction is present in the time component of the four-momentum and in the Lorentz boost; the point form, where all components of the four-momentum operator depend on the interaction; and the light-front form, where the interaction appears in one component of the four-momentum and in the transverse rotations. In each of these forms the Poincaré invariance can be broken in the number of constituents (by the creation of $q\bar{q}$ pairs) or by the use of approximate current operators. Although most of these calculations correctly describe the EMFF behaviour at large Q^2-values, effective degrees of freedom, such as a pion cloud and/or a finite size of the constituent quarks, are introduced to correctly describe the behaviour at lower Q^2-values.

Miller (2002a) uses an extension of the cloudy bag model (Théberge *et al.* 1981), three relativistically moving (in light-front kinematics) constituent quarks, surrounded by a pion cloud. He chose a spatial wavefunction, as derived by Schlumpf (1994), whose parameters (and those of the pion cloud) are chosen to describe the magnetic moments, the neutron charge radius, and the EMFF behaviour at large Q^2-values. Cardarelli and Simula (2000) also use light-front kinematics, but they calculate the nucleon wavefunction by solving the three-quark Hamiltonian in the Isgur-Capstick one-gluon-exchange potential. In order to get good agreement with the EMFF data they introduce a finite size of the constituent quarks in agreement (Petronzio *et al.* 2003) with recent DIS data. The results of Wagenbrunn *et al.* (2001) and Boffi *et al.* (2002) are calculated in a covariant manner in the point-form spectator approximation (PFSA). In addition to a linear confinement, the quark-quark interaction is based on Goldstone-boson exchange dynamics. The PFSA current is effectively a three-body operator (in the case of the nucleon as a three-quark system) because of its relativistic nature. It is still incomplete but it leads to surprisingly good results for the electric radii and magnetic moments of the other light and strange baryon ground states beyond the nucleon. Although Desplanques and Theussl (2003) have criticised the use of the point form in its introduction of two-body currents in the form of a neutral boson exchange, Coester and Riska (2003) obtain a reasonable representation of empirical form factors in this frame. Giannini *et al.* (2003), De Sanctis *et al.* (2000) and Ferraris M *et al.* (1995) explicitly introduced a three-quark interaction in the form of a gluon-gluon interaction in a hypercentral model, which successfully describes various static baryon properties. Relativistic effects are included by boosting the three quark states to the Breit frame and by introducing a relativistic quark current. All previously described RCQM calculations used a non-relativistic treatment of the quark dynamics, supplemented by a relativistic calculation of the electromagnetic current matrix elements. Merten *et al.* (2002) have solved the Bethe-Salpeter equation with instantaneous forces, inherently respecting relativistic covariance. In addition to a linear confinement potential, they used an effective flavour-dependent two-body interaction. For static properties this approach yields results (Van Cauteren *et al.* 2004) similar to those obtained by Wagenbrunn *et al.* (2001) and Boffi *et al.* (2002). The results of these five calculations are compared to the EMFF data in Figure 14. The calculations of Miller (2002a) do well for all EMFFs, except for G_M^n at low Q^2-values. Those of Cardarelli and Simula (2000), Giannini *et al.* (2003), Wagenbrunn *et al.* (2001) and Boffi *et al.* (2002) are in reasonable agreement with the data, except for that of Wagenbrunn *et al.* for G_M^p, while the results of Merten *et al.* (2002) provide the poorest description of the data.

Before Jefferson Laboratory polarisation-transfer data on G_E^p/G_M^p became available, Holzwarth (1996, 2004) predicted a linear drop in a chiral soliton model. In such a model the quarks are bound in a nucleon by their interaction with chiral fields. In the bare version quarks are eliminated and the nucleon becomes a skyrmion with a spatial extension, but the Skyrme model provided an inadequate description of the EMFF data. Holzwarth's extension introduced one vector-meson propagator for both isospin channnels in the Lagrangian and a relativistic boost to the Breit frame. His later calculations used separate isovector and isoscalar vector-meson form factors. He obtained excellent agreement for the proton data, but only a reasonable description of the neutron data. Christov *et al.* (1995) and Kim *et al.* (1996) used an SU(3) Nambu-Jona-Lasinio Lagrangian, an effective theory that incorporates spontaneous chiral symmetry breaking. This procedure is comparable to the inclusion of vector mesons into the Skyrme model, but it involves many fewer free parameters (which are fitted to the masses and decay constants of pions and kaons). The calculations are limited to $Q^2 \leq 1$ GeV^2 because the model is restricted to Goldstone bosons and because higher-order terms, such as recoil corrections, are neglected. A constituent quark mass of 420 MeV provided a reasonable description of the EMFF data (Figure 13).

In the asymptotically free limit, QCD can be solved perturbatively, providing predictions for the EMFF behaviour at large Q^2-values. Brodsky and Farrar (1975) derived a scaling law for the Pauli and Dirac form factors based on a dimensional analysis, that entailed counting propagators and the number of scattered constituents:

$$F_1 \propto (Q^2)^{-2}, \quad F_2 \propto (Q^2)^{-3}, \quad F_2/F_1 \propto Q^{-2} . \tag{17}$$

Brodsky and Lepage (1981) later reached the same asymptotic behaviour based on a more detailed theory that assumed factorisation and hadron helicity conservation. The recent polarisation transfer data clearly do not follow this pQCD prediction (which the Rosenbluth data unfortunately do). Miller (2002b) was the first to observe that imposing

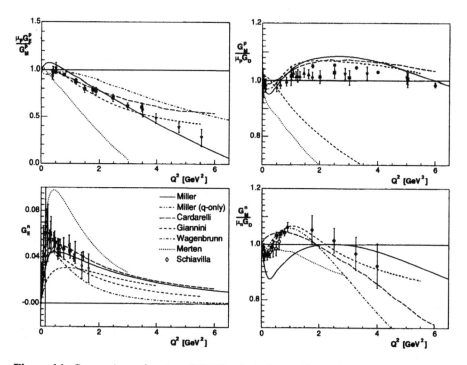

Figure 14. *Comparison of various RCQM calculations with available EMFF data, similar to the comparison in Figure 13. The calculations shown are from Miller (2002a), Cardarelli and Simula (2000), Giannini et al. (2003), Wagenbrunn et al. (2001), Boffi et al. (2002) and Merten et al. (2002). Miller (q-only) denotes a calculation by Miller (2002a) in which the pion cloud has been suppressed. Where applicable, the calculations have been normalised to the calculated values of $\mu_{p,n}$.*

Poincaré invariance removes the pQCD condition that the transverse momentum must be zero, and introduces a quark OAM component in the wavefunction of the proton, thus violating hadron helicity conservation. His model predicts a $1/Q$ behaviour for the ratio of the Dirac and Pauli form factors at intermediate Q^2-values, in excellent agreement with the polarisation transfer data for $Q^2 \geq 3\,\text{GeV}^2$. Iachello and Wan (2004) and others have pointed out that this $1/Q$ behaviour is accidental and only valid in an intermediate Q^2-region. Ralston and Jain (2004) have generalised this issue to conclude that the Q^2-behaviour of the Jefferson Lab data signals substantial quark OAM in the proton. Recently, Brodsky *et al.* (2004) and Belitsky *et al.* (2003) have independently revisited the pQCD domain. Belitsky *et al.* derive the following large Q^2-behaviour:

$$\frac{F_2}{F_1} \propto \frac{\ln^2 Q^2/\Lambda^2}{Q^2}, \tag{18}$$

where Λ is a soft scale related to the size of the nucleon. Even though the Jefferson Lab data follow this behaviour (Figure 15), Belitsky *et al.* warn that this could very well be precocious, since pQCD is not expected to be valid at such low Q^2-values. Brodsky *et al.* (2004) argue that a non-zero OAM wavefunction should contribute to both F_1 and

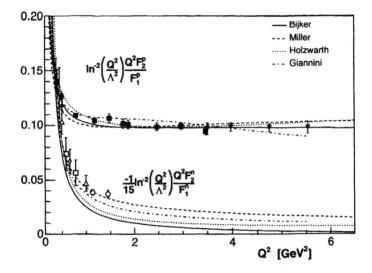

Figure 15. *The ratio* $(Q^2 F_2/F_1)/\ln^2(Q^2/\Lambda^2)$ *as a function of* Q^2 *for the polarisation-transfer data and the calculations of Bijker and Iachello (2004), Miller (2002a), Holzwarth (1996, 2004) and Giannini et al. (2003). The same ratio, scaled by a factor* $-1/15$, *is shown for the neutron with open symbols. For* Λ *a value of 300 MeV has been used.*

F_2 and that thus $Q^2 F_2/F_1$ should still be asymptotically constant.

Once enough data have been collected on generalised parton distributions, it will become possible to construct a three-dimensional picture of the nucleons, with the three dimensions being the two transverse spatial coordinates and the longitudinal momentum. Miller (2003) has further investigated the information that can be extracted from form-factor data by themselves. His colourful images of the proton should be interpreted as three-dimensional pictures of the proton as a function of the momentum of the quark, probed by the virtual photon, and for different orientations of the spin of that quark relative to that of the proton. Ji (2003) has derived similar images from generalised parton distributions using Wigner correlation functions for the quark and gluon distributions.

However, all theories described until now are at least to some extent effective (or parameterisations). They use models constructed to focus on certain selected aspects of QCD. Only lattice gauge theory can provide a truly *ab initio* calculation, but accurate lattice QCD results for the EMFFs are still several years away. One of the most advanced lattice calculations of EMFFs has been performed by the QCDSF collaboration (Göckeler *et al.* 2003). The technical state of the art limits these calculations to the quenched approximation (in which sea-quark contributions are neglected), to a box size of 1.6 fm and to a pion mass of 650 MeV. Ashley *et al.* (2004) have extrapolated the results of these calculations to the chiral limit, using chiral coefficients appropriate to full QCD. The agreement with the data (Figure 13) is poorer than that of any of the other calculations, a clear indication of the technology developments required before lattice QCD calculations can provide a stringent test of experimental EMFF data.

6 Summary, outlook and conclusions

Recent advances in polarised electron sources, polarised nucleon targets and nucleon recoil polarimeters have enabled accurate measurements of the spin-dependent elastic electron-nucleon cross section. New data on nucleon electro-magnetic form factors with unprecedented precision have (and will continue to) become available in an ever increasing Q^2-domain. Highly accurate measurements with the Rosenbluth technique have established that the discrepancy between results on G_E^p/G_M^p with the Rosenbluth techniques and with polarisation transfer is not an instrumentation problem. Recent advances on two-photon exchange contributions make it highly likely that the application of TPE corrections will resolve that discrepancy. However, a continuing strong effort, both experimental and theoretical, is needed to firmly establish the applicability of TPE corrections.

The two magnetic form factors G_M^p and G_M^n closely follow the simple dipole form factor G_D. G_E^p/G_M^p drops linearly with Q^2 and G_E^n appears to drop at the same rate as G_E^p from $\sim 1\,\text{GeV}^2$ onwards. The Q^2-behaviour of G_E^n has provided a signal of substantial non-zero orbital angular momentum in the proton. Only scant data are available in the time-like region. The full EMFF data set forms tight constraints on models of nucleon structure. So far, all available theories are at least to some extent effective (or parameterisations). Still, only few of these adequately describe all four EMFFs. Only lattice gauge theory can provide a truely *ab initio* calculation, but accurate lattice QCD results for the EMFFs are still several years away. A scaling prediction has been developed for the ratio of the Pauli and Dirac form factors, which the data appear to follow even at a Q^2-value as low as $1\,\text{GeV}^2$. Novel procedures allow a visualisation of the nucleon structure as a function of the momentum of the struck quark. A fully three-dimensional picture of the nucleon will become available when future exclusive data have allowed the determination of the Generalised Parton Distributions.

I would like to conclude with a forty-year-old quote from the review by Wilson and Levinger (1964) which is still fully appropriate now: "Although the major landmarks of this field of study are now clear, we are left with the feeling that much is yet to be learned about the nucleon by refining and extending both measurement and theory."

Acknowledgments

The work presented was supported by the U.S. Department of Energy (DOE) contract DE-AC05-84ER40150 Modification NO. M175, under which the Southeastern Universities Research Association operates the Thomas Jefferson National Accelerator Facility.

References

Akhiezer AI, Rozentsweig LN and Shmushkevich IM, 1958, *Sov Phys JETP* **6** 588.
Alcorn J *et al.* , 2004, *Nucl Instrum Methods A* **522** 294.
Andivahis L *et al.* , 1994, *Phys Rev D* **50** 5491.
Anklin H *et al.* , 1994, *Phys Lett B* **336** 313.
Anklin H *et al.* , 1998, *Phys Lett B* **428** 248.

Arnold R, Carlson C and Gross F, 1981, *Phys Rev C* **23** 363.

Arnold R *et al.* , 1988, *Phys Rev Lett* **61** 806.

Arrington J, 2003, *Phys Rev C* **68** 034325.

Arrington J, 2004a, *Phys Rev C* **69** 022201R.

Arrington J, 2004b, *Phys Rev C* **69** 032201R, and references therein.

Ashley J D *et al.* , 2004 *Eur Phys Jour A* **19** (Suppl 1) 9.

Aulenbacher K, 2002, *Proc. 9th Int. Workshop on Polarised Sources and Targets*, (World Scientific, Singapore) 141.

Averett T D *et al.* , 1999, *Nucl Instrum Methods A* **427** 440.

Bartel W, *et al.* , 1966, *Phys Rev Lett* **17** 608.

Baylac M *et al.* , 2001, *Phys Lett B* **459** 412.

Becker J *et al.* , 1999, *Eur Phys Jour A* **6** 329.

Belitsky A V, Ji X and Yuan F, 2003, *Phys Rev Lett* **91** 092003.

Berger C, *et al.* , 1971, *Phys Lett B* **35** 87.

Bermuth J *et al.* , 2003, *Phys Lett B* **564** 199.

Bijker R and Iachello F, 2004, *Phys Rev C* **69** 068201.

Blunden P G, Melnitchouk W and Tjon J A, 2003, *Phys Rev Lett* **91** 142304.

Boffi S *et al.* , 2002, *Eur Phys Jour A* **14** 17.

Borkowski F *et al.* , 1975, *Nucl Phys B* **93** 461.

Bosted P E *et al.* , 1990, *Phys Rev C* **42** 38.

Bosted P E *et al.* , 1995, *Phys Rev C* **51** 409.

Brash E *et al.* , 2002, *Phys Rev C* **65** 051001.

Brodsky S J and Farrar G, 1975, *Phys Rev D* **11** 1309.

Brodsky S J and Lepage G P, 1981, *Phys Rev D* **24** 2848.

Brodsky S J *et al.* , 2004, *Phys Rev D* **69** 076001.

Brooks W and Vineyard M F, 1994, spokespersons, Jefferson Lab experiment E94-017; private communication.

Brooks W *et al.* , 2004, spokespersons, Jefferson Lab experiment E04-116.

Bruins E E W *et al.* , 1995, *Phys Rev Lett* **75** 21.

Cardarelli F and Simula S, 2000, *Phys Rev C* **62** 065201.

Cardman LS *et al.* , eds, 2003, *Pre-Conceptual Design Report for the Science and Experimental Equipment for the 12 GeV Upgrade of CEBAF*, http://www.jlab.org/div_dept/physics_division/pCDR_public/pCDR_final.

Chen Y C *et al.* , 2004, *Phys Rev Lett* **93** 122301.

Christov C V *et al.* , 1995, *Nucl Phys A* **592** 513.

Christy M E *et al.* , 2004, *Phys Rev C* **70** 015206.

Chudakov E and Luppov V, 2004, *IEEE Trans on Nucl Science*, **51** 1533.

Clasie B *et al.* , 2003, *Presented at X-th Workshop on Polarised Sources and Targets*, Novosibirsk.

Coester F and Riska D O, 2003, *Nucl Phys A* **728** 439.

Crabb D *et al.* , 1995, *Nucl Instrum Methods A* **356** 9.

Derbenev Y S and Kondratenko A M, 1973, *Sov Phys JETP* **37** 968.

De Sanctis M *et al.* , 2000, *Phys Rev C* **62** 025208.

Desplanques B and Theussl L, 2003, hep-ph/0307028.

Dieterich S *et al.* , 2001, *Phys Lett B* **500** 47.

Donnelly T W and Raskin A S, 1986, *Ann Phys, NY* **169** 247.

Dubnicka S, Dubnickova A Z and Weisenpacher P, 2003, *J Phys G: Nucl Part Phys* **29** 405.

Eden T *et al.* , 1994, *Phys Rev C* **50** R1749.

Estermann I and Stern O, 1933, *Z Phys* **85** 17.

Ferraris M *et al.* , 1995, *Phys Lett B* **364** 231.

Friedrich J and Walcher T, 2003, *Eur Phys Jour A* **17** 607.

Frisch R and Stern O, 1933, *Z Phys* **85** 4.

Fuchs T, Gegelia J and Scherer S, 2004, *Eur Phys Jour A* **19** (Suppl 1) 35 and references therein.

Galster S *et al.*, 1971, *Nucl Phys B* **32** 221.

Gao H *et al.*, 1994, *Phys Rev C* **50** R546.

Gao H, Calarco J R and Kolster H, 2001, spokespersons, MIT-Bates proposal 01-01.

Gari M F and Krümpelmann W, 1985, *Z Phys A* **322** 689.

Gari M F and Krümpelmann W, 1992, *Phys Lett B* **274** 159.

Gayou O *et al.*, 2002, *Phys Rev Lett* **88** 092301.

Giannini M, Santopinto E and Vassallo A, 2003, *Prog Part Nucl Phys* **50** 263.

Glazier D I *et al.*, 2005, *Eur Phys Jour A* **24** 101.

Göckeler M *et al.*, 2003, hep-lat/0303019.

Golak J *et al.*, 2001, *Phys Rev C* **63** 034006.

Guichon P A M and Vanderhaeghen M, 2003, *Phys Rev Lett* **91** 142303.

Hammer H W, Meissner U G and Drechsel D, 1996, *Phys Lett B* **385** 343.

Hammer H W and Meissner U G, 2004, *Eur Phys Jour A* **20** 469.

Hammer H W, Drechsel D and Meissner U G, 2004, *Phys Lett B* **586** 291.

Hand L N, Miller D G and Wilson R, 1963, *Rev Mod Phys* **35** 335.

Hauger M *et al.*, 2001, *Nucl Instrum Methods A* **462** 382.

Herberg C *et al.*, 1999, *Eur Phys Jour A* **5** 131.

Höhler G *et al.*, 1976, *Nucl Phys B* **114** 505.

Holzwarth G, 1996, *Z Phys A* **356** 339.

Holzwarth G, 2004, hep-ph/0201138.

Iachello F, Jackson A and Lande A, 1973, *Phys Lett B* **43** 191.

Iachello F and Wan Q., 2004, *Phys Rev C* **69** 055204.

Isgur N, 1999, *Phys Rev Lett* **83** 272.

Janssens T, *et al.*, 1966, *Phys Rev* **142** 922.

Ji X, 2003, *Phys Rev Lett* **91** 062001.

Jones M K *et al.*, 2000, *Phys Rev Lett* **84** 1398.

Jones-Woodward C E *et al.*, 1991, *Phys Rev C* **44** R571.

Kelly J J, 2002, *Phys Rev C* **66** 065203.

Kim H C *et al.*, 1996, *Phys Rev D* **53** 4013.

Kopecky S *et al.*, 1997, *Phys Rev C* **56** 2229.

Kubon G *et al.*, 2002, *Phys Lett B* **524** 26.

Litt J, *et al.*, 1970, *Phys Lett B* **31** 40 and references therein.

Lomon E L, 2001, *Phys Rev C* **64** 035204.

Lomon E L, 2002, *Phys Rev C* **66** 045501.

Lung A F *et al.*, 1993, *Phys Rev Lett* **70** 718.

Madey R *et al.*, 2003, *Phys Rev Lett* **91** 122002.

Markowitz P *et al.*, 1993, *Phys Rev C* **48** R5.

Maximon L C and Tjon J A, 2000, *Phys Rev C* **62** 054320.

Melnikov K and van Ritbergen T, 2000, *Phys Rev Lett* **84** 1673.

Melnitchouk W, 2003, private communication.

Mergell P, Meissner UG and Drechsel D, 1996, *Nucl Phys A* **596** 367.

Merten D *et al.*, 2002, *Eur Phys Jour A* **14** 477.

Milbrath B *et al.*, 1998, *Phys Rev Lett* **80** 452; erratum, 1999, *Phys Rev Lett* **82** 2221.

Miller G A, 2002a, *Phys Rev C* **66** 032001R.

Miller G A, 2002b, *Phys Rev C* **65** 065205.

Miller G A, 2003, *Phys Rev C* **68** 022201R.

Mo L W and Tsai Y S, 1969, *Rev Mod Phys* **41** 205.

Offermann EAJM *et al.*, 1991, *Phys Rev C* **44** 1096.

Ostrick M *et al.*, 1999, *Phys Rev Lett* **83** 276.

Passchier I *et al.*, 1999, *Phys Rev Lett* **82** 4988.

Perdrisat C F *et al.* , 2001, spokespersons, Jefferson Lab experiment E01-109.

Petronzio R, Simula S and Ricco G, 2003, *Phys Rev D* **67** 094004.

Platchkov S *et al.* , 1990, *Nucl Phys A* **510** 740.

Pospischil T *et al.* , 2001, *Eur Phys Jour A* **12** 125.

Punjabi V *et al.* , 2003, submitted to *Phys Rev C*.

Qattan I A *et al.* , 2004, nucl-ex/0410010.

Ralston J P and Jain P, 2004, *Phys Rev D* **69** 053008.

Rock S *et al.* , 1992, *Phys Rev D* **46** 24.

Rohe D *et al.* , 1999, *Phys Rev Lett* **83** 4257.

Rosenbluth M N, 1950, *Phys Rev* **79** 615.

Sargsian M, 2001, *Int J Mod Phys E* **10** 405.

Schiavilla R and Sick I, 2001, *Phys Rev C* **64** 041002.

Schlumpf F, 1994, *J Phys G: Nucl Part Phys* **20** 237.

Sick I, 2004, *Phys Lett B* **576** 62; private communication.

Sill A F *et al.* , 1993, *Phys Rev D* **48** 29.

Steffens E and Haeberli W, 2003, *Rep Prog Phys* **66** 1887.

Steigerwald M, 2001, *Proc 14th Int Workshop on Polarised Sources, AIP Conf Proc* **570,** 935
    (American Institute of Physics, New York).

Surkau R *et al.* , 1997, *Nucl Instrum Methods A* **384** 444.

Théberge S, Thomas A W and Miller G A, 1981, *Phys Rev D* **24** 216.

Thomas A W and Weise W, 2001, *The Structure of the Nucleon*, (Wiley-VCH, Berlin).

Van Cauteren T *et al.* , 2004, *Eur Phys Jour A* **20** 283.

Wagenbrunn R F *et al.* , 2001, *Phys Lett B* **511** 33.

Walker R C *et al.* , 1994, *Phys Rev D* **49** 5671.

Warren G *et al.* , 2004, *Phys Rev Lett* **92** 042301.

Wilson R R and Levinger J S, 1964, *Ann Rev Nucl Science* **14** 135.

Wojtsekhowski B *et al.* , 2002, spokespersons, Jefferson Lab experiment E02-013.

Xu W *et al.* , 2000, *Phys Rev Lett* **85** 2900.

Xu W *et al.* , 2003, *Phys Rev C* **67** 012201R.

Zhu H *et al.* , 2001, *Phys Rev Lett* **87** 081801.

Hard scattering processes

Dirk Ryckbosch

Department of Subatomic and Radiation Physics
University of Gent, Belgium

1 Introduction

The ultimate goal of hadronic physics is to understand the structure of hadrons in terms of quark and gluon degrees of freedom. The interaction of these particles is known to be described by Quantum ChromoDynamics (QCD). However, due to the increasing strength of this interaction at length scales typical of hadronic radii the theory is not quite tractable and one has to resort to various models. There are only a few reactions which give information which is directly sensitive to the partonic degrees of freedom. Among these, the hard scattering processes like Deep Inelastic Scattering (DIS), stand out. In the case of DIS the reaction directly probes the quark structure of the nucleon through a determination of the Parton Distribution Functions (PDF).

The unpolarised PDF $q(x, Q^2)$ gives essentially the probability of finding a parton with a momentum fraction x in a nucleon. The polarised PDF's describe the spin structure of the nucleon. In particular the longitudinal PDF $\Delta q(x, Q^2)$ gives the helicity distribution of partons in the nucleon. Unpolarised PDF's have been measured in great detail over a wide range of x and Q^2. The information on polarised PDF's is much more sketchy. A reasonably accurate picture of the quark helicity structure has emerged in recent years, but other aspects of the spin structure remain virtually untouched. Among these the lack of experimental information on the gluon polarisation and the transverse spin structure of the nucleon are particularly obvious. Several experiments are gearing up to address these issues in the near future.

In these lectures the main features of hard electron scattering are reviewed. The focus is on experiments determining the spin structure of the nucleon, and the topics are illustrated with recent results, mainly from the HERMES experiment.

2   The HERMES experiment

HERMES is a fixed target experiment (Ackerstaff *et al.* 1998) on the 27.6 GeV HERA e-ring. The electrons in the ring are transversely polarised by the Sokolov-Ternov effect, and longitudinal polarisation is achieved by a spin rotator in front of the HERMES target. A second rotator behind the experiment restores the transverse polarisation. Typical beam polarisations are between 50 and 60%. The polarisation of the beam is continuously monitored by two independent Compton backscattering polarimeters. The beam traverses the HERMES target consisting of a 40 cm long windowless storage cell which confines the injected polarised target gas to the region around the beam. The polarised atomic H and D are provided by an atomic beam source using Stern-Gerlach separation. A small sample of the target gas is continuously analysed in a Breit-Rabi polarimeter. Typical polarisations are 85%.

In 1996 and 1997 data were taken with a polarised proton target, while in 1998, 1999 and 2000 a large data sample was collected using a polarised deuterium target. The scattered electrons, and some of the hadrons produced in the reaction, are detected in the HERMES spectrometer. This is a typical forward magnetic spectrometer consisting of two symmetric halves above and below the plane of the accelerator. Scattered leptons and produced hadrons are detected and identified within the angular acceptance of ±170 mrad horizontally and 40 − 140 mrad vertically. The hadron-lepton separation is done on the basis of the signals in a transition radiation detector, an electromagnetic calorimeter and a pair of scintillator hodoscopes where the second one is preceded by a lead sheet preshower.

The identification of pions, kaons and (anti-)protons has been performed since 1998 using a Ring Imaging CHerenkov detector (RICH) with clear aerogel and C_4F_{10} gas as radiators (Akopov *et al.* 2002). The hadrons can be separated over almost the full momentum range of HERMES, *ie* between 2 and 15 GeV. Prior to 1998 a threshold Cerenkov was used to give pion identification in a limited momentum range.

The typical kinematic domain for the analysis of polarised DIS in HERMES is $0.1 < Q^2 < 15$ GeV^2, $x > 0.02$, with Q^2 the (negative) momentum transfer squared, and $x = Q^2/2M\nu$ the Bjorken scaling variable. To avoid regions with very large contributions from resonance production or large radiative corrections further conditions, $W > 2$ GeV, $y = \nu/E > 0.85$, are usually imposed, where $W^2 = 2M\nu + M^2 - Q^2$ is the invariant mass (squared) of the photon-nucleon system and y is the fraction of the initial beam energy E transferred. In semi-inclusive reactions, where in addition to the scattered lepton a produced hadron is also detected, the fraction of the energy of the virtual photon carried by the hadron is given by $z = E_h/\nu$. A cut at $z > 0.2$ is usually used to avoid the region dominated by target fragmentation.

3   Inclusive scattering

Most of the information on the PDF's has up to now come from inclusive DIS: experiments where only the scattered lepton is detected. Fig. 1 shows the world data on the polarised structure function $g_1(x)$ for deuterium.

It is obvious from this figure that all experiments are basically in agreement with

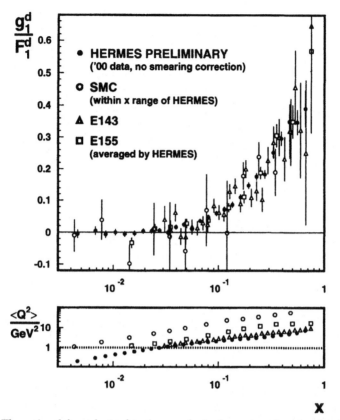

Figure 1. *The ratio of the polarised to the unpolarised structure function for deuterium g_1^d/F_1^d. The bottom panel shows the Q^2 values corresponding to each x-bin for the different experiments.*

each other. The recent preliminary HERMES data considerably improve the statistical accuracy. However, there is a difference of about one order of magnitude in the scale of the various experiments, with the SMC results being taken at much higher beam energy than the data from SLAC and HERMES. The fact that the ratio is similar in the different experiments thus implies an evolution for g_1 that is very similar to the one for F_1. Just as was done succesfully in the case of unpolarised structure functions one can analyse the scaling violations observed in $g_1(x, Q^2)$ for proton and neutron targets, and deduce information on the polarisation of the different quark flavours and the gluon polarisation.

Such a Next to Leading Order (NLO) QCD analysis was recently undertaken by HERMES (De Nardo 2002). The results confirm the facts already known about the quark polarisation: the quark spin contributes only 20-30% to the spin of the nucleon; most of this contribution comes from the valence quarks; the light quark sea is only a little (and negatively) polarised. The gluon polarisation that can be derived from the analysis of the presently available data, shown in Fig. 2.

This figure shows that the present data do not really constrain the gluon polarisation.

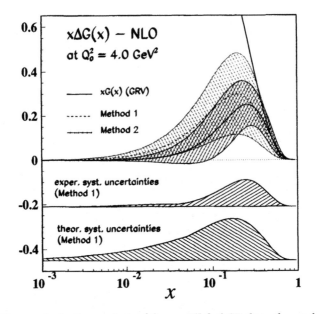

Figure 2. *Gluon polarisation as derived from a NLO QCD fit to the world data on g_1 (De Nardo 2002). "Method 1" and "Method 2" refer to two different methods to solve the evolution equations. The full line gives the positivity limit.*

The only strong suggestion is that it should be positive, but no value can be given. More direct methods to determine the gluon contribution to the nucleon spin are necessary. This is one of the aims of the COMPASS experiment at CERN which is now starting up. Also the spin physics programme at RHIC intends to determine the gluon polarisation independently.

4   Semi-inclusive scattering

4.1   Data and analysis

The HERMES experiment was specifically designed to perform accurate measurements of semi-inclusive reactions, where in addition to the scattered lepton some of the produced hadrons are also detected. The idea is that these hadrons contain extra information on the quark that took part in the scattering process. This technique of flavour tagging allows the determination of the polarisation of individual quark flavours directly from the spin asymmetries observed in hadron production.

The quantity of interest here is the photon-nucleon asymmetry A_1^h when a hadron of type h is produced. This can be derived from the experimental semi-inclusive spin asymmetry $A_\parallel^h$:

$$A_1^h = \frac{A_\parallel^h}{D(1+\eta\gamma)} = \frac{1}{D(1+\eta\gamma)} \frac{N_h^{\uparrow\downarrow} - N_h^{\uparrow\uparrow}}{N_h^{\uparrow\downarrow} + N_h^{\uparrow\uparrow}}, \tag{1}$$

where D is the virtual photon depolarisation, η and γ are kinematic factors, and $N_h^{\uparrow\uparrow}$ ($N_h^{\uparrow\downarrow}$) are the number of semi-inclusive events (properly taking into account polarisation of target and beam, and relative luminosity) for target polarisation parallel (antiparallel) to the beam polarisation.

An example of the asymmetries obtained at HERMES is given in Fig. 3. Similar asymmetries were derived for positive and negative pions and on a proton target. A striking feature of the asymmetries in this figure is the fact that the asymmetry for the K^- is compatible with zero. Since this is an all-sea object this already indicates that the polarisation of the quark sea will be very small.

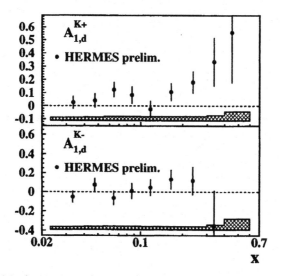

Figure 3. *Semi-inclusive spin asymmetries for kaons produced on a deuteron target, plotted as a function of x.*

In leading order QCD and assuming factorisation of the cross section one can write this asymmetry in terms of the PDF and fragmentation functions:

$$A_1^h(x, z) \quad = \frac{\int_{z_m}^1 dz \sum_q e_q^2 \Delta q(x) D_q^h(z)}{\int_{z_m}^1 dz \sum_q e_q^2 q(x) D_q^h(z)} \tag{2}$$

$$= \sum_q P_q^h(x) \frac{\Delta q(x)}{q(x)}. \tag{3}$$

In the last equation the *purities* P_q^h are introduced. These are spin-independent quantities which give the probability that a hadron of type h, observed in the final state, came from a struck quark of flavour q. In that sense they are the inverse of fragmentation functions. The purities depend on the unpolarised quark densities and the fragmentation functions. The former were measured to high precision in unpolarised DIS experiments. Information on the fragmentation functions is, however, less precise. There is some information for pions, but those data were taken at quite different kinematics from HERMES. Hence, the purities were calculated using a Monte Carlo simulation of the entire scattering process. Standard unpolarised parton distribution parametrisations (Lai *et al.* 2000) were used, while the fragmentation was modelled in the LUND string model as implemented in JETSET (Sjostrand 1994). The LUND model fully describes the fragmentation from the current as well as the target fragmentation region. The parameters of the model were tuned to fit the hadron multiplicities measured at HERMES in order to achieve a good description of the fragmentation process at our kinematics. The resulting purities also include effects of the acceptance of the spectrometer.

The analysis applies Eq. 3 in matrix form:

$$\vec{A_1}(x) = \mathcal{P}(x) \cdot \vec{Q}(x) \tag{4}$$

where the elements of vector $\vec{A_1}$ are the measured inclusive and semi-inclusive (Born) asymmetries, the elements of matrix $\mathcal{P}$ are the effective integrated purities for the proton and neutron targets and the vector $\vec{Q}$ contains the (anti-)quark polarisations. This is an over-constrained system of equations which is solved by a minimisation procedure.

4.2 Results

The purity formalism has been used in the HERMES analysis (Airapetian *et al.* 2004) to make a flavour decomposition of the quark polarisations for $u, \bar{u}, d, \bar{d}$, and $s + \bar{s}$. A symmetric strange sea polarisation $\Delta s/s = \Delta \bar{s}/\bar{s}$ was assumed. The measured asymmetries were integrated over the z-range from 0.2 to 0.8, and over $Q^2 > 1 \text{GeV}^2$. The resulting spin densities are shown in Fig. 4.

The polarisation of the u-quark is positive over the entire measured range in x, with the largest polarisation at high x where the valence quarks dominate. The polarisation of the down-quark is negative, particularly in the region of the valence quarks. The polarisation of the light sea flavours $\bar{u}$ and $\bar{d}$, and the polarisation of the strange sea are consistent with zero.

The data are compared to polarised PDF's derived from fits to inclusive data. Good agreement is found between the spin densities directly measured here and the results of NLO QCD analyses, in particular for the valence distributions. The tendency of the parameterisations to yield a small negative sea polarisation is not confirmed by the data; neither is it ruled out with the present statistical accuracy.

5 Exclusive reactions

Exclusive DIS reactions, where the target nucleon remains in or close to its ground state, can be described in terms of the Generalised Parton Distributions (GPD) that were introduced a few years ago. These GPD's form a natural off-forward extension of the standard Parton Distribution Functions which are well determined from (semi-)inclusive DIS reactions. They form a connection between the PDF's and the form-factors of hadrons.

The cleanest example of the appearance of GPD's in an exclusive reaction is that of Deeply Virtual Compton Scattering (DVCS). The main diagram for this reaction is shown in Fig. 5.

There are 4 (flavour sets of) GPD's: 2 unpolarised functions (H and E) and 2 polarised ($\tilde{H}$ and $\tilde{E}$). The two functions E and $\tilde{E}$ correspond to helicity flip operators and

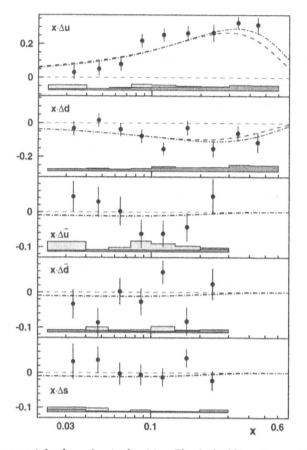

Figure 4. *The x-weighted quark spin densities. The dashed line shows a GRSV param-eterisation (Glück et al. 2001), and the dash-dotted curve an alternate parameterisation due to Blümlein and Böttcher (Blümlein 2002).*

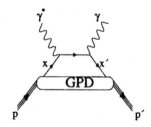

Figure 5. *Handbag diagram describing DVCS.*

have no direct analogue in the forward PDF's. The functions H and $\tilde{H}$ have as their forward limit the standard unpolarised and polarised PDF's F_1 and g_1, respectively. The GPD's depend on 3 independent kinematic variables: $H(x, \xi, t)$ where ξ is related to the usual Bjorken variable by $2\xi = \frac{x_B}{1 - x_B/2}$. (Note that there are other equivalent choices possible for the variables.)

The importance of GPD's in the context of spin physics is embodied by the sum rule derived by Ji (1997) for the second moment of the GPD's:

$$\frac{1}{2} \int_{-1}^{+1} x[H^q(x,\xi,t=0) + E^q(x,\xi,t=0)]dx = J_q, \tag{5}$$

where J_q is the *total* angular momentum contribution of quarks to the spin of the nucleon. Up to now there is experimental information only on the contribution from the spins of quarks (and gluons). Combined with this sum rule GPD's and thus exclusive reactions are the only way presently known that may lead to information on the orbital angular momentum contribution, thus providing a complete picture of the nucleon spin structure.

The main problem in the experimental determination of the GPD's is the fact that they do not appear as simple factors in the expressions for cross sections, asymmetries *etc.* They always appear in convolutions over the x and ξ variables, making a direct determination impossible. Finding the best deconvolution procedure to lead from the experimental data to the GPD is a challenge for the future. To do this a good understanding of the structure of the GPD is essential to develop adequate models.

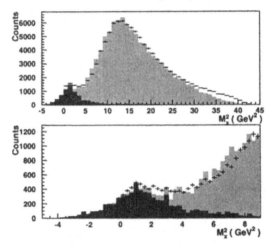

Figure 6. *Missing mass distribution for electroproduction of real photons (Airapetian et al. 2001). The top plot shows the full range of M_x^2, wheras the bottom plot is an expanded view of the region round the ground state mass. The darker histogram shows the simulated contribution of exclusive photonproduction, while the lighter histogram corresponds to photons produced in fragmentation processes (mainly π^0 production).*

Fig. 6 shows the missing mass (squared) spectrum for the DVCS reaction at HERMES. A clear peak at the ground state mass can be seen, thus establishing the observation of exclusive DVCS.

However, with relatively low beam energy the main source of hard photons in the HERMES kinematics is not DVCS but the Bethe-Heitler process where the photon is radiated from either the incoming or the outgoing electron. It is in principle impossible to disentangle these two processes. On the other hand the presence of two processes will also lead to interference between them and this can be exploited to access the DVCS amplitude. The interference terms can be projected out by measuring azimuthal asymmetries in single photon electroproduction. It was shown (Diehl *et al.* 1997, Belitsky *et al.* 2002) that beam-spin (A_{LU}), target-spin (A_{UL}) and beam-charge (A_C) azimuthal asymmetries provide access to different combinations of the real and imaginary parts of the interfering amplitudes. It should be noted that at present HERA is the only place where beam charge asymmetries can be studied because of the availability of both electron and positron beams.

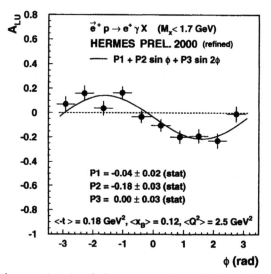

Figure 7. *The beam-spin azimuthal asymmetry for exclusive real photon production on a hydrogen target.*

In Fig. 7 the beam-spin azimuthal asymmetry for a proton target is shown, while Fig. 8 presents the results for the beam-charge asymmetry on the proton. The $\sin\phi$ moment of the former one is related to the imaginary part of the interference amplitude; the $\cos\phi$ moment of the latter is related to the real part of the interference amplitude.

It is obvious from the results presented in these figures that the DVCS-BH interference indeed dominates the distributions. The extracted $\sin\phi$ and $\cos\phi$ moments are in encouraging agreement with theoretical calculations based on models of GPD's. In the next step the kinematic distributions of the various moments of the azimuthal asymmetries are extracted and to be compared to more detailed calculations.

From Fig. 6 it can be seen that the experimental resolution of the HERMES spectrometer is insufficient to allow a clear separation of exclusive and fragmentation processes. At present only the scattered electron and the real photon are detected in the forward spectrometer. To really establish exclusivity on the event level would require the detection of the recoil target proton. Kinematics dictate that this proton recoils with low momentum at high angles relative to the beam line, outside the present acceptance of the HERMES spectrometer. A major project at HERMES now is the construction of a recoil detector to detect and identify such recoil particles. The detector will consist of three

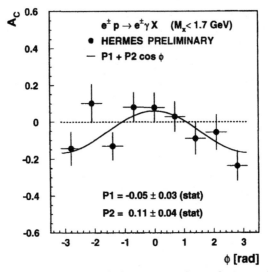

Figure 8. *The beam-charge azimuthal asymmetry for exclusive real photon production on a hydrogen target.*

active detector parts. A silicon detector around the target cell inside the beam vacuum, a scintillating fibre tracker in a longitudinal magnetic field and a layer of scintillator strips with interspersed W-sheets as preshower material.

It is planned to install the recoil detector sometime in 2004 or 2005 and to operate it for about 2 years with an unpolarised hydrogen target and polarised electron and positron beams. The higher target densities that are possible for unpolarised targets will enable HERMES to obtain a large data sample on DVCS with unprecedented statistical and systematic accuracy for both beam-spin and beam-charge asymmetries. An added advantage of the recoil detector is that the determination of the momentum of the recoil proton immediately yields an accurate measurement of the momentum transfer $-t$ to the target. This will improve the resolution in this important kinematic quantity (see *eg* Eq. 5) by about an order of magnitude.

6 Transversity

Up to now we have only discussed the helicity structure of the nucleon. In fact there are three different structure functions which describe the parton structure of the nucleon at leading twist. These are the unpolarised structure function F_1, the helicity structure function g_1 and a new quantity, transversity h_1. The related PDF is denoted $h_1^q(x)$ and gives the probability of finding a transversely polarised quark in a transversely polarised nucleon. Mainly because, as a chiral-odd object, it cannot be measured through inclusive DIS it is experimentally unknown. However, in combination with another chiral-odd object, *eg* a fragmentation function, transversity becomes accessible. In semi-inclusive reactions it leads to single-spin azimuthal asymmetries in *eg* pion and kaon production on a transversely polarised target.

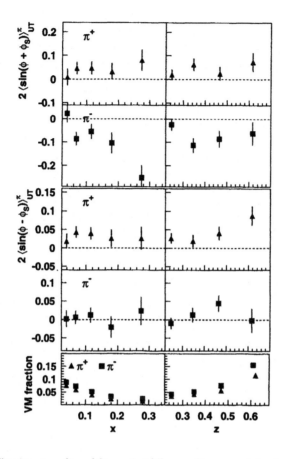

Figure 9. *Collins (top panel) and Sivers (middle panel) asymmetries for charged pions. Error bars are statistical only. There is an additional 8 % scale uncertainty. The lower panel gives an estimate of the contribution from pions coming from exclusive vector meson production and decay.*

Such a transversely polarised target has been installed at HERMES since 2001. The first data were taken with it after HERA became operational again following a major luminosity upgrade for the collider experiments. These data allow a simultaneous measurement of the Collins and Sivers asymmetries. The Collins asymmetry is related to the transversity distribution, while the Sivers asymmetry involves a new unknown distribution function $f_{1T}^\perp$. This T-odd distribution function, which is related to the orbital angular momentum of the quarks, was believed to be zero, but recent work has shown that it may arise from final state interactions involving soft gluons (Brodsky 2002).

Figure 9 shows first data from HERMES on transverse single spin asymmetries, in particular the Collins and Sivers asymmetries (Airapetian *et al.* 2005). The data show a clear non-zero signal for both asymmetries indicating that the distribution (and fragmentation) functions involved are significant. The large and negative Collins asymmetry for negative pions is surprising: from u quark dominance arguments one would expect this asymmetry to be similar or smaller than the asymmetry for positive pions. The present data seem to suggest that the unfavoured fragmentation function involved is surprisingly large. The Sivers asymmetries show for the first time that it may be possible to access this new non-integrated distribution function, which may conceivably lead to an alternative access to the elusive orbital angular momentum contribution of quarks.

7 Conclusions

Some recent results from the HERMES experiment at HERA have been presented. The analysis of the large data set on longitudinally polarised proton and deuteron targets, collected in the first run of HERMES from 1995 through 2000, is nearing completion. Results of a NLO QCD analysis of the inclusive scattering show indications for a positive and possibly large gluon polarisation. The results from the semi-inclusive analysis are consistent with those from the inclusive data and moreover give for the first time a flavour decomposition of the quark polarisations in 5 components. The valence quarks are seen to dominate the contribution of the quark spins to the nucleon spin, while the sea appears to have a negligible polarisation.

The observation of azimuthal asymmetries in exclusive reactions, here exemplified by the DVCS reaction, promises access to the new GPD's. The implementation of the recoil detector now under construction at HERMES will give a much enhanced accuracy for such studies. The main thrust of the data taking of HERMES at the present time is with a transversely polarised proton target. On the basis of the single-spin azimuthal asymmetries observed now it is expected that the full data set which will become available after the end of this programme in 2005 will give valuable new insights into the transverse spin structure of the nucleon.

References

Ackerstaff K *et al.* , 1998, *Nucl Instrum Methods A* **417** 230.
Airapetian A *et al.* , 2001, *Phys Rev Lett* **87** 182001.
Airapetian A *et al.* , 2004, *Phys Rev Lett* **92** 012005.
Airapetian A *et al.* [HERMES collaboration], 2005, *Phys Rev Lett* **94** 012002.
Akopov A *et al.* , 2002, *Nucl Instrum Methods A* **479** 511.
Belitsky A V *et al.* , 2002, *Nucl Phys B* **629** 323.
Blümlein J, and Böttcher H, 2002, *Nucl Phys B* **636** 225.
Brodsky S J, Hwang D S, and Schmidt I, 2002, *Phys Lett B* **530** 99.
De Nardo L, 2002, Dissertation, University of Alberta.
Diehl M *et al.* , 1997, *Phys Lett B* **411** 192.
Glück M *et al.* , 2001, *Phys Rev D* **63** 094005.
Ji X, 1997, *Phys Rev Lett* **78** 610.
Lai H *et al.* , 2000, *Eur Phys Jour C* **12** 375.
Sjostrand T, 1994, *Comp Phys Comm* **82** 74.

Selected topics in baryon spectroscopy and structure

Volker D Burkert

Jefferson Laboratory,
12000 Jefferson Avenue, Newport News, VA 23606, USA

1 Introduction

At the N*2000 conference held at Jefferson Laboratory Isgur (2000) gave his last public presentation on the subject "Why N*'s are important?". He gave a 3-fold answer to this question that I would like to repeat here:

- Nucleons represent the real world, therefore they must be at the centre of any discussion on why the world is the way it is.

- Nucleons represent the simplest system where the non-abelian character of QCD is manifest.

- Nucleons are complex enough to reveal physics hidden from us in mesons.

Nucleons, and baryons in general, have played an important role in the development of the quark model and of QCD. Gell-Mann (1964) and Zweig (1964) were led to the $3 \otimes 3 \otimes 3$ symmetry of the quark model by the systematics of the baryon spectrum. The excitation of the $\Delta(1232)$ resonance, first seen in pion-proton scattering experiments conducted by Fermi and collaborators (Anderson *et al.* 1952) at the Institute for Nuclear Studies at the University of Chicago, provided the first experimental hints at the proton substructure. The excitation of the doubly-charged $\Delta^{++}(1232)$, which represents the largest hadronic cross section in pion-proton scattering, was prohibited within the symmetric quark model. The state with the quark configuration (u, u, u) and total quark spin $J_{3q} = \frac{3}{2}$ did not fit into the symmetric quark model, as its ground state wavefunction in the quark model was symmetric while the overall wavefunction for fermions must be anti-symmetric. A new degree of freedom for the quarks was needed, which is now called colour. It was Greenberg (1964) who introduced coloured quarks first as parafermions.

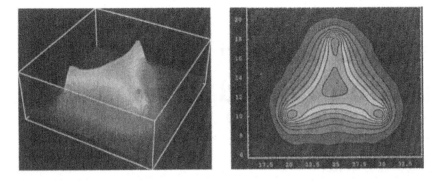

Figure 1. *Lattice QCD calculation of the 3D colour flux distribution for a baryon with static quarks. The calculation was done to study the abelian colour-flux distribution in a static 3-quark system. The "Y-shape" configuration is evident, indicating the presence of a genuine 3-body force. The graph shows high density at the quark locations and in the centre. The Δ-shaped flux configuration, characteristic of 2-body forces would have a depletion in the centre.*

Recent QCD calculations on the lattice (Ichie *et al.* 2003) show evidence for the "Y-shape" colour flux indicating a genuine 3-body force for baryons as shown in Figure 1. A dominant 2-body force would generate Δ-shape colour flux. The 3-body force is a unique feature of the baryon system in QCD. Studying nucleon excitations versus the distance scale will teach us a great deal about the interquark interaction, and how confinement works in nucleons. Studying the nucleon ground state in elastic electron scattering allows us to determine the charge and magnetic moment distribution in the nucleon. To reveal the substructure of the nucleon we need to study the excitation spectrum as a function of the distance scale. Only then can we hope to more fully understand the force that confines the quarks in nucleons, and that allows nucleons to form stable atomic nuclei, the fundamental building blocks of the matter we observe in our daily lives.

The lectures are organised as follows: In section 2 we briefly review the basic classification of baryons in $SU(6) \otimes O(3)$ multiplets. For a detailed discussion of the quark model see the lectures by Close (2004). In section 3 recent studies of electromagnetic transitions of non-strange baryon resonances in single pion or eta electroproduction will be discussed, and in section 4 searches for so-called "missing states" in $p\pi\pi$ and $K\Lambda$ final states measured in photo- and electroproduction will be presented. Finally, some aspects of searches for pentaquark baryons will be addressed in section 5.

2   Baryon multiplets

Properties of the ground state baryons with 3 quarks (u, d, s) quarks, are well described within the flavour symmetry group $SU(3)$. They are organised in flavour multiplets of

$$\mathbf{3} \otimes \mathbf{3} \otimes \mathbf{3} = \mathbf{10} \oplus \mathbf{8} \oplus \mathbf{8} \oplus \mathbf{1}. \tag{1}$$

When quark spins are included in $SU(2)$, a rich structure within spin-flavour group $SU(6)$ is obtained,

$$6 \otimes 6 \otimes 6 = 56_s \oplus 70_{ms} \oplus 70_{ma} \oplus 20_a. \qquad (2)$$

with symmetric, mixed symmetric, mixed antisymmetric, and antisymmetric representations. The $SU(6)$ representation decomposes into flavour multiplets with defined total quark spin:

$$\begin{aligned}
56 &= {}^4 10 \oplus {}^2 8 \\
70 &= {}^2 10 \oplus {}^4 8 \oplus {}^2 8 \oplus {}^2 1 \qquad (3) \\
20 &= {}^2 8 \oplus {}^4 1.
\end{aligned}$$

In addition to these flavour multiplets, the 3-quark system may also have orbital angular momentum $\vec{L}$, so that the total angular momentum is given by

$$\vec{J} = \vec{L} + \sum_{i=1}^{3} \vec{s}_i. \qquad (4)$$

The symmetry properties of baryons are described within $SU(6) \otimes O(3)$, where $O(3)$ represents the orbital angular momentum excitations. The full spectrum of baryons, made from u, d, and s quarks, is then obtained by designing a Hamiltonian for the interquark interaction and the orbital excitation that breaks the mass degeneracy of the baryon multiplets. Within $SU(6) \otimes O(3)$ baryons fall into supermultiplets of $[SU(6), L^P]_N$, where $P = (-1)^L$ is the parity of the 3 quark system. Most of the lowest lying baryon states can be associated with states within the 4 supermultiplets $[56, 0^+]_0$, $[70, 1^-]_1$, $[56, 0^+]_2$, and $[56, 2^+]_2$. The lowest supermultiplet $[56, 0^+]_0$ contains the ground state nucleon with $I = \frac{1}{2}$, and the Δ with $I = \frac{3}{2}$. The $[70, 1^-]_1$ supermultiplet contains the $S_{11}(1535)$, $D_{13}(1520)$, $S_{11}(1650)$, $S_{31}(1620)$, $D_{13}(1700)$, $D_{33}1700)$, and $D_{15}(1675)$, all well known states.

In the following section the electromagnetic excitation of these resonances from the ground state nucleon will be discussed.

3 Electromagnetic excitations of S=0 baryon resonances

In the past several decades, the non-strange baryon resonance spectrum has been explored mostly with pion beams. We will focus on the study of baryon resonances using electromagnetic probes. The electromagnetic interaction probes the internal structure of the excited state and samples the nucleon wavefunction. A number of electron accelerators are currently in use at Jefferson Lab, ELSA, GRAAL, SPrinG-8, MAMI, Bates, and LEGS at BNL. They are producing large amounts of high statistics data in many channels relevant to the study of excited baryons.

In experimental studies of baryon resonances there are two main goals:

(1) The measurement of electromagnetic transition amplitudes and their dependence on the distance scales being probed, *ie* the transition form factors. Since in most experiments the hadronic final state is measured, the extraction of the transition form factors requires knowledge of the hadronic vertices. The excitation of an isolated N^* resonance in s-channel proceeds through the two-step process $\gamma N \rightarrow N^*$, and $N^* \rightarrow Nm$, where m is the final state meson. In order to extract the information on the electromagnetic vertex $\gamma N N^*$, knowledge of the hadronic vertex $N^* Nm$ is needed. This information has to come from hadronic processes, *eg* from $\pi N \rightarrow N^* \rightarrow Nm$.

Electromagnetic transition form factors can also be computed in models of baryon structure, for example in various versions of dynamical quark models, and in other microscopic models. Also, Lattice QCD is now making predictions for these form factors, especially for the $N\Delta(1232)$ transition (Alexandrou *et al.* 2004, 2005). From such comparisons we learn about the basic degrees of freedom that underly the baryon structure, and how QCD "works" in the real world of nucleons and nuclei.

(2) The other main thrust of the excited baryon program is the search for undiscovered states, so-called "missing" states. These are excited baryon states that are predicted by symmetric quark models with $SU(6) \otimes O(3)$ symmetry but have not been observed in πN scattering. Since most of the experiments studying N^* excitations have been carried in elastic πN scattering, many higher mass states with only small coupling to the πN channel, may have been missed in the partial wave analysis of elastic scattering data. Many of the "missing" states are predicted to couple more strongly to $N\pi\pi$, $N\omega$, or $K\Lambda$ (Capstick and Roberts 1994). Some of these states may have significant couplings to photons and may thus be searched for most efficiently in the photo- or electroproduction of these final states.

In the following subsection I will discuss recent measurements of electroproduction multipoles in the $\Delta(1232)$ region, and helicity amplitudes for states in the 2nd resonance region, the Roper $P_{11}(1440)$, $S_{11}(1535)$, and $D_{13}(1520)$. Using symmetry relations from the single quark transition model we can make predictions for many states in the 3rd resonance region with masses near 1.7 GeV.

3.1   The $N\Delta(1232)$ quadrupole transition and the shape of the nucleon

The possibility of determining the shape of the Δ ground state has been a main motivation for experiments to measure the transition amplitudes from the nucleon to the $\Delta(1232)$. This has become an achievable goal due to recent advances in experimental techniques, and the development of unitary models for the analysis single pion electroproduction data (Burkert and Lee 2004).

In models with $SU(6)$ spherical symmetry the transition from the nucleon ground state to the $\Delta(1232)$ is due to a quark spin flip mediated by the magnetic dipole transition M_{1+}, while the quadrupole transitions $E_{1+} = 0$, and $S_{1+} = 0$. (Here I use here the multipole notations for the electroproduction of pions M_{1+}, E_{1+}, S_{1+} which correpond to the photon multipoles $M1$, $E2$, $C2$.) Non-zero values for E_{1+} and S_{1+} would indicate deformation of the $N - \Delta$ system. Dynamically, such deformation could arise through interaction of the photon with the pion cloud (Sato and Lee 2001, Kamalov and

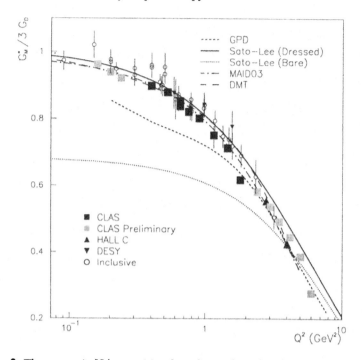

Figure 2. *The magnetic $N\Delta$ transition form factor determined from the M_{1+} multipole.*

Yang 1999), and through the one-gluon exchange mechanism at short distances (Koniuk and Isgur 1980). At asymptotically high momentum transfer, a model-independent prediction of helicity conservation (Carlsson 1986) requires $R_{EM} = E_{1+}/M_{1+} \rightarrow +1$, and $R_{SM} = S_{1+}/M_{1+} \rightarrow constant$. An interpretation of R_{EM} and R_{SM} in terms of quadrupole contributions to Δ shape can thus only be valid at small or moderate momentum transfers. In the following we show the multipole expansion that is often used in the analysis of pion electroproduction in the region of the $\Delta(1232)$.

The differential cross section for the electroproduction of pions is given by:

$$\frac{d\sigma}{d\Omega_M^*} = \frac{d\sigma_T}{d\Omega_M^*} + \epsilon\frac{d\sigma_L}{d\Omega_M^*} + \epsilon\frac{d\sigma_{TT}}{d\Omega_M^*}\cos 2\phi_M + \sqrt{2\epsilon(1+\epsilon)}\frac{d\sigma_{LT}}{d\Omega_M^*}\cos\phi_M, \qquad (5)$$

where σ_T, σ_L, σ_{TT}, σ_{LT} are the transverse, longitudinal, polarisation, and interference response functions. The response functions can be expanded in terms of Legendre polynomials:

$$\frac{d\sigma_T}{d\Omega_M^*} + \epsilon\frac{d\sigma_L}{d\Omega_M^*} = \sum_{\ell=0}^{\infty} A_\ell P_\ell(\cos\theta_M^*)$$

$$\sqrt{\epsilon(1+\epsilon)}\frac{d\sigma_{LT}}{d\Omega_M^*} = \sum_{\ell=1}^{\infty} B_\ell P_\ell'(\cos\theta_M^*) \qquad (6)$$

$$\frac{\epsilon\, d\sigma_{TT}}{d\Omega_M^*} = \sum_{\ell=2}^{\infty} C_\ell P_\ell''(\cos\theta_M^*).$$

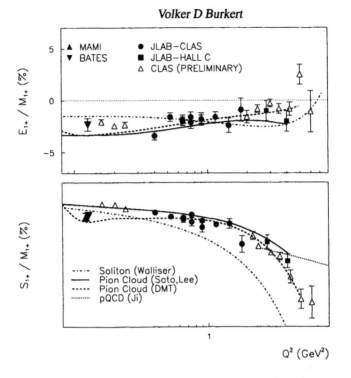

Figure 3. *Multipole ratios R_{EM} and R_{SM} for the $N\Delta(1232)$ transition vs photon virtuality Q^2.*

In the approximation that only terms containing the dominant multipole M_{1+} are retained the coefficients in the Legendre expansion are related to $|M_{1+}|$, and its projections onto the other s-and p-wave multipoles E_{1+}, S_{1+}, M_{1-}, E_{0+}, S_{0+}:

$$
\begin{aligned}
|M_{1+}|^2 &= A_0/2 \\
Re(E_{1+}M_{1+}^*) &= (A_2 - 2C_2/3)/8 \\
Re(M_{1-}M_{1+}^*) &= -[A_2 + 2(A_0 + 2(A_0 + C_2/3)]/8 \\
Re(E_{0+}M_{1+}^*) &= A_1/2 \\
Re(S_{0+}M_{1+}^*) &= B_1 \\
Re(S_{1+}M_{1+}^*) &= B_2/6.
\end{aligned}
\tag{7}
$$

To the degree that M_{1+} is the dominant contribution at the Δ mass it can be directly extracted using this approximation. This is the case for small values of Q^2. However, multipole M_{1+} drops more rapidly with increasing Q^2 than other multipoles, and the approximation may no longer be valid. In this case additional corrections will be neccesary using, for example, a unitary isobar model that takes into account higher partial waves resulting from background amplitudes and from higher mass resonances that may leak strength into the $\Delta(1232)$ region. Alternatively, one can use the unitary isobar model to fit the cross section directly, and extract the resonant multipoles. This latter approach will introduce additional model-dependency and increase the overall uncertainty. A summary of our current knowledge on the Q^2 dependence of the multipoles for the $N\Delta(1232)$ transition is shown in Figures 2 and 3.

We see that the electric quadrupole ratio R_{EM} is small and negative, and remains so over a large Q^2 range. R_{SM} is also negative, but rises in magnitude with Q^2. The progress made during the past few years is very impressive considering that just five years ago, even the sign of R_{EM} was not known beyond the photon point. Dynamical models that include dressed vertices and coupling to the pion cloud can describe both multipole ratios quite well. However, this agreement should be taken with some caution, as the models are fitted to the data in some range of Q^2. For example, the dynamical models have been fitted to the photon point, and to the highest Q^2 points at 2.8 and 4 GeV^2. The predictions only cover the ranges outside of these points. Nonetheless, the good simultaneous description of both R_{EM} and R_{SM} gives some credence to the claim that the quadrupole contributions are largely due to meson effects in the wavefunction. Such effects are not explicitly included in dynamical quark model, which do not describe the data well.

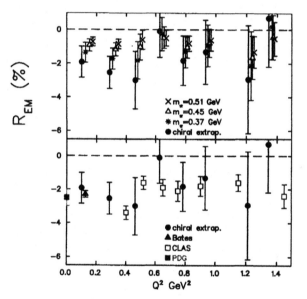

Figure 4. *Comparison of LQCD results in quenched approximation and experimental data for R_{EM}. The upper panels shows the Lattice results for different quark masses and the chiral extrapolation to realistic quark masses (filled circles). The lower panels show the comparison of the Lattice results with the data.*

Ultimately, we want to come to a QCD description of these important structure quantities. There has been considerable progress in this regard in the past few years. The first full calculations on the lattice (LQCD) are now available (Alexandrou *et al.* 2004), and the Q^2 dependence of both $R_{EM}(Q^2)$ and $R_{SM}(Q^2)$ have been computed in the lower Q^2 range within quenched LQCD (Alexandrou *et al.* 2005). While the results for R_{EM}, shown in Figure 4, still have large uncertainities, they agree in sign and magnitude with the experimental results from CLAS (Joo *et al.* 2002) and Bates (Mertz *et al.* 2001). The Lattice results also agree with the trend of the CLAS data for R_{SM}, but show a significant discrepancy with the low Q^2 point from Bates, as shown in Figure 5. The results confirm the interpretation of the quadrupole contributions in terms of an oblate deviation of the Δ from spherical symmetry. Such a deformation was recently discussed in a model analysis (Buchmann and Henley 2000), in which two-body $q\bar{q}$ currents are largely responsible for the deformation. This suggests that its origin lies in the baryon's $q\bar{q}$ cloud. Dynamical models also attribute the significant quadrupole transition amplitudes to meson cloud effects (Sato and Lee 2001, Kamalov and Yang 1999, Kamalov and Yang 2001).

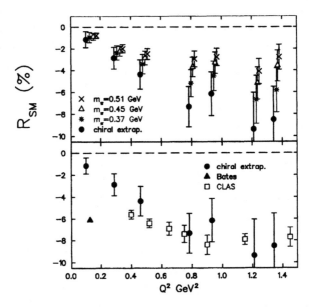

Figure 5. *Comparison of LQCD results in quenched approximation and experimental data for R_{SM}. The upper panels shows the Lattice results for different quark masses and the chiral extrapolation to realistic quark masses (filled circles). The lower panels show the comparison of the Lattice results with the data.*

If the nucleon and Delta deformations are mainly caused by meson cloud effects, we should expect a diminishing of this contribution with increasing Q^2 when the interior of the nucleon is probed. Figure 2 shows the magnetic transition form factor $G_M^*(Q^2)$ as determined from the M_{1+} multipole in comparison with dynamical model calculations. Meson cloud effects, which in dynamical models contribute more than 30% to the total G_M^* strength at $Q^2 = 0$, are reduced with increasing Q^2, and contribute less than 10% at $Q^2 > 4$ GeV^2. The R_{EM} values in Figure 4 also appear to drop in magnitude with increasing Q^2, in comparison with $R_{EM} \approx 0.025$ at small Q^2.

With the preliminary CLAS data, the $N\Delta$ transition measurements have now been extended to $Q^2 = 6$ GeV^2. An extension to even higher Q^2 may require additional experimental information to help reduce the model uncertainty in the analysis. The biggest uncertainties in the analysis are due to background amplitudes, some of which are unknown at high Q^2. For example, the pion form factor $F_\pi(Q^2)$, which is unknown at high Q^2, enters into the $N\Delta$ analysis through the pion rescattering. The simultaneous analysis of the $ep \to en\pi^+$ channel can eliminate this uncertainty. Also, as will be discussed in the following section, measurements of the polarised beam response function $\sigma_{LT'}$ can provide information about background amplitudes.

3.2 Electroproduction of single pseudoscalar mesons, beyond the $\Delta(1232)$

In the higher mass region above the $\Delta(1232)$ there are increasing numbers of partial waves and corresponding multipoles contributing to the production strength. A multipole analysis, as described for the Δ region, will be much more difficult and requires measurement of many polarisation observables. There are no resonances at higher masses that dominate the cross section to the same degree as the $\Delta(1232)$ does. Resonances overlap and background amplitudes become more important. However, as resonances couple to different final states with different isospin composition, more experimental information may be accessed. Beside single pions, many other final states such as $p\eta$, $N\pi\pi$, $p\omega$, $K\Lambda$, Σ, $p\eta'$ can be employed to study resonance transitions at higher W. Moreover, theoretical constraints on meson production amplitudes can be used to reduce ambiguities in the absence of sufficient data needed for a completely model-independent extraction of amplitudes. The most extensively used phenomenological approaches in the analysis of pion electroproduction data are the unitary isobar model, and fixed-t dispersion relations. Detailed descriptions of various phenomenological approaches are given in a recent review article (Burkert and Lee 2004). We here describe briefly the unitary isobar model of which MAID is the best known representative.

3.2.1 Unitary Isobar Model

Unitary isobar models (UIM) such as MAID (Drechsel *et al.* 1999), and later extensions such as JANR (Aznauryan 2003a, 2003b) are designed for the analysis of single channel processes such as single pion production. They relate the scattering amplitude T to the K matrix:

$$T_{\pi N, \gamma N} = e^{i\delta_{\pi N}} \cos \delta_{\pi N} K_{\pi N, \gamma N}, \tag{8}$$

where $\delta_{\pi N}$ is the pion-nucleon scattering phase shift taken from elastic πN scattering. If one assumes that K can be expressed as a sum of background and resonant terms $K = V = v^{bg} + v^R$ one obtains the following relation:

$$T_{\pi N, \gamma N}(UIM) = e^{\delta_{\pi N}} \cos \delta_{\pi N} [v_\pi^{bg} N, \gamma N] + \sum_{N_i^*} T_{\pi N, \gamma N}^{N_i^*}(E), \tag{9}$$

where the second term contains all the resonance contributions. In this approximation, non-resonant multi-channel effects such as $\gamma N \to (\rho N, \Delta\pi) \to N\pi\pi$, which could be important at higher masses, are neglected. The resonances are parameterised using a relativistic Breit-Wigner form with energy-dependent width

$$T_{\pi N, \gamma N}^{N_i^*}(E) = f_{\pi N}^i(E) \frac{\Gamma_{tot} M_i e^{i\Phi}}{M_i^2 - E^2 - iM_i\Gamma^{tot}} f_{\gamma N}^i(E) \bar{A}^i, \tag{10}$$

where $f^i_{\gamma N}(E)$ and $f^i_{\pi N}(E)$ are the form factors describing the decays of N^*, Γ_{tot} is the total decay width and $\bar{A}^i$ is the $\gamma N \to N^*$ excitation strength. The phase is determined by the unitarity condition and by the assumption that the phase ψ of the total amplitude is related to the πN phase shift $\delta_{\pi N}$ and inelasticity $\eta_{\pi N}$ by

$$\psi(E) = \tan^{-1}\left[\frac{1 - \eta_{\pi N}(E)\cos 2\delta_{\pi N}(E)}{\eta_{\pi N}(E)\sin 2\delta_{\pi N}(E)}\right]. \tag{11}$$

In JANR the high energy behaviour is parameterised using a Regge dependence that smoothly connects to the lower energy regime. In all UIM the background amplitudes are parameterised by the pion Born terms, and ρ and ω exchange terms that contribute in the higher mass region. The pion Born terms are parameterised by the nucleon and pion form factors.

3.3 The nature of the Roper resonance $P_{11}(1440)$

In the $SU(6) \otimes O(3)$ symmetry scheme discussed in section 2, the Roper resonance has traditionally been assigned to the $[56, 0^+]_2$ supermultiplet. However, the non-relativistic constituent quark model has difficulties in describing its photo- and electro-coupling amplitudes. This has led to alternate approaches, such as modelling the Roper on the light-cone (Capstick and Keister 1995), or as a hybrid baryon with a large gluonic component (Li *et al.* 1992), as a baryon with a small quark core and a large meson cloud (Cano and Gonzalez 1998), and as a nucleon-sigma molecule (Krewald *et al.* 2000). Recent quenched LQCD calculations (Lee *et al.* 2003, Lee 2004) show a clear signal for the Roper as a 3-quark state indicating that the state must have a significant 3-quark component, while other components are not ruled out. Also, in this calculation the mass was found close to the experimental value. The aforementioned models make distinct predictions for the photocoupling amplitudes, especially their Q^2 evolution. While the photocoupling amplitude $A_{1/2}$ is quite well known, there has been a lack of reliable data on the Q^2 dependence. Also, the longitudinal amplitude $S_{1/2}(Q^2)$, which is especially sensitive to the spatial structure, is poorly known. One analysis (Gerhardt *et al.* 1980) of $p\pi^0$ electroproduction data gave results consistent with $S_{1/2}(Q^2) = 0$. In electroproduction of π^0 from protons the resonance is masked by the large contribution from the Δ. As an isospin $\frac{1}{2}$ state the P_{11} couples more strongly to the $n\pi^+$ channel than to $p\pi^0$ ($\pi^+n/\pi^0p = 2$), while the opposite is the case for the Δ ($\pi^+n/\pi^0p = 1/2$). The high statistics data from CLAS on $n\pi^+$ electroproduction (Smith 2004) have now been included in the analysis of pion electroproduction data. Moreover, there are now data on the beam-helicity dependent response function $\sigma_{LT'}$ for both $n\pi^+$ and $p\pi^0$ available. The response function corresponds to the imaginary part of an LT interference term

$$\sigma_{LT'} \propto Im(L) * Re(T) + Re(L) * Im(T). \tag{12}$$

In the presence of a large (real) background amplitude σ'_{LT} is highly sensitive to even small (imaginary) resonance amplitudes. This helps in the extraction of a small resonance signal within a large background. This is somewhat analogous to the extraction of the small S_{1+} in the $\Delta(1232)$ region, described earlier. In the response function

$$\sigma_{LT} \propto Re(L) * Re(T) + Im(L) * Im(T), \tag{13}$$

the Δ contributes a large $Im(T)$ resonant amplitude through the M_{1+} multipole. The interference boosts the small resonant $Im(L)$ amplitude, *ie* S_{1+}.

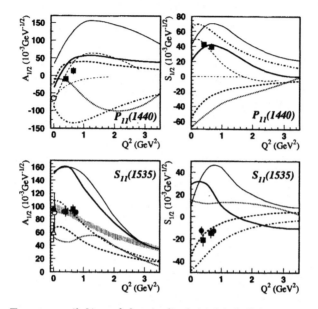

Figure 6. *Transverse (left) and longitudinal (right) helicity amplitudes for the $P_{11}(1440)$. Bold and thin solid lines correspond to relativistic and non-relativistic quark model calculations (Capstick and Keister 1995). Dashed lines correspond to light-front calculations (Pace et al. 1999). Dotted, dashed-dotted and thin dashed lines correspond to quark models (Warns et al. 1990, Aiello et al. 1998, Cano and Gonzalez 1998). The shaded band indicates the averages from previous analyses.*

Using the unitary isobar model JANR03, as well as fixed-t dispersion relations, the $P_{11}(1440)$ transverse and longitudinal helicity amplitudes $A_{1/2}$ and $S_{1/2}$ have been extracted from a combined fit to the π^+n and π^0p cross sections, and to the beam polarisation data for these channels. The results for two values of Q^2 are shown in Figure 6. We see a rapid change of the $A_{1/2}$ helicity amplitude with Q^2. Instead of increasing in magnitude as predicted by nonrelativistic constituent quark models, $A_{1/2}$ drops rapidly in magnitude from its value at the photon point, and appears to change sign near $Q^2 = 0.5$ GeV^2. The longitudinal helicity amplitude $S_{1/2}$ is large and positive. Both amplitudes show a behaviour that is consistent with what is expected from a model that includes large vector meson contributions and a small quark core (Cano and Gonzales 1998). The large $S_{1/2}$ amplitude effectively excludes an interpretation of this state as a hybrid baryon. Ironically, the hybrid baryon model was used to explain what was seen as lack of longitudinal strength for the Roper excitation resulting from the aforementioned analysis (Gerhardt *et al.* 1980) of older $p\pi^0$ cross section data. The results of that analysis are clearly not supported by the analysis of the new precise, and more complete data. It also illustrates the importance of including both final states $p\pi^0$ and $n\pi^+$ as well as polarisation observables in the resonance analysis. The comparison with the light cone model (Capstick and Keister 1995) shows that relativity may also be important. Interestingly, a large meson component is also found in the chiral quark model (Risko 2004), which assigns only a 20-30% 3-quark component to the Roper.

3.4 The hard transition form factor of the $S_{11}(1535)$

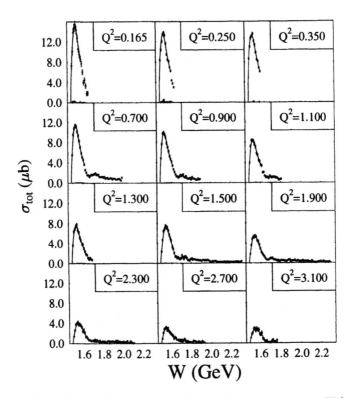

Figure 7. *Total cross section for η electroproduction from protons versus W for different* Q^2.

The $S_{11}(1535)$ is known to have a large branching fraction to the $p\eta$ final state. There is no other state close by in mass with significant coupling to $N\eta$. This channel can then used to isolate the $S_{11}(1535)$ from other isospin $\frac{1}{2}$ resonances, especially the nearby $P_{11}(1440)$ and $D_{13}(1520)$. In addition to a strong $p\eta$ coupling the $N\pi$ coupling is also significant. The two channels can then used to study a possible channel-dependence of the analysis results. Earlier comparisons of results from the analysis of $p\eta$ and $N\pi$ channels in photoproduction showed significant disagreement (Arndt *et al.* 1996). The $S_{11}(1535)$ shows up prominently and for all Q^2 values in the total cross section (Thompson *et al.* 2001, Denizli 2004) of $\gamma^*p \to p\eta$. This is shown in Figure 7. Even at the highest $Q^2 = 3.1$ GeV^2, there is almost no background under the resonance peak.

Similar to the procedure used for the multipole analysis of the $N\Delta$ transition described in section 3.1, the differential cross section for $p\eta$ production can be expanded in terms of the Legendre polynomials using Equation 7. If we limit $\ell \leq 2$ we retain

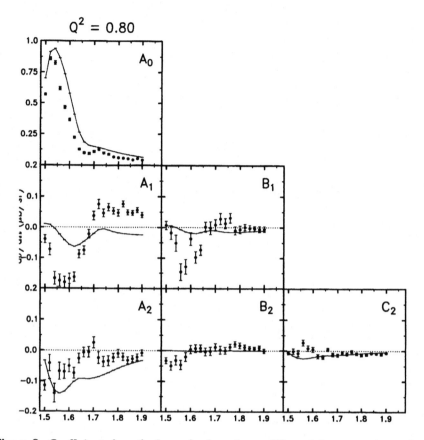

Figure 8. *Coefficients from the Legendre fit to the pη differential cross section at fixed Q^2.*

the six Legendre coefficients shown in Figure 8. The amplitude A_0 is dominated by the resonant s-wave multipole $|E_{0+}|^2$ which can be extracted directly from A_0. In contrast to the Δ region, where the interference term $Re(S_{1+}M_{1+}^*)$ enhances the small longitudinal multipole, there is no interference term in the σ_{LT} response function for the s-wave that could boost the longitudinal multipole S_{0+}. In leading order, S_{0+} only interferes with p-wave multipoles such as M_{1-} and M_{1+}. In η production there are also no large resonant p-wave multipoles nearby making the extraction of S_{0+} through interference in that channel less sensitive. In pion production, the high energy tail of the dominant M_{1+} from the Δ or the M_{1-} of the nearby Roper $P_{11}(1440)$ may generate sizeable interference responses. This emphasises the importance of using both the $p\eta$ and the $N\pi$ channels in the analysis.

An analysis of the $N\pi$ channel in photoproduction (Arndt *et al.* 1996) yielded an $A_{1/2}(0)$ amplitude of only about 60% of the corresponding $p\eta$ analysis. This discrepancy is absent from the electroproduction analysis (Aznauryan *et al.* 2005). The latter results are shown in Figure 6. The $p\eta$ and $N\pi$ channel give consistent results for both Q^2 values. Moreover, the isobar model and the dispersion relation analysis also agree within their uncertainties. There is no visible Q^2 dependence of $A_{1/2}$ in the range $Q^2 = 0. - 0.75$ GeV^2. The very slow fall off of $A_{1/2}$ at higher Q^2 is indicated by the shaded band in the graph. The hard transition form factor has been a nagging problem for quark models for decades. However, progress has been made recently. The quark model with hypercentral potential (Aiello *et al.* 1998) gets the magnitude near the photon point right, while the light-front model (Pace *et al.* 1999) has approximately the correct Q^2 dependence but is low in magnitude. The two models also get the correct sign for the longitudinal amplitude $S_{1/2}$, and predict the correct magnitude. Lattice QCD calculations (Lee 2004) show a solid 3-quark component for this state, and estimates within the chiral quark model give a nearly 100% 3-quark component for this state (Riska 2004). This is qualitatively consistent with the hard form factor discussed here. The $S_{11}(1535)$ appears to have all the characteristics of the lowest, pure 3-quark state in the nucleon spectrum. To come to more definite conclusions, full LQCD calculations of the electro-coupling amplitudes for the $\gamma_v p S_{11}(Q^2)$ transition will be needed.

3.5   The helicity structure of the $D_{13}(1520)$

The $D_{13}(1520)$ resonance is excited predominantly through the helicity amplitude $A_{3/2}$. This behaviour can qualitatively be accomodated in the non-relativistic constituent quark model with harmonic oscillator potential. In the simplest version of the model only spin flip and orbit flip transitions acting on a single quark occur. The ratio of the two helicity amplitudes is given by (Copley *et al.* 1969):

$$\frac{A_{1/2}^{D13}}{A_{3/2}^{D13}} = \frac{1}{\sqrt{3}}\left(1 - \frac{\vec{Q}^2}{\alpha^2}\right). \tag{14}$$

Obviously, $A_{1/2}$ will become dominant at sufficiently high Q^2. This prediction was one of the early successes of the dynamical quark model (Close and Gilman 1972). The data confirm this predictions as can be seen in Figure 9. The trend of older analyses is shown with the shaded band. The most recent data confirm the low Q^2 behaviour, and indicate that the cross over may occur in the range $Q^2 = 0.45 - 0.65$ GeV^2.

3.6   Single Quark Transition Model (SQTM)

To the degree that resonance transitions can be described by a single quark transition, the electromagnetic current operator for radiative transitions between the $[56, 0^+]_0$ and $[70, 1^-]_1$ supermultiplets is given by (Hey and Weyers 1974, Cottingham and Dunbar 1979):

$$J_{em}^+ = AL^+ + B\sigma^+ L_z + C\sigma_z L^+, \tag{15}$$

where σ is the quark Pauli spin operator, and the terms with B and C in front operate on the quark spatial wavefunction changing the component of orbital angular momentum along the direction of the momentum transfer (z axis). The A term corresponds to a quark orbit flip with $\Delta L_z = +1$, term B to a quark spin flip with $\Delta L_z = 0$, and term C corresponds to a simultaneous quark orbit and quark spin flip with $\Delta L_z = +1$. The relations between the SQTM amplitudes and the $A_{1/2}$ and $A_{3/2}$ helicity amplitudes are

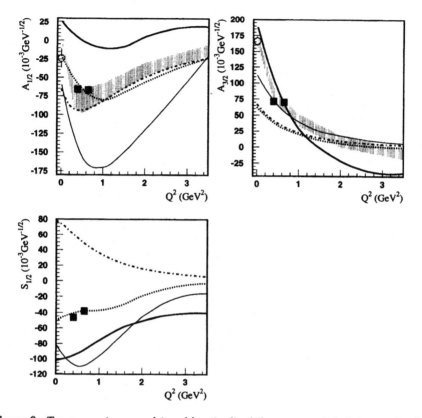

Figure 9. *Transverse (top panels) and longitudinal (bottom panel) helicity amplitudes for the $\gamma p D_{13}(1520)$ transition. The curves are explained in Fig. 6.*

given in Table 1. The Moorhouse (1966) selection rule, derived from $SU(6)$ symmetry, forbids excitation of the $S_{11}(1650)$, $D_{13}(1700)$, and $D_{15}(1675)$ from the proton. Configuration mixing between states within $[70, 1^-]$ breaks this symmetry. The degree of symmetry breaking is parameterised by the mixing angle.

Using the experimentally measured $A_{1/2}$ and $A_{3/2}$ amplidues for the $S_{11}(1535)$ and the $D_{13}(1520)$, the SQTM amplitudes can be determined (Burkert *et al.* 2003). They are shown in Figure 10. Amplitudes for all other states in the same multiplet can then be predicted. Figure 11 shows predictions for the helicity amplitudes in other states in comparison with data.

The SQTM predictions show remarkable agreement with the data at the photon point, and where electroproduction data are available, *eg* for the $S_{11}(1650)$, there is also good agreement. For most of the states, the predicted Q^2 dependence cannot be tested due to the lack of data. It is worth noting that the SQTM predicts the $S_{11}(1650)$ and $D_{13}(1700)$ states to be excited from proton targets only because of configuration mixing of two $J^P = \frac{1}{2}^-$ quark configuration states with total quark spin $S_{3q} = \frac{3}{2}$ and $S_{3q} = \frac{1}{2}$. For the two D_{13} states the mixing angle is small, and excitation from protons is predicted

State	Proton target	Neutron target
$S_{11}(1535)$	$A_{1/2}^+ = \frac{1}{6}(A + B - C)\cos\theta$	$A_{1/2}^\circ = -\frac{1}{6}(A + \frac{1}{6}B - \frac{1}{3}C)$
$D_{13}(1520)$	$A_{1/2}^+ = \frac{1}{6\sqrt{2}}(A - 2B - C)$	$A_{1/2}^\circ = -\frac{1}{18\sqrt{2}}(3A - 2B - C)$
	$A_{3/2}^+ = \frac{1}{2\sqrt{6}}(A + C)$	$A_{3/2}^\circ = \frac{1}{6\sqrt{6}}(3A - C)$
$S_{11}(1650)$	$A_{1/2}^+ = \frac{1}{6}(A + B - C)\sin\theta$	$A_{1/2}^\circ = \frac{1}{18}(B - C)$
$D_{13}(1700)$	$A_{1/2}^+ = 0$	$A_{1/2}^\circ = \frac{1}{18\sqrt{5}}(B - 4C)$
	$A_{3/2}^+ = 0$	$A_{3/2}^\circ = \frac{1}{6\sqrt{15}}(3B - 2C)$
$D_{15}(1675)$	$A_{1/2}^+ = 0$	$A_{1/2}^\circ = -\frac{1}{6\sqrt{5}}(B + C)$
	$A_{3/2}^+ = 0$	$A_{3/2}^\circ = -\frac{1}{6}\sqrt{\frac{2}{5}}(B + C)$
$D_{33}(1700)$	$A_{1/2}^+ = \frac{1}{6\sqrt{2}}(A - 2B - C)$	same
	$A_{3/2}^+ = \frac{1}{2\sqrt{6}}(A + C)$	same
$S_{31}(1620)$	$A_{1/2}^+ = \frac{1}{18}(3A - B + C)$	same

Table 1. *Helicity amplitudes for the electromagnetic transition from the ground state* $[56, 0^+]$ *to the* $[70, 1^-]$ *multiplet as a function of the SQTM amplitudes.* θ *is the mixing angle relating two* $J^P = \frac{1}{2}^-$ *states with* $s_{3q} = \frac{3}{2}$ *and* $s_{3q} = \frac{1}{2}$. *There is also a small mixing angle for the two* $\frac{3}{2}^-$ *states resulting in the physical states* $D_{13}(1520)$ *and* $D_{13}(1700)$. *We have not included the latter mixing angle in the table.*

to be very small. In case of the $D_{15}(1675)$ only the configuration with $s_{3q} = \frac{3}{2}$ exists, and there is no mixing possible. Within the SQTM this state cannot be photoexcited from protons. The Particle Data Group (PDG: Eidelmann *et al.* 2004) gives small but non-zero values of $A_{1/2} = (19\pm8) \times 10^{-3}\,\text{GeV}^{-1/2}$, and $A_{3/2} = (15\pm9) \times 10^{-3}\,\text{GeV}^{-1/2}$. If this indicates a small violation of the SQTM remains to be seen with new, more precise photoproduction data. For neutron targets, all states within $[70, 1^-]$ may be excited. While the $D_{13}(1700)$ should only be weakly excited from neutrons, the $D_{15}(1675)$ should have sizeable photocoupling amplitudes for neutrons. This is indeed seen in the data.

Given its simplicity, the SQTM appears to represent the main features of phototransition from the nucleon to the $[70, 1^-]_1$ supermultiplet. It will be interesting to see if this remains to be the case for electroproduction, especially when the short distance behaviour of resonance transition is probed at high photon virtuality Q^2. One also would like to test this model for other supermultiplets, *eg* the transition to the $[56, 2^+]_2$. This will be possible when the photo- and electrocoupling amplitudes for at least two states in that muliplet have been measured. To date, electroproduction data are available only for the $F_{15}(1680)$. Since most states in that multiplet couple strongly to the $N\pi\pi$ channel we have to wait until data are available for that channel. The prospects for that are good, as several experiments are currently analysing photo- and electroproduction data of $p\pi^0\pi^0$ and $p\pi^+\pi^-$ channels, respectively.

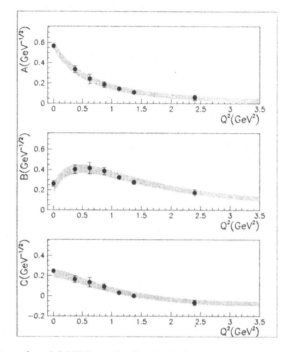

Figure 10. *The reduced SQTM amplitudes A, B, C extracted from the $S_{11}(1535)$ and $D_{13}(1520)$ helicity amplitudes.*

4 Searching for new baryon states

The symmetric quark model predicts many states with masses near 2 GeV that have not been found experimentally. This has been called the problem of the "missing resonances". As the quark model predicts so many states with presumably large widths, it may not be surprising that some excited states have escaped detection, especially if they have small coupling to the initial or final state. Why should we care about the "missing resonances"? The answer is that $SU(6) \otimes O(3)$ is not the only possible symmetry group representing the spectrum of excited states. Other symmetries have recently been discussed that represent a different baryon spectrum (Kirchbach 2000). In some models, two quarks form a cluster that acts as a single diquark. For example, the diquark cluster model (Carroll *et al.* 1968) predicts a different baryon spectrum. While the supermultiplets $[56, 0^+]$, $[70, 1^-]$, $[56, 2^+]$ also exist in this scheme, the diquark model predicts no states belonging to supermultiplets $[70, 0^+]$, $[70, 2^+]$ and $[20, 1^-]$. Hence, many P_{11}, P_{13}, and F_{15} states are not predicted in this model. Nonetheless, all states with 4-star ratings by the PDG can be accomodated in that scheme, while only one 3-star state, the $P_{11}(1710)$, is not predicted. For all other states that are not predicted in the diquark model there is also no solid experimental evidence.

The absence of some states that are predicted in the symmetric model may also have a much simpler explanation: they haven't been found yet. Indeed most of the known states have been observed in πN scattering. If the state does not couple to πN, or couples

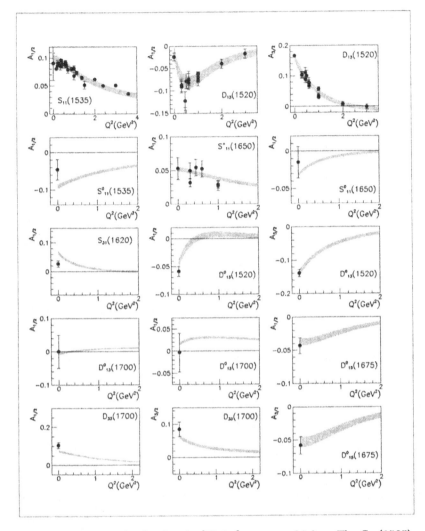

Figure 11. *SQTM amplitudes for the $[70, 1^-]_1$ supermultiplet. The $S_{11}(1535)$ and $D_{13}(1520)$ amplitudes on the proton are used as input. All other amplitudes are predictions of the SQTM.*

only weakly, it should not be surprising that it hasn't been seen. This conjecture is supported by calculations of decay properties of baryon resonances (Koniuk and Isgur 1980, Capstick and Roberts 1994), and many of the unobserved states are indeed predicted to have a small πN widths. On the other hand, many states are predicted to have strong $N\pi\pi$ couplings, and some with significant $p\omega$ or $K\Lambda$ widths. A number of states are also predicted with significant γN coupling. These predictions let to systematic searches for "missing" states in the channels $\gamma p \rightarrow p\pi^+\pi^-$, $\gamma p \rightarrow p\pi^0\pi^0$, $\gamma p \rightarrow K^+\Lambda$, and $\gamma p \rightarrow p\omega$. While many hints of new resonant states are clearly present in these data sets, no new state has yet been convincingly identified in electromagnetic interactions. This, however, seems to be just a matter of time until the new data include measurements of polarisation observables, and the analysis procedures are sufficiently sophisticated to deal with large and complex background amplitudes that are present in many of these channels. In the following, we focus on the $p\pi^+\pi^-$ channel where new data are available (Ripani *et al.* 2003, Thoma 2003, Bellis 2004) and are currently being analysed.

4.1 The $\gamma^*p \to p\pi^+\pi^-$ channel

The photoproduction of two pions is a complicated channel to analyze. Besides the expected resonance contributions, large background contributions are expected. For example, the channel $\gamma p \to p\rho$ will have a large background component due to diffractive scattering, especially for real photons, and at high W will dominate the forward angle region. With increasing photon virtuality, Q^2, this background component should decrease, which makes electroproduction possibly a more sensitive probe for resonances in that channel. To extract resonance information one has to deal with complex amplitude interferences. Analysis procedures can be divided into two distinctly different approaches. The historically most frequently used approach is based on an energy-dependent description of the reaction within an isobar model representation. The second approach uses a partial wave formulation. The two methods are described briefly in the following subsections.

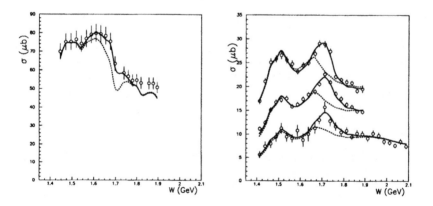

Figure 12. *Total cross section for $p\pi^+\pi^-$ in photoproduction (left panel), and electroproduction (right panel) at $Q^2 = 0.65$, 0.95, 1.3 GeV² values (from the top). The curves are described in the text.*

4.1.1 Isobar model

In this model, amplitudes are parameterised as a sum of resonant and non-resonant amplitudes. Non-resonant amplitudes are described by Born terms and other terms that may contribute to the production cross section, and whose strengths must be determined by fits to the data. Resonances are usually parameterised as relativistic Breit-Wigner forms. Model parameters are fitted to one-dimensional projections of the cross section on invariant mass distributions and angular distributions. To search for new resonances s-channel Breit-Wigner forms are introduced in some partial waves and included in the fit. This approach makes maximum use of accumulated knowledge from hadronic processes, and can provide a good description of the projected data. A certain disadvantage of this approach is that due to the use of one-dimensional projections, correlations within events are lost. The method has been used successfully in the analysis of CLAS electroproduction data (Ripani *et al.* 2003). In this analysis a significant discrepancy between the data and the resonance parameterisations, implemented in the fit model, was found near a mass of 1.7 GeV for the $p\pi^+\pi^-$ system. The discrepancy is attributed to either inaccurate hadronic couplings for the well known $P_{13}(1720)$ determined in previous analysis of hadronic experiments, or to an additional resonance with $J^P = \frac{3}{2}^+$ with undetermined isospin. The discrepancy is visible in the total cross section for electroproduction, shown in Figure 12 (right). The dotted line shows the model predictions using resonance parameters from single pion electroproduction and from the analysis of $\pi N \to N\pi\pi$. The solid line represents the fit when the hadronic couplings of the $P_{13}(1720)$ to $\Delta\pi$ and $N\rho$ are allowed to vary significantly beyond the ranges established in the analysis of hadronic data. Alternatively, a new state was introduced with hadronic couplings extracted from the fit while keeping the parameters of the known $P_{13}(1720)$ at the previously established values. In either case, the fit requires a resonance with hadronic couplings that are significantly different from the ones of $P_{13}(1720)$ listed by PDG.

Another, at first surprising, aspect is that the total photoproduction cross section in Figure 12 (left) shows a W dependence that is very different from the electroproduction data (right). The peak seen at 1.72 GeV is completely invisible. The isobar analysis shows that this behaviour is largely due to much higher background contribution from non-resonant $\Delta^{++}\pi^-$ production in the real photon sector. Also, diffractive $p\rho^0$ production increases rapidly with W above threshold. Both data sets are consistent with a strong resonance near $W = 1.72$ GeV in the $\frac{3}{2}^+$ partial wave (Mokeev *et al.* 2004).

4.1.2 Partial wave analysis

The second approach (Bellis 2004) is based on a partial wave formulation starting from the T-matrix at a given photon energy E:

$$T_{fi}(E) = <p\pi^+\pi^-; \tau_f|T|\gamma p; E> \tag{16}$$

$$= \sum_\alpha <p\pi^+\pi^-; \tau_f|\alpha><\alpha|T|\gamma p; E> \tag{17}$$

$$= \sum_\alpha \psi^\alpha(\tau_f)V^\alpha(E), \tag{18}$$

where α denotes all the intermediate states, and τ_f characterises the final state kinematics. The decay amplitude $\psi^\alpha(\tau_f)$ is calculated using an isobar model for specific decay channels, *eg* $\Delta^{++}\pi^-$, $\Delta^0\pi^+$, or $p\rho^0$. The production amplitude V^α is then determined at fixed energy using a maximum likelihood fit. In contrast to the isobar model formulation, no assumptions are made on the energy dependence of intermediate resonances in an attempt to avoid introducing any bias. Each energy bin is then fitted independently.

This method makes use of all information contained in the data, and takes into account correlations within individual events. Since the energy dependence is not constrained the method is more prone to ambiguities if the number of partial waves needed is large and only unpolarised cross section data are available.

4.2 Strangeness production and resonances

Kaon production from nucleons has long been recognised as a potentially very sensitive tool in the search for excited baryon states (Capstick and Roberts 1994). Analyses of the $K\Lambda$ and $K\Sigma$ channels include the isospin selectivity; the $K\Lambda$ final state selects isospin $\frac{1}{2}$, similar to the $p\eta$ channel, while $K\Sigma$ couples to both N^* and Δ^* resonances. An important tool in resonance studies is the measurement of polarisation observables. The self-analysing power of the weak decay $\Lambda \to \pi^- p$ can be utilised to measure the recoil Λ polarisation. New high statistics data on cross sections and polarisation observables covering the nucleon resonance region are now available from CLAS (McNabb *et al.* 2004, Niculescu 2004) and SAPHIR (Glander 2004).

The high statistics of these data allow structures in the differential cross sections to be identified that hint at the presence of previously unobserved resonances. These new states emerge largely through the interference with non-resonant amplitudes. The presence of s-channel resonances is particularly evident in the W-dependence shown in Figure 13. At the most forward angles (upper panel), two resonance-like structures are visible at $W = 1.7$ GeV, and at $W = 1.95$ GeV. The structure at 1.7 GeV could be accomodated by the known states $S_{11}(1650)$, $P_{11}(1710)$, and $P_{13}(1720)$, if the $K\Lambda$ strength of these states are allowed to vary. At intermediate angles (middle panel), the data show a smoother falloff with W, while at backward angles (lower panel) another resonance-like structure near 1.87 GeV emerges, overlapping with the structure at higher mass. These distributions reveal that complex interference processes are at work likely involving several resonances.

A similar pattern is seen in the electroproduction of $K^+\Lambda$. Integrated $K^+\Lambda$ cross sections are shown in Figure 14 for the forward and backward hemispheres. The forward cross section shows no pronounced structures, while the backward cross section clearly shows resonance-like structures near 1.7 GeV and near 1.87 GeV. Likely, these represent excitation of the same resonances also seen in photoproduction.

Searches for baryon resonances are also underway in $p\pi^0\pi^0$ photoproduction (Thoma 2003) and in $p\omega$ photo- and electroproduction (Cole 2004).

5 Flavour exotic baryons - Pentaquarks

If we scatter K^- mesons with quark content ($\bar{u}s$) off a target of protons with quark content (uud) we have a system with quark content ($uud\bar{u}s$). A state with such a 5-quark configuration would have the same quantum numbers as a Λ hyperon (uds). Indeed, the well known $\Lambda(1520)$ is generated in such a way through s-channel resonance formation. The state could have both 3-quark and 5-quark (or baryon-meson) components. The latter component would not be "flavour exotic", and we will not consider such states here. If we scatter K^+ mesons with quark content ($\bar{s}u$) from neutrons with quark content

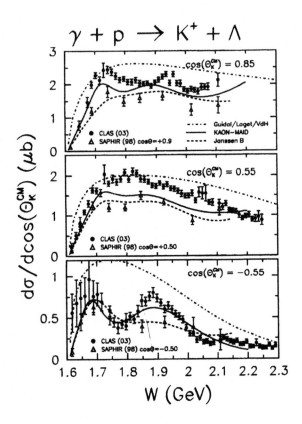

Figure 13. *Differential cross sections for the photoproduction of $K^+\Lambda$. Structures visible at different angular ranges indicate the excitations of resonances interfering with background amplitudes. Various model calculations are included (Janssen et al. 2001, Guidal et al. 2003, Mart et al. 2000).*

(udd) we have a system with quark content of $(uudd\bar{s})$. Such a system does not have flavour quantum numbers that can be represented by a 3-quark baryon. It would be a baryon with minimal quark content of 5-quark, or a "flavour exotic" baryon. Can such a pentaquark baryon exist? QCD does not present any restriction that would preclude such an object from existing, and searches for pentaquarks have been undertaken for decades. No convincing evidence has been found, until recently.

The interest in pentaquark baryons was rekindled with a seminal paper by Diakonov *et al.* (1997). Using a chiral soliton model (χSM) the Θ^+ would be an isosinglet with strangeness $= +1$ in an anti-decuplet of ten 5-quark states with spin $s = \frac{1}{2}$. The anti-decuplet is shown in Figure 15. Only the Θ^+ with strangeness $S = +1$, and the Ξ^{--} and Ξ^+ with $S = -2$ have flavour exotic quantum numbers. The other members of the anti-decuplet have non-exotic quantum numbers that are also observed for 3-quark states. In the original prediction of the Θ^+, the mass was predicted based on the equal

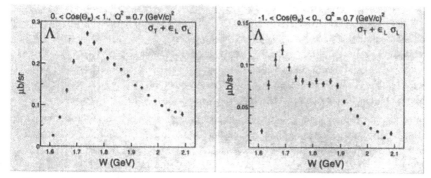

Figure 14. *The electroproduction of $K^+\Lambda$ integrated over the forward (left) and backward (right) hemispheres.*

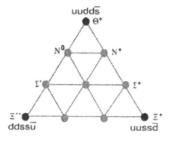

Figure 15. *The $\overline{10}$ as predicted in the χSM.*

180 MeV mass spacing between states with $\Delta S = \pm 1$ resulting from the symmetry of the χSM. The exact splitting depends on the magnitude of the $\Sigma_{\pi N}$ term, which is not well known and has changed significantly over the past decade. By assigning the mass of the N^0 and N^+ members of the anti-decuplet to the known $N'(1710)$, the mass of the Θ^+ was predicted by Diakonov et al. (1997) to be 1530 MeV with a width of less than 15 MeV. The low mass and the narrow width made the Θ^+ an interesting object for experimentalists to search for. Recent theoretical developments are discussed in Oka (2004).

The first claim of the observation of the Θ^+ came from the LEPS experiment (Nakano et al. 2003) at the SPring-8 accelerator. LEPS studied the reaction $\gamma C \rightarrow K^+K^-(n)$, where a peak with about 20 events is observed in the missing mass of $\gamma C \rightarrow K^-X$ after corrections have been made for the Fermi momentum in the carbon nucleus. This observation was followed by several confirmations, and to date there are ten published observations of a Θ^+ signal. Most observations are made in electromagnetic interactions (Stepanyan et al. 2003, Koubarovsky et al. 2004, Barth et al. 2003, Chekanov et al. 2004, Airapetian et al. 2004), but signals have also been seen in hadronic processes (Barmin et al. 2003, Abdel-Bary et al. 2004, Aleev et al. 2004), and in neutrino interactions (Asratyan et al. 2004). The width of the experimental signal is limited by the experimental resolution. However, theoretical analyses of K^+d and K^+A scattering data (Cahn and Trilling 2004, Gibbs 2004a, Arndt *et al.* 2003) indicate that the natural widths of the observed state cannot be much larger than 1 MeV. This is a very narrow width for a hadronic decay, and a generally accepted explanation is still lacking. There are also a number of experimental data mostly from high energy experiments that show no signal (Hicks 2005). At least for some of these non observations it is questionable if the measurements had sufficient sensitivity to reveal a signal for the Θ^+.

In the following I will discuss as examples the data from CLAS, both on deuterium and on hydrogen. Other experiments, including the NA49 results on another exotic candidate Ξ^{--} and the negative searches, are discussed in the lectures by Rosner (2004).

5.1 Photoproduction of Θ^+ on deuterium

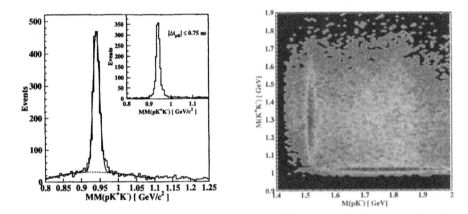

Figure 16. *Left: Missing mass M_X of $\gamma d \rightarrow pK^-K^+X$. The peak is due to the neutron. The inset shows how the background vanishes for tighter timing cuts. Right: Dalitz plot for the masses of K^+K^- versus K^-p. The ϕ and $\Lambda(1520)$ bands are clearly visible.*

In the absence of low energy kaon beams, photoproduction maybe the most efficient way of exciting the Θ^+. The generalised vector dominance picture (Sakurai and Schildknecht 1972) tells us that high energy photons have a certain probablity to fluctuate into virtual vector mesons (ρ, ω, and ϕ), and interact hadronically. Through the ϕ content, high energy photons contain a significant amount of ($s\bar{s}$). In a hadronic picture, the ϕ meson would couple to virtual $K\bar{K}$ pairs, and thus the photon has the ingredients needed to form a $udud\bar{s}$ configuration in an s-channel process on neutrons $"K^+" + n \rightarrow \Theta^+$. This provides photoproduction experiments with a significant advantage over experiments where the Θ^+ has to be produced through quark fragmentation, such as in central e^+e^- collisions.

The CLAS experiment claims to have observed the Θ^+ signal on deuterium in the reaction $\gamma d \rightarrow K^-pK^+(n)$. In order to observe the process in CLAS a second scattering is required to give the spectator proton sufficient momentum (300 MeV/c) to be detected.

The neutron is then observed in missing mass from the over determined kinematics. Figure 16 (left) shows the missing mass distribution M_X for $K^- p K^+ X$. The Dalitz plot in Figure 16 (right) shows bands in the $K^+ K^-$ and the $p K^-$ invariant masses indicating the presence of ϕ and $\Lambda(1520)$ production. These events are eliminated from the final data sample as they will only contribute to the background in the $K^+ n$ invariant mass distribution where the Θ^+ signal is expected. The Dalitz plot also shows broader band from higher mass Λs near 1.7 and 1.8 GeV. While these resonances will also contribute to the background, they cannot be eliminated by simple mass cuts, and are therefore kept in the data sample. The only additional cut that is applied is to require $p_n > 100$ MeV/c to ensure that the neutron is not a spectator but participated in the interaction. Note that the neutron momentum in deuterium peaks near 50 MeV/c.

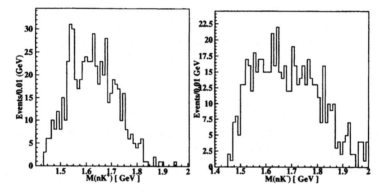

Figure 17. *Left: Mass spectrum of nK^+. Right: Mass spectrum of nK^-. Both figures contain the same pnK^+K^- events.*

In Figure 17 (left) the nK^+ spectrum obtained in the CLAS experiment is shown. A narrow peak near 1540 MeV is visible, although a recent reanalysis of this reaction, using a different analysis technique, indicates a lower significance for the peak.

One may ask if the observed peak can be due to statistical fluctuations. As the background under the peak is not well defined, there is a certain probablity that fluctuations can generate the observed signal. Other effects such as kinematical reflections of high mass mesons decaying into $K^+ K^-$ have also been discussed. While such effects have been claimed (Dzierba *et al.* 2004) to contribute to the structure seen in the M_{nK+} mass spectrum, it has been shown recently (Oh *et al.* 2004, Hicks *et al.* 2004) that this particular calculation is incorrect and not applicable to the CLAS results. For comparison, Figure 17 (right) shows the M_{nK-} spectrum of the same final state events used in the M_{nK+} sample. One would expect kinematical reflections to also show up in this spectrum. However, there is no structure visible, indicating that kinematical reflections are not generating narrow structures in this spectrum.

5.2 Photoproduction on hydrogen

The CLAS collaboration also studied the process $\gamma p \rightarrow \pi^+ K^+ K^- n$ using data measured at photon energies from 3 to 5.4 GeV. Events are selected with the π^+ at forward

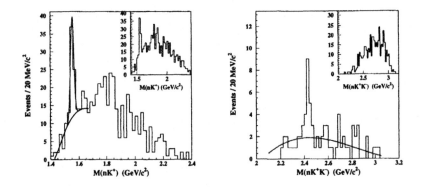

Figure 18. *Left: Invariant mass distribution of $M(nK^+)$ after all cuts. The inset shows the nK^+ mass distribution with only the $\cos\theta^*_{\pi^+} > 0.8$ cut applied. Right: Mass distribution $M(K^-nK^+)$ for events selected in the peak region of the graph on the left. The inset shows the distribution for events outside the Θ^+ region.*

angles and the K^+ at large angles. Such a selection will reduce t-channel contributions to K^+ production and favour t-channel proceses for the π^+. The final mass spectrum is shown in Figure 18 (left). A significant peak is seen at a mass of 1555 ± 10 MeV which has been associated with the Θ^+. In this analysis, background processes were subjected to a partial wave analysis allowing for a precise determination of the background shape under the Θ^+ peak. Possible contributions include the production of Θ^+ through an intermediate N^*. This assumption can be tested by selecting the events in the region of the Θ^+ and plotting the nK^+K^- mass distribution, which is shown in Figure 18 (right). While the data hint at a narrow structure near 2.4 GeV, the statistics are too poor to allow more definite conclusions. Further studies with higher statistics are clearly needed to get a better handle on possible production mechanisms of the Θ^+ at different kinematics.

5.3 Towards a quantitative comparison of pentaquark production

The $\Theta^+(1540)$ signal has been seen in experiments with different beams and in different final states. Nonetheless, the physics community is not fully convinced that what has been seen is a real particle state. This may be partly due to results from a number of high energy experiments that did not observe a signal. Unfortunately, it is unclear if the various results are compatible or if they contradict each other. In the following, I give arguments for two cases, the BaBaR and Belle e^+e^- experiments, that the results may not conflict. For example, often the $K^0_s p$ mass distribution is compared with the K^-p distribution. In the latter the $\Lambda(1520)$ is clearly visible, while in the former the $\Theta^+(1540)$ may not. Only a quantitative estimate of what this ratio should be for a specific experiment can reveal possible discrepancies between different experiments. In the Belle experiment the process $K^+n \rightarrow pK^0_s$ is measured in second scattering processes. The total width of the Θ has been estimated in analyses of K^+Xe scattering (Cahn and Trilling 2004), and K^+d scattering (Gibbs 2004a), giving $\Gamma_{\Theta^+} = 0.9 \pm 0.3$ MeV. This result can directly be compared with the Belle measurement. A straightforward estimate

(Gibbs 2004b) using the respective widths of $\Lambda(1520)$ and $\Theta^+(1540)$, and Clebsch-Gordon coeffients for the decays, shows that one expects a ratio for the *formation* of the Θ^+ and the $\Lambda(1520)$ with subsequent decays to the observed final states as $\approx 1 : 60$ (Gibbs 2004b). Such a ratio is indeed compatible with the limit given by the Belle experiment (Abe *et al.* 2004).

An even greater reduction of 5-quark baryons could be expected for e^+e^- annihilation processes that require baryon production through $q\bar{q}$ creation from the vacuum, which should be a much more rare process for a 5-quark system than for a 3-quark system. Data from BaBaR (Halyo 2004) indicate that the rate of pseudoscalar meson production, requiring a single $q\bar{q}$ from the vacuum, drops as $10^{-2}/\text{GeV}/c^2$ mass. That means a meson with a mass of 2.0 GeV/c^2 would be produced with a 100 times lower rate than a meson of 1.0 GeV/c^2 mass. For a 3-quark baryon, requiring two $q\bar{q}$ pairs, the rate drops like $10^{-4}/\text{GeV}/c^2$. This suggests that for every quark pair created from the vacuum a drop in rate by $10^{-2}/\text{GeV}/c^2$ is observed. If we assume pentaquark production in e^+e^- annihilation processes proceeds through the creation of four $q\bar{q}$ pairs, this would yield a rate reduction of $10^{-8}/\text{GeV}/c^2$. Such a rapid drop in rate could make pentaquark production in e^+e^- annihilation a very rare process in comparison to photoproduction from nuclear targets, where all quark ingredients necessary for Θ^+ formation and production are already present in the initial state.

The two examples illustrate that the searches for pentaquark baryons must be elevated to a more quantitative level. Cross sections or upper limits are needed for any future publication on new results. In the following, I briefly describe the current program with CLAS at the Jefferson Laboratory to study the $\Theta^+(1540)$ and other 5-quark states with higher precision.

5.4 Search for the Θ^+ and other pentaquark baryons at Jefferson Laboratory

The analysis of existing data shows the capabilities of CLAS to select exclusive final states with high multiplicity (Burkert *et al.* 2004). The reaction channels described above were cleanly identified with small background due to misidentified particles. Concurrent reactions decaying to the same final states were seen and rejected from the final event sample. However, the numbers of Θ^+ events in the peaks are rather small with sizeable backgrounds, and do not allow detailed checks of systematic dependencies to be performed.

To obtain a definitive answer on the existence of pentaquark states, four dedicated experiments were recently approved and are currently being carried out using CLAS. The goals and experimental conditions of these experiments are summarised in Table 2.

5.5 Search for the Θ^+ and excited states

The g10 experiment (Hicks *et al.* 2003), which has taken data during the spring of 2004, aims at studying the production channels $\gamma d \rightarrow pK^-\Theta^+$, $\gamma d \rightarrow pK^0X$, and $\gamma d \rightarrow \Lambda\Theta^+$ with an order of magnitude improved statistics over the previous g2a run. The g11 experiment studies $\gamma p \rightarrow \Theta^+\bar{K}^0$ and $\gamma p \rightarrow \Theta^+K^-\pi^+$, and two decay modes, $\Theta^+ \rightarrow$

Run	Beam	Energy	Target	Reaction	Status
g10	γ	3.8 GeV	LD$_2$	$\gamma d \to \Theta^+ K^- p$	Completed
				$\gamma d \to \Theta^+ \Lambda^0$	
g11	γ	4.0 GeV	LH$_2$	$\gamma p \to \Theta^+ \bar{K}^0$	Completed
				$\gamma p \to \Theta^+ K^- \pi^+$	
eg3	γ	5.7 GeV	LH$_2$	$\gamma p \to \Xi^{--} X$	In progress
				$\gamma p \to \Xi^+ X$	
g12	γ	5.7 GeV	LH$_2$	$\gamma p \to \Theta^+ K^- \pi^+$	2006
				$\gamma p \to \Theta^+ \bar{K}^0$	
				$\gamma p \to K^+ K^- \Xi^-$	

Table 2. *New experiments proposed in Hall B for the search for pentaquark states.*

nK^+ and $\Theta^+ \to pK^0$, increasing by an order of magnitude the statistics of the previous data. Both experiments have similar experimental setups and beam conditions to that used in the g2a and g1c runs, respectively. The g11 experiment (Battaglieri *et al.* 2004) used photons generated from a 4 GeV primary electron beam and a gold radiator incident on a 40-cm long hydrogen target. Photons from 1.6 GeV up 3.8 GeV are tagged and the data acquisition is triggered by events with at least two tracks to maximize the efficiency for the reaction of interest. The total expected integrated luminosity is approximately 20 times larger than in the previous run.

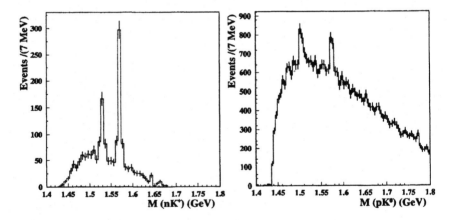

Figure 19. *Expected statistical accuracy of the mass spectra for the reactions $\gamma p \to \Theta^+(\Theta^{+*})\bar{K}^0$, with $\Theta^+(\Theta^{+*})$ decaying into $K^+ n$ (left) and pK^0 (right). A production cross section of 10nb is assumed.*

If the Θ^+ can be established with certainty, the new data will allow us to make progress in establishing the phenomenology of the Θ^+ spectrum, *eg* determining in which production channels the Θ^+ is seen and what higher mass states are excited. The expected statistical accuracy is shown in Figure 19, where the background was estimated based on the existing data and the signal was simulated assuming a production cross section of ~ 10 nb. If the existence of the Θ^+ is confirmed or new states are seen, these data will provide accurate measurements of the mass position. In addition, the large acceptance of the CLAS detector will allow us to measure both the production and decay angular distribution, providing information on the production mechanism and spin. Expected statistical accuracy for the measurement of the production and decay angular distributions are shown in Figure 20 for different assumptions on the production angular distribution and for the spin and parity of the pentaquark state.

This measurement will provide a solid foundation for a long term plan for the investigation of the pentaquark spectrum and properties.

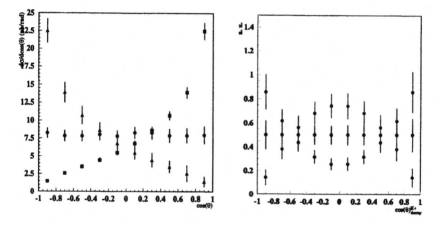

Figure 20. *Expected statistical accuracy for the measurement of the production and decay angular distribution. A total cross section of ~ 10 nb was assumed. The left plot shows the expected error bars for the production angular distribution in the assumption of a t-channel, s-channel, and u-channel production mechanism. The right plot shows the expected error bars for the decay angular distribution for different assumption on the spin and parity of the state and 100% polarisation.*

5.6 Search for Ξ^{--} and Ξ^- baryons

The anti-decuplet predicted by the χSM or quark-cluster models (Jaffe and Wilczek 2003, Karliner and Lipkin 2003) for 5-quark states also contains Ξ_5 states, two of them of exotic nature, the Ξ^{--} and the Ξ^+. Evidence for such states has so far been seen in only one experiment (Alt *et al.* 2004). This makes it urgent to confirm or refute these claims. Two new experiments with CLAS will search for Ξ baryons.

The eg3 experiment (Smith *et al.* 2004) uses an untagged photon beam of 5.75 GeV impinging on a liquid-deuterium target. The process $\gamma n \rightarrow \Xi^{--} X$ will be searched for by measuring the decay chain $\Xi^{--} \rightarrow \pi^- \Xi^- \rightarrow \pi^- \Lambda \rightarrow \pi^- p$. One proton and three π^- emerging from three different vertices have to be reconstructed. The experiment is taking data in the winter 2004/2005.

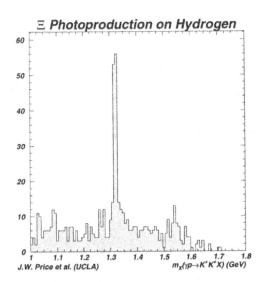

Figure 21. *Missing mass M_X for the reaction $\gamma p \rightarrow K^+ K^+ X$ for photon energies in the range 3.2 to 3.9 GeV. The narrow peak observed for the ground state $\Xi(1320)$ illustrates the excellent mass resolution that can be obtained using this method (Price et al. 2004). For the pentaquark Ξ_5 search, higher energies and much higher statistics are needed.*

The second experiment (Price *et al.* 2004) is part of g12. It uses the missing mass method to search for the Ξ^- in the exclusive reaction $\gamma p \rightarrow K^+ K^+ X^-$. If the NA49 results are correct, the Ξ^- would be seen in the missing mass spectrum as a peak at 1862 MeV. Excellent missing mass resolution is required for such a measurements which is provided by CLAS. Figure 21 illustrates the method with data taken at photon energies between 3.2 - 3.9 GeV. The $\Xi(1320)$ ground state is observed as a narrow spike. Limitations in beam energy and/or in the statistics did not allow the observation of higher mass $\Xi' s$ in this measurement.

In conclusion of this section, CLAS is currently pursuing high statistics searches for the Θ^+ on hydrogen and deuterium in various final states. We are also searching for possible excited states of the Θ^+. The experiments are conducted under similar kinematical conditions to previous measurements. The much higher statistics will allow more definite conclusions as to the existence of the Θ^+ in the exclusive channel $\gamma d \rightarrow K^- p K^+ (n)$. In addition, the existence of a possible excited state with mass about 50 MeV above the Θ^+ will be clarified. Moreover, an experiment is underway to search for the Ξ_5^{--} in the mass range where the NA49 experiment at CERN claimed evidence for the observation of the $\Xi_5(1862)$ in various charge channels. If evidence for the pentaquark states remains and is considerably strengthened, another high statistics experiment (g12) will be able to study the spectroscopy of pentaquark states.

References

Abe K *et al.*, 2004, hep-ex/0411005.
Abdel-Bary M *et al.*, 2004, *Phys Lett B* **595** 127.
Aiello M *et al.*, 1998, *J Phys G: Nucl Part Phys* **24** 753.
Airapetian A *et al.*, 2004, *Phys Lett B* **585** 213.
Aleev A *et al.* [SVD collaboration], 2004, hep-ex/0401024.
Alexandrou C *et al.*, 2004, *Phys Rev D* **69** 114506.
Alexandrou C *et al.*, 2005, *Phys Rev Lett* **94** 021601.
Alt C *et al.*, 2004, *Phys Rev Lett* **92** 042003.
Anderson H L *et al.*, 1952, *Phys Rev* **85** 936.
Arndt R A *et al.*, 1996, *Phys Rev C* **53** 430.
Arndt R A *et al.*, 2003, *Phys Rev C* **68** 042201; Erratum, 2004, *Phys Rev C* **69** 019901.
Asratyan A E *et al.*, 2004, *Phys Atom Nucl* **67** 682; *Yad Fiz* **67** 704.
Aznauryan I G, 2003a, *Phys Rev C* **67** 015209.
Aznauryan I G, 2003b, *Phys Rev C* **68** 065204.
Aznauryan I G *et al.*, 2005, *Phys Rev C* **71** 015210.
Barmin V V *et al.*, 2003, *Phys Atom Nucl* **66** 1715; *Yad Fiz* **66** 1763.
Barth J *et al.*, 2003, *Phys Lett B* **572** 127.
Battaglieri M *et al.*, 2004, JLab experiment E-04-021.
Bellis M, 2004, *proc N*2004*, Grenoble, ed V Kuznetsov, World Scientific, to be published.
Buchmann A and Henley E, 2000, *Phys Rev C* **63** 015202.
Burkert V D *et al.*, 2003, *Phys Rev C* **67** 035204.
Burkert V D and Lee T S, 2004, *Int J Mod Phys E* **13** 1035.
Burkert V D *et al.*, 2004, *proc N*2004*, Grenoble, ed V Kuznetsov, World Scientific, to be published.
Cahn R N and Trilling G H, 2004, *Phys Rev D* **69** 011501.
Cano F and Gonzales P, 1998, *Phys Lett B* **431** 270.
Capstick S and Roberts W, 1994, *Phys Rev D* **49** 4570.
Capstick S and Keister B, 1995, *Phys Rev D* **51** 3598.
Carlsson C E, 1986, *Phys Rev D* **34** 2704.
Carroll J *et al.*, 1968, *Phys Rev* **174** 1681.
Chekanov S *et al.*, 2004, *Phys Lett B*, **591** 7.
Close F E and Gilman F J, 1972, *Phys LettB* **38** 541.
Close F E, 2004, *contribution to these proceedings*.
Cole P, 2004, *proc N*2004*, Grenoble, ed V Kuznetsov, World Scientific, to be published.
Copley L A *et al.*, 1969, *Phys Lett B* **29** 117.
Cottingham W N and Dunbar I H, 1979, *Z Phys C* **2** 41.
Denizli H, 2004, *proc BARYONS 2002*, ed C Carlson and B Mecking, World Scientific.
Diakonov D *et al.*, 1997, *Z Phys A* **359** 305.
Drechsel D *et al.*, 1999, *Nucl Phys B* **645** 145.
Dzierba A R *et al.*, 2004, *Phys Rev D* **69** 051901.
Eidelmann S *et al.* [Review of Particle Properties], 2004, *Phys Lett B* **592** 1.
Gell-Mann M, 1964, *Phys Lett* **8,** 214.
Gerhardt B *et al.*, 1980, *Z Phys C* **7** 11.
Gibbs W, 2004a, *Phys Rev C* **70** 045208.
Gibbs W, 2004b, private communication.
Glander K H *et al.*, 2004, *Eur Phys J A* **19** 251.
Greenberg O W, 1964, *Phys Rev Lett* **13** 598..
Guidal M *et al.*, 2003, *Phys Rev C* **68** 058201.
Halyo V, 2004, Pentaquark 2004, SPring-8.

Hey A J G and Weyers J, 1974, *Phys Lett B* **48** 69.

Hicks K *et al.* , 2003, Jlab experiment E-03-113.

Hicks K, 2005, *Int J Mod Phys A* **20** 219.

Hicks K *et al.* , 2004, hep-ph/0411265.

Ichie H *et al.* , 2003, *Nucl Phys Proc Suppl* **119** 751.

Isgur N, 2000, *Excited nucleons and hadronic structure*, ed V D Burkert *et al.* , World Scientific, 403.

Jaffe R and Wilczek F, 2003, *Phys Rev Lett* **91** 232003.

Joo K *et al.* , 2002, *Phys Rev Lett* **88** 122001.

Kamalov S S and Yang S N, 1999, *Phys Rev Lett* **83** 4494.

Kamalov S S and Yang S N, 2001, *Proc NSTAR 2000*, Ed V Burkert *et al.* , World Scientific.

Karliner M and Lipkin H, 2003, *Phys Lett B* **575** 249.

Kirchbach M, 2000, *Int J Mod Phys A* **15** 1435.

Koniuk R and Isgur N, 1980, *Phys Rev D* **21** 1868.

Koubarovsky V *et al.* , 2004, *Phys Rev Lett* **92** 032001.

Krewald S *et al.* , 2000, *Phys Rev C* **62** 025207.

Janssen S *et al.* , 2001, *Phys Rev C* **65** 015201.

Lee F X *et al.* , 2003, *Nucl Phys Proc Suppl* **119** 296.

Lee F X, 2004, *proc N*2004*, Grenoble, ed V Kuznetsov, World Scientific, to be published.

Li Z *et al.* , 1992, *Phys Rev D* **46** 70.

Mart T *et al.* , 2000, *KaonMAID*, www.kph-uni-mainz.de/MAID/kaon/kaonmaid.html.

McNabb J W C *et al.* , 2004, *Phys Rev C* **69** 042201.

Mertz C *et al.* , 2001, *Phys Rev Lett* **86** 2963.

Mokeev V *et al.* , 2004, *proc N*2004*, Grenoble, ed V Kuznetsov, World Scientific, to be published.

Moorhouse R G, 1966, *Phys Rev Lett* **16** 771.

Nakano T *et al.* , 2003, *Phys Rev Lett* **91** 012002.

Niculescu G, 2004, *proc BARYONS2002*, ed C Carlson and B Mecking, World Scientific.

Oh Y *et al.* , 2004, hep-ph/0412363.

Oka M, 2004, *Prog Theor Phys* **112** 1.

Pace E *et al.* , 1999, *Few Body Syst Suppl* **10** 407.

Price J *et al.* , 2004, Jlab experiment E-04-017.

Ripani M *et al.* , 2003, *Phys Rev Lett* **91** 022002.

Riska D O, 2004, *Talk at MENU2004*, Beijing.

Rosner G, 2004, *Lectures given to SUSSP58 Hadron Physics Summer School, St Andrews.*

Sakurai J J and Schildknecht D, 1972, *Phys Lett B* **40** 121.

Sato T and Lee T S, 2001, *Phys Rev C* **63** 055201.

Smith E *et al.* 2004, JLab experiment E-04-010.

Smith L C, 2004, *proc N*2004*, Grenoble, ed V Kuznetsov, World Scientific, to be published.

Stepanyan S *et al.* , 2003, *Phys Rev Lett* **91** 252001.

Thoma U [CB-ELSA and CLAS collaborations], 2003, *proc NSTAR 2002*, Pittsburgh, ed S A Dytman and E S Swansan, World Scientific, 21.

Thompson R *et al.* , 2001, *Phys Rev Lett* **86** 1702.

Warns M *et al.* , 1990, *Z Phys C* **45** 627.

Zweig G, 1964, *CERN Report 8182/TH-412* (unpublished).

Hadrons in the nuclear medium: recent results and perspectives

Volker Metag

II. Physikalisches Institut, University of Giessen,
D-35392 Giessen, Germany

1 Introduction

From astrophysical observations we learn that only 5% of the energy/ matter in the universe is visible (baryonic) matter; the rest is dark matter and dark energy which we know very little about. However, the mass of baryonic matter is also not fully understood; it cannot be explained by the masses of the elementary building blocks of matter: quarks and leptons. Quarks are not observed as free particles; they are bound in hadrons such as protons, neutrons or mesons. In case of other composite systems like atoms or nuclei, the mass of the total system is given by the sum of the masses of the constituents apart from very small binding energy effects. However, the nucleon mass of 938 MeV is much larger than the sum of the current quark masses, estimated to be of the order of 15 MeV. The mass of the nucleon and of other light hadrons is explained in terms of energy stored in the motion of quarks and in the gluon fields. Thereby one obtains a massive system out of almost massless constituents: *mass without mass* (Wilczek 2002a, 2002b).

Since quarks are Fermions they show a quantum mechanical resistance to localisation: being confined to the dimensions of a nucleon they exhibit an uncertainty in the momentum according to Heisenberg's uncertainty relation. The associated kinetic energy has to be overcome by the interactions among the quarks. As in any other bound system the potential energy has to be larger than the kinetic energy of the constituents. (In case of the hydrogen atom the potential energy is twice as large as the kinetic energy). For small distances among quarks, probed at large momentum transfers, the strong interaction is weak, so that it can be treated perturbatively like Quantum Electrodynamics (QED). This is the reason for the high predictive power of Quantum Chromdodynamics (QCD) in the perturbative regime. At larger distances among the quarks, approaching the dimensions of a nucleon, the interaction becomes so strong (see Figure 1) that it can overcome the kinetic energy of the quarks. This sets the scale for the size of the nucleon and for the energy associated with quark motion and colour gluon fields which determines the order of magnitude of the nucleon mass.

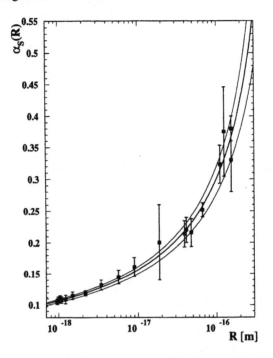

Figure 1. *Variation of the strong interaction coupling strength with the distance between quarks.*

While this dynamic effect accounts for the bulk of the nucleon mass it is not the only source of mass in nature. An important role is played by the spontaneous breaking of chiral symmetry which is a fundamental symmetry of QCD in the limit of vanishing quark masses. For massless quarks, the interaction through gluon exchange occurs in two completely separated sectors of left-handed and right-handed quarks: left-handed quarks, with their spin and momentum pointing in opposite directions, remain left-handed, and correspondingly, right-handed quarks remain right-handed, *ie* quarks retain their chirality while interacting through gluon exchange. If chiral symmetry were to hold also in the hadronic sector we would expect the existence of mass degenerate chiral partners, *ie* hadronic states with the same spin but opposite parity should have the same mass. This is not realised in nature; chiral symmtery is broken: The ground state $1/2^+$ of the nucleon has a mass of 938 MeV while the lowest $1/2^-$ state, the $S_{11}(1535)$ nucleon resonance, has a mass of 1535 MeV. The mass split between these two states is of the same order as the nucleon mass, demonstrating the impact of chiral symmetry breaking on hadron masses. Similar mass gaps are observed in the meson sector: vector mesons: ρ ($J^\pi = 1^-$, 776 MeV) and a_1 (1^+, 1260 MeV); scalar modes: $\pi(0^-, 135$ MeV) and σ (0^+, 600 MeV).

Various theoretical considerations predict at least a partial restoration of chiral symmetry in compressed or heated nuclear matter (Brown and Rho 1991). The restoration of a symmetry is associated with the vanishing of an order parameter. In classical physics, a well known example is the transition from the ferromagnetic to the paramagnetic state of matter with increasing temperature. This corresponds to the restoration of full rotational symmetry, *ie* the transition from a state with rotational symmetry around one axis - all spins in the ferromagnet are aligned along one direction - to a state with full rotational symmetry - all spin directions are equally likely in a paramagnet. The order parameter of this transition is the magnetisation which - to first order - vanishes when the full rotational symmetry is restored.

In Quantum Chromodynamics the order parameter of the spontaneously broken symmetry is the quark condensate $<\bar{q}q>$ which vanishes in the limit of chiral symmetry restoration (see Figure 2).

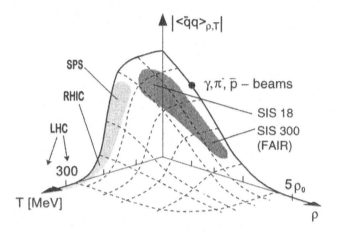

Figure 2. *The chiral condensate calculated within the Nambu-Jona-Lasinio model as a function of temperature and baryon density (adapted from Klimt et al. 1990). Parameter ranges accessible at different accelerators in heavy-ion collsions and elementary reactions are indicated.*

Its behaviour as a function of temperature and baryon density is therefore of prime interest. If chiral symmetry breaking has an impact on hadron masses, the restoration of this symmetry will influence hadron masses as well. Hadron masses are, however, not directly related to the chiral condensate which is not a direct physical observable. The link between hadron masses and the $<\bar{q}q>$ and higher order condensates is provided via QCD sum rules (Hatsuda and Lee 1992, Leupold *et al.* 1998). A more detailed account of these considerations has been given in a recent review by Weise (2003).

The key to the generation of the mass of elementary particles may have been found once the Higgs particle is discovered in planned experiments at high energy accelerators. This, however, does not answer the quest for the mass of matter around us and in the visible part of the universe. Understanding the mass of matter is a challenge for hadron physics. Experiments are being performed at several accelerators in the few GeV energy range trying to clarify whether hadron masses are indeed changed if produced in the nuclear medium. The main motivation for this experimental program is to achieve a better understanding of the origin of hadron masses in the context of spontaneous chiral symmetry breaking in QCD and their modification due to chiral dynamics and the partial restoration of chiral symmtery in a hadronic environment.

In these lectures an overview over the current status of this field is given. Specific theoretical predictions, experimental approaches and first evidence for medium modifications of hadrons will be presented in the following sections. An outlook on the possible future development of this field is also given.

2 In-medium behaviour of vector mesons

In-medium modifications of vector mesons have been investigated theoretically by a number of groups. As an example, predictions by the Munich group (Renk *et al.* 2002) are shown in Figure 3.

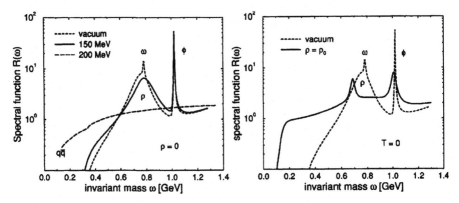

Figure 3. *Vector meson spectral functions, calculated by Renk et al. (2002), for (left) different temperatures at vanishing baryon density and (right) vanishing baryon density and normal nuclear matter density at temperature T=0.*

These calculations have been performed in the framework of an extended vector meson dominance model which incorporates the relevant features of chiral dynamics in many body systems of interacting hadrons (Klingl *et al.* 1996). For vector mesons at rest in the nuclear medium, the vacuum spectral distribution (dashed curves) is modified either by heating or compressing the nuclear matter. While the ρ meson is dissolved at sufficiently high temperatures and densities the ω meson retains its character as a quasiparticle but exhibits a strong mass shift and broadening; due to its weak interaction with the medium the Φ meson, on the other hand, is expected to show only some broadening and almost no mass shift.

These predictions can be checked experimentally in a variety of reactions. The properties of vector mesons as a function of temperature T can be studied in relativistic and ultra-relativistic heavy ion reactions with increasing incident energy. Medium modifications at T=0 and normal nuclear matter density are experimentally accessible in photonuclear or elementary hadronic reactions. Here, the medium effects are only in the 10-30% range. They can, however, be measured quite accurately and theoretically interpreted in a relatively straight forward way. In heavy-ion reactions, the expected medium modifications are much stronger but the experiment integrates over the complex space-time history of the reaction which makes the interpretation more complicated since the density in the collision zone varies strongly with time.

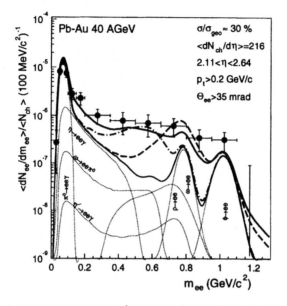

Figure 4. *Invariant mass spectrum of e^+e^- pairs obtained by the CERES collaboration in Pb + Au collisions at 40 AGeV (Wessels et al. 2003). The thin dotted curves give the contributions of individual hadronic sources to the total dilepton yield. The thick solid (modified spectral function) and the dashed-dotted (dropping mass only) curves represent the result of calculations (Rapp 2003) taking in-medium modifications of vector mesons into account.*

The best approach to study in-medium properties of mesons is to observe their decay within cold nuclei. Lepton pairs as decay products have the advantage that they do not undergo strong final state interactions within the medium. The invariant mass of the meson within the medium can be determined from the measured and undistorted 4-momentum vectors of the leptons from

$$m_{meson} = \sqrt{(p_1 + p_2)^2}. \tag{1}$$

If this invariant mass differs from the in-vacuum mass of the meson, *eg* as listed by the Particle Data Group (PDG: Eidelmann *et al.* 2004), an in-medium effect has been observed. Lepton spectroscopy is thus the preferred experimental tool despite the extremely small branching ratios, of the order of $10^{-5} - 10^{-4}$, of vector mesons into dileptons. One has, however, to ensure that the observed decays really occur in the nuclear medium. This can be achieved by requiring sufficiently low recoil momenta of the respective mesons which can be deduced from the measured dilepton momenta.

Experimental investigations of in-medium properties of vector mesons have been pioneered by the CERES collaboration (Agakichiev *et al.* 1995, 1997) at CERN who studied dilepton emission in ultra-relativistic nucleus-nucleus collisions. A more recent result from their work is shown in Figure 4.

The observed dilepton yield is compared to the sum of hadron decays and to model calculations which take, or do not take, medium effects into account. Calculations assuming only a broadening or a shift and broadening of the ρ meson spectral function are in good agreement with the data while estimates ignoring these medium effects fail to reproduce the experimental spectrum. These results together with the initial work (Agakichiev *et al.* 1995, 1997) provided the first evidence for medium modifications of vector mesons in heated nuclear matter.

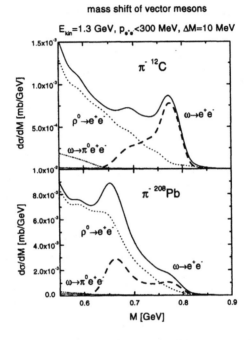

Figure 5. *Predicted invariant mass spectrum for e^+e^- pairs in the ω mass range for π^- induced reactions on C and Pb nuclei at $p_\pi = 1.3$ GeV/c. To enhance in-medium ω decays a cut of less than 300 MeV/c has been placed on the dilepton pair momentum. The assumed mass resolution of 10 MeV corresponds to the resolution achievable with the HADES spectrometer. The calculation has been performed by Effenberger et al. (1999).*

Experiments to search for medium modifications of vector mesons at normal nuclear matter density have been performed or proposed at KEK, JLab, GSI, and ELSA. At KEK dilepton decays of ρ, ω, and Φ mesons have been investigated in proton-nucleus reactions at 12 GeV (Ozawa *et al.* 2001). Evidence for medium modifications of the ρ meson has been deduced from a shift in the dilepton invariant mass spectrum for different nuclear mass numbers (Muto *et al.* 2004). A corresponding experiment using a photon beam has been performed at Jlab and is presently being analysed. At GSI, it has been proposed to use a pion beam for almost recoilless production of ω mesons in a nuclear target (Schön *et al.* 1996).

The expected dilepton invariant mass spectra of ω mesons produced on C and Pb targets, respectively, are shown in Figure 5. The in-medium modification leads to a double structure in the mass distribution: a peak at the nominal meson mass of 783 MeV/c^2 for ω mesons decaying outside of the nucleus and a shifted peak for in-medium decays. The latter peak becomes stronger for larger nuclei because more omegas are produced inside the nucleus and decay at densities close to normal nuclear matter density. It should be noted that the ρ meson - which is also expected to undergo medium modifications - contributes to the dilepton spectrum with similar strength which makes it more difficult to isolate the ω in-medium effect. The corresponding experiment will be performed with the dilepton spectrometer HADES as soon as sufficiently intense pion beams will be available at GSI.

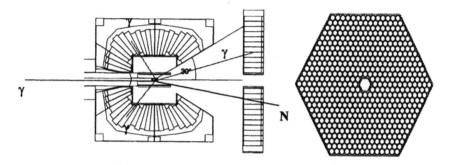

Figure 6. *Experimental setup at the ELSA accelerator in Bonn: the Crystal Barrel detector consisting of 1290 CsI (Tl) scintillators is combined with the photon spectrometer TAPS to make an almost 4π photon detector system. A typical $\omega \to \pi^0\gamma \to \gamma\gamma\gamma$ decay is indicated. The right hand part of the figure shows the granularity of the TAPS forward wall (528 BaF$_2$ scintillators).*

An alternative experimental approach to study the in-medium properties of the ω meson is the investigation of the decay branch $\omega \to \pi^0\gamma$. The advantage of this decay channel is the branching ratio of 9% which is three orders of magnitude larger than the dilepton decay. Furthermore, it is larger by two orders of magnitude than the $\rho \to \pi^0\gamma$ decay. This makes it easier to identify medium modifications of the ω meson in this decay mode. The disadvantage of the $\omega \to \pi^0\gamma$ channel is a possible rescattering of the π^0 within the nuclear medium which changes the pion momentum and thus distorts the deduced ω invariant mass. This effect was studied in a detailed simulation (Messchendorp *et al.* 2001) including recent parameterisations of the pion-nucleus interaction. Any pion rescattering proceeds predominantly through the formation of an intermediate Δ resonance. The momentum of the re-emitted pion is small, governed by the Δ-decay kinematics and, smeared by Fermi motion, mainly leads to small invariant $\pi^0\gamma$ masses far below the range of interest between 600 and 800 MeV. According to Messchendorp *et al.* (2001) the contribution to this mass range is of the order of a few percent which can be further reduced by removing all events with π^0 mesons with kinetic energies less than 150 MeV.

This consideration has been applied in the analysis of a corresponding experiment performed at the electron accelerator ELSA in Bonn (Trnka *et al.* 2004). The photoproduction of ω mesons on the proton and on Nb has been investigated in a comparative study. Bremsstrahlung photons in the range 800 - 2600 MeV have been produced by the continuous wave electron beam from the electron stretcher ring ELSA impinging on a radiator wire. The photon energy $E_\gamma = E_e - E'_e$ is deduced event-by-event from the electron beam energy E_e and the energy E'_e of the scattered electron measured with a magnetic spectrometer (tagger). ω mesons have been identified via the $\omega \rightarrow \pi^0\gamma \rightarrow \gamma\gamma\gamma$ decay in an almost 4π photon detector system consisting of the Crystal Barrel (1290 Cs(Tl) modules) as target detector and the photon spectrometer TAPS (528 BaF$_2$ modules) in a forward wall configuration (see Figure 6).

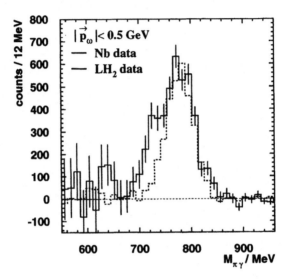

Figure 7. $\pi^0\gamma$ *invariant mass distributions near the ω mass after background subtraction measured for a Nb target (solid histogram) and a LH$_2$ target (dashed histogram) for events with ω momenta < 500 MeV/c (Trnka et al. 2004).*

Figure 7 shows the $\pi^0\gamma$ invariant mass spectrum in the ω mass range measured for a Nb and, for comparison, a LH$_2$ target. Only events with ω meson momenta less than 500 MeV/c have been selected to enhance the fraction of in-medium decays.

For long-lived recoiling mesons, like π^0, η, and η', which decay outside of the nucleus, no difference in the lineshape is observed for the two data sets. However, a deviation from the LH$_2$ target lineshape is found for ω mesons produced on Nb. The structure on the low mass side of the ω signal is attributed to in-medium decays of ω mesons of reduced mass.

Suppressing further π^0 rescattering by removing events with π^0 mesons of less than 150 MeV, the experimental $\pi^0\gamma$ invariant mass spectrum shown in Figure 8 has been obtained. Here, the fraction of in-medium decays has been enhanced by selecting only events with incident photon energies in the range of 800 - 1200 MeV. The data points for the Nb target (Trnka *et al.* 2004) are to be compared to the dashed curve representing the result of a transport model calculation (Mühlich *et al.* 2004b) taking π^0 rescattering in the Nb nucleus into account but ignoring any in-medium modification of the ω meson.

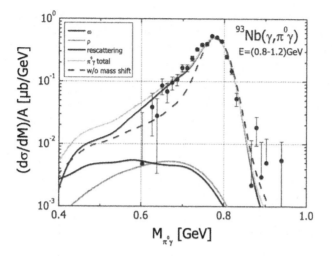

Figure 8. $\pi^0\gamma$ *invariant mass spectrum for photonuclear reactions on a Nb target. Preliminary experimental data (Trnka et al. 2004) are compared to calculations within a coupled-channel transport model (Mühlich et al. 2004b).*

The full calculation is given by the solid curve, assuming in addition a drop of the ω mass according to

$$m = m_0(1 - 0.16\rho/\rho_0) \tag{2}$$

and an in-medium broadening of the ω meson of 40 MeV. Furthermore, the calculated contribution of remaining events with π^0 rescattering in the medium and the calculated ρ meson contribution to the invariant mass distribution are indicated.

For the photon energy range 800 - 1200 MeV good agreement with the experimental data is achieved. This is taken as first evidence for a medium modification of the ω meson as expected theoretically. Moreover, the experimental result is of special theoretical importance since Zschocke *et al.* (2004) have pointed out that the ω mass is particularly sensitive to the density dependence of higher order quark condensates. The experimentally determined in-medium ω mass is thus a critical testing ground for our current understanding of the origin of hadron masses.

3 In-medium behaviour of scalar mesons

Chiral partners - hadronic states with the same spin but opposite parity - are not degenerate in mass as one would expect if chiral symmetry were to hold in the hadronic sector. If chiral symmetry became partially restored with increasing nuclear matter density the masses of chiral partners should approach each other. This is illlustrated in Figure 9 which shows the result of a calculation within the Nambu-Jona-Lasinio model (Bernard *et al.* 1987). While the mass of the pion ($J^\pi = 0^-$) is almost unaffected, the mass of the chiral partner the σ meson ($J^\pi = 0^+$) is predicted to decrease almost linearly with nuclear density, as also found by Lutz *et al.* (1992). At densities much larger than normal nuclear matter density both masses become degenerate. The sigma meson is an elusive particle for which the most recent PDG evaluation (Eidelmann *et al.* 2004) quotes a mass of about 600 MeV and a width of about 300 MeV in vacuum. Whether this particle is a genuine meson state of predominant $q\bar{q}$ structure or rather a resonance dynamically generated in $\pi\pi$ scattering, is still under debate. The current status has been reviewed by Ishida (2000).

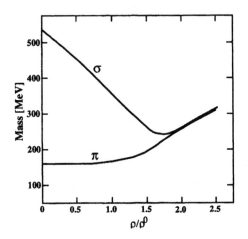

Figure 9. *Dependence of the π and σ masses on the baryon density as calculated by Bernard et al. (1987) within the Nambu-Jona-Lasinio model. The meson masses become degenerate when chiral symmetry is restored at high baryon densities.*

The σ meson decays into $\pi^0\pi^0$ or $\pi^+\pi^-$. As a consequence of the decreasing σ mass the phase space for two pion decays shrinks; the 2π strength gets more and more concentrated near the 2π threshold, as studied quantitatively by Hatsuda *et al.* (1999). Their theoretical prediction is given in Figure 10.

A strong concentration of 2π strength at low masses is also predicted within the chiral unitary model of the Valencia group (Chiang *et al.* 1998, Vicente Vacas and Oset 2002). They provide a comparative study of s-wave I=0 $\pi\pi$ correlations in vacuum and in the nuclear medium. Here, in-medium modifications in the σ channel are induced by the strong p- wave coupling of pions to particle-hole and Δ-hole excitations. As shown in Figure 11 the σ mass is also predicted to drop to values close to the 2π threshold at normal nuclear matter density. Because of the decreasing 2π phase space the width is reduced accordingly.

On the experimental side, several results have been reported showing a strong enhancement in the 2π invariant mass distribution close to the 2π mass with increasing nuclear mass number. Pioneering experiments were performed by the CHAOS collaboration (Bonutti *et al.* 1996) studying the $A(\pi^+, \pi^+\pi^-)$ reaction. They found a concentration of strength in the $\pi^+\pi^-$ invariant mass distribution near $2m_\pi$ while in the $\pi^+\pi^+$ channel, which does not couple to the neutral σ meson, an invariant mass spectrum consistent with a phase space distribution has been observed. Similarly, the Crystal Ball collaboration (Starostin *et al.* 2000) observed a shift in the $2\pi^0$ invariant mass distribu-

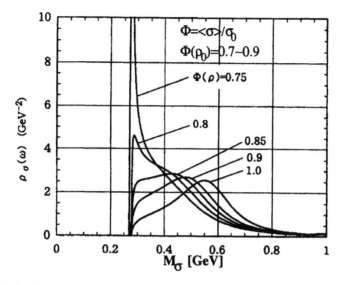

Figure 10. *Predicted spectral function of the σ meson for different values of Φ which represents the ratio of the chiral condensate at finite nuclear matter density ρ to that in vacuum. From Hatsuda et al. (1999).*

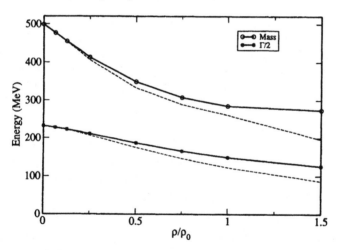

Figure 11. *Mass and width of the σ meson as a function of the nuclear density, calculated with the chiral unitary model (Vicente Vacas and Oset 2002). The dashed lines include two-particle two-hole contributions to the π self energy.*

tion towards the 2π threshold with increasing nuclear mass (see Figure 12).

Pion induced reactions are, however, affected by strong initial state interactions leading to absorption of the incoming pions near the nuclear surface with densities in the range of 1/4 - 1/2 ρ_0. This prevents the study of $\pi\pi$ correlations at close to normal nuclear densities. Since photons below 1 GeV penetrate the whole nucleus photon induced

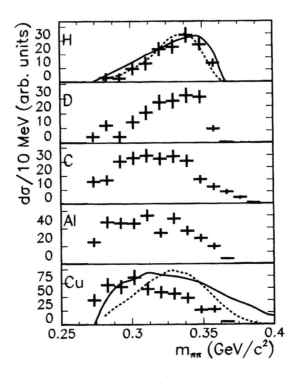

Figure 12. *Invariant mass distributions of $\pi^0\pi^0$ pairs produced in π^- induced reations on p, d, C, Al, and Cu targets (Starostin 2000) at a pion momentum of 408 MeV/c. The solid and dashed curves represent calculations by Rapp et al. (1999) and Vicente Vacas and Oset (1999).*

double pion production on nuclei appears to be a more favourable approach to study in-medium $\pi\pi$ correlations. Corresponding experiments have been performed with the Glasgow tagged photon spectrometer at MAMI, using the photon spectrometer TAPS for the identification of the 2π channel (see Figure 13).

For incident photon energies of $E_\gamma = 400 - 500$ MeV preliminary $\pi^0\pi^0$ invariant mass distributions are shown in Figure 14 for a variety of targets. A shift in strength towards small $m_{\pi^0\pi^0}$ is observed with increasing nuclear mass number. The peak in the $m_{\pi^0\pi^0}$ distribution moves from about 340 MeV for the proton target to 320 MeV for C, Ca and to approximately 270 MeV for Pb. The experimentally observed angular distributions in the $A(\gamma, \pi^0\pi^0)$ reaction are found to be isotropic in the $\pi^0\pi^0$ centre-of-mass frame and are compatible with J=0, supporting the conclusion that a significant A-dependence has been observed in the $\pi^0\pi^0$ I=J=0 channel.

The question arises whether the observed shift in the $\pi^0\pi^0$ invariant mass distribution is not simply the result of pion absorption. Energetic pions which mainly contribute to the high mass part of the $\pi^0\pi^0$ invariant mass distribution have a higher chance with increasing nuclear size of exciting a Δ resonance in a pion-nucleon collision. They are more strongly absorbed than low energy ones, leading to a reduced yield at higher

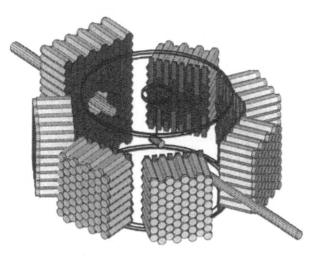

Figure 13. *Arrangement of the TAPS photon spectrometer in a 6 block configuration with forward wall. This detector setup has been used in the studies of pion pair photo-production off nuclei at the electron accelerator MAMI, Mainz. The photon beam comes along the pipe from the right.*

masses.

This hypothesis has been tested experimentally by comparing the invariant mass distributions in the $\pi^0\pi^0$ channel to those in the $\pi^\pm\pi^0$ channel. The neutral σ meson can not decay into the latter. In the experiments of Messchendorp *et al.* (2002) this final state has been measured concurrently, exploiting the capability of the TAPS spectrometer to identify charged pions via a time-of-flight and energy measurement (see Figure 16).

Positive and negative pions can, however, not be distinguished because of the lack of a magnetic field. The preliminary $\pi^\pm\pi^0$ invariant mass distributions measured for p, C, Ca, and Pb are shown in the right panel of Figure 14. In contrast to the $\pi^0\pi^0$ channel, the $\pi^\pm\pi^0$ invariant mass distributions do not show a variation in shape with increasing nuclear mass. This becomes even clearer from the ratio of distributions measured for different target nuclei shown in Figure 15. The ratio of the $\pi^\pm\pi^0$ invariant mass distributions for Pb an C targets is approximately constant, indicating no shape variation, while the corresponding ratio for the $\pi^0\pi^0$ channel shows a strong dependence on the invariant mass, emphasising the shift of the $\pi^0\pi^0$ strength towards the 2π threshold for increasing nuclear mass. This comparison of isospin channels shows that the observed shift is a genuine feature of the $\pi^0\pi^0$ channel and cannot be ascribed to absorption processes which would be similar for all isospin channels. The experimental observation is thus consistent with the theoretically predicted in-medium mass modifications of the σ meson. At least, it has been shown experimentally that the $\pi^0\pi^0$ interaction in the nuclear medium is different from the one in vacuum.

The importance of final state interactions on the reported experimental results has recently been revisited by Mühlich *et al.* (2004c). More detailed experimental information is needed to clarify the issue. Higher statistics data will allow for a clean selection of

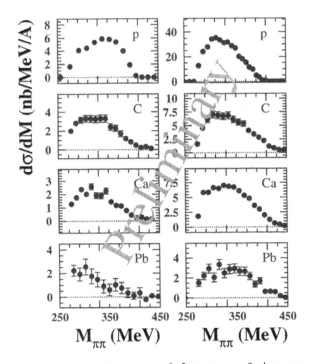

Figure 14. *Invariant mass distributions of $\pi^0\pi^0$ (left) and $\pi^0\pi^\pm$ (right) pairs produced in photonuclear reactions on p, C, Ca, and Pb targets (Messchendorp 2002, Bloch 2004, Schadmand 2004) for the incident photon energy range of 400 - 500 MeV.*

quasi-free production events, for cuts on the momentum of the $\pi\pi$ pair, and for detailed angular distributions, *etc.* To achieve this goal, an improved experiment with almost 4π coverage for photon detection is under way at the electron accelerator MAMI, using the combined Crystal Ball and TAPS photon spectrometers.

4   In-medium properties of charmed mesons

So far only in-medium properties of hadrons built of light quarks have been discussed. What will happen when one extends these studies into the charm sector? As charmonium ($c\bar{c}$) states are dominated by the large mass of the charm quark pair, rather little sensitivity to changes in the quark condensate is expected. The in-medium mass of these states would primarily be affected by a modification of the gluon condensate. Calculations (Klingl *et al.* 1999) indicate only small mass reductions of the order of a few MeV for the J/Ψ and η_c in nuclear matter. For higher excited $c\bar{c}$ states, Lee and Ko (2003) calculate reduced masses in nuclear matter. Sizable medium effects are expected in the open charm sector, *eg* for D mesons. Built of a heavy c quark and a light antiquark, the D meson is the QCD analogue to the hydrogen atom. Thereby, D mesons provide the unique opportunity to study the in-medium dynamics of a system with a single light quark.

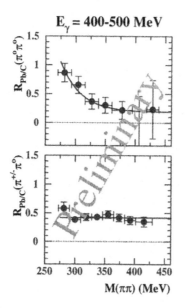

Figure 15. *Ratio of differential cross sections for the $A(\gamma, \pi^0\pi^0)$ (top panel) and $A(\gamma, \pi^0\pi^\pm)$ (bottom panel) reactions on Pb and C targets (Schadmand 2004).*

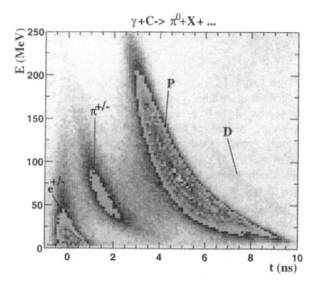

Figure 16. *Energy of charged particles deposited in TAPS as a function of their time-of-flight. Separated bands of protons and charged pions can be identified.*

Recent model calculations (Hayashigaki 2000) predict a substantial lowering of both the D^+ and D^- meson masses in the nuclear medium by about 50 MeV as a result of the decrease of the light quark condensate. The resulting drop of the $D\bar{D}$ threshold in the medium by about 100 MeV would allow the Ψ' state of charmonium to decay into this channel as indicated in Figure 17.

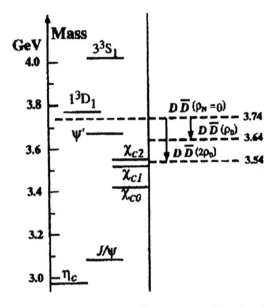

Figure 17. *Excitation energy spectrum of the $c\bar{c}$ system. The thresholds for decay into $D\bar{D}$ pairs in vacuum and for normal and twice the normal nuclear matter density are indicated (Hayashigaki 2000).*

While in vacuum this state has a width of of 0.3 MeV as it is 52 MeV below the free $D\bar{D}$ threshold its width would dramatically increase in the medium. A sensitive probe for studying this phenomenon would be e^+e^- spectroscopy of charmonium states produced via $\bar{p}$ annihilation in nuclei. Because of the increased width an appreciable fraction of the excited charmonium states is expected to decay inside the nucleus. This would lead to an observable shift of the peak position in the dilepton invariant mass spectrum if the mass is indeed modified in the nuclear medium as predicted by Lee and Ko (2003).

Moreover, in vacuum the $\Psi(3770)$ state is 31 MeV above the $D\bar{D}$ threshold. Its decay is completely dominated by the $D\bar{D}$ channel leading to a width of 24 MeV. The branching ratio into electron pairs is of the order of 10^{-5}.

An in-medium lowering of the D-meson masses would increase the phase space for the $D\bar{D}$ decay and thereby lead to a suppression of the e^+e^- decay of this resonance in comparison to the decay in vacuum, as predicted by Golubeva *et al.* (2003) and shown in Figure 18. A suppression of the e^+e^- yield would, however, be much more difficult to detect than an enhancement as observed in heavy-ion reactions. The feasibility of these experiments is under investigation.

The behaviour of heavy quark systems in a nuclear medium is also of considerable interest because of the ongoing discussion of J/Ψ suppression in ultra-relativistic heavy-ion collisions as a possible signal for quark-gluon plasma formation. To clarify this issue, detailed knowledge about the in-medium interaction of $c\bar{c}$ systems under "normal", non-

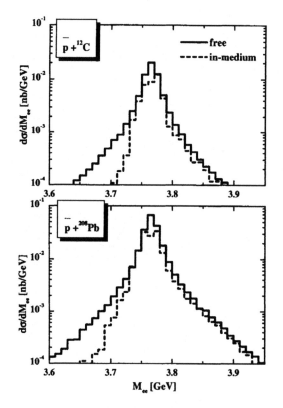

Figure 18. *Predicted invariant mass spectrum of e^+e^- pairs emitted in the decay of the $\Psi(3770)$ state in C and Pb (dashed histograms) in comparison to the decay in vacuum (solid histogram). The figure is taken from Golubeva et al. (2003).*

plasma conditions is required.

These investigations are part of the planned physics program at the High Energy Storage Ring (HESR) which will provide cooled $\bar{p}$ beams up to momenta of 15 GeV/c at luminosities of up to $L = 10^{32}cm^{-2}s^{-1}$. This storage ring is one of the major components of the future Facility for Antiproton and Ion Research (FAIR) (2001) at GSI which is expected to go into operation around 2012. The experiments will be performed with the internal target detector PANDA (2004) which will allow for neutral and charged particle detection and identification at high event rates. The detector concept is presently being worked out. It is envisaged to have a central detector around the target and a forward spectrometer, both of modular design and optimised for the specific kinematics of the antiproton-nucleon annihilation process. The current layout of the storage ring and the detector configuration is shown in Figure 19.

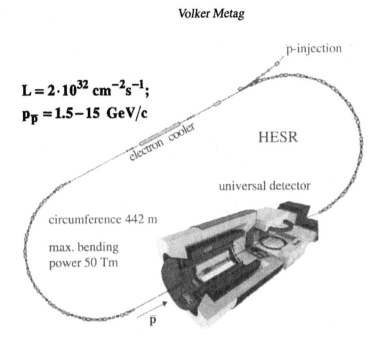

$$L = 2 \cdot 10^{32} \text{ cm}^{-2}\text{s}^{-1};$$
$$p_{\bar{p}} = 1.5 - 15 \text{ GeV/c}$$

Figure 19. *Current layout of the planned High Energy Storage Ring (HESR) with electron cooler and the PANDA detector at the future Facility for Antiproton and Ion Research (FAIR) at GSI, Darmstadt (Metag 2003).*

5 The search for η-mesic nuclei

The interaction of mesons with nuclei is not only essential for understanding the in-medium properties of mesons but also governs the possible existence of bound systems of mesons and nuclei. The existence of pionic atoms is well established and described in text books (see *eg* Ericson and Weise 1988). Deeply bound pionic states have recently been studied by Geissel *et al.* (2002a, 2002b). A detailed account is given in a recent review by Kienle and Yamazaki (2004). The superposition of a large repulsive s-wave π^--nucleus interaction at low pion momenta and an attractive Coulomb interaction leads to a potential pocket at the surface of the nucleus which gives rise to pion-nucleus bound states with a halo-like pion distribution around the nucleus.

These states have been populated in (d,^{3}He) reactions in almost recoil-free kinematics where sufficient energy for pion production, but little momentum, is transferred to the nucleus. A beautiful example of the high resolution spectroscopy in the (d,^{3}He) reaction is shown in Figure 20.

The deeply bound pionic states are clearly visible above some background. The energy and widths of these states have been carefully analysed and interpreted by Kienle and Yamazaki (2004) as evidence for a lowering of the in-medium pion decay constant f_π. This observation is attributed to a partial restoration of chiral symmetry in the nuclear medium, demonstrating the direct link of these investigations to the study of in-medium hadron properties, discussed in the previous sections.

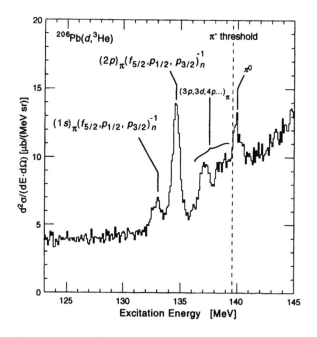

Figure 20. *Double differential cross section for the ^{206}Pb $(d,^3He)$ reaction at an incident deuteron energy of 604.3 MeV as a function of the excitation energy above the ^{205}Pb ground state (Geissel et al. 2002a).*

In the case of the neutral η meson an attractive, strong η-nucleon interaction is required for the formation of η-mesic nuclei. First predictions for such states were based on investigations of the η nucleon scattering length. This scattering length has been estimated from coupled channel analyses, performed by Bahlerao and Liu (1985), indicating the possible existence of η-mesic nuclei with mass numbers A > 10. From near threshold cross section enhancements observed in η photoproduction on the deuteron (Krusche *et al.* 1995, Weiss *et al.* 2001), Sibirtsev *et al.* (2002) have extracted a scattering length of $a_{\eta N} = (0.42 + i0.34)$ fm. This value is consistent with a weakly bound η-nucleus system even for nuclei as light as ^{3}He.

Sokol *et al.* (2001) searched for the existence of η-mesic Boron in the reaction $\gamma +^{12} C \rightarrow p +_{\eta}^{11} B \rightarrow \pi^+ + n + X$ proceeding via an intermediate $S_{11}(1535)$ resonance which decays into $\pi^+ n$. The correlated energies and momenta of the pion and neutron have been used as signature for this reaction channel. A peak, observed in the kinetic energy distribution of $\pi^+ n$ pairs below the threshold of the elemenatry $\pi n \rightarrow \eta n$ reaction, is taken as evidence for the formation and decay of an intermediate η-mesic nucleus.

In a recent experiment at MAMI, the TAPS/A2 collaboration has searched for the existence of η-mesic ^{3}He (Pfeiffer *et al.* 2004). After removing quasifree η production events by a missing energy analysis the coherent η photoproduction cross section in Figure 21 has been obtained. The peak-like behaviour of the near threshold cross section is interpreted as a first sign for the existence of an η bound state. This is corroborated by the

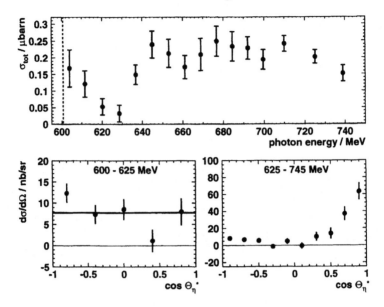

Figure 21. *Top: Cross section for coherent η photoproduction off* ^{3}*He. Bottom: Angular distribution of η mesons in the γ^{3}He centre-of mass system for incident photon energies of 600 - 625 MeV (left) and 625 - 745 MeV (right). From Pfeiffer et al. (2004).*

analysis of the near threshold η angular distribution which is almost isotropic in the γ^{3}He centre-of-mass system in contrast to an expected characteristic forward rise which is actually observed at higher incident photon energies above 625 MeV (see Figure 21). The observations are consistent with the assumption of an η-mesic nucleus which isotropically decays into the coherent η channel above the η threshold. Earlier measurements of the η production off ^{3}He in a proton induced reaction on the deuteron (Mayer *et al.* 1996, Bilger *et al.* 2002) showed a similar energy dependence of the production amplitude after dividing out phase space factors.

Additional evidence for a possible η-mesic state can be deduced by studying another decay channel which is open also below the η threshold, *eg* the decay into a correlated π^0p pair. In the decay of the η-mesic state, this reaction may occur via the excitation of a nucleon to the S_{11} nucleon resonance which has a high branching into the π^0p channel. Because of the limited phase space the S_{11} resonance will have a very low momentum in the γ^{3}He centre-of-mass system. As a consequence, the π^0 and proton have to be emitted almost back-to-back in this frame. This signature has been found in the data as shown in Figure 22.

After background subtraction, a structure near the η threshold is observed in the excitation function of π^0p production for opening angles between the π^0 and the proton of 170° - 180°. A simultaneous fit to both data sets leads to a resonance energy of 1481±4.2 MeV and a width of 25.3±6.1 MeV (Pfeiffer *et al.* 2004). This corresponds to a binding energy of 4.4± 4.2 MeV. The experimental data thus indicate the existence of an extremely weakly bound η mesic state which, because of its large width, extends into the continuum and cannot only decay into the π^0 p but also into the η^{3}He chan-

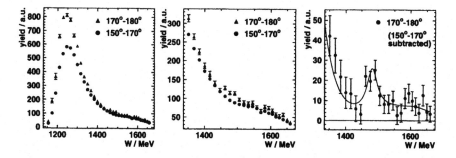

Figure 22. *Left and centre: Excitation function of π^0 proton production for opening angles of 170° - 180° (triangles) compared to opening angles of 150° - 170° (circles) in the $\gamma-^3He$ centre-of-mass system. Right: difference of both distributions with a background plus Breit-Wigner fit to the data. (Pfeiffer et al. 2004).*

nel. An independent analysis (Sibirtsev *et al.* 2004) of the same data, however, yields a binding energy of 0.4 ± 2.9 MeV which is consistent within errors with the results of Pfeiffer *et al.* (2004) but also compatible with zero. This raises the question whether the observed structure is due to a bound or rather to a virtual state. According to Hanhart (2005), this ambiguity can be solved if data of higher statistics and finer energy binning are provided. This would allow a proper treatment of both decay channels within the Flatté formalism (Flatté 1976). Furthermore, data from hadronic reactions should also be included in the analysis. In response to Hahnhart (2005), Pfeiffer *et al.* (2005) have performed a Flatté fit to the existing data in finer energy binning (see Figure 23) using the following expressions:

$$\sigma_{fi} = \left(\frac{16\pi}{s}\right)\left(\frac{\rho_f}{\rho_i}\right)|\hat{T}_{fi}|^2 \tag{3}$$

$$\hat{T}_{p\pi^0,\eta} = \frac{m_0\Gamma_0 \cdot A_{\pi^0,\eta}}{m_0^2 - m^2 - im_0(\rho_{p\pi^0}\Gamma_0\gamma_{p\pi^0}^2 + \rho_{p\eta}\Gamma_0\gamma_{p\eta}^2)} \tag{4}$$

with $m_0, \Gamma_0\gamma_{p\pi^0}^2, \Gamma_0\gamma_{p\eta}^2, A_{\pi^0,\eta}$ as fit parameters. This formalism takes both decay channels of the tentative η-mesic state into account, the $\pi^0 p$ channel and the η^3He channel, and allows for the opening of the η^3He channel within the broad resonance. Again, only a weakly bound state is deduced. Much higher statistics data are expected from an improved photoproduction experiment, using the Crystal Ball/TAPS detector system, which is in preparation at MAMI.

It will be interesting to extend the search for η-mesic states to heavier nuclei for which more strongly bound states are expected to exist (Garcia-Recio *et al.* 2002, Roy *et al.* 2004). An alternative future direction is to search for bound states of nuclei with still heavier mesons *eg*, ω mesons as proposed by Marco and Weise (2001).

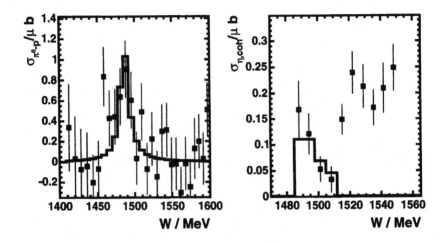

Figure 23. *Comparison of the decay channels of η mesic ^3He into $\pi^0 p + X$ (left) and η^3He (right). The solid histograms represent a simultaneous fit to both data sets, using a Flatté parameterisation (Flatté 1976) of the cross sections (Pfeiffer et al. 2005).*

6 Summary

An overview of the current status of experiments on in-medium modifications of hadrons has been given. The motivation for these studies is to achieve a deeper understanding of the origin of hadron masses. Dilepton spectroscopy in ultra-relativistic heavy-ion reactions has provided first evidence for an in-medium lowering of the mass and broadening of the ρ meson. These results have been corroborated in proton-nucleus collisions. In-medium properties of the ω meson have been studied in the $\pi^0\gamma$ decay channel in photonuclear reactions. A dropping of the ω mass by about 7% at effective nuclear densities of about 0.6 ρ_0 has been observed. Double pion photoproduction experiments have revealed a difference in the in-medium modification for the $\pi^0\pi^\pm$ and $\pi^0\pi^0$ channels, consistent with the scenario of a dropping sigma meson mass in the nuclear medium. All experimental observations are in line with theoretical predictions and thus support our understanding of hadron masses in the context of breaking and restoration of chiral symmetry. Antiproton beams provided by the High Energy Storage Ring (HESR) at the future Facility for Antiproton and Ion Research (FAIR) will open up the possibility to extend the study of in-medium properties of hadrons to the charm sector. The physics potential of meson-nucleus bound states and their importance for understanding in-medium properties of hadrons has been highlighted.

Acknowledgments

It is a pleasure to thank the many PhD students and colleagues from the CBELSA/TAPS collaboration in Bonn and the A2 collaboration in Mainz who have performed and analysed the experiments which are the basis of the physics program presented in these lectures. In particular, I would like to acknowledge contributions from Frederic Bloch, Silke Janssen, Johan Messchendorp, Marco Pfeiffer, Susan Schadmand, and David Trnka. Illuminating discussions on the planning and theoretical interpretation of the measurements with Wolfgang Cassing, Stefan Leupold, Ulrich Mosel, Eulogio Oset, Jochen Wambach, and Wolfram Weise are highly appreciated. I would like to thank Ulrike Thoma for critical reading of the manuscript. This work has been supported by Deutsche Forschungsgemeinschaft through Schwerpunktprogram 1034 *"Untersuchung der hadronischen Struktur von Nukleonen und Kernen mit elektromagnetischen Sonden"* and through the Sonderforschungsbereich/Transregio 16 *"Subnuclear structure of matter"* as well as by Bundesministerium für Bildung and Forschung and Gesellschaft für Schwerionenforschung.

References

Agakichiev G *et al.* , 1995, *Phys Rev Lett* **75** 1272.
Agakichiev G *et al.* , 1995, *Phys Lett B* **422** 405.
Bernard V *et al.* , 1987, *Phys Rev Lett* **59** 996.
Bhalereo R S and Liu L C, 1985, *Phys Rev Lett* **54** 865.
Bilger R *et al.* , 2002, *Phys Rev C* **65** 044608.
Bloch F (Univ. Basel), 2004, *priv com.*
Bonutti F *et al.* , 1996, *Phys Rev Lett* **77** 603.
Brown G E and Rho M, 1991, *Phys Rev Lett* **66** 2720.
Chiang H C *et al.* , 1998, *Nucl Phys A* **644** 77.
Effenberger M *et al.* , *Phys Rev C* **60** 044614.
Eidelmann S *et al.* [Review of Particle Properties], *Phys Lett B* **592** 1.
Ericson T E O and Weise W, 1988, *Pions in Nuclei*, Clarendon Press, Oxford.
FAIR, 2001, *An International Accelerator Facility for Beams of Ions and Antiprotons, Conceptual Design Report*, GSI.
Flatté S M, 1976, *Phys Lett B* **63** 224.
Garcia-Recio C *et al.* , 2002, *Phys Lett B* **550** 47.
Geissel H *et al.* , 2002a, *Phys Rev Lett* **88** 122301.
Geissel H *et al.* , 2002b, *Phys Lett B* **549** 64.
Golubeva Ye S *et al.* , 2003, *Eur Phys J A* **17** 275.
Hanhart C, 2005, *Phys Rev Lett* **94** 049101.
Hayashigaki A, 2000, *Phys Lett B* **487** 96.
Hatsuda T and Lee S H, 1992, *Phys Rev C* **46** R34.
Hatsuda T *et al.* , 1999, *Phys Rev Lett* **82** 2840.
Kienle P and Yamazyki T, 2004, *Prog Part Nucl Phys* **52** 85.
Klimt S *et al.* , 1990, *Phys Lett B* **249** 386.
Klingl F *et al.* , 1996, *Z Phys A* **356** 193.
Klingl F *et al.* , 1999, *Phys Rev Lett* **82** 3396.
Krusche B *et al.* , 1995, *Phys Lett B* **358** 40.
Ishida S *et al.* , 2000, in *proc Sigma Meson Workshop*, Kyoto, Japan.
Lee S H and Ko C M, 2003, *Phys Rev C* **67** 038202.
Leupold S *et al.* , 1998, *Nucl Phys A* **628** 311.
Lutz M *et al.* , 1992, *Nucl Phys A* **542** 521.
Marco E and Weise W, 2001, *Phys Lett B* **502** 59.
Mayer B *et al.* , 1996, *Phys Rev C* **53** 2068.
Metag V, 2003, in *proc MESON 2002*, World Scientific, Singapore.

Messchendorp J G *et al.* , 2001, *Eur Phys J A* **11** 95.

Messchendorp J G *et al.* , 2002, *Phys Rev Lett* **89** 222302.

Mühlich P *et al.* , 2004a, *Eur Phys J A* **20** 409.

Mühlich P *et al.* , 2004b, *Phys Lett B* **595** 216.

Mühlich P *et al.* , 2004c, *proc 32nd Int Workshop on Gross Properties of Nuclei and Nuclear Excitation*, Hirschegg, Austria, 113.

Muto R *et al.* , 2004, *J Phys G: Nucl Part Phys* **30** S1023.

Ozawa K *et al.* , 2001, *Phys Rev Lett* **86** 5019.

PANDA, 2004, *Letter of Intent*, GSI.

Pfeiffer M *et al.* , 2004, *Phys Rev Lett* **92** 252001.

Pfeiffer M *et al.* , 2005, *Phys Rev Lett* **94** 049102.

Rapp R *et al.* , 1999, *Phys Rev C* **59** 1237.

Rapp R, 2003, *Pramana* **60** 675.

Renk R *et al.* , 2002, *Phys Rev C* **66** 014902.

Roy B J *et al.* , 2004, COSY-proposal Nr 50.

Schadmand S, 2004, *priv com.*

Schön *et al.* , 1996, *Acta Phys Pol B* **27** 2959,

Sibirtsev A *et al.* , 2002, *Phys Rev C* **65** 044007.

Sibirtsev A *et al.* , 2004, *Phys Rev C* **70** 047001.

Sokol *et al.* , 2001, nucl-ex/0011005.

Starostin A *et al.* , 2000, *Phys Rev Lett* **85** 5539.

Trnka D *et al.* , 2004, PhD thesis *Uni Giessen*; *priv com.*

Vicente Vacas M J and Oset E, 1999, *Phys Rev C* **60** 064621.

Vicente Vacas M J and Oset E, 2002, nucl-th/0204055.

Weise W, 2003, in *proc Int School Physics "Enrico Fermi"*, IOS Press, Amsterdam

Weiss J *et al.* , 2001, *Eur Phys J A* **11** 371.

Wessels J P *et al.* , 2003, *Nucl Phys B* **715** 262.

Wilczek F, 2002a, *Phys Today August* 10.

Wilczek F, 2002b, *proc Fundamental physics - Heisenberg and beyond*, Munich 2001 79.

Zschocke *et al.* , 2003, *Phys Lett B* **562** 57.

Topical aspects of hyperon physics

Stanislav Belostotski

Petersburg Nuclear Physics Institute, High Energy Physics Division,
St Petersburg, Gatchina, Russia

1 Hyperon spin structure

It is well known that the spin structure of the proton, neutron and other baryons is non-trivial. All the baryons are apparently 3-quark states (qqq). Existence of more complex states, like pentoquarks ($ud ud\bar{s}$), is still subject to intensive experimental investigation. Using the lowest mass flavours (u, d and s quarks), one may build an SU(3)-flavour symmetric wavefunction of any member of the spin-$\frac{1}{2}$ baryon octet or spin-$\frac{3}{2}$ baryon decuplet (see *eg* Martin and Halzen (1984)). In the naive Constituent Quark Model (CQM) all quarks are assumed to be in an S-state and the spin of a baryon is composed of its constituent quark spins. Thus one may write for the proton, neutron and Λ^0 hyperon, the lowest mass states of the spin-$\frac{1}{2}$ baryon octet:

$$|p \Uparrow> \ = \ \frac{1}{\sqrt{18}}(u \uparrow u \downarrow d \uparrow +u \uparrow u \uparrow d \downarrow) + cycl\ perm \ ,$$

$$|n \Uparrow> \ = \ \frac{1}{\sqrt{18}}(d \uparrow d \downarrow u \uparrow +d \uparrow d \uparrow u \downarrow) + cycl\ perm \ ,$$

$$|\Lambda \Uparrow> \ = \ \frac{1}{\sqrt{18}}[(u \uparrow d \downarrow +u \downarrow d \uparrow) - (u \downarrow d \uparrow +u \uparrow d \downarrow)]s \uparrow +cycl\ perm \ . \quad (1)$$

Here $\Uparrow$ indicates that a baryon is in a pure spin state, and $\uparrow, \downarrow$ denote spin states of the constituent quarks. Using these wavefunctions one can directly calculate the quark polarisation P_q in a fully polarised baryon:

$$< p \Uparrow |\sigma_z^u|p \Uparrow> = P_u = \frac{2}{3} \quad ; \quad < p \Uparrow |\sigma_z^d|p \Uparrow> = P_d = -\frac{1}{3}$$

$$< n \Uparrow |\sigma_z^u|n \Uparrow> = P_u = -\frac{1}{3} \quad ; \quad < n \Uparrow |\sigma_z^d|n \Uparrow> = P_d = \frac{2}{3}$$

$$< \Lambda \Uparrow |\sigma_z^{u,d}|\Lambda \Uparrow> = P_{u,d} = 0 \quad ; \quad < \Lambda \Uparrow |\sigma_z^s|\Lambda \Uparrow> = P_s = 1 . \tag{2}$$

According to this simple model the u-quark is strongly polarised along the proton spin and the d-quark is polarised in the opposite direction. Since the proton and neutron are isotopically symmetric states quark polarisations in the neutron are obtained by interchanging u and d quarks. As for the Λ baryon, its spin is entirely carried by the s quark, while the ud pair is in a spinless (singlet) state.

In the CQM approach the following sum rule must be fulfilled for a member of the barion spin-$\frac{1}{2}$ octet:

$$\Sigma = \Delta\Sigma_q = \sum_f n_f < \sigma_z^f > = \sum_f n_f P_f = 1, \tag{3}$$

where n_f is number of quarks with flavour f, $\frac{1}{2}\Sigma$ is the baryon spin, and $\Delta\Sigma_q$ is contribution from the quark polarisations. Thus for the proton in a pure spin state $\Delta_q\Sigma = 2 \cdot \frac{2}{3} - \frac{1}{3} = 1$.

It is important that the spin stucture of the proton and neutron can be directly studied in polarised deep inelastic lepton scattering (DIS) experiments. As it follows from *inclusive* DIS experiments (Antony *et al.* 1997, Aveda *et al.* 1998, Airapetiain *et al.* 1998) that the quark spins account for only a fraction of the nucleon spin, the sum rule Equation 3 is not confirmed by the experiment. A similar result for the neutron has been obtained by measuremnets on the target containing polarised neutrons (Abe *et al.* 1995, Anthony *et al.* 1996, Ackersfatt *et al.* 1997). These experiments allowed a high precision verification of the fundamental Bjorken sum rule (Bjorken 1970).

Information about specific contributions to the nucleon spin from different quark flavours has been obtained in *semi-inclusive* DIS experiments in which a final state hadron is detected in coincidence with the scattered lepton (Adams *et al.* 1998, Ackerstaff *et al.* 1999, Airapetian *et al.* 2004). In the Quark Parton Model (QPM) these contributions are given by the first moments of the helicity dependent (polarised) parton distributions:

$$\Delta q_f(x) = q_f(x) \uparrow - q_f(x) \downarrow \quad ; \quad \Delta q_f(x) = \Delta u(x), \Delta d(x)... \; , \tag{4}$$

where x represents the Bjorken scaling variable. The overall quark polarisation discussed above is defined as

$$P_f = \frac{\Delta q_f}{n_f} = \frac{\int \Delta q_f(x)dx}{\int q_f(x)dx} \quad , \tag{5}$$

where $q_f(x) = q_f(x) \uparrow + q_f(x) \downarrow$ represents helicity independent (unpolarised) parton distributions, and $n_f = \int q_f(x)dx$.

Recent semi-inclusive DIS data (Airapetian *et al.* 2004) obtained by the HERMES collaboration have provided the direct measurements of $\Delta q_f(x)$ in the proton. Integrated over the measured kinematic range $0.023 < x < 0.6$, the following moments are obtained:

$$\Delta u + \Delta \bar{u} = 0.599 \pm 0.022(stat) \pm 0.065(syst),$$
$$\Delta d + \Delta \bar{d} = -0.280 \pm 0.026(stat) \pm 0.057(syst),$$
$$\Delta s = -0.028 \pm 0.033(stat) \pm 0.009(syst), \tag{6}$$

with $\Delta\Sigma = 0.347 \pm 0.024 \pm 0.066$.

With the help of SU(3) rotation, for members of the spin-$\frac{1}{2}$ octet one obtains:

$$\Delta u_p = \Delta u, \quad \Delta d_p = \Delta d, \quad \Delta s_p = \Delta s;$$
$$\Delta u_n = \Delta d, \quad \Delta d_n = \Delta u, \quad \Delta s_n = \Delta s;$$
$$\Delta u_\Lambda = \Delta d_\Lambda = \frac{2}{3}\Delta d + \frac{1}{6}(\Delta u + \Delta s), \quad \Delta s_\Lambda = -\frac{1}{3}\Delta d + \frac{2}{3}(\Delta u + \Delta s);$$
$$\Delta u_{\Sigma^+} = \Delta u, \quad \Delta d_{\Sigma^+} = \Delta s, \quad \Delta s_{\Sigma^+} = \Delta d;$$
$$\Delta u_{\Sigma^0} = \Delta d_{\Sigma^0} = \frac{1}{2}(\Delta u + \Delta s), \quad \Delta s_{\Sigma^0} = \Delta d;$$
$$\Delta u_{\Sigma^-} = \Delta s, \quad \Delta d_{\Sigma^-} = \Delta u, \quad \Delta s_{\Sigma^-} = \Delta d;$$
$$\Delta u_{\Xi^0} = \Delta d, \quad \Delta d_{\Xi^0} = \Delta s, \quad \Delta s_{\Xi^0} = \Delta u;$$
$$\Delta u_{\Xi^-} = \Delta s, \quad \Delta d_{\Xi^-} = \Delta d, \quad \Delta s_{\Xi^-} = \Delta u. \tag{7}$$

The results of calculations made using Equation 7 and the experimentally found numbers (Equation 6) are listed in Table 1. Here we assumed that $\Delta s_p \equiv 0$. The numbers corresponding to the constituent quark model are listed for comparison in the right part of the table.

	qqq	S	I_3	Δu_{exp}	Δd_{exp}	Δs_{exp}	Δu_{QM}	Δd_{QM}	Δs_{QM}
p	uud	0	1/2	0.6	-0.28	0.	4/3	-1/3	0
n	udd	0	-1/2	-0.28	0.6	0.	-1/3	4/3	0
Λ^0	uds	-1	0	-0.086	0	0.493	0	0	1
Σ^+	uus	-1	1	0.6	0.	-0.28	4/3	0	-1/3
Σ^0	uds	-1	0	0.3	0.3	-0.28	2/3	2/3	-1/3
Σ^-	dds	-1	-1	0.	0.6	-0.28	0	4/3	-1/3
Ξ^0	uss	-2	1/2	-0.28	0.	0.6	-1/3	0	4/3
Ξ^-	dss	-2	-1/2	0	-0.28	0.6	0	-1/3	4/3

Table 1. *Hyperon spin structure according to SU(3)-flavour symmetry. Values of $\Delta u_{exp}, \Delta d_{exp},$ and Δs_{exp} are obtained using Equation 7 and the experimental HERMES results for the proton (Airapetian et al. 2004).*

One experimental possibility to probe hyperon spin structure is to measure the spin transfer coefficient from a polarised quark to the hyperon produced in the process of fragmentation. It has been shown that the longitudinal spin transfer is sensitive to the spin structure of the final hyperon (Jaffe 1996). In this respect the Λ^0 hyperon is of particular interest. The polarisation of final-state Λ^0 hyperons can be measured via the weak decay $\Lambda^0 \to p\pi^-$, through the angular distribution of the final state particles:

$$\frac{dN_p}{d\Omega} \propto 1 + \alpha \vec{P}_\Lambda \cdot \hat{k} \ . \tag{8}$$

Here $\alpha = 0.642 \pm 0.013$ is the asymmetry parameter of the parity-violating weak decay, $\vec{P}_\Lambda$ is the polarisation of the Λ^0, and $\hat{k}$ is the unit vector along the proton momentum in the rest frame of the Λ^0. Since this decay is parity-violating, the decay amplitude includes the term $\vec{\sigma}_\Lambda \cdot \hat{k}$, which gives the possibility of detecting the Λ^0 polarisation by the angular asymmetry of the emitted proton with respect to the Λ polarisation vector.

In a similar way the polarisation of heavier hyperons may be studied. For example the polarisation of the Ξ^- hyperon can be measured via its parity-violating weak decay $\Xi^- \to \Lambda^0 + \pi^-$ with the decay asymmetry parameter $\alpha = -0.458 \pm 0.012$.

The polarisation of Λ^0 s, emitted in the decay of a polarised hyperon, provides another opportunity to measure hyperon polarisation. Thus in the case of the $\Sigma^0 \to \Lambda^0 + \gamma$ decay, the Λ^0 polarisation is directly connected to the polarisation P_{Σ^0} of the the parent Σ^0 (Gatto 1958):

$$\vec{P}_{\Lambda^0} = P_{\Sigma^0}(\vec{n}_{\Sigma^0} \cdot \vec{k}_{\Lambda^0})\vec{k}_{\Lambda^0}, \tag{9}$$

where $\vec{n}_{\Sigma^0}$ and $\vec{k}_{\Lambda^0}$ are the unit vectors along the Σ^0 polarisation and Λ^0 momentum, respectively.

It should be noted that in all the above mentioned cases of hyperon polarisation measurements, the detector acceptance is usually strongly involved. The acceptance correction may be obtained in principle using Monte Carlo simulations. Unfortunately, systematic uncertainties in these simulations are often at the level of the measured asymmetries. The acceptance correction does not practically affect the polarisation (spin transfer) extacted from a "helicity balanced data set" provided a possiblity exists to flip the primary beam polarisation (as discussed in the next section).

In these lectures existing data on the longitudinal spin transfer to the Λ^0 hyperon are reviewed and related possibilities to study hyperon spin structure are discussed. The transverse polarisation of Λ^0 and $\bar{\Lambda}^0$ hyperons, produced inclusively in quasi-real photonnucleon interactions, are presented, and future prospects for hyperon spin structure studies are outlined.

2 Longitudinal spin transfer to the Λ^0 hyperon

2.1 Overview of experiment results

Longitudinal spin transfer was first studied by the LEP experiments OPAL and ALEPH at an energy corresponding to the Z^0 pole (Ackerstaff *et al.* 1998a, Buskulic *et al.* 1996). In these experiments the $\Lambda(\bar{\Lambda})$ hyperons are predominantly produced via the decay $Z^0 \to s\bar{s}$, in which the primary strange quarks initiating the hadronisation process are negatively polarised at the level of -91%. The OPAL and ALEPH (LEP) data show a significant polarisation transfer from the $s(\bar{s})$ quark to the $\Lambda(\bar{\Lambda})$, thus confirming that Δs^{Λ} dominates the Λ spin. The Λ^0 polarisation was found to be -0.32 ± 0.04 at $z > 0.3$, where $z = \frac{E_{\Lambda}}{E_q}$, and E_q is the energy of the primary quark.

The experiments succeeded in describing the data using a Lund Monte Carlo model with the following hypotheses:

- The primary quarks produced in the Z^0 decay retain their helicity throughout the fragmentation process.

- The quarks produced from colour-string breaking have no preferred spin direction.

- The spin structure of the produced hyperons can be adequately described by the naive CQM.

The interpretation of these data is not unique, however, as the measured polarisation is dominated by the strange-quark contribution with rather little sensitivity to possible contributions from u and d quarks.

In contrast to the LEP experiments, production of Λ^0 hyperons in DIS originates predominately from the u and d quarks in the proton. In the NOMAD experiment (Astier *et al.* 2000), the production of Λ^0 hyperons was studied in ν_μ charged-current interactions. Also in contrast to the LEP experiments, the NOMAD data are concentrated in the kinematic domain corresponding to the target fragmentation region. A negative longitudinal Λ polarisation $P_\Lambda = -0.21 \pm 0.04$ was observed.

Using charged lepton beams, only one measurement of longitudinal Λ^0 polarisation in DIS has been reported to date. The E665 collaboration (Adams *et al.* 1994) measured a negative polarisation using polarised muon beams of 470 GeV. Unfortunately, the statistical quality of the E665 experiment is rather limited as only 750 Λ^0 events were identified.

The HERMES spectrometer (Ackerstaff *et al.* 1998b) has provided preliminary measurements of the longitudinal spin transfer to the Λ^0 in DIS using the 27.5 GeV polarised positron beam of the HERA collider and unpolarised gas target (Airapetian *et al.* 2001). This first result was based on limited statistics of about 1800 Λ^0 events. Since then the statistical accuracy was essentially improved by the HERMES experiment. The final analysis has been recently performed with a data sample containing 8000 unpolarised target events with a cut $Q^2 > 0.8$ GeV^2, where Q^2 is negative four-momentum transfer squared. A polarised target data sample with more than $\simeq 3500$ Λ^0 events is still to be analysed.

2.2 Longitudinal spin transfer in the Quark Parton Model (QPM)

The dominant mechanism for semi-inclusive production of longitudinally polarised Λ^0 hyperons in polarised DIS is depicted in Figure 1. A longitudinally polarised electron or positron emits a polarised virtual photon (denoted γ^*) which is absorbed by a quark of opposite spin direction in the target proton. As indicated by the arrows in Figure 1, this fixes the spin orientation of the struck quark: after the spin-1 photon is absorbed, the outgoing quark has the same helicity as the virtual photon.

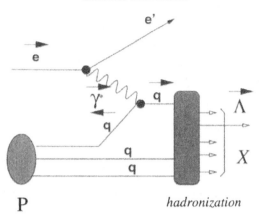

Figure 1. *The single-quark scattering mechanism leading to Λ^0 production. The arrows indicate spin orientations of the participating particles. Produced Λ spin orientation is conditionally shown.*

If the longitudinal polarisation of the beam is given by P_b and the target is unpolarised, the struck quark will acquire a polarisation $P_q = P_b D(y)$. Here $y = (E - E')/E$ is the fractional energy carried by the photon, E and E' represent the energy of the primary and scattered electron, and $D(y) = [1 - (1 - y)^2]/[1 + (1 - y)^2]$ gives the "depolarisation" of the virtual photon as compared to the incident electron. The polarisation of the Λ^0 hyperon is then given by

$$P_\Lambda = P_b D(y) D^\Lambda_{LL'}, \qquad (10)$$

where $D^\Lambda_{LL'}$ is the spin transfer coefficient describing the probability that the polarisation of the struck quark will be transfered to the Λ^0 .

It is not specified in Figure 1 whether the final Λ^0 is related to the struck quark (current fragmentation) or whether it is produced via a remnant di-quark (target fragmentation). At high energy these two mechanisms are usually separated with the help of Feynman variable $x_F = p_\parallel / p_{max}$, where $p_\parallel$ is the hadron momentum component along the virtual-photon direction in the $\gamma^* p$ centre-of-mass system. Current fragmentation is selected by the cut $x_F > 0$, while target fragmentation is selected by $x_F < 0$. This separation, however, is very conditional. According to the Lund fragmentation model a substantial fraction of the Λs are produced from the central part of the "fragmentation string" which is neither related to the struck quark nor to the remnant diquark.

Results of MC studies for HERMES kinematics (Makins 2004) are presented in Figure 2. The open histograms show the distribution of Λs in x_F accepted by the HERMES detector. It must be remembered that a substantial fraction of Λs (about 60%) are produced via decay of the Σ^0 or Σ^*_{1385} resonances. The filled histograms in Figure 2 indicate those Λs where either the struck quark or the target remnant (*ie*, one of the ends of the string) were actually present in the Λ^0 or its parent hyperon. As expected, those containing the struck quark are concentrated at high values of x_F (current fragmentation region), while those containing the remnant appear at negative x_F (target fragmentation region). However, even in the forward region $x_F > 0$, Λs containing the target remnant account for a significant fraction of the events.

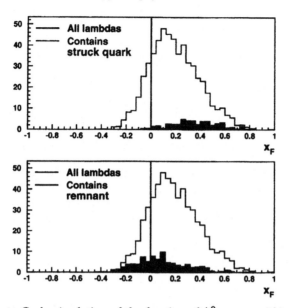

Figure 2. *Monte Carlo simulation of the fraction of Λ^0 events within the HERMES acceptance where the Λ^0 (or its parent hyperon) contains either the struck quark or the target remnant.*

Therefore one can conclude that in the kinematic domain covered by the HERMES spectrometer only a fraction of Λ^0 events F can "remember" the struck quark polarisation. At $z > 0.3$ the factor F is ≈ 0.15, *ie* "dilution" due to the unpolarised part of the fragmentation $(1 - F)$ is very large. It is not clear at the moment, however, how large the systematic uncertainty of the MC calculations of the factor F is, particularly at the relatively low primary beam energy of the HERMES experiment. For comparison, in the ALEPH experiment at the same cut $z > 0.3$ the factor F is found to be 0.49 ± 0.09 (Buskulic *et al.* 1996).

In the Quark-Parton Model (QPM), spin-transfer integrated over x (the Bjorken variable) in the current-fragmentation region is expressed as follows (Jaffe 1996):

$$D_{LL'}^{\Lambda}(z, Q^2) = \sum_f \omega_f^{\Lambda}(z, Q^2) D_{LL',f}^{\Lambda}(z, Q^2), \tag{11}$$

where the purity ω_f^{Λ} is the probability that a Λ^0 was produced in a fragmentation process originating from a struck quark of flavour f, and $z = E_\Lambda / E_q$. (In the DIS case $E_q = \nu$, where $\nu = E - E'$ is the energy transfered from the scattered lepton to the virtual photon. The latter is absorbed by the struck quark which acquires the energy ν.)

The quantity $D_{LL',f}^{\Lambda}$ in Equation 11 is the "partial" spin transfer from a quark of flavour q to a Λ^0 hyperon. The $D_{LL',f}^{\Lambda}$ coefficient is read

$$D_{LL',f}^{\Lambda}(z) = \frac{\Delta D_f^{\Lambda}(z)}{D_f^{\Lambda}(z)} \equiv \frac{D_{f+}^{\Lambda+}(z) - D_{f+}^{\Lambda-}(z)}{D_{f+}^{\Lambda+}(z) + D_{f+}^{\Lambda-}(z)}. \tag{12}$$

Here $D_f^\Lambda(z)$ and $\Delta D_f^\Lambda(z)$ are the spin-independent and spin-dependent fragmentation functions for Λ^0 production from a primary quark q_f, $D_{f+}^{\Lambda+}(z)$ and $D_{f+}^{\Lambda-}(z)$ are used to denote the fragmentation functions for a quark of helicity + to produce a Λ^0 of helicity + or − respectively. Because of the weak dependence of $D_{LL'}^\Lambda$ on Q^2, this dependence is omitted hereafter in the text.

At present it is still not possible to evaluate the polarised or unpolarised fragmentation functions from first princples. On the other hand, it is apparent that the spin transfer coefficient must be closely related to the spin structure of the produced Λ hyperon. Under conditions that the produced Λ contains the struck quark q_f, and the original helicity of the quark is preserved during the hadronisation process, the spin state of the Λ will be a direct consequence of its helicity substructure. For example, if u-quark with helicity + fragments to a Λ, then according to Table 1 the produced Λ will have polarisation −0.086, or zero in CQM. For the s-quark the Λ polarisation is expected to be 0.493 and 1, respectively. In other words, the net $D_{LL'}^\Lambda$ coefficient for Λs containing the primary struck quark can be written as

$$D_{LL',f}^\Lambda \simeq \frac{\Delta q_f^\Lambda}{q_f^\Lambda}. \tag{13}$$

This property of $D_{LL',f}^\Lambda$ is based on the assumption that the struck quark q_f preserves its helicity in the produced Λ. Assume that $N_+^{\Lambda+}$ is number of Λ events with helicity +. In this sample the spin of q_f is parallel to the Λ spin. In the sample $N_+^{\Lambda-}$ the spin of q_f is antiparallel to the Λ spin. Since $N_+^{\Lambda-} = N_-^{\Lambda+}$ one may write

$$D_{LL',f}^\Lambda \equiv \frac{N_+^{\Lambda+} - N_+^{\Lambda-}}{N_+^{\Lambda+} + N_+^{\Lambda-}} = \frac{N_+^{\Lambda+} - N_-^{\Lambda+}}{N_+^{\Lambda+} + N_-^{\Lambda+}} \equiv \frac{\Delta q_f^\Lambda}{q_f^\Lambda}. \tag{14}$$

One should note that $D_{LL',f}^\Lambda$ is a function of z, while $\frac{\Delta q_f^\Lambda}{q_f^\Lambda}$ is a function of x^Λ, where x^Λ is the Bjorken variable for the Λ^0 hyperon. The problem of the practical application of Equation 14 can be solved with the help of so called "reciprocity relation" (Gribov and Lipatov 1971). Proceeding from that it was suggested in Schmidt *et al.* (2000) and Ma *et al.* (2002) that Equation 14 may be used assuming that $z = x^\Lambda$.

2.3   Longitudinal spin transfer measured by HERMES

In the HERMES experiment the 27.5 GeV polarised positron (or electron) beam of the HERA *ep*-collider is passed through an open-ended tubular storage cell into which polarised or unpolarised hydrogen is injected. The experimental setup is described in detail in Ackerstaff *et al.* (1998b).

The data presented here were recorded during a five-year period from 1996 to 2000 using longitudinally polarised positron or electron beams from the HERA ring. Most of the data taking involved positron scattering off unpolarised hydrogen and deuterium targets, but in order to increase statistics, data collected from unpolarised ^{3}He and ^{14}N targets were also included in the analysis. The data from all targets were combined in one data set.

The scattered positrons and the Λ^0 decay products were detected in coincidence by the HERMES spectrometer (Ackerstaff *et al.* 1998b) in the polar angular range from 40 to 220 mrad. The Λ^0 hyperons were identified through their $p\pi^-$ decay channel. Two spatial vertices were reconstructed for each event. The primary (production) vertex was determined from the intersection of the beamline and the scattered lepton, while the secondary (decay) vertex was determined from the intersection of the proton and pion tracks. All tracks were required to satisfy a series of fiducial-volume cuts designed to avoid the inactive edges of the detector. For tracks fulfilling these requirements the invariant mass of the hadron pair was evaluated, revealing a clear Λ^0 peak even without background suppression cuts. These spectra are displayed in Figure 3.

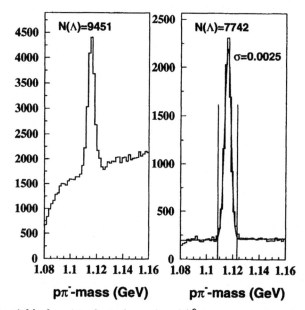

Figure 3. *The yield of semi-inclusively produced Λ^0 hyperons in deep inelastic scattering. The left (right) panel shows the invariant mass spectrum before (after) the application of background suppression cuts.*

In order to suppress background, a vertex separation cut $z_2 - z_1 > 10$ cm was used together with particle identification information allowing rejection of $\pi^+\pi^-$ pairs contribution to the combinatorial background.

With all requirements imposed, the Λ^0 sample contained 7,700 hyperon events for all the data obtained from unpolarised targets. These numbers were obtained by integrating the Λ^0 peak between the boundaries shown in Figure 3 and subtracting the background.

An average beam polarisation of around 55% was typical during the data taking. Reversal of the polarisation direction was performed three times during the 1996-1997 data taking period, but more frequently thereafter. This history is displayed in Figure 4 where the average beam polarisation recorded in each data-taking run is plotted versus (sequential) run number. As shown in the figure, similar amounts of data were recorded in each helicity state.

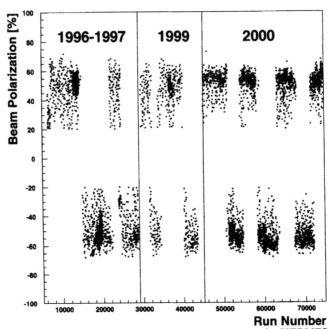

Figure 4. *The polarisation of the HERA positron beam versus the HERMES data-taking run number in the years 1996-2000.*

Two independent polarimeters were used to measure the beam polarisation, using similar techniques based on laser Compton backscattering. The transverse polarimeter (TPOL) measures the polarisation outside the HERMES spin rotators, while the longitudinal polarimeter (LPOL) is located near the HERMES interaction point and inside the spin rotators. The systematic uncertainty on the polarisation is approximately 2%.

Data analysis

The coefficient $D_{LL'}^{\Lambda}$ describes the correlation between the longitudinal polarisations of the struck quark and produced Λ^0 hyperon. The spin-quantisation axis L for the quark is the momentum direction of the virtual photon which it absorbs.

Unfortunately the QPM does not define how the polarisation vector of the produced hyperon (L') is directed. In general case for an orthogonal coordinate frame xyz the beam induced Λ polarisation vector (where the target is unpolarised) is read as

$$P_\Lambda^j = P_B D(y) D_{LL'}^{\Lambda}{}^{,j}, \qquad j = x, y, z. \tag{15}$$

Here $D_{LL'}^{\Lambda}{}^{,j}$ are the components of the polarisation transfer vector to be found in the experiment.

Extraction of the $D_{LL'}^{\Lambda}{}^{,j}$ components is considered in the Addendum at the end of these lecture notes. Here for simplicity we assume two possibilities for the L' direction:

	Years 96-97	Years 99-00	All data
$\langle D_{LL'}^\Lambda \rangle$, axis 1	0.18 ± 0.16	0.08 ± 0.12	0.11 ± 0.10
$\langle D_{LL'}^\Lambda \rangle$, axis 2	0.15 ± 0.15	0.10 ± 0.11	0.11 ± 0.09
N_Λ	2,930	4,820	7,770
$\langle z \rangle$	0.438	0.456	0.449
$\langle x_F \rangle$	0.287	0.306	0.299

Table 2. *Results for $D_{LL'}^\Lambda$ after averaging over kinematics with cut $x_F > 0$.*

along the Λ momentum (axis 1); and along the virtual photon momentum in the Λ rest frame (axis 2).

As the HERMES spectrometer is a forward detector, its acceptance for the reconstruction of Λ hyperons is limited and strongly depends on $\cos \Theta_{pL'}$, where $\Theta_{pL'}$ is the angle between the proton momentum direction and L' axis in the Λ rest frame. To minimise acceptance effects, the spin transfer to the Λ has been determined by combining the two data sets measured with opposite beam helicities in such a way that the luminosity-weighted average beam polarisation for the selected data sample is zero. A detailed derivation is presented in the Addendum, and leads to the following extraction formula:

$$D_{LL'}^\Lambda = \frac{1}{\alpha \overline{P_b^2}} \cdot \frac{\sum_{i=1}^{N_\Lambda} P_{b,i} \, D(y_i) \cos \Theta_{pL'}^i}{\sum_{i=1}^{N_\Lambda} D^2(y_i) \cos^2 \Theta_{pL'}^i}. \tag{16}$$

Here, $\overline{P_b^2} \equiv (\frac{1}{L}) \int P_b^2 \, dL$ is the luminosity-weighted average of the square of the beam polarisation ($L = \int dL$ is the total luminosity). The indicated sums are over the Λ events, selected so as to form a helicity-balanced sample with $\overline{P_b} = 0$.

Table 2 presents the average results for $D_{LL'}^\Lambda$. As shown in the table, the choice of coordinate system does not affect the result: the spin-transfer along the virtual photon direction (axis 1) is the same as that along the Λ momentum direction (axis 2).

As the measured value for $D_{LL'}^\Lambda$ shows no significant dependence on the choice of longitudinal spin-quantisation axis, the results can be summarised by a single value:

$$D_{LL'}^\Lambda = 0.12 \pm 0.10 \,(\text{stat}) \pm 0.03 \,(\text{syst}) \tag{17}$$

This represents the spin-transfer to the Λ^0 along the Λ^0 momentum direction, averaged over the kinematic region $x_F > 0$. The average fractional energy of the Λ^0 hyperons in this sample is $\langle z \rangle = 0.45$. The net spin-transfer is consistent with zero.

Discussion

The dependence of $D_{LL'}^\Lambda$ on the energy fraction z is presented in Figure 5 (left). A cut of $x_F > 0$ was imposed on the data.

Superimposed on the data are the phenomenological model calculations of Schmidt *et al.* (2000) (pQCD and quark-diquark models) which predict a pronounced rise of the

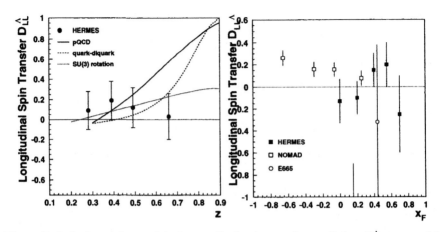

Figure 5. *Left: Dependence of the longitudinal spin-transfer coefficient $D_{LL'}^{\Lambda}$ on z with cut $x_F > 0$ imposed. The curves represent various theoretical calculations described in the text. Right: Compilation of data on longitudinal spin transfer to the Λ^0 in lepton deep inelastic scattering.*

spin transfer at high z values, and of Ma *et al.* (2002) (SU(3)-flavour rotation of proton values) which predicts a more gradual increase. Although the HERMES data extend to the highest values of z yet explored in DIS, they display no evidence of a kinematic dependence. The measurements are fully consistent with zero but they are also not in contradiction with a modest variation with z.

In Figure 5 (right) the HERMES data are presented as a function of x_F together with results obtained by the NOMAD (Astier *et al.* 2000) and E665 (Adams *et al.* 1994) experiments.

The NOMAD neutrino beam of 43 GeV places the experiment in an energy range similar to that of HERMES (27.6 GeV positron beam).

The neutrino is "left-handed", *ie* its helicity is -1 which formally corresponds to $P_B = -1$ in Equation 10. The depolarisation factor D_y can be taken equal to 1 for (ν_μ, μ^-) scattering. Intermediate W^+ boson may be absorbed only by d, s, and $\bar{u}$ producing a u quark in the final state, however absorbtion by d dominates. The longitudinal Λ^0 polarisation measured in a (ν_μ, μ^-) experiment may be interpreted in the QPM as follows:

$$P_\Lambda(x,y,z) = -\frac{d(x)\,\Delta D_u^\Lambda(z) - (1-y)^2\,\bar{u}(x)\,\Delta D_{\bar{d}}^\Lambda(z)}{d(x)\,D_u^\Lambda(z) + (1-y)^2\,\bar{u}(x)\,D_{\bar{d}}^\Lambda(z)} \approx \frac{\Delta D_u^\Lambda(z)}{D_u^\Lambda(z)}\ . \qquad (18)$$

Here for simplicity the Cabibbo suppressed processes and contributions from the strange quarks in the target are neglected. As Λ production at HERMES is also dominated by u quark fragmentation, $-P_\Lambda$ from NOMAD can be qualitatively compared to $D_{LL'}^\Lambda$ from HERMES (Makins 2004).

As shown in Figure 5 (right) the NOMAD and HERMES results are compatible in the kinematic region of overlap $-0.1 < x_F < 0.3$. In the $x_F > 0$ domain the NOMAD experiment has obtained $D_{LL'}^\Lambda = 0.09 \pm 0.06(stat) \pm 0.03(syst)$, which is close to the HERMES result $(0.12 \pm 0.1(stat) \pm 0.03(syst))$. Both results are compatible with zero. A trend towards positive spin transfers is observed in the transition from positive to negative x_F. In the region of negative x_F the NOMAD experiment has obtained a statistically significant positive longitudinal spin transfer:

$$D_{LL'}^\Lambda = 0.21 \pm 0.04(stat) \pm 0.02(syst).$$

Theoretical interpretation of this result has not yet been accomplished. A Monte Carlo simulation is needed to take into account the above mentioned hyperon decays and to extract the net spin transfer.

The new HERMES data presented here are the most precise measurements to date of spin transfer in DIS at $x_F > 0$. The spin transfer is found to be consistent with zero which is in marked contrast with the large Λ polarisation observed in e^+e^- annihilation at OPAL and ALEPH.

This result could be qualitatively predicted. In the e^+e^- annihilation case a strongly polarised s quark fragments to a Λ and, according to SU(3) rotation, the s quark carries a substantial part of the Λ spin. Therefore a sizable $D_{LL'}^\Lambda$ would be expected. On the other hand, in deep inelastic lepton scattering Λs are predominantly produced via u or d quarks, carrying just a small fraction of the Λ spin. This would result in a $D_{LL'}^\Lambda$ close to zero.

In order to understand this result quantitatively a complex Monte Carlo study must be done. If preliminary MC estimates of the F factor for HERMES kinematics (Figure 2) are realistic then the $D_{LL'}^\Lambda$ might be trivially equal to zero due to a large unpolarised background. In order to enlarge the spin-dependent component of $D_{LL'}^\Lambda$ the events with highest possible x_F, or z must be accumulated to a statistically significant data sample. This is, however, a task for future experiments.

3 Transverse hyperon polarisation

3.1 "Self-polarised" hyperons in hadron collisions

The transverse (perpendicular to the production plane) hyperon polarisation was observed and investigated in many high-energy scattering experiments, with a wide variety of hadron beams and kinematic settings (Heller 1997, Lach 1996).

The transverse polarisation of Λ^0 hyperons (P_n^Λ) is almost always found to be negative. One notable exception is the positive polarisation measured in the K^-p interaction. The kinematic behaviour of P_n^Λ has also been investigated. It increases almost linearly with transverse momentum p_T up to a value of about 1 GeV, where a plateau is reached, and rises slowly with x_F.

Possible mechanisms for the origin of this polarisation have been reviewed, for example by Panagiotou (1990) and by Soffer and Törnqvist (1992).

3.2   General formalism of P_n^Λ extraction

Because of the parity-conserving nature of the strong interaction, any final-state hadron polarisation in a reaction with unpolarised beams and targets must point along a pseudo-vector direction. In the case of inclusive hyperon production, the only available direction of this type is the normal $\hat{n}$ to the scattering plane formed by the cross-product of the vectors along the laboratory-frame momenta of the beam $(\vec{p}_B)$ and the Λ $(\vec{p}_\Lambda)$:

$$\hat{n} = \frac{\vec{p}_B \times \vec{p}_\Lambda}{|\vec{p}_B \times \vec{p}_\Lambda|}. \tag{19}$$

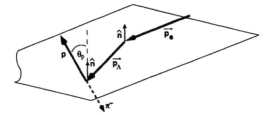

Figure 6. *Schematic diagram of Λ production plane and Λ decay. The angle θ_p of the decay proton with respect to the normal $\hat{n}$ to the scattering plane is defined in the Λ rest frame.*

A kinematic diagram of inclusive Λ production and decay is given in Figure 6. The Λ decay is shown in the Λ rest frame, where θ_p is the angle of proton emission relative to the axis given by the normal $\hat{n}$ to the scattering plane. Although $\hat{n}$ is defined in Equation 19 using vectors in the laboratory frame, it is important to note that the direction is unaffected by the boost into the Λ rest frame. The value of $cos\theta_p$ in the Λ rest frame can be explicitly expressed through the x and y components of the decay proton and pion momenta $p_{px,y}$ and $p_{\pi x,y}$ directly reconstructed in the laboratory frame:

$$cos\theta_p = \frac{-p_{px}p_{\pi y} + p_{py}p_{\pi x}}{q^\Lambda p_T^\Lambda}, \tag{20}$$

where q^Λ is the Λ cm decay momentum and

$$p_T^\Lambda = \sqrt{(p_{px} + p_{\pi x})^2 + (p_{py} + p_{\pi y})^2}$$

is transverse Λ momentum.

The laboratory coordinate system is defined as follows: axis z is directed along the beam, axis x is perpendicular to the beam in horizontal plane, and axis y is perpendicular to the beam in vertical plane.

An important feature of Equation 20 is its "mirror symmetry", *ie* $cos\theta_p \to -cos\theta_p$ if $y \to -y$ or $x \to -x$. This property, as will be shown, is very important for the extraction of polarisation from the data by minimising false asymmetries due to the spectrometer acceptance.

A moment method, similar to that used for longitudinal spin transfer (see Addendum), can be evaluated for the transverse Λ polarisation extraction. Since only the normal component of Λ polarisation may be non-zero the polarised Λ decay distribution is written as

$$\frac{dN}{d\Omega_p} = \frac{dN_0}{d\Omega_p}(1 + \alpha P_n^\Lambda \cos\theta_p), \tag{21}$$

where θ_p denotes the angle of proton emission relative to the $\hat{n}$ axis in the Λ rest frame (as shown in Figure 6). As the decay of unpolarised Λ's is isotropic, $dN_0/d\Omega_p$ is simply a normalisation factor in a Λ kinematic bin, independent of $\cos\theta_p$. In the case of limited spectrometer acceptance $dN_0/d\Omega_p$ must be multiplied by a factor ε, the acceptance function in a kinematic bin, which gives the probability of detecting a Λ hyperon produced in a given kinematics and event topology. It is obvious that ε might have a strong dependence on $\cos\theta_p$ resulting in modification of the decay distribution $dN/d\Omega_p$. Here one should remember that ε is a function of Λ kinematics *and* $\cos\theta_p$ while $dN_0/d\Omega_p$, being also a function of Λ kinematics, does *not* depend on $\cos\theta_p$.

For the case of limited acceptance given by the function $\varepsilon(\cos\theta_p)$ let us define the following moments ($n = 1, 2, ...$)

$$\langle\cos^n\theta_p\rangle = \frac{\int \cos^n\theta_p \frac{dN}{d\Omega_p}(\cos\theta_p)d\Omega_p}{\int \frac{dN}{d\Omega_p}(\cos\theta_p)d\Omega_p} \equiv \frac{\int \cos^n\theta_p(1 + \alpha P_n^\Lambda \cos\theta_p)\varepsilon(\cos\theta_p)d\Omega_p}{\int(1 + \alpha P_n^\Lambda \cos\theta_p)\varepsilon(\cos\theta_p)d\Omega_p}, \tag{22}$$

$$\langle\cos^n\theta_p\rangle_0 \equiv \frac{\int \cos^n\theta_p \varepsilon(\cos\theta_p)d\Omega_p}{\int \varepsilon(\cos\theta_p)d\Omega_p}. \tag{23}$$

Here the symbol $\langle...\rangle$ represents an average over an actual data sample, while $\langle...\rangle_0$ denotes an average over a hypothetical purely unpolarised sample of Λ particles.

Combining Equations 22 and 23 one obtains

$$\langle\cos^n\theta_p\rangle = \frac{\langle\cos^n\theta_p\rangle_0 + \alpha P_n^\Lambda \langle\cos^{n+1}\theta_p\rangle_0}{1 + \alpha P_n^\Lambda \langle\cos\theta_p\rangle_0}. \tag{24}$$

Experimentally, the moment $\langle\cos^n\theta_p\rangle$ can be determined by taking an average over the data set:

$$\langle\cos^n\theta_p\rangle = \frac{1}{N_\Lambda}\sum_{i=1}^{N_\Lambda}\cos^n\theta_{p,i}, \tag{25}$$

where N_Λ is the number of Λ events.

In principle the polarisation P_n^Λ can be found from Equation 24 if unpolarisaed moments $\langle\cos^n\theta_p\rangle_0$ are known. These moments could be calculated with the help of a Monte Carlo simulation of the spectrometer acceptance, but systematic uncertainties in these calculations are often found to be of the same level as the measured effect.

Fortuitously, extraction of the transverse Λ polarisation from the data, even in case of very limited acceptance, is greatly simplified if the detector has up/down symmetry, as in the case of the HERMES spectrometer. It can be readily shown that this geometric symmetry leads to the relations

$$\langle \cos^n \theta_p \rangle_0 = 0 \quad n = 1, 3, \ldots$$
$$\langle \cos^n \theta_p \rangle = \langle \cos^n \theta_p \rangle_0 \quad n = 2, 4, \ldots$$

Using Equation 24 for $n = 1$ it readily follows that

$$P_n^\Lambda = \frac{\langle \cos \theta_p \rangle}{\alpha \langle \cos^2 \theta_p \rangle} \equiv \frac{1}{\alpha} \frac{\sum\limits_{i=1}^{N_\Lambda} \cos \theta_{p,i}}{\sum\limits_{i=1}^{N_\Lambda} \cos^2 \theta_{p,i}}, \tag{26}$$

which gives the possibility to find P_n^Λ using *only* the experimental data set in which a value for $\cos \theta_p$ is calculated for each event.

It is sometimes convenient to calculate $\langle \cos \theta_p \rangle$ and $\langle \cos^2 \theta_p \rangle$ separately for the *top* and *bottom* data sample to acount for a possible normalisation difference in the overall efficiency of each half of the detector. Neglecting very small higher order terms one obtains from Equation 26

$$P_n^\Lambda = \frac{1}{\alpha} \frac{\frac{1}{N_\Lambda^{top}} \sum\limits_{i=1}^{N_\Lambda^{top}} \cos \theta_{p,i} + \frac{1}{N_\Lambda^{bot}} \sum\limits_{j=1}^{N_\Lambda^{bot}} \cos \theta_{p,j}}{\frac{2}{N_\Lambda} \sum\limits_{m=1}^{N_\Lambda} \cos^2 \theta_{p,m}}. \tag{27}$$

3.3 Transverse Λ^0 and $\overline{\Lambda}^0$ polarisation measured by HERMES

While transverse hyperon polarisation has been studied extensively in hadron-hadron interactions, very little experimental information exists about this effect in photo- and electro-production. Transverse polarisation in the inclusive photoproduction of neutral strange particles was investigated 20 years ago at CERN (Aston *et al.* 1982) and at SLAC (Abe *et al.* 1984). However, the statistical accuracy of these data is indecisive. The CERN measurements, for incident tagged photons with energies between 25 and 70 GeV, resulted in an average polarisation of 0.06 ± 0.04. At SLAC, the overall polarisation was observed to be 0.09 ± 0.07 for Λ hyperons produced using a 20 GeV photon beam.

The HERMES experiment has measured the transverse polarisation of Λ and $\overline{\Lambda}$ hyperons produced inclusively in quasi-real photon-nucleon interactions at a positron beam energy of 27.6 GeV. The average transverse polarisations were found to be $P_n^\Lambda = 0.058 \pm 0.005(stat) \pm 0.006(syst)$ and $P_n^{\overline{\Lambda}} = -0.042 \pm 0.012(stat) \pm 0.005(syst)$ for Λ and $\overline{\Lambda}$, respectively. The data analysis is not yet finalised and the results presented here are preliminary.

The HERMES spectrometer (Ackerstaff *et al.* 1998b) offers a good opportunity to detect inclusively produced hyperons. The yield of inclusive Λ^0 and $\overline{\Lambda}^0$ hyperon production is $\simeq 50$ times larger then that for the semi-inclusive DIS case. In contrast to the semi-inclusive DIS when relatively large Q^2 is defined by detection of the scattered positron in coincidence with a hadron, the inclusively detected hadrons are mostly produced due to interaction of quasi-real photons ($Q^2 \simeq 0$) with the target. Most of these photons are emitted by the beam positrons nearly along their momenta, and are thus strongly peaked at $Q^2 = 0$.

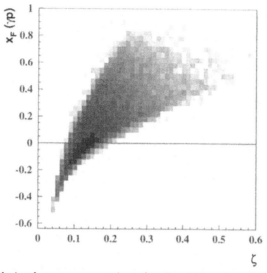

Figure 7. *Correlation between x_F, evaluated in the $\gamma^* N$ centre-of-mass frame, and the light-cone fraction ζ determined in the eN frame, as determined from a PYTHIA Monte Carlo.*

Since the scattered positrons are not detected the momenta of emitted photons are not tagged. Therefore only kinematic variables related to the eN system are available on an event-by-event basis. It is convenient to analyse the data using the kinematic variable $\zeta \equiv (E_\Lambda + p_{z\Lambda})/(E_e + p_e)$, where E_e, p_e are the energy and momentum of the positron beam. This variable is the light-cone momentum fraction of the beam positron carried by the outgoing Λ or $\bar{\Lambda}$. As shown in Figure 7, a simulation of the reaction using the PYTHIA program reveals a correlation between ζ and x_F. In particular, all events at $\zeta \geq 0.25$ are produced in the kinematic region $x_F > 0$.

An indication that the dominant production mechanism changes at ζ values around 0.25 can be observed in the ratio of Λ to $\bar{\Lambda}$ yields displayed in Figure 8. Above $\zeta \approx 0.25$, a constant ratio of approximately 4 is seen. At lower values the ratio increases significantly, likely indicating the influence of the nucleon target remnant in Λ formation.

It may be speculated that the different behaviours of P_n^Λ and $P_n^{\bar{\Lambda}}$ in the low and high ζ regions are related to the different hadron-formation mechanisms as suggested by the ratio of Λ to $\bar{\Lambda}$ yields in Figure 8.

In Figure 9 the transverse Λ and $\bar{\Lambda}$ polarisations are shown versus p_T for the two intervals $\zeta < 0.25$ and $\zeta > 0.25$. Similar to the hadron-hadron collision case, in both regimes the Λ polarisation rises with p_T, particularly for the forward-going hyperons ($\zeta > 0.25$) where this behaviour is more pronounced. In the forward region the $\bar{\Lambda}$ polarisation is consistent with zero, also in agreement with hadronic reactions. In the backward region, however, the measured $\bar{\Lambda}$ polarisation favours a negative value.

Very few theoretical models of the kinematic dependence of Λ polarisation in photo-

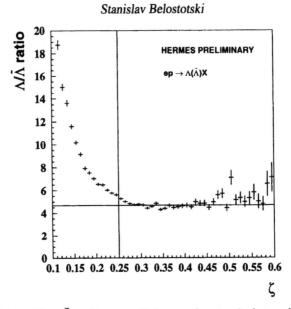

Figure 8. *Ratio of Λ to Λ̄ yields versus light-cone fraction ζ observed in the data.*

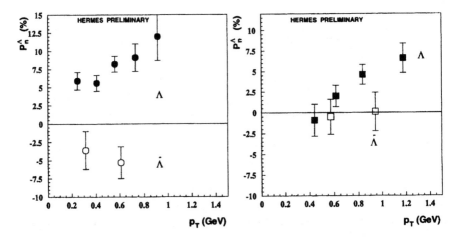

Figure 9. *Transverse polarisations P_n^Λ and $P_n^{\bar\Lambda}$ as a function of p_T for the two intervals $\zeta < 0.25$ (left) and $\zeta > 0.25$ (right). The error bars represent the statistical uncertainties, the systematic uncertainties are given by the error bands.*

or electro-production are available for comparison with the data. One set of curves for the photoproduction case is provided by Nakajima *et al.* (1999), based on the Quark-Recombination Model. As only the dependence on x_F in the γN frame is calculated in this work, no direct comparison with the present data is possible. However the model does predict a negative polarisation in the $x_F^{\gamma N} > 0$ regime, in contradiction with the positive value measured at HERMES for $\zeta > 0.25$.

One may speculate on the reason for the positive Λ polarisation in $\gamma^* N \to \Lambda X$.

In the model of DeGrand and Miettinen (1985), for example, forward-going Λ particles produced in proton-proton scattering are formed from the recombination of a high-momentum spin and isospin singlet (ud) diquark from the beam with a strange sea quark from the target. The Λ polarisation then arises from the acceleration of the strange quark, via the Thomas precession effect.

Conversely, the positive Λ polarisation observed with K^- beams is indicative of the deceleration of strange quarks from the beam. The positive polarisation observed in the HERMES quasi-real photoproduction data might therefore indicate that the $\gamma \to s\bar{s}$ hadronic component of the beam plays a significant role in inclusive Λ production.

It should also be noted that some fraction of the observed Λ particles must arise from the decay of heavier hyperons such as Σ, Σ^* and Ξ. The transverse polarisations of these hyperons, produced in γN reactions at high energies, are unknown and therefore it is currently impossible to evaluate their contribution to the present result.

3.4 Hyperon yields

A substantial part of the Λ^0 and $\overline{\Lambda}^0$ hyperons detected by HERMES originate from the decay of Σ^0, $\Sigma(1385)$, Ξ, and their antiparticles. As it has been discussed above a Monte Carlo simulation must be done to estimate these contributions. Uncertainties in these Monte Carlo calculations can be reduced by measurement of the hyperon yields and (where possible) hyperon polarisations. As an example, typical invariant mass spectra obtained at HERMES by detection of a Λ^0 ($\overline{\Lambda}^0$) in coincidence with a pion are shown in Figure 10.

The Ξ and $\Sigma(1385)$ hyperons decaying to $\Lambda\pi$ are clearly identified in these spectra. The hyperon yields accumulated during data taking periods of the years 1996, 1997, 1999 and 2000 are listed in Table 3.

hyperon	*decay mode,* %	*hyperon yield*	*antihyp. yield*
Λ^0 (1116)	$p\pi^-$ (63.9)	386000	72000
Σ^0(1193)	$\Lambda^0\gamma$(100)	19000	5200
Ξ^-(1321)	$\Lambda\pi^-$ (99.)	2500	650
Σ^+(1393)	$\Lambda\pi^+$(88.)	5700	820
Σ^+(1388)	$\Lambda\pi^-$(88.)	6300	1200

Table 3. *Hyperon yields without correction for HERMES acceptance.*

4 Conclusion and outlook

As shown in these lectures, hyperon spin structure and production mechanisms can be studied by measuring final state hyperon polarisation. The interpretation of existing ex-

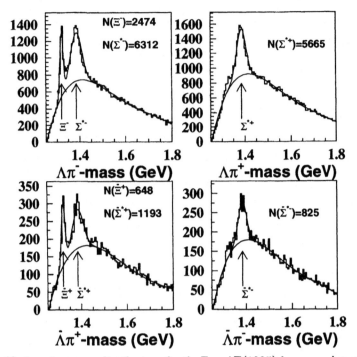

Figure 10. *Invariant mass distributions for the* Ξ *and* $\Sigma(1385)$ *hyperons detected by the HERMES spectrometer.*

perimental data are complex due to insufficient statistical precision, limited acceptance of the experiments, and also because of uncertainties in the MC simulations.

In future experiments, measurements of the longitudinal spin transfer from a struck quark to the Λ^0, or a heavier hyperon, in the current fragmentation region need to focus on the high z kinematic domain, since the data obtained at moderate z are unfortunately not very conclusive. The Λ production rate must be increased by at least a factor of 100 as compared, for example, to that suggested by the present version of the HERMES detector. HERMES is a forward spectrometer with a relatively low efficiency ($\simeq$ 1%) for detecting decaying hyperons. The Λ hyperons are mostly detected at $x_F > 0$, and negative x_F kinematics are practically unreachable due to the limit at $x_F \approx -0.25$.

A substantial increase of the hyperon rate at HERMES is expected with the Recoil Detector (RD) due to be installed in 2005. The RD together with the LW forward silicon detector (HERMES 1997) will provide practically full coverage for the detection of hyperon decay products, *eg* pions emitted from the target at large angles. Very slow hyperons coming out of the target may also be detected by the RD thus giving a full access to the target fragmentation region. The importance of the exclusive channel for strangeness production has been pointed out by Frankfurt *et al.* (1997) and Strikman *et al.* (2000). The RD provides an excellent opportunity to measure polarisation observables in exclusive reactions with strangeness, like $\gamma^* p \Rightarrow K^+ \Lambda^0$ or $\gamma^* p \Rightarrow K^+ \Sigma^0$.

A high Λ production rate is expected by the COMPASS collaboration in CERN (COMPASS 1996).

The polarised beam of the Jefferson laboratory (JLAB), in particular after its planned 11-12 GeV upgrade, suggests an excellent opportunity of measuring $D^\Lambda_{LL'}$ with the kinematic constraints mentioned above (x_F or z larger than ~ 0.8). Furthermore, a very high Λ production rate is expected in the target fragmentation region, and also for exclusive reactions. Spin transfer to the Λ is presently under study at JLAB at lower beam energies, *eg* the CLAS collaboration has recently reported spin transfer measurements in exclusive Λ production (Carman *et al.* 2003). A very interesting conclusion has already been made in this paper about the anti-aligned variant of the spin orientation in $s\bar{s}$ quark pairs.

To conclude, the study of polarisation phenomena in hyperon production reactions provides very interesting new insights into the nature of fragmentation processes and the structure of hadrons. Longitudinal spin transfer is still poorly investigated, and transverse polarisation remains a puzzle to be solved in future experiments.

Addendum: Extracting $D^\Lambda_{LL'}$ from helicity-balanced data

The self-analysing (parity violating) decay $\Lambda^0 \to p + \pi^-$ provides the means to measure the polarisation of Λ^0 hyperons. The polarisation can be determined from the asymmetry in the angular distribution of the emitted protons, which in the Λ^0 rest frame is given by

$$\frac{dN}{d\Omega} = \frac{dN_0}{d\Omega}\left(1 + \alpha \vec{P}_\Lambda \cdot \hat{k}\right) = \frac{dN_0}{d\Omega}\left(1 + \alpha P^i_\Lambda k_i\right). \tag{28}$$

In this equation, α is the analysing power of the parity violating decay ($\alpha = 0.642$), and $\vec{P}_\Lambda$ is the vector representing the three polarisation components of the Λ^0. $\hat{k}$ is the unit vector along the momentum of the decay proton, which is emitted within the solid angle $d\Omega$. The index $i = x, y, z$ refers to each component of the given vector in the chosen coordinate system, and summation over repeated indices is assumed throughout. Finally $dN/d\Omega$ represents the differential distribution of detected Λ^0 events, while $dN_0/d\Omega$ is that same distribution for the special case of zero Λ^0 polarisation. In this latter case, the protons are emitted isotropically, and so $dN_0/d\Omega$ is a measure of the experimental acceptance.

Introducing the moments

$$\langle k_i \rangle = \frac{\int k_i \frac{dN}{d\Omega} d\Omega}{\int \frac{dN}{d\Omega} d\Omega}, \quad \langle k_i \rangle_0 = \frac{\int k_i \frac{dN_0}{d\Omega} d\Omega}{\int \frac{dN_0}{d\Omega} d\Omega}, \quad \langle k_i k_j \rangle_0 = \frac{\int k_i k_j \frac{dN_0}{d\Omega} d\Omega}{\int \frac{dN_0}{d\Omega} d\Omega}, \tag{29}$$

one can rewrite Equation 28 in the form

$$\langle k_i \rangle = \frac{\langle k_i \rangle_0 + \alpha P^j_\Lambda \langle k_j k_i \rangle_0}{1 + \alpha P^j_\Lambda \langle k_j \rangle_0}, \quad \text{where } i, j = x, y, z. \tag{30}$$

In the simple case of a uniform 4π acceptance, $dN_0/d\Omega$ is just a constant and the zero-polarisation moments in Equation 30 have simple values. For example, $\langle k_i \rangle_0 = 0$ for each component i, because the decay protons are emitted isotropically when $\vec{P}_\Lambda$ is zero.

It can also be shown that the matrix $\langle k_j k_i \rangle_0$ is diagonal, and that $\langle k_x k_x \rangle_0 = \langle k_y k_y \rangle_0 = \langle k_z k_z \rangle_0 = 1/3$. Equation 30 can then be trivially inverted, and the polarisation vector $\vec{P}_\Lambda$ can be obtained directly from the measured moments

$$P_\Lambda^i = \frac{3}{\alpha} \langle k_i \rangle. \tag{31}$$

Experimentally, the moments $\langle k_i \rangle$ are obtained by evaluating the following average over all recorded Λ^0 events:

$$\langle k_i \rangle = \frac{1}{N_\Lambda} \sum_{\nu=1}^{N_\Lambda} k_{i,\nu} \tag{32}$$

where N_Λ is the total number of Λ^0 events and the symbol $k_{i,\nu}$ represents component i of the proton unit-vector k for event ν.

If the experiment has a limited acceptance, however, the differential distribution $dN_0/d\Omega$ for a zero-polarisation sample will be non-trivial, as will the zero-polarisation moments $\langle k_i \rangle_0$ and $\langle k_i k_j \rangle_0$ appearing in Equation 30. Moreover, the matrix $\langle k_i k_j \rangle_0$ may no longer be diagonal, resulting in correlations between the three components of $\vec{P}_\Lambda$.

The zero-polarisation moments, reflecting the experimental acceptance, can be found with the help of a Monte Carlo simulation based on the geometry and materials of the spectrometer. Unfortunately, the systematic uncertainty of such computations is easily similar to the size of the measured polarisation: a small miscalculated acceptance effect will directly influence the value of the difference $\langle k_i \rangle - \langle k_i \rangle_0$, which is (to first approximation) also the size of the measured Λ^0 polarisation. This large dependence of $\vec{P}_\Lambda$ on acceptance effects can be reduced significantly by reversing the helicity of the lepton beam. The basic idea is that the zero-polarisation moments can be evaluated directly from the data by combining balanced samples taken with opposite beam helicity.

The final goal of the analysis is to determine the spin-transfer coefficient $D_{LL'}^\Lambda$. Generalising to three components (D_{Lx}^Λ , D_{Ly}^Λ , D_{Lz}^Λ), $D_{LL'}^\Lambda$ is related to the Λ^0 polarisation components P_Λ^i via the photon depolarisation factor $D(y)$ and the longitudinal beam polarisation P_b:

$$P_\Lambda^i = P_b D(y) D_{Li}^\Lambda, \quad i = x, y, z. \tag{33}$$

Substituting into Equation 28, one can evaluate the following P_b-weighted moments:

$$\langle P_b D(y) k_i \rangle = \frac{\overline{P_b} \langle D(y) k_i \rangle_0 + D_{Lj}^\Lambda \alpha \overline{P_b^2} \langle D^2(y) k_j k_i \rangle_0}{1 + D_{Lj}^\Lambda \alpha \overline{P_b} \langle D(y) k_j \rangle_0}. \tag{34}$$

Here,

$$\overline{P_b} \equiv \left(\frac{1}{L} \right) \int P_b \, dL \tag{35}$$

is the luminosity-weighted beam polarisation ($L = \int dL$ is the total luminosity). Similarly, $\overline{P_b^2} \equiv (\frac{1}{L}) \int P_b^2 \, dL$ is the luminosity-weighted average of the beam polarisation squared. The moments on the left-hand side of Equation 34 can be evaluated experimentally as in Equation 32:

$$\langle P_b D(y) k_i \rangle = \frac{1}{N_\Lambda} \sum_{\nu=1}^{N_\Lambda} P_{b,\nu} D(y)_\nu k_{i,\nu}. \tag{36}$$

If one now requires that the sign of the beam polarisation is changed during the data-taking period such that

$$\overline{P_b} = 0, \tag{37}$$

the terms proportional to $\overline{P_b}$ in Equation 34 vanish. We refer to a data set satisfying Equation 37 as a "helicity-balanced" sample. Further, as such a data set has no net beam polarisation, it cannot have any net Λ^0 polarisation. Thus, $dN/d\Omega = dN_0/d\Omega$ and the remaining zero-polarisation moment $\langle D^2(y)k_jk_i\rangle_0 = \langle D^2(y)k_jk_i\rangle$, which can be determined from the data in the manner of Equation 36.

Finally, one obtains the following formula for a helicity-balanced data set:

$$\langle P_b\, D(y)\, k_i\rangle = D_{Lj}^\Lambda\, \alpha\, \overline{P_b^2}\, \langle D^2(y)k_jk_i\rangle. \tag{38}$$

Using this system of equations ($i, j = x, y, z$), the 3 components D_{Li}^Λ of the spin transfer may be extracted from the experimental data without the need for a Monte Carlo simulation the acceptance function.

In the specific case of the HERMES spectrometer, the matrix $\langle k_ik_j\rangle$ is practically diagonal and $\langle D^2(y)k_ik_j\rangle \approx \langle D^2(y)\rangle\langle k_ik_j\rangle$. Consequently, correlations are small between the various components of the spin transfer. As a result, Equation 38 can be simplified to the one-dimensional form

$$\langle P_b\, D(y)\, k_i\rangle = D_{Li}^\Lambda\, \alpha\, \overline{P_b^2}\, \langle D^2(y)k_i^2\rangle. \tag{39}$$

In one dimension case only the spin transfer $D_{LL'}^\Lambda$ along the longitudinal direction (labelled z) is considered. With $k_z = \cos\Theta_{pL'}$, one obtains

$$\langle P_b\, D(y)\, \cos\Theta_{pL'}\rangle = D_{LL'}^\Lambda\, \alpha\, \overline{P_b^2}\, \langle D^2(y)\, \cos^2\Theta_{pL'}\rangle \tag{40}$$

which immediately yields the $D_{LL'}^\Lambda$ extraction formula, Equation 16.

Acknowledgments

I gratefully acknowledge my colleagues from the HERMES Collaboration who were involved in the analysis of Λ polarisation and hyperon production for their valuable contribution to this field.

References

Abe K *et al.* , 1984, *Phys Rev D* **29** 1877.

Abe K *et al.* [E143 experiment], 1995, *Phys Rev Lett* **75** 25.

Ackerstaff K *et al.* [HERMES collaboration], 1997, *Phys Lett B* **404** 383.

Ackerstaff K *et al.* [OPAL collaboration], 1998a, *Eur Phys J C* **2** 49.

Ackerstaff K *et al.* , 1998b, *Nucl Inst Meth A* **417** 230.

Ackerstaff K *et al.* [HERMES collaboration], 1999, *Phys Lett B* **464** 123.

Adams M R *et al.* , 1994, *Z Phys C* **61** 539.

Adams D *et al.* [SMC collaboration], 1998, *Phys Lett B* **240** 180.

Airapetian A *et al.* [HERMES collaboration], 1998, *Phys LettB* **442** 484.

Airapetian A *et al.* , 2001, *Phys Rev D* **64** 112005.

Airapetian A *et al.* [HERMES collaboration], 2004, *Phys Rev Lett* **92** 012005.

Anthony P L *et al.* [E142 experiment], 1996, *Phys Rev D* **54** 6620.

Anthony P L *et al.* , 1997, *Phys Rev Lett* **72** 26.

Astier P *et al.* , 2000, *Nucl Phys B* **588** 3.

Aston D *et al.* , 1982, *Nucl Phys B* **195** 189.

Aveda B *et al.* , 1998, *Phys Rev D* **58** 11201.

Bjorken J D, 1970, *Phys Rev D* **1** 1376.

Buskulic D *et al.* [ALEPH collaboration], 1996, *Phys Lett B* **374** 319.

Carman D S *et al.* [CLAS Collaboration], 2003, *Phys Rev Lett* **90** 131804.

COMPASS Collaboration, 1996, *Proposal CERN-SPSLC-96-14.*

DeGrand T A and Miettinen H I, 1985, *Phys Rev D* **32** 2445.

Frankfurt L *et al.* , 1997, *Phys Rev D* **56** 2982.

Gatto R, 1958, *Phys Rev* **109** 610.

Gribov V N and Lipatov L N, 1971, *Phys Lett B* **37** 78.

Heller K, 1997, *proc 12th Int Symp on High Energy Spin Physics (Spin96)*, 23.

HERMES, 1997, *internal report* 97-032.

Jaffe R L, 1996, *Phys Rev D* **54** 6581.

Lach J, 1996, *Nucl PhysProc Suppl B* **50** 216.

Ma J S B-Q *et al.* 2002, *Phys Rev D* **65** 034004.

Makins N C R, 2004, *priv com.*

Martin A D and Halzen F, 1984, *Quarks and Leptons: An Introductory Course in Modern Particle Physics*, John Wiley and Sons Inc.

Nakajima N *et al.* , 1999, *proc APCTP Workshop on Strangeness in Nuclear Physics*, Seoul, Korea, 214.

Panagiotou A D, 1990, *Int J Mod Phys A* **5** 1197.

Schmidt I *et al.* , 2000, *Phys Lett B* **477** 107

Soffer J and Törnqvist N A, 1992, *Phys Rev Lett* **68** 907.

Strikman M *et al.* , 2000, *Phys Rev Lett* **84** 2589.

Light-front QCD

Stanley J Brodsky

Stanford Linear Accelerator Center, Stanford University,
Stanford, California 94309, USA

1 Introduction

In principle, quantum chromodynamics (QCD) provides a fundamental description of hadronic and nuclear structure and dynamics in terms of their elementary quark and gluon degrees of freedom. The theory has extraordinary properties such as colour confinement (Greensite 2003), asymptotic freedom (Gross and Wilczek 1973, Politzer 1973), a complex vacuum structure, and it predicts an array of new forms of hadronic matter such as gluonium and hybrid states (Klempt 2000). The phase structure of QCD (Rajagopal and Wilczek 2000) implies the formation of a quark-gluon plasma in high energy heavy-ion collisions (Rischke 2003) as well insight into the evolution of the early universe (Schwarz 2003). Its non-Abelian Yang Mills gauge theory structure provides the foundation for the electroweak interactions and the eventual unification of the electrodynamic, weak, and hadronic forces at very short distances.

The asymptotic freedom property of QCD explains why the strong interactions become weak at short distances, thus allowing hard processes to be interpreted directly in terms of the perturbative interactions of quark and gluon quanta. This in turn leads to factorisation theorems (Collins *et al.* 1989, Bodwin 1984) for both inclusive and exclusive processes (Brodsky and Lepage 1989) which separate the hard scattering sub-processes which control the reaction from the non-perturbative physics of the interacting hadrons.

QCD becomes scale free and conformally symmetric in the analytic limit of zero quark mass and zero β function. Conversely, one can start with the conformal prediction and systematically incorporate the non-zero β function contributions into the scale of the running coupling. This "conformal correspondence principle" determines the form of the expansion polynomials for distribution amplitudes and the behaviour of non-perturbative wavefunctions which control hard exclusive processes at leading twist. The conformal template also can be used to derive commensurate scale relations which connect observables in QCD without scale or scheme ambiguity.

Recently, a remarkable duality has been established between supergravity string theory in 10 dimensions and conformal supersymmetric extensions of QCD (Maldacena 1998, Polchinski and Strassler 2001, Brower and Tan 2002, Andreev 2002). This Anti-de Sitter/Conformal Field Theory (AdS/CFT) correspondence is now leading to a new understanding of QCD at strong coupling and the implications of its near-conformal structure. As I will discuss here, the AdS/CFT correspondence of large N_C supergravity theory in higher-dimensional anti-de Sitter space with supersymmetric QCD in 4-dimensional space-time has important implications for hadron phenomenology in the conformal limit, including the non-perturbative derivation of counting rules for exclusive processes and the behaviour of structure functions at large x_{bj}. String/gauge duality also predicts the QCD power-law fall-off of light-front Fock-state hadronic wavefunctions with arbitrary orbital angular momentum at high momentum transfer.

The Lagrangian density of QCD (Fritzsch *et al.* 1973) has a deceptively simple form:

$$\mathcal{L} = \overline{\psi}(i\gamma_\mu D^\mu - m)\psi - \frac{1}{4}G^2_{\mu\nu} , \tag{1}$$

where the covariant derivative is $iD_\mu = i\partial_\mu - gA_\mu$ and where the gluon field strength is $G_{\mu\nu} = \frac{i}{g}[D_\mu, D_\nu]$. The structure of the QCD Lagrangian is dictated by two principles: (i) local $SU(N_C)$ colour gauge invariance – the theory is invariant when a quark field is rotated in colour space and transformed in phase by an arbitrary unitary matrix $\psi(x) \rightarrow U(x)\psi(x)$ locally at any point x^μ in space and time; and (ii) renormalisability, which requires the appearance of dimension-four interactions. In principle, the only parameters of QCD are the quark masses and the QCD coupling determined from a single observable at a single scale.

Solving QCD from first principles is extremely challenging because of the non-Abelian three-point and four-point gluonic couplings contained in its Lagrangian. The analytic problem of describing QCD bound states is compounded not only by the physics of confinement, but also by the fact that the wavefunction of a composite of relativistic constituents has to describe systems of an arbitrary number of quanta with arbitrary momenta and helicities. The conventional Fock state expansion based on equal-time quantisation quickly becomes intractable because of the complexity of the vacuum in a relativistic quantum field theory. Furthermore, boosting such a wavefunction from the hadron's rest frame to a moving frame is as complex a problem as solving the bound state problem itself. The Bethe-Salpeter bound state formalism, although manifestly covariant, requires an infinite number of irreducible kernels to compute the matrix element of the electromagnetic current even in the limit where one constituent is heavy.

The description of relativistic composite systems using light-front quantisation is in contrast remarkably simple. The Heisenberg problem for QCD can be written in the form

$$H_{LC}|H\rangle = M^2_H|H\rangle , \tag{2}$$

where $H_{LC} = P^+P^- - P^2_\perp$ is the mass operator. The operator $P^- = P^0 - P^3$ is the generator of translations in the light-front time $x^+ = x^0 + x^3$. Its form is predicted from the QCD Lagrangian. The quantities $P^+ = P^0 + P^3$ and $P_\perp$ play the role of the conserved three-momentum. The simplicity of the light-front Fock representation relative to that in equal-time quantisation arises from the fact that the physical vacuum state has a much simpler structure on the light cone. Indeed, kinematical arguments suggest that the light-front Fock vacuum is the physical vacuum state. This means that all constituents in a physical eigenstate are directly related to that state, and not disconnected vacuum fluctuations. In the light-front formalism the parton model is literally true.

Formally, the light-front expansion is constructed by quantising QCD at fixed light-front time (Dirac 1949) $\tau = t + z/c$ and forming the invariant light-front Hamiltonian: $H_{LF}^{QCD} = P^+ P^- - \vec{P}_\perp^2$ where $P^\pm = P^0 \pm P^z$ (Brodsky *et al.* 1998a). The momentum generators P^+ and $\vec{P}_\perp$ are kinematical; *ie* they are independent of the interactions. The generator $P^- = i\frac{d}{d\tau}$ generates light-front time translations, and the eigenspectrum of the Lorentz scalar H_{LF}^{QCD} gives the mass spectrum of the colour-singlet hadron states in QCD together with their respective light-front wavefunctions.

Each hadronic eigenstate $|H\rangle$ of the QCD light-front Hamiltonian can be expanded on the complete set of eigenstates $\{|n\rangle\}$ of the free Hamiltonian which have the same global quantum numbers: $|H\rangle = \sum \psi_n^H(x_i, k_\perp, \lambda_i)|n\rangle$. For example, the proton state satisfies: $H_{LF}^{QCD}|\psi_p\rangle = M_p^2|\psi_p\rangle$. This equation can be written as a Heisenberg matrix eigenvalue problem by introducing a complete set of free Fock states. The Fock expansion begins with the colour singlet state $|uud\rangle$ of free quarks, and continues with $|uudg\rangle$ and the other quark and gluon states that span the degrees of freedom of the proton in QCD. The Fock states $\{|n\rangle\}$ are built on the free vacuum by applying the free light-front creation operators. The summation is over all momenta $(x_i, k_{\perp i})$ and helicities λ_i satisfying momentum conservation $\sum_i^n x_i = 1$ and $\sum_i^n k_{\perp i} = 0$ and conservation of the projection J^3 of angular momentum.

The light-front wavefunctions (LFWFs) $\psi_{n/H}(x_i, \vec{k}_{\perp i}, \lambda_i)$ of hadrons are the central elements of QCD phenomenology, encoding the bound state properties of hadrons in terms of their fundamental quark and gluon degrees of freedom at the amplitude level. It is the probability amplitude that a proton of momentum $P^+ = P^0 + P^3$ and transverse momentum $P_\perp$ consists of n quarks and gluons with helicities λ_i and physical momenta $p_i^+ = x_i P^+$ and $p_{\perp i} = x_i P_\perp + k_{\perp i}$. The wavefunctions $\{\psi_n^p(x_i, k_{\perp i}, \lambda_i)\}, n = 3, \ldots$ thus describe the proton in an arbitrary moving frame. The variables $(x_i, k_{\perp i})$ are internal relative momentum coordinates. The fractions $x_i = p_i^+/P^+ = (p_i^0 + p_i^3)/(P^0 + P^3)$, $0 < x_i < 1$, are the boost-invariant light-front momentum fractions; $y_i = \log x_i$ is the difference between the rapidity of the constituent i and the rapidity of the parent hadron. The appearance of relative coordinates is connected to the simplicity of performing Lorentz boosts in the light-front framework. This is another major advantage of the light-front representation.

For example, take the eigensolution $|\psi_p\rangle$ of the QCD light-front Hamiltonian for the proton expanded on the colour-singlet $B = 1$, $Q = 1$ eigenstates $\{|n\rangle\}$ of the free Hamiltonian $H_{LF}^{QCD}(g = 0)$. This defines the light-front Fock expansion:

$$\left|\psi_p(P^+, \vec{P}_\perp)\right\rangle = \sum_n \prod_{i=1}^n \frac{dx_i\, d^2\vec{k}_{\perp i}}{\sqrt{x_i}\, 16\pi^3}\, 16\pi^3\, \delta\left(1 - \sum_{i=1}^n x_i\right) \delta^{(2)}\left(\sum_{i=1}^n \vec{k}_{\perp i}\right) \tag{3}$$

$$\times\, \psi_{n/H}(x_i, \vec{k}_{\perp i}, \lambda_i)\left|n;\, x_i P^+, x_i \vec{P}_\perp + \vec{k}_{\perp i}, \lambda_i\right\rangle.$$

The light-front momentum fractions $x_i = k_i^+/P^+$ and $\vec{k}_{\perp i}$ represent the relative momentum coordinates of the QCD constituents. The physical transverse momenta are

$\vec{p}_{\perp i} = x_i \vec{P}_\perp + \vec{k}_{\perp i}$. The λ_i label the light-front spin projections S^z of the quarks and gluons along the quantisation direction z. Each Fock component has the invariant mass squared

$$\mathcal{M}_n^2 = (\sum_{i=1}^n k_i^\mu)^2 = \sum_{i=1}^n \frac{k_{\perp i}^2 + m_i^2}{x_i}. \tag{4}$$

The physical gluon polarisation vectors $\epsilon^\mu(k, \lambda = \pm 1)$ are specified in light-cone gauge by the conditions $k \cdot \epsilon = 0$, $\eta \cdot \epsilon = \epsilon^+ = 0$. The gluonic quanta which appear in the Fock states thus have physical polarisation $\lambda = \pm 1$ and positive metric. Since each Fock particle is on its mass shell in a Hamiltonian framework, $k^- = k^0 - k^z = \frac{k_\perp^2 + m^2}{k^+}$. The dominant configurations in the wavefunction are generally those with minimum values of $\mathcal{M}^2$. Note that, except for the case where $m_i = 0$ and $k_{\perp i} = 0$, the limit $x_i \to 0$ is an ultraviolet limit, *ie* it corresponds to particles moving with infinite momentum in the negative z direction: $k_i^z \to -k_i^0 \to -\infty$.

LFWFs have the remarkable property of being independent of the hadron's four-momentum. In contrast, in equal-time quantisation, a Lorentz boost mixes dynamically with the interactions, so that computing a wavefunction in a new frame at fixed t requires solving a non-perturbative problem as complicated as the Hamiltonian eigenvalue problem itself. The LFWFs are properties of the hadron itself; they are thus universal and process independent.

The central tool which will be used in these lectures are the light-front Fock state wavefunctions which encode the bound-state properties of hadrons in terms of their quark and gluon degrees of freedom at the amplitude level. Given these frame-independent wavefunctions, one can compute a large array of hadronic processes ranging from the generalised parton distributions measured in deep inelastic scatterings, hard exclusive reactions, and the weak decays of hadrons. As I will review below, the quantum fluctuations contained in the LFWFs lead to the prediction of novel QCD phenomena such as colour transparency, intrinsic charm, sea quark asymmetries, and hidden colour in nuclear wavefunctions.

Given the light-front wavefunctions $\{\psi_n(x_i, k_{\perp i}, \lambda_i)\}$ one can compute the electromagnetic and weak form factors from a simple overlap of light-front wavefunctions, summed over all Fock states (Drell and Yan 1970, Brodsky and Drell 1980). Form factors are generally constructed from hadronic matrix elements of the current $\langle p|j^\mu(0)|p + q\rangle$, where in the interaction picture we can identify the fully interacting Heisenberg current J^μ with the free current j_μ at the spacetime point $x^\mu = 0$. In the case of matrix elements of the current $j^+ = j^0 + j^3$, in a frame with $q^+ = 0$, only diagonal matrix elements in particle number $n' = n$ are needed. In contrast, in the equal-time theory one must also consider off-diagonal matrix elements and fluctuations due to particle creation and annihilation in the vacuum. In the non-relativistic limit one can make contact with the usual formulae for form factors in Schrödinger many-body theory.

One of the important aspects of fundamental hadron structure is the presence of non-zero orbital angular momentum in the bound-state wavefunctions. The evidence for a "spin crisis" in the Ellis-Jaffe sum rule signals a significant orbital contribution in the proton wavefunction (Jaffe and Manohar 1990, Ji 2003). The Pauli form factor of nucleons is computed from the overlap of LFWFs differing by one unit of orbital angular momentum $\Delta L_z = \pm 1$. Thus the fact that the anomalous moment of the proton is non-zero requires non-zero orbital angular momentum in the proton wavefunction (Brodsky and Drell 1980). In the light-front method, orbital angular momentum is treated explicitly; it includes the orbital contributions induced by relativistic effects, such as the spin-orbit effects normally associated with the conventional Dirac spinors. Angular momentum conservation for each Fock state implies

$$J^z = \sum_i^n S_i^z + \sum_i^{n-1} L_i^z, \qquad (5)$$

where L_i^z is one of the $n-1$ relative orbital angular momenta.

The quark and gluon probability distributions of a hadron are constructed from integrals over the absolute squares $|\psi_n|^2$, summed over n. In the far off-shell domain of large parton virtuality, one can use perturbative QCD or conformal arguments to derive the asymptotic fall-off of the Fock amplitudes, which then in turn leads to the QCD evolution equations for distribution amplitudes and structure functions. More generally, one can prove factorisation theorems for exclusive and inclusive reactions which separate the hard and soft momentum transfer regimes, thus obtaining rigorous predictions for the leading power behaviour contributions to large momentum transfer cross sections. One can also compute the far off-shell amplitudes within the light-front wavefunctions where heavy quark pairs appear in the Fock states. Such states persist over a time $\tau \simeq P^+/\mathcal{M}^2$ until they are materialised in hadron collisions. As I shall discuss, this leads to a number of novel effects in the hadroproduction of heavy quark hadronic states.

2 Light-front statistical physics

As shown by Raufeisen and Brodsky (2004), one can construct a "light-front density matrix" from the complete set of light-front wavefunctions which is a Lorentz scalar. This form can be used at finite temperature to give a boost-invariant formulation of thermodynamics. At zero temperature the light-front density matrix is directly connected to the Green's function for quark propagation in the hadron as well as deeply virtual Compton scattering. One can also define a light-front partition function Z_{LF} as an outer product of light-front wavefunctions. The deeply virtual Compton amplitude and generalised parton distributions can then be computed as the trace $Tr[Z_{LF}\mathcal{O}]$, where $\mathcal{O}$ is the appropriate local operator (Raufeisen and Brodsky 2004). This partition function formalism can be extended to multi-hadronic systems and systems in statistical equilibrium to provide a Lorentz-invariant description of relativistic thermodynamics (Raufeisen and Brodsky 2004).

3 AdS/CFT and hadron phenomenology

Maldacena (1998) has shown that there is a remarkable correspondence between large N_C supergravity theory in a higher dimensional anti-de Sitter space and supersymmetric QCD in 4-dimensional space-time. String/gauge duality provides a framework for predicting QCD phenomena based on the conformal properties of the AdS/CFT correspondence. For example, Polchinski and Strassler (2001) have shown that the power-law fall-off of hard exclusive hadron-hadron scattering amplitudes at large momentum transfer can be derived without the use of perturbation theory by using the scaling properties of the hadronic interpolating fields in the large-r region of AdS space. Thus one can use the Maldacena correspondence to compute the leading power-law fall-off of exclusive processes such as high-energy fixed-angle scattering of gluonium-gluonium scattering in supersymmetric QCD. The resulting predictions for hadron physics effectively coincide (Polchinski and Strassler 2001, Brower and Tan 2002, Andreev 2002) with QCD dimensional counting rules (Brodsky and Farrar 1973, Matveev *et al.* 1973, Brodsky and Farrar 1975, Brodsky 2002). Polchinski and Strassler (2001) have also derived counting rules for deep inelastic structure functions at $x \to 1$ in agreement with perturbative QCD predictions (Brodsky *et al.* 1994a) as well as Bloom-Gilman exclusive-inclusive duality. An interesting point is that the hard scattering amplitudes which are normally of order α_s^p in perturbative QCD (PQCD) appear as order $\alpha_s^{p/2}$ in the supergravity predictions. This can be understood as an all-orders resummation of the effective potential (Maldacena 1998, Rey and Yee 2001). The near-conformal scaling properties of light-front wavefunctions thus lead to a number of important predictions for QCD which are normally discussed in the context of perturbation theory.

Brodsky and de Teramond (2004) have shown how one can use the scaling properties of the hadronic interpolating operator in the extended AdS/CFT space-time theory to determine the form of QCD wavefunctions at large transverse momentum $k_\perp^2 \to \infty$ and at $x \to 1$. The angular momentum dependence of the light-front wavefunctions also follow from the conformal properties of the AdS/CFT correspondence. The scaling and conformal properties of the correspondence leads to a hard component of the light-front Fock state wavefunctions of the form:

$$\psi_{n/h}(x_i, \vec{k}_{\perp i}, \lambda_i, l_{zi}) \sim \frac{(g_s N_C)^{\frac{1}{2}(n-1)}}{\sqrt{N_C}} \prod_{i=1}^{n-1} (k_{i\perp}^\pm)^{|l_{zi}|} \qquad (6)$$

$$\times \left[\frac{\Lambda_o}{M^2 - \sum_i \frac{\vec{k}_{\perp i}^2 + m_i^2}{x_i} + \Lambda_o^2} \right]^{n + \sum_i |l_{zi}| - 1} ,$$

where g_s is the string scale and Λ_o represents the basic QCD mass scale. The scaling predictions agree with the perturbative QCD analysis given in Ji *et al.* (2003), but the AdS/CFT analysis is performed at strong coupling without the use of perturbation theory. The form of these near-conformal wavefunctions can be used as an initial ansatz for a variational treatment of the light-front QCD Hamiltonian. The same ansatz leads to predictions for the hadron spectrum, which I will discuss in the conclusions.

4 Light-front wavefunctions & hadron phenomenology

Even though QCD was motivated by the successes of the parton model, QCD predicts many new features which go well beyond the simple three-quark description of the proton. Since the number of Fock components cannot be limited in relativity and quantum mechanics, the non-perturbative wavefunction of a proton contains gluons and sea quarks, including heavy quarks at any resolution scale. Thus there is no scale Q_0 in deep inelastic lepton-proton scattering where the proton can be approximated by its valence quarks. The non-perturbative Fock state wavefunctions contain intrinsic gluons, strange quarks, charm quarks, *etc.*, at any scale. The internal QCD interactions lead to asymmetries such as $s(x) \neq \bar{s}(x)$, $\bar{u}(x) \neq \bar{d}(x)$ and intrinsic charm and bottom distributions at large x since this minimises the invariant mass and off-shellness of the higher Fock state. As discussed above, the Fock state expansion for nuclei contains hidden colour states which cannot be classified in terms of nucleonic degrees of freedom. However, some leading-twist phenomena such as the diffractive component of deep inelastic scattering, single-spin asymmetries, nuclear shadowing and anti-shadowing cannot be computed from the LFWFs of hadrons in isolation.

4.1 The strange quark asymmetry

In the simplest treatment of deep inelastic scattering, non-valence quarks are produced via gluon splitting and Dokshitzer-Gribov-Lipatov-Altarelli-Parisi (DGLAP) evolution. However, in the full theory, heavy quarks are multiply connected to the valence quarks (Brodsky *et al.* 1980a). Although the strange and anti-strange distributions in the nucleon are identical when they derive from gluon-splitting $g \rightarrow s\bar{s}$, this is not the case when the strange quarks are part of the intrinsic structure of the nucleon – the multiple interactions of the sea quarks produce an asymmetry of the strange and anti-strange distributions in the nucleon due to their different interactions with the other quark constituents. A Quantum Electrodynamic (QED) analogy is the distribution of τ^+ and τ^- in a higher Fock state of muonium $\mu^+ e^-$. The τ^- is attracted to the higher momentum μ^+ thus asymmetrically distorting its momentum distribution. Similar effects will happen in QCD. If we use the diquark model $|p\rangle \sim \left| u_{3_c} (ud)_{\bar{3}_C} \right\rangle$, then the Q_{3_C} in the $|u(ud)Q\bar{Q}\rangle$ Fock state will be attracted to the heavy diquark and thus have higher rapidity than the $\bar{Q}$. An alternative model is the $|K\Lambda\rangle$ fluctuation model for the $|uuds\bar{s}\rangle$ Fock state of the proton (Brodsky and Ma 1996). The s quark tends to have higher x.

Empirical evidence also continues to accumulate that the strange and anti-strange quark distributions are not symmetric in the proton (Brodsky and Ma 1996, Kretzer 2004, Portheault 2004). The experimentally observed asymmetry appears to be small but positive: $\int dx x[s(x) - \bar{s}(x)] > 0$. The results of a recent CTEQ collaboration analysis (Olness *et al.* 2004) of neutrino-induced dimuon data are shown in Figure 1.

The fit is constrained so that the number of s and $\bar{s}$ quarks in the nucleon are equal. The shape of the strangeness asymmetry is consistent with the ΛK fluctuation model (Brodsky and Ma 1996). Kretzer (2004) has noted that a significant part of the NuTeV anomaly could be due to this asymmetry, The $\bar{s}(x) - s(x)$ asymmetry can be studied in detail in $p\bar{p}$ collisions by searching for antisymmetric forward-backward strange quark distributions in the $\bar{p} - p$ CM frame.

4.2 Intrinsic heavy quarks

The probability for Fock states of a light hadron such as the proton to have an extra heavy quark pair decreases as $1/m_Q^2$ in non-Abelian gauge theory (Franz *et al.* 2000, Brodsky *et al.* 1984). The relevant matrix element is the cube of the QCD field strength

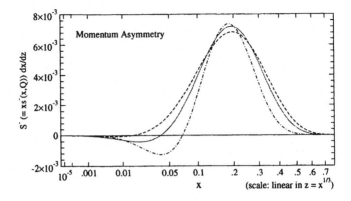

Figure 1. *Representative results of the strangeness asymmetry analysis by the CTEQ collaboration.*

$G_{\mu\nu}^3$. This is in contrast to Abelian gauge theory where the relevant operator is $F_{\mu\nu}^4$ and the probability of intrinsic heavy leptons in QED bound state is suppressed as $1/m_\ell^4$. The intrinsic Fock state probability is maximised at minimal off-shellness. It is useful to define the transverse mass $m_{\perp i} = \sqrt{k_{\perp i}^2 + m_i^2}$. The maximum probability then occurs at $x_i = m_\perp^i / \sum_{j=1}^n m_\perp^j$; ie, when the constituents have minimal invariant mass and equal rapidity. Thus the heaviest constituents have the highest momentum fractions and the highest x_i. Intrinsic charm thus predicts that the charm structure function has support at large x_{bj} in excess of DGLAP extrapolations (Brodsky *et al.* 1980a); this is in agreement with the EMC measurements (Harris *et al.* 1996).

4.3 Diffractive dissociation and intrinsic heavy quark production

Diffractive dissociation is particularly relevant to the production of leading heavy quark states. The projectile proton can be decomposed as a sum over all of its Fock state components. The diffractive dissociation of the intrinsic charm $|uudc\bar{c}>$ Fock state of the proton on a nucleus can produce a leading heavy quarkonium state at high $x_F = x_c + x_{\bar{c}}$ in $pA \rightarrow J/\psi X A'$ since the c and $\bar{c}$ can readily coalesce into the charmonium state. Since the constituents of a given intrinsic heavy-quark Fock state tend to have the same rapidity, coalescence of multiple partons from the projectile Fock state into charmed hadrons and mesons is also favoured. For example, as illustrated in Figure 2, one can produce leading Λ_c at high x_F and low p_T from the coalescence of the udc constituents of the projectile IC Fock state. A similar coalescence mechanism was used in atomic physics to produce relativistic anti-hydrogen in $\bar{p}A$ collisions (Munger *et al.* 1994). This phenomena is important not only for understanding heavy-hadron phenomenology, but also for understanding the sources of neutrinos in astrophysics experiments (Halzen 2004).

The charmonium state will be produced at small transverse momentum and high x_F with a characteristic $A^{2/3}$ nuclear dependence. This forward contribution is in addition to the A^1 contribution derived from the usual perturbative QCD fusion contribution at small x_F. Because of these two components, the cross section violates perturbative QCD factorisation for hard inclusive reactions (Hoyer *et al.* 1990). This is consistent with the

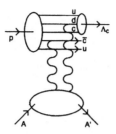

Figure 2. *Production of forward heavy baryons by diffractive dissociation.*

observed two-component cross section for charmonium production observed by the NA3 collaboration at CERN (Badier *et al.* 1981).

The diffractive dissociation of the intrinsic charm Fock state leads to leading charm hadron production and fast charmonium production in agreement with measurements (Anjos *et al.* 2001). Intrinsic charm can also explain the $J/\psi \rightarrow \rho\pi$ puzzle (Brodsky and Karliner 1997), and it affects the extraction of suppressed CKM matrix elements in B decays (Brodsky and Gardner 2002). Intrinsic charm can also enhance the production probability of Higgs bosons at hadron colliders from processes such as $gc \rightarrow Hc$. It is thus critical for new experiments (HERMES, HERA, COMPASS) to definitively establish the phenomenology of the charm structure function at large x_{bj}.

The production cross section for the double-charm Ξ_{cc}^+ baryon (Ocherashvili *et al.* 2004) and the production of J/ψ pairs appears to be consistent with the diffractive dissociation and coalescence of double IC Fock states (Brodsky *et al.* 2004a, Vogt and Brodsky 1995). It is unlikely that the appearance of two heavy quarks at high x_F could be explained by the "colour drag model" used in PYTHIA simulations (Andersson *et al.* 1983) in which the heavy quarks are accelerated from low to high x by the fast valence quarks. These observations provide compelling evidence for the diffractive dissociation of complex off-shell Fock states of the projectile and contradict the traditional view that sea quarks and gluons are always produced perturbatively via DGLAP evolution. It is also conceivable that the observations (Bari *et al.* 1991) of Λ_b at high x_F at the ISR in high energy pp collisions could be due to the diffractive dissociation and coalescence of the "intrinsic bottom" $|uudb\bar{b}>$ Fock states of the proton.

4.4 Colour transparency

The small transverse size fluctuations of a hadron wavefunction with a small colour dipole moment will have minimal interactions in a nucleus (Bertsch *et al.* 1981, Brodsky and Mueller 1988).

This has been verified in the case of diffractive dissociation of a high energy pion into dijets $\pi A \rightarrow q\bar{q}A'$ in which the nucleus is left in its ground state (Ashery 2002). As discussed in the next subsection, when the hadronic jets have balancing but high transverse momentum, one studies the small size fluctuation of the incident pion. The diffractive dissociation cross section is found to be proportional to A^2 in agreement with the colour transparency prediction.

Colour transparency has also been observed in diffractive electroproduction of ρ mesons (Borisov *et al.* 2002) and in quasi-elastic $pA \rightarrow pp(A - 1)$ scattering (Aclander *et al.* 2004) where only the small size fluctuations of the hadron wavefunction enters the hard exclusive scattering amplitude. In the latter case an anomaly occurs at $\sqrt{s} \simeq 5$ GeV, most likely signaling a resonance effect at the charm threshold (Brodsky and de Teramond 1988).

4.5   Diffraction dissociation as a tool to resolve hadron substructure

Diffractive multi-jet production in heavy nuclei provides a novel way to measure the shape of light-front Fock state wavefunctions and test colour transparency (Brodsky and Mueller 1988). For example, consider the reaction (Bertsch *et al.* 1981, Frankfurt *et al.* 2000) $\pi A \rightarrow \text{Jet}_1 + \text{Jet}_2 + A'$ at high energy where the nucleus A' is left intact in its ground state. The transverse momenta of the jets balance so that $\vec{k}_{\perp 1} + \vec{k}_{\perp 2} = \vec{q}_\perp < R^{-1}{}_A$. The light-front longitudinal momentum fractions also need to add to $x_1 + x_2 \sim 1$. Diffractive dissociation on a nucleus also requires that the energy of the beam has to be sufficiently large such that the momentum transfer to the nucleus $\Delta p_L = \frac{\Delta M^2}{2E_{lab}}$ is smaller than the inverse nuclear size R_A^{-1}. The process can then occur coherently in the nucleus.

Because of colour transparency, the valence wavefunction of the pion with small impact separation will penetrate the nucleus with minimal interactions, diffracting into jet pairs (Bertsch *et al.* 1981). The $x_1 = x$, $x_2 = 1 - x$ dependence of the di-jet distributions will thus reflect the shape of the pion valence light-front wavefunction in x; similarly, the $\vec{k}_{\perp 1} - \vec{k}_{\perp 2}$ relative transverse momenta of the jets gives key information on the second transverse momentum derivative of the underlying shape of the valence pion wavefunction (Frankfurt *et al.* 2000, Nikolaev *et al.* 2001). The diffractive nuclear amplitude extrapolated to $t = 0$ should be linear in nuclear number A if colour transparency is correct. The integrated diffractive rate will then scale as $A^2/R_A^2 \sim A^{4/3}$. This is in fact what has been observed by the E791 collaboration at FermiLab for 500 GeV incident pions on nuclear targets (Aitala *et al.* 2001a). The measured momentum fraction distribution of the jets is found to be approximately consistent with the shape of the pion asymptotic distribution amplitude, $\phi_\pi^{\text{asympt}}(x) = \sqrt{3}f_\pi x(1 - x)$ (Aitala *et al.* 2001b). Data from CLEO (Gronberg *et al.* 1998) for the $\gamma\gamma^* \rightarrow \pi^0$ transition form factor also favour a form for the pion distribution amplitude close to the asymptotic solution to its perturbative QCD evolution equation (Lepage and Brodsky 1979a, Efremov and Radyushkin 1980a, Lepage and Brodsky 1980).

Colour transparency, as evidenced by the FermiLab measurements of diffractive dijet production, implies that a pion can interact coherently throughout a nucleus with minimal absorption, in dramatic contrast to traditional Glauber theory based on a fixed $\sigma_{\pi n}$ cross section. Colour transparency gives direct validation of the gauge interactions of QCD.

4.6   Diffractive dissociation and hidden colour in nuclear wavefunctions

The concept of high energy diffractive dissociation can be generalised to provide a tool to materialise the individual Fock states of a hadron, nucleus or photon. For example, the diffractive dissociation of a high energy proton on a nucleus $pA \rightarrow XA'$ where the diffractive system is three jets $X = qqq$ can be used to determine the valence light-front wavefunction of the proton.

In the case of a deuteron projectile, one can study diffractive processes such as $dA \rightarrow pnA'$ or $dA \rightarrow \pi^- pp$ to measure the mesonic Fock state of a nuclear wavefunction. At small hadron transverse momentum, diffractive dissociation of the deuteron should be controlled by conventional nuclear interactions; however at large relative k_T, the diffractive system should be sensitive to the hidden colour components of the deuteron wavefunction. The theory of hidden colour is reviewed below.

5 Discrete Light Cone Quantisation (DLCQ) solutions

The entire spectrum of hadrons and nuclei and their scattering states is given by the set of eigenstates of the light-front Hamiltonian H_{LC} for QCD. In principle it is possible to compute the light-front wavefunctions by diagonalising the QCD light-front Hamiltonian on the free Hamiltonian basis. In the case of QCD in one space and one time dimensions, the application of discretised light-front quantisation (Pauli and Brodsky 1985a) provides complete solutions of the theory, including the entire spectrum of mesons, baryons, and nuclei, and their wavefunctions. In the DLCQ method, one uses periodic boundary conditions in x^- and $b_\perp$ to discretise the light-front momentum space. One then diagonalizes the light-front Hamiltonian for QCD on a discretised Fock state basis. The DLCQ solutions can be obtained for arbitrary parameters including the number of flavours and colours and quark masses. Exact solutions are known for $QCD(1+1)$ at $N_C \rightarrow \infty$ by 't Hooft (1974). The one-space one-time theory can be solved numerically to any precision at finite N_C for any coupling strength and number of quark flavours using discretised light-front quantisation (DLCQ) (Pauli and Brodsky 1985a, Hornbostel *et al.* 1990, Burkardt 1989, Brodsky *et al.* 1998a). One can use DLCQ to calculate the entire spectrum of virtually any 1+1 theory, its discrete bound states as well as the scattering continuum. The main emphasis of the DLCQ method applied to QCD is the determination of the wavefunctions of the hadrons from first principles.

A large number of studies have been performed of model field theories in the LF framework. This approach has been remarkably successful in a range of toy models in 1+1 dimensions: Yukawa theory (Pauli and Brodsky 1985b), the Schwinger model (for both massless and massive fermions) (Eller *et al.* 1987, McCartor 1991, McCartor 1994), ϕ^4 theory (Harindranath and Vary 1987, Harindranath and Vary 1988), QCD with various types of matter (Burkardt 1989, Hornbostel *et al.* 1990, Demeterfi *et al.* 1994, Dalley and Klebanov 1993), and the sine-Gordon model (Burkardt 1993). It has also been applied with promising results to theories in 3+1 dimensions, in particular QED (Krautgartner *et al.* 1992, Kaluza and Pauli 1992) and Yukawa theory (Brodsky *et al.* 2003a) in a truncated basis. In all cases agreement was found between the LC calculations and results obtained by more conventional approaches, for example, lattice gauge theory.

The extension of this program to physical theories in 3+1 dimensions is a formidable computational task because of the much larger number of degrees of freedom; however, progress is being made. Analyses of the spectrum and light-front wavefunctions of positronium in QED_{3+1} are given in Krautgartner *et al.* (1992).

5.1 A DLCQ example: QCD_{1+1} with fundamental matter

This theory was originally considered by 't Hooft in the limit of large N_c ('t Hooft 1974). Later Burkardt (1989), and Hornbostel *et al.* (1990) gave essentially complete numerical solutions of the theory for finite N_c, obtaining the spectra of baryons, mesons, and nucleons and their wavefunctions. The DLCQ results are consistent with the few other calculations available for comparison, and are generally much more efficiently obtained. In particular, the mass of the lowest meson agrees to within numerical accuracy with lattice Hamiltonian results (Hamer 1982). For $N_c = 4$ this mass is close to that obtained by 't Hooft in the $N_c \to \infty$ limit ('t Hooft 1974). Finally, the ratio of baryon to meson mass as a function of N_c agrees with the strong-coupling results of Date *et al.* (1987).

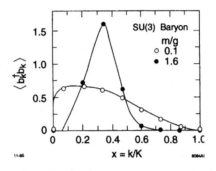

Figure 3. *Valence contribution to the baryon structure function in QCD_{1+1}, as a function of the light-front longitudinal momentum fraction. The gauge group is SU(3), m is the quark mass, and g is the gauge coupling. (From Hornbostel et al. (1990).)*

In addition to the spectrum, one obtains the wavefunctions. These allow direct computation of, *eg*, structure functions. As an example, Figure 3 shows the valence contribution to the structure function for an SU(3) baryon, for two values of the dimensionless coupling m/g. As expected, for weak coupling the distribution is peaked near $x = 1/3$, reflecting that the baryon momentum is shared essentially equally among its constituents. For comparison, the contributions from Fock states with one and two additional $q\bar{q}$ pairs are shown in Figure 4. Note that the amplitudes for these higher Fock components are quite small relative to the valence configuration. The lightest hadrons are nearly always dominated by the valence Fock state in these super-renormalisable models; higher Fock wavefunctions are typically suppressed by factors of 100 or more. Thus the light-front quarks are much more like constituent quarks in these theories than equal-time quarks would be. As discussed above, in an equal-time formulation even the vacuum state would be an infinite superposition of Fock states. Identifying constituents in this case, three of which could account for most of the structure of a baryon, would be quite difficult.

6 Light-front wavefunctions and hadron observables

Light-front Fock state wavefunctions $\psi_{n/H}(x_i, \vec{k}_{\perp i}, \lambda_i)$ play an essential role in QCD phenomenology, generalising Schrödinger wavefunctions $\psi_H(\vec{k})$ of atomic physics to

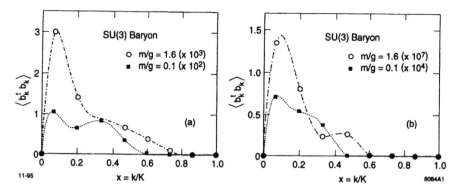

Figure 4. *Contributions to the baryon structure function from higher Fock components: (a) valence plus one additional $q\bar{q}$ pair; (b) valence plus two additional $q\bar{q}$ pairs. (From Hornbostel et al. (1990).)*

relativistic quantum field theory. Given the $\psi_{n/H}^{(\Lambda)}$, one can construct any spacelike electromagnetic, electroweak, or gravitational form factor or local operator product matrix element of a composite or elementary system from the diagonal overlap of the LFWFs (Brodsky and Drell 1980). Exclusive semi-leptonic B-decay amplitudes involving timelike currents such as $B \to A\ell\bar{\nu}$ can also be evaluated exactly in the light-front formalism (Brodsky and Hwang 1999). In this case, the timelike decay matrix elements require the computation of both the diagonal matrix element $n \to n$ where parton number is conserved and the off-diagonal $n + 1 \to n - 1$ convolution such that the current operator annihilates a $q\bar{q}'$ pair in the initial B wavefunction. This term is a consequence of the fact that the time-like decay $q^2 = (p_\ell + p_{\bar{\nu}})^2 > 0$ requires a positive light-front momentum fraction $q^+ > 0$. Conversely for space-like currents, one can choose $q^+ = 0$, as in the Drell-Yan-West representation of the space-like electromagnetic form factors. The light-front Fock representation thus provides an exact formulation of current matrix elements of local operators. In contrast, in equal-time Hamiltonian theory, one must evaluate connected time-ordered diagrams where the gauge particle or graviton couples to particles associated with vacuum fluctuations. Thus even if one knows the equal-time wavefunction for the initial and final hadron, one cannot determine the current matrix elements. In the case of the covariant Bethe-Salpeter formalism, the evaluation of the matrix element of the current requires the calculation of an infinite number of irreducible diagram contributions.

One can also prove directly from the LFWF overlap representation that the anomalous gravitomagnetic moment $B(0)$ vanishes for any composite system (Brodsky et al. 2001a). This property follows directly from the Lorentz boost properties of the light-front Fock representation and holds separately for each Fock state component.

Given the light-front wavefunctions, one can define positive-definite probability distributions, such as the quark and gluon distributions $q(x, Q)$, $g(x, Q)$ which enter deep inelastic scattering and other hard inclusive reactions. These include all spin-dependent distributions such as quark transversity. The resulting distributions obey DGLAP evolution; the moments defined as the matrix elements of the operator product expansion have the correct anomalous dimensions. In addition one can compute the unintegrated distributions in x and $k_\perp$ which underlie the generalised parton distributions for non-zero skewness. For example, the polarised quark distributions at resolution Λ correspond to

$$q_{\lambda_q/\Lambda_p}(x, \Lambda) = \sum_{n, q_a} \int \prod_{j=1}^n dx_j d^2 k_{\perp j} \sum_{\lambda_i} |\psi_{n/H}^{(\Lambda)}(x_i, \vec{k}_{\perp i}, \lambda_i)|^2 \qquad (7)$$

$$\times \, \delta\left(1 - \sum_i^n x_i\right) \delta^{(2)}\left(\sum_i^n \vec{k}_{\perp i}\right) \delta(x - x_q)$$

$$\times \, \delta_{\lambda_a, \lambda_q} \Theta(\Lambda^2 - \mathcal{M}_n^2) \, ,$$

where the sum is over all quarks q_a which match the quantum numbers, light-front momentum fraction x, and helicity of the struck quark.

Brodsky *et al.* (2001b) have shown how to represent virtual Compton scattering $\gamma^* p \to \gamma p$ at large initial photon virtuality Q^2 and small momentum transfer squared t in the handbag approximation in terms of the light-front wavefunctions of the target proton. Thus the generalised parton distributions which enter virtual Compton scattering and the two-photon exchange contribution to lepton-proton scattering are given by overlaps of the LFWFs with $n = n'$ and $n - n' = \pm 2$. One can verify that the skewed parton distributions $H(x, \zeta, t)$ and $E(x, \zeta, t)$ which appear in deeply virtual Compton scattering are the integrands of the Dirac and Pauli form factors $F_1(t)$ and $F_2(t)$ and the gravitational form factors $A_q(t)$ and $B_q(t)$ for each quark and anti-quark constituent. We have given an explicit illustration of the general formalism for the case of deeply virtual Compton scattering on the quantum fluctuations of a fermion in quantum electrodynamics at one loop. The absolute square of the LFWFs define the unintegrated parton distributions. The integrals of the unintegrated parton distributions over transverse momentum at zero skewness provide the helicity and transversity distributions measurable in polarised deep inelastic experiments (Lepage and Brodsky 1980).

The relationship of QCD processes to the hadron LFWFs is illustrated in Figures 5 and 6. Other applications include two-photon exclusive reactions, and diffractive dissociation into jets. The universal light-front wavefunctions and distribution amplitudes control hard exclusive processes such as form factors, deeply virtual Compton scattering, high momentum transfer photoproduction, and two-photon processes.

Hadronisation phenomena such as the coalescence mechanism for leading heavy hadron production can also be computed from LFWF overlaps. Diffractive jet production provides another phenomenological window into the structure of LFWFs. However, as shown recently (Brodsky *et al.* 2002a) some leading-twist phenomena such as the diffractive component of deep inelastic scattering, single spin asymmetries, nuclear shadowing and anti-shadowing cannot be computed from the LFWFs of hadrons in isolation.

Given the LFWFs, one can also compute the hadronic distribution amplitudes $\phi_H(x_i, Q)$ which control hard exclusive processes as an integral over the transverse momenta of the valence Fock state LFWFs (Lepage and Brodsky 1980). The hadron distribution amplitudes are obtained by integrating the $n-$parton valence light-front wavefunctions:

$$\phi(x_i, Q) = \int^Q \Pi_{i=1}^{n-1} d^2 k_{\perp i} \, \psi_{\text{val}}(x_i, k_\perp) \, . \qquad (8)$$

The distribution amplitudes are gauge-invariant vacuum to hadron matrix elements and

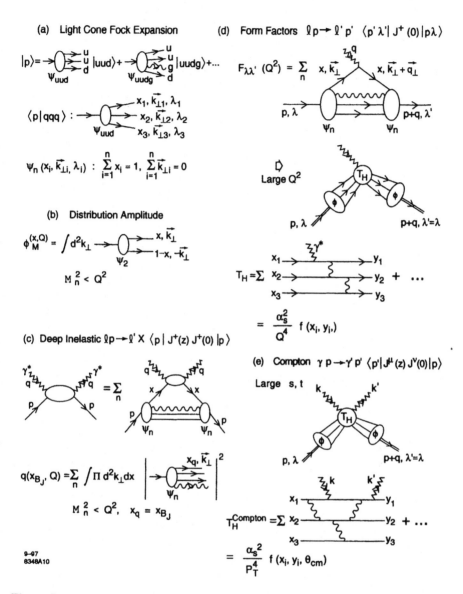

Figure 5. *Representation of QCD hadronic processes in the light-front Fock expansion. (a) The valence uud and higher Fock uudg contributions for the proton. (b) The distribution amplitude* $\phi(x, Q)$ *of a meson expressed as an integral over its valence light-front wavefunction restricted to invariant masses less than Q. (c) Representation of DIS and the quark distributions* $q(x, Q)$ *as probabilistic measures of the light-front Fock wavefunctions. The sum is over the Fock states with invariant mass less than Q. (d) Exact representation of spacelike proton form factors in the light-front Fock basis. The sum is over all Fock components. (e) Leading-twist factorisation of the Compton amplitude at large momentum transfer.*

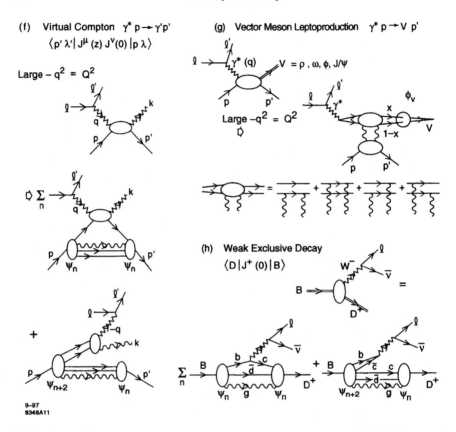

(f) Virtual Compton $\gamma^* p \rightarrow \gamma' p'$

$\langle p' \lambda' | J^\mu (z) J^\nu(0) | p \lambda \rangle$

Large $- q^2 = Q^2$

(g) Vector Meson Leptoproduction $\gamma^* p \rightarrow V p'$

$V = \rho, \omega, \phi, J/\psi$

Large $-q^2 = Q^2$

(h) Weak Exclusive Decay

$\langle D | J^+ (0) | B \rangle$

9-97
8348A11

Figure 6. (f) *Representation of deeply virtual Compton scattering in the light-front Fock expansion at leading twist. Both diagonal n $\rightarrow$ n and off-diagonal n + 2 $\rightarrow$ n contributions are required.* (g) *Diffractive vector meson production at large photon virtuality Q^2 and longitudinal polarisation. The high energy behaviour involves two gluons in the t channel coupling to the compact colour dipole structure of the upper vertex. The bound-state structure of the vector meson enters through its distribution amplitude.* (h) *Exact representation of the weak semileptonic decays of heavy hadrons in the light-front Fock expansion. Both diagonal n $\rightarrow$ n and off-diagonal pair annihilation n + 2 $\rightarrow$ n contributions are required.*

they obey evolution equation as dictated by the OPE. Leading-twist PQCD predictions for hard exclusive amplitudes (Lepage and Brodsky 1980) are written in a factorised form as the product of hadron distribution amplitudes $\phi_I(x_i, Q)$ for each hadron I convoluted with the hard scattering amplitude T_H obtained by replacing each hadron with collinear on-shell quarks with light-front momentum fractions $x_i = k_i^+ / P^+$. The logarithmic evolution equations for the distribution amplitudes require that the valence light-front wavefunctions fall-off asymptotically as the nominal power $[\frac{1}{k_\perp^2}]^{n-1}$, where n is the number of elementary fields in the minimal Fock state.

The light-front Fock representation thus provides an exact formulation of current matrix elements of local and bi-local operators. In contrast, in equal-time Hamiltonian theory, one must evaluate connected time-ordered diagrams where the gauge particle or graviton couples to particles associated with vacuum fluctuations. Thus even if one knows the equal-time wavefunction for the initial and final hadron, one cannot determine the current matrix elements. In the case of the covariant Bethe-Salpeter formalism, the evaluation of the matrix element of the current requires the calculation of an infinite number of irreducible diagram contributions.

7 General structure of light-front wavefunctions

Even without explicit solutions, much is known about the explicit form and structure of LFWFs. They can be matched to non-relativistic Schrödinger wavefunctions at soft scales. At high momenta, the LFWFs at large $k_\perp$ and $x_i \to 1$ are constrained by arguments based on conformal symmetry, the operator product expansion, or perturbative QCD. The pattern of higher Fock states with extra gluons is given by ladder relations (Antonuccio *et al.* 1997).

The structure of Fock states with non-zero orbital angular momentum is also constrained by the Karmanov-Smirnov operator (Karmanov and Smirnov 1992). One can define the light-front Fock expansion using a covariant generalisation of light-front time: $\tau = x \cdot \omega$. The four-vector ω, with $\omega^2 = 0$, determines the orientation of the light-front plane; the freedom to choose ω provides an explicitly covariant formulation of light-front quantisation (Carbonell *et al.* 1998): all observables such as matrix elements of local current operators, form factors, and cross sections are light-front invariants – they must be independent of ω_μ. In recent work, Brodsky *et al.* (2004b) have studied the analytic structure of LFWFs using the explicitly Lorentz-invariant formulation of the front form. Eigensolutions of the Bethe–Salpeter equation have specific angular momentum as specified by the Pauli–Lubanski vector. The corresponding LFWF for an n-particle Fock state evaluated at equal light-front time $\tau = \omega \cdot x$ can be obtained by integrating the Bethe-Salpeter solutions over the corresponding relative light-front energies. The resulting LFWFs $\psi_n^I(x_i, k_{\perp i})$ are functions of the light-front momentum fractions $x_i = k_i \cdot \omega/p \cdot \omega$ and the invariant mass of the constituents $\mathcal{M}_n$, each multiplying spin-vector and polarisation tensor invariants which can involve ω^μ. They are eigenstates of the Karmanov–Smirnov kinematic angular momentum operator (Karmanov and Smirnov 1992, Carbonell *et al.* 1998).

$$\vec{J} = -i[\vec{k} \times \partial/\partial\vec{k}] - i[\vec{n} \times \partial/\partial\vec{n}] + \frac{1}{2}\vec{\sigma} , \qquad (9)$$

where $\vec{n}$ is the spatial component of ω in the constituent rest frame ($\vec{\mathcal{P}} = \vec{0}$). Although this form is written specifically in the constituent rest frame, it can be generalised to an arbitrary frame by a Lorentz boost.

Normally the generators of angular rotations in the LF formalism contain interactions, as in the Pauli–Lubanski formulation; however, the LF angular momentum operator can also be represented in the kinematical form (Equation 9) without interactions. The key term is the generator of rotations of the LF plane $-i[\vec{n} \times \partial/\partial\vec{n}]$ which replaces the interaction term; it appears only in the explicitly covariant formulation, where the dependence on $\vec{n}$ is present. Thus LFWFs satisfy all Lorentz symmetries of the front form, including boost invariance, and they are proper eigenstates of angular momentum.

In principle, one can solve for the LFWFs directly from the fundamental theory using methods such as DLCQ (Pauli and Brodsky 1985b), the transverse lattice (Bardeen *et al.* 1980, Dalley 2004, Burkardt and Dalley 2002), lattice gauge theory moments (DelDebbio *et al.* 2000), Dyson-Schwinger techniques (Maris and Roberts 2003), and Bethe–Salpeter techniques (Brodsky *et al.* 2004b). DLCQ has been remarkably successful in determining the entire spectrum and corresponding LFWFs in one space-one time field theories (Gross *et al.* 1998), including QCD(1+1) (Hornbostel *et al.* 1990) and Supersymmetric Quantum Chromodynamics (SQCD)(1+1) (Harada *et al.* 2004). There are also DLCQ solutions for low sectors of Yukawa theory in physical space-time dimensions (Brodsky *et al.* 2003a). The DLCQ boundary conditions allow a truncation of the Fock space to finite dimensions while retaining the kinematic boost and Lorentz invariance of light-front quantisation.

One can also project known solutions of the Bethe–Salpeter equation to equal light-front time, producing hadronic light-front Fock wavefunctions (Brodsky *et al.* 2004b). Recently new methods have been developed (Bakker *et al.* 2003, van Iersel and Bakker 2004) to find solutions to bound-state light-front equations in the ladder approximation. Pauli (2004) has shown how one can construct an effective light-front Hamiltonian which acts within the valence Fock state sector alone. Another possible method is to construct the $q\bar{q}$ Green's function using light-front Hamiltonian theory, DLCQ boundary conditions and Lippmann-Schwinger resummation. The zeros of the resulting resolvent projected on states of specific angular momentum J_z can then generate the meson spectrum and their light-front Fock wavefunctions. As emphasised by Weinstein (2004) and Zhan *et al.* (2004) new effective operator methods, which have been developed for Hamiltonian theories in condensed matter and nuclear physics, could also be applied advantageously to light-front Hamiltonians. Reviews of non-perturbative light-front methods may be found in Brodsky *et al.* (1998), Carbonell *et al.* (1998), Dalley (2002), and Brodsky (2003).

Other important non-perturbative QCD methods are Dyson-Schwinger techniques (Maris and Roberts 2003) and the transverse lattice (Dalley 2004). The transverse lattice method combines DLCQ for one-space and the light-front time dimensions with lattice theory in transverse space. It has recently provided the first computation of the generalised parton distributions of the pion (Dalley 2004).

Currently the most important computational tool for making predictions in strong-coupling QCD(3+1) is lattice gauge theory (Wilson 1976) which has made enormous progress in recent years, particularly in computing mass spectra and decay constants. Lattice gauge theory can only provide limited dynamical information because of the difficulty of continuing predictions from Euclidean to Minkowski space. At present, results are limited to large quark and pion masses such that the ρ meson is stable (DeGrand 2004). In contrast to lattice gauge theory path integral methods, Light-front Hamiltonian methods are frame-independent, formulated in Minkowski space, only two physical polarisation gluonic degrees of freedom appear as quanta, and there are no complications from fermions. The known DLCQ solutions for 1+1 quantum field theories could provide a powerful test of lattice methods.

The Hamiltonian approach is in fact the method of choice in virtually every area of physics and quantum chemistry. It has the desirable feature that the output of such a calculation is immediately useful: the spectrum of states and wavefunctions. Furthermore, it allows the use of intuition developed in the study of simple quantum systems, and also the application of, *eg*, powerful variational techniques. The one area of physics where it is *not* widely employed is relativistic quantum field theory. The basic reason for this is that in a relativistic field theory quantised at equal time ("the Instant Form") one has particle creation/annihilation in the vacuum. Thus the true ground state is in general extremely complicated, involving a superposition of states with arbitrary numbers of bare quanta, and one must understand the complicated structure of this state before excitations can be considered. Furthermore, one must have a non-perturbative way of separating out disconnected contributions to physical quantities, which are physically irrelevant. Finally, the truncations that are required inevitably violate Lorentz covariance and, for gauge theories, gauge invariance. These difficulties (along with the development of covariant Lagrangian techniques) eventually led to the almost complete abandonment of fixed-time Hamiltonian methods in relativistic field theories.

Light-front quantisation provides an alternative to the usual formulation of field theories in which these problems appear to be tractable. This raises the prospect of developing a practical Hamiltonian approach to solving field theories non-perturbatively based on diagonalising LC Hamiltonians.

8 Consequences of near-conformal field theory

One of the most exciting recent developments is the AdS/CFT correspondence (Maldacena 1998, Polchinski and Strassler 2001, Brower and Tan 2002, Andreev 2002) between superstring theory in 10 dimensions and supersymmetric Yang Mills theory in 3+1 dimensions. As discussed below, one can use this connection to establish the form of QCD wavefunctions at large transverse momentum $k_\perp^2 \to \infty$ and at $x \to 1$ (Brodsky and de Teramond 2004). The AdS/CFT correspondence has important implications for hadron phenomenology in the conformal limit, including an all-orders demonstration of counting rules (Brodsky and Farrar 1973, Matveev *et al.* 1973, Brodsky and Farrar 1975) for hard exclusive processes (Polchinski and Strassler 2001), as well as determining essential aspects of hadronic light-front wavefunctions (Brodsky and de Teramond 2004).

8.1 The conformal correspondence principle

The classical Lagrangian of QCD for massless quarks is conformally symmetric. Since it has no intrinsic mass scale, the classical theory is invariant under the $SO(4, 2)$ translations, boosts, and rotations of the Poincaré group, plus the dilatations and other transformations of the conformal group. Scale invariance and therefore conformal symmetry is destroyed in the quantum theory by the renormalisation procedure which introduces a renormalisation scale as well as by quark masses. Conformal symmetry is thus broken in physical QCD; nevertheless, we can still recover the underlying features of the conformally invariant theory by evaluating any expression in QCD in the analytic limit of zero quark mass and zero β function (Parisi 1972):

$$\lim_{m_q \to 0, \beta \to 0} \mathcal{O}_{QCD} = \mathcal{O}_{\text{conformal QCD}} \,. \tag{10}$$

This conformal correspondence limit is analogous to Bohr's correspondence principle where one recovers predictions of classical theory from quantum theory in the limit of zero Planck constant. The contributions to an expression in QCD from its non-zero β-function can be systematically identified (Brodsky *et al.* 2001c, Rathsman 2001, Grunberg 2001) order-by-order in perturbation theory using the Banks-Zaks (1982) procedure.

The "conformal correspondence principle" provides a new tool, the conformal template (Brodsky 2004, Brodsky 2003b), which is very useful for theory analyses, such as the expansion polynomials for distribution amplitudes (Brodsky *et al.* 1980b, Brodsky *et al.* 1986a, Brodsky *et al.* 1986b, Braun *et al.* 2003), the non-perturbative wavefunctions which control exclusive processes at leading twist (Lepage and Brodsky 1979a, Brodsky and Lepage 1989, Brodsky 2001).

8.2 Commensurate scale relations

The near-conformal behaviour of QCD is the basis for commensurate scale relations (Brodsky and Lu 1995) which relate observables to each other without renormalisation scale or scheme ambiguities (Brodsky *et al.* 2001c, Rathsman 2001). One can derive the commensurate scale relation between the effective charges of any two observables by first computing their relation in conformal gauge theory; the effects of the non-zero QCD β function are then taken into account using the BLM method (Brodsky *et al.* 1983a) to set the scales of the respective couplings. An important example is the generalised Crewther relation (Brodsky *et al.* 1996):

$$\left[1 + \frac{\alpha_R(s^*)}{\pi}\right]\left[1 - \frac{\alpha_{g_1}(Q^2)}{\pi}\right] = 1 \qquad (11)$$

where the underlying form at zero β function is determined by conformal symmetry (Crewther 1972). Here $\alpha_R(s)/\pi$ and $-\alpha_{g_1}(Q^2)/\pi$ represent the entire radiative corrections to $R_{e^+e^-}(s)$ and the Bjorken sum rule for the $g_1(x, Q^2)$ structure function measured in spin-dependent deep inelastic scattering, respectively. The relation between s^* and Q^2 can be computed order by order in perturbation theory using the BLM method (Brodsky *et al.* 1983a). The ratio of physical scales guarantees that the effect of new quark thresholds is commensurate. Commensurate scale relations are renormalisation-scheme independent and satisfy the group properties of the renormalisation group. Each observable can be computed in any convenient renormalisation scheme such as dimensional regularisation. The $\overline{MS}$ coupling can then be eliminated; it becomes only an intermediary (Brodsky and Lu 1995). In such a procedure there are no further renormalisation scale (μ) or scheme ambiguities.

The effective charge (Brodsky *et al.* 1998b) defined from the ratio of elastic pion and photon-to-pion transition form factors $\alpha_s^{\text{exclusive}}(Q^2) = F_\pi(Q^2)/4\pi Q^2 F_{\gamma\pi^0}^2(Q^2)$ can also be connected to other effective charges and observables by commensurate scale relations. Its magnitude, $\alpha_s^{\text{exclusive}}(Q^2) \sim 0.8$ at small Q^2, is sufficiently large as to explain the observed magnitude of exclusive amplitudes such as the pion form factor using the asymptotic distribution amplitude. An analytic effective charge such as the pinch scheme (Cornwall 1982) provides a method to unify the electroweak and strong couplings and forces.

8.3 Fixed point behaviour

Although the QCD coupling decreases logarithmically at high virtuality due to asymptotic freedom, theoretical (von Smekal *et al.* 1997, Zwanziger 2004, Howe and Maxwell 2002, Howe and Maxwell 2004, Furui and Nakajima 2004a, Furui and Nakajima 2004b, Badalian and Veselov 2004, Ackerstaff *et al.* 1999) and phenomenological (Mattingly and Stevenson 1994, Brodsky *et al.* 2003b, Baldicchi and Prosperi 2002) evidence is now accumulating that the QCD coupling becomes constant at small virtuality; *ie*, $\alpha_s(Q^2)$ develops an infrared fixed point in contradiction to the usual assumption of singular growth in the infrared. If QCD running couplings are bounded, the integration over the running coupling is finite and renormalisation resummations are not required. If the QCD coupling becomes scale-invariant in the infrared, then elements of conformal theory (Braun *et al.* 2003) become relevant even at relatively small momentum transfers.

Brodsky *et al.* (2003b) have presented a definition of a physical coupling for QCD which has a direct relation to high precision measurements of the hadronic decay channels of the $\tau^- \rightarrow \nu_\tau H^-$. Let R_τ be the ratio of the hadronic decay rate to the leptonic one. Then $R_\tau \equiv R_\tau^0 \left[1 + \frac{\alpha_\tau}{\pi}\right]$, where R_τ^0 is the zeroth order QCD prediction, defines the effective charge α_τ. The data for τ decays is well-understood channel by channel, thus allowing the calculation of the hadronic decay rate and the effective charge as a function of the τ mass below the physical mass. The vector and axial-vector decay modes can be studied separately. Using an analysis of the τ data from the OPAL collaboration (Ackerstaff *et al.* 1999), we have found that the experimental value of the coupling $\alpha_\tau(s) = 0.621 \pm 0.008$ at $s = m_\tau^2$ corresponds to a value of $\alpha_{\overline{MS}}(M_Z^2) = (0.117$-$0.122) \pm 0.002$, where the range corresponds to three different perturbative methods used in analysing the data. This result is in good agreement with the world average $\alpha_{\overline{MS}}(M_Z^2) = 0.117 \pm 0.002$. However, one also finds that the effective charge only reaches $\alpha_\tau(s) \sim 0.9 \pm 0.1$ at $s = 1\,\text{GeV}^2$, and it even stays within the same range down to $s \sim 0.5\,\text{GeV}^2$. The effective coupling is close to constant at low scales, suggesting that physical QCD couplings become constant or "frozen" at low scales.

Figure 7 shows a comparison of the experimentally determined effective charge $\alpha_\tau(s)$ with solutions to the evolution equation for α_τ at two-, three-, and four-loop order normalised at m_τ. At three loops the behaviour of the perturbative solution drastically changes, and instead of diverging, it freezes to a value $\alpha_\tau \simeq 2$ in the infrared. The infrared behaviour is not perturbatively stable since the evolution of the coupling is governed by the highest order term. This is illustrated by the widely different results obtained for three different values of the unknown four loop term $\beta_{\tau,3}$ which are also shown. The values of $\beta_{\tau,3}$ used are obtained from the estimate of the four loop term in the perturbative series of R_τ, $K_4^{\overline{MS}} = 25 \pm 50$ (Le Diberder and Pich 1992). It is interesting to note that the central four-loop solution is in good agreement with the data all the way down to $s \simeq 1\,\text{GeV}^2$.

The results for α_τ resemble the behaviour of the one-loop "time-like" effective coupling (Beneke and Braun 1995, Ball *et al.* 1995, Dokshitzer *et al.* 1996)

$$\alpha_{\text{eff}}(s) = \frac{4\pi}{\beta_0} \left\{ \frac{1}{2} - \frac{1}{\pi} \arctan\left[\frac{1}{\pi} \ln \frac{s}{\Lambda^2}\right] \right\}, \tag{12}$$

which is finite in the infrared and freezes to the value $\alpha_{\text{eff}}(s) = 4\pi/\beta_0$ as $s \rightarrow 0$. It is

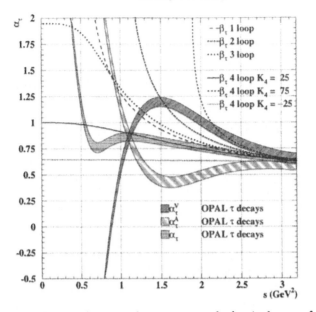

Figure 7. *The effective charge α_τ for non-strange hadronic decays of a hypothetical τ lepton with $m_{\tau'}^2 = s$ compared to solutions of the fixed order evolution equation for α_τ at two-, three-, and four-loop order. The error bands include statistical and systematic errors.*

instructive to expand the "time-like" effective coupling for large s,

$$
\begin{aligned}
\alpha_{\text{eff}}(s) &= \frac{4\pi}{\beta_0 \ln(s/\Lambda^2)} \left\{ 1 - \frac{1}{3} \frac{\pi^2}{\ln^2(s/\Lambda^2)} + \frac{1}{5} \frac{\pi^4}{\ln^4(s/\Lambda^2)} + \dots \right\} \\
&= \alpha_s(s) \left\{ 1 - \frac{\pi^2 \beta_0^2}{3} \left(\frac{\alpha_s(s)}{4\pi} \right)^2 + \frac{\pi^4 \beta_0^4}{5} \left(\frac{\alpha_s(s)}{4\pi} \right)^4 + \dots \right\} .
\end{aligned}
\tag{13}
$$

This shows that the "time-like" effective coupling is a resummation of $(\pi^2 \beta_0^2 \alpha_s^2)^n$ - corrections to the usual running couplings. The finite coupling α_{eff} given in Equation 12 obeys standard PQCD evolution at LO. Thus one can have a solution for the perturbative running of the QCD coupling which obeys asymptotic freedom but does not have a Landau singularity.

The near constancy of the effective QCD coupling at small scales helps explain the empirical success of dimensional counting rules for the power law fall-off of form factors and fixed angle scaling. As shown in Brodsky *et al.* (1998b) and Melic *et al.* (2002), one can calculate the hard scattering amplitude T_H for such processes (Lepage and Brodsky 1980) without scale ambiguity in terms of the effective charge α_τ or α_R using commensurate scale relations. The effective coupling is evaluated in the regime where the coupling is approximately constant, in contrast to the rapidly varying behaviour from powers of α_s predicted by perturbation theory (the universal two-loop coupling). For example, the nucleon form factors are proportional at leading order to two powers of α_s evaluated at low scales in addition to two powers of $1/q^2$; The pion photoproduction amplitude at fixed angles is proportional at leading order to three powers of the QCD coupling. The essential variation from leading-twist counting-rule behaviour then only arises from the anomalous dimensions of the hadron distribution amplitudes.

8.4 The Abelian correspondence principle

Another important guide to QCD predictions is consistency in a limit where the theory becomes Abelian. One can consider QCD predictions as functions of analytic variables of the number of colours N_C and flavours N_F. At $N_C \to \infty$ at fixed $N_C\alpha_s$, calculations in QCD greatly simplify since only planar diagrams enter. However, the $N_C \to 0$ limit is also very interesting. Remarkably, one can show at all orders of perturbation theory (Brodsky and Huet 1998) that PQCD predictions reduce to those of an Abelian theory similar to QED at $N_C \to 0$ with $C_F\alpha_s$ and $\frac{N_F}{T_F C_F}$ held fixed, where $C_F = \frac{N_C^2 - 1}{2N_C}$ and $T_F = 1/2$. The resulting theory corresponds to the group $1/U(1)$ which means that light-by-light diagrams acquire a particular topological factor. The $N_C \to 0$ limit provides an important check on QCD analyses; QCD formulae and phenomena must match their Abelian analog. The renormalisation scale is effectively fixed by this requirement. Commensurate scale relations obey the Abelian Correspondence principle, giving the correct Abelian relations between observables in the limit $N_C \to 0$.

9 Perturbative QCD and exclusive processes

Exclusive processes provide an important window on QCD and the structure of hadrons. There has been considerable progress analysing exclusive and diffractive reactions at large momentum transfer from first principles in QCD. Rigorous statements can be made on the basis of asymptotic freedom and factorisation theorems which separate the underlying hard quark and gluon sub-process amplitude from the non-perturbative physics of the hadronic wavefunctions. The leading-power contribution to exclusive hadronic amplitudes such as quarkonium decay, heavy hadron decay, and scattering amplitudes where hadrons are scattered with large momentum transfer can often be factorised as a convolution of distribution amplitudes $\phi_H(x_i, \Lambda)$ and hard-scattering quark/gluon scattering amplitudes T_H integrated over the light-front momentum fractions of the valence quarks (Lepage and Brodsky 1980):

$$\mathcal{M}_{\text{Hadron}} = \int \prod \phi_H^{(\Lambda)}(x_i, \lambda_i) T_H^{(\Lambda)} dx_i \ . \tag{14}$$

Here $T_H^{(\Lambda)}$ is the underlying quark-gluon sub-process scattering amplitude in which each incident and final hadron is replaced by valence quarks with collinear momenta $k_i^+ = x_i p_H^+$, $\vec{k}_{\perp i} = x_i \vec{p}_{\perp H}$. The invariant mass of all intermediate states in T_H is evaluated above the separation scale $\mathcal{M}_n^2 > \Lambda^2$. The essential part of the hadronic wavefunction is the distribution amplitude (Lepage and Brodsky 1980), defined as the integral over transverse momenta of the valence (lowest particle number) Fock wavefunction; *eg* for the pion

$$\phi_\pi(x_i, Q) \equiv \int d^2 k_\perp \, \psi_{q\bar{q}/\pi}^{(Q)}(x_i, \vec{k}_{\perp i}, \lambda) \ , \tag{15}$$

where the separation scale Λ can be taken to be order of the characteristic momentum transfer Q in the process. It should be emphasised that the hard scattering amplitude T_H is evaluated in the QCD perturbative domain where the propagator virtualities are above the separation scale.

The leading power fall-off of the hard scattering amplitude as given by dimensional counting rules follows from the nominal scaling of the hard-scattering amplitude: $T_H \sim 1/Q^{n-4}$, where n is the total number of fields (quarks, leptons, or gauge fields) participating in the hard scattering (Matveev *et al.* 1973, Brodsky and Farrar 1975). Thus the reaction is dominated by sub-processes and Fock states involving the minimum number of interacting fields. In the case of $2 \to 2$ scattering processes, this implies

$$\frac{d\sigma}{dt}(AB \to CD) = F_{AB \to CD}(t/s)/s^{n-2} \, , \tag{16}$$

where $n = N_A + N_B + N_C + N_D$ and n_H is the minimum number of constituents of H.

In the case of form factors, the dominant helicity conserving amplitude has the nominal power-law fall-off $F_H(t) \sim (1/t)^{n_H-1}$. The complete predictions from PQCD modify the nominal scaling by logarithms from the running coupling and the evolution of the distribution amplitudes. In some cases, such as large angle $pp \to pp$ scattering, there can be "pinch" contributions (Landshoff 1974) when the scattering can occur from a sequence of independent near-on shell quark-quark scattering amplitudes at the same CM angle. After inclusion of Sudakov suppression form factors, these contributions also have a scaling behaviour close to that predicted by constituent counting.

The constituent counting rules were originally derived (Matveev *et al.* 1973, Brodsky and Farrar 1975) before the development of QCD in anticipation that the underlying theory of hadron physics would be renormalisable and close to a conformal theory. The factorised structure of hard exclusive amplitudes in terms of a convolution of valence hadron wavefunctions times a hard-scattering quark scattering amplitude was also proposed (Brodsky and Farrar 1975). Upon the discovery of the asymptotic freedom in QCD, there was a systematical development of the theory of hard exclusive reactions, including factorisation theorems, counting rules, and evolution equations for the hadronic distribution amplitudes (Brodsky and Lepage 1979, Lepage and Brodsky 1979a, Lepage and Brodsky 1979b, Efremov and Radyushkin 1980a).

The distribution amplitudes which control leading-twist exclusive amplitudes at high momentum transfer can be related to the gauge-invariant Bethe-Salpeter wavefunction at equal light-front time $\tau = x^+$. The logarithmic evolution of the hadron distribution amplitudes $\phi_H(x_i, Q)$ with respect to the resolution scale Q can be derived from the perturbatively-computable tail of the valence light-front wavefunction in the high transverse momentum regime. The DGLAP evolution of quark and gluon distributions can also be derived in an analogous way by computing the variation of the Fock expansion with respect to the separation scale. Other key features of the perturbative QCD analyses are: (a) evolution equations for distribution amplitudes which incorporate the operator product expansion, renormalisation group invariance, and conformal symmetry (Lepage and Brodsky 1980, Brodsky *et al.* 1980b, Muller 1995, Ball and Braun 1999, Braun *et al.* 1999); (b) hadron helicity conservation which follows from the underlying chiral structure of QCD (Brodsky and Lepage 1981a); (c) colour transparency, which eliminates corrections to hard exclusive amplitudes from initial and final state interactions at leading power and reflects the underlying gauge theoretic basis for the strong interactions (Brodsky and Mueller 1988) and (d) hidden colour degrees of freedom in nuclear wavefunctions, which reflect the colour structure of hadron and nuclear wavefunctions (Brodsky *et al.* 1983b). There have also been recent advances eliminating renormalisation scale ambiguities in hard-scattering amplitudes via commensurate scale relations (Brodsky and Lu 1995) which connect the couplings entering exclusive amplitudes to the α_V coupling which controls the QCD heavy quark potential.

Exclusive processes such as $\bar{p}p \rightarrow \bar{p}p$, $\bar{p}p \rightarrow K^+K^-$ and $\bar{p}p \rightarrow \gamma\gamma$ provide a unique window for viewing QCD processes and hadron dynamics at the amplitude level (Brodsky and Lepage 1981b, Brodsky and Lepage 1989, Brodsky 2001). New tests of theory and comprehensive measurements of hard exclusive amplitudes can also be carried out for electroproduction at Jefferson Laboratory and in two-photon collisions at CLEO, Belle, and BaBar (Brodsky 2001b). Hadronic exclusive processes are closely related to exclusive hadronic B decays, processes which are essential for determining the CKM phases and the physics of CP violation. The universal light-front wavefunctions which control hard exclusive processes such as form factors, deeply virtual Compton scattering, high momentum transfer photoproduction, and two-photon processes, are also required for computing exclusive heavy hadron decays (Beneke *et al.* 2000, Keum *et al.* 2001, Szczepaniak *et al.* 1990, Brodsky 2001c), such as $B \rightarrow K\pi$, $B \rightarrow \ell\nu\pi$, and $B \rightarrow K p\bar{p}$ (Chua *et al.* 2002). The same physics issues, including colour transparency, hadron helicity rules, and the question of dominance of leading-twist perturbative QCD mechanisms enter in both realms of physics.

The data for virtually all measured hard scattering processes appear to be consistent with the conformal predictions of QCD. For example, one also sees the onset of the predicted perturbative QCD scaling behaviour for exclusive nuclear amplitudes such as deuteron photodisintegration (Here $n = 1 + 6 + 3 + 3 = 13$.) $s^{11}\frac{d\sigma}{dt}(\gamma d \rightarrow pn) \sim$ constant, at fixed CM angle. The measured deuteron form factor and the deuteron photodisintegration cross section appear to follow the leading-twist QCD predictions at large momentum transfers in the few GeV region (Holt 1990, Bochna *et al.* 1998, Rossi *et al.* 2005). A comparison of the data with the QCD predictions is shown in Figure 8.

Another application to exclusive nuclear processes is the approach to scaling of the deuteron form factor $[Q^2]^5 \sqrt{A(Q^2)} \rightarrow$ constant, observed at SLAC and Jefferson Laboratory at high Q^2. These scaling laws reflects the underlying scaling of the nucleon-nucleon interaction and the nuclear force at short distances. The phenomenological successes provide further evidence for the dominance of leading-twist quark-gluon subprocesses and the near conformal behaviour of the QCD coupling. As discussed above, the evidence that the running coupling has constant fixed-point behaviour, which together with BLM scale fixing, could help explain the near conformal scaling behaviour of the fixed-CM angle cross sections. The angular distribution of hard exclusive processes is generally consistent with quark interchange, as predicted from large N_C considerations.

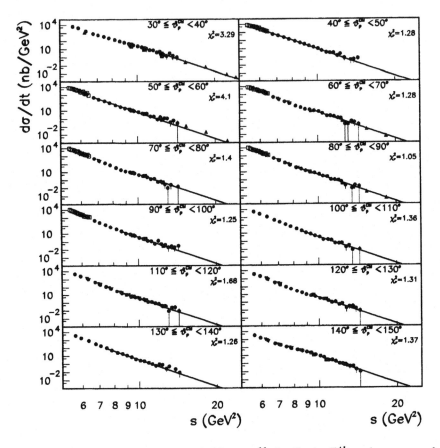

Figure 8. *Fits of the cross sections $d\sigma/dt$ to s^{-11} for $P_T \geq P_T^{th}$ and proton angles between 30° and 150° (solid lines). Data are from CLAS (full circles), Mainz (open squares), SLAC (full downward triangles), Jefferson Laboratory Hall A (full squares) and Hall C (full upward triangles). Also shown in each panel is the χ_ν^2 value of the fit. (From Rossi et al. (2005).)*

10 The evolution of the deuteron distribution amplitude and hidden colour

In this section I will review an analysis by Brodsky *et al.* (1983b) which shows how the asymptotic behaviour of the deuteron form factor at large momentum transfer, and the evolution of the deuteron six-quark distribution amplitude at short distances, can be computed systematically as an expansion in $\alpha_s(Q^2)$. The results agree with the operator product expansion as well as the conformal scaling implied by the AdS/CFT correspondence. As we shall see, the QCD predictions appear to be in remarkable agreement with experiment for $Q^2 \gtrsim 1$ GeV^2 particularly when expressed in terms of the deuteron reduced form factor. This provides a good check on the six-quark description of the deuteron at short distances as well as the scale invariance of the elastic quark-quark scattering amplitude. I will also discuss how the dominance of the hidden colour amplitudes at short distances also provides an explanation for the repulsive behaviour of the nucleon-nucleon potential at small inter-nucleon separation.

Hadronic form factors in QCD at large momentum transfer $Q^2 = \vec{q}^2 - q_0^2$ can be written in a factorised form where all non-perturbative effects are incorporated into process-independent distribution amplitudes $\phi_H(x_i, Q)$, computed from the equal $\tau = t + z$, six-quark valence wavefunction at small relative quark transverse separation $b_\perp^i \sim O(1/Q)$. The $x_i = (k^0 + k^3)_i / (p^0 + p^3)$ are the light-front longitudinal momentum fractions with $\sum_{i=1}^n x_i = 1$. In the case of the deuteron, only the six-quark Fock state needs to be considered for the purpose of computing a hard scattering amplitude since in a physical gauge any additional quark or gluon forced to absorb large momentum transfer yields a power law suppressed contribution to the form factor. The deuteron form factor can then be written as a convolution

$$F_d(Q^2) = \int_0^1 [dx][dy] \phi_d^\dagger(y, Q) \, T_H^{6_q + \gamma^* \to 6_q}(x, y, Q) \phi_d(x, Q) \,, \tag{17}$$

where the hard scattering amplitude

$$T_H^{6_q + \gamma^* \to 6_q} = \left[\frac{\alpha_s(Q^2)}{Q^2} \right]^5 t(x, y) \left[1 + O(\alpha_s(Q^2)) \right] \tag{18}$$

gives the probability amplitude for scattering six quarks collinear with the initial to the final deuteron momentum and

$$\phi_d(x_i, Q) \propto \int^{k_{\perp i} < Q} [d^2 k_\perp] \, \psi_{qqq\,qqq}(x_i, \vec{k}_{\perp i}) \tag{19}$$

gives the probability amplitude for finding the quarks with longitudinal momentum fractions x_i in the deuteron wavefunction collinear up to the scale Q. Because the coupling of the gauge gluon is helicity-conserving and the fact that $\phi_d(x_i, Q)$ is the $L_z = 0$ projection of the deuteron wavefunction, hadron helicity is conserved: The dominant form factor corresponds to $\sqrt{A(Q^2)}$; ie, $h = h' = 0$.

The distribution amplitude $\phi_d(x_i, Q)$ is the basic deuteron wavefunction which controls high momentum transfer exclusive reactions in QCD. The logarithmic Q^2 dependence of ϕ_d is determined by an evolution equation computed from perturbative quark-quark scattering kernels at large momentum transfer, or equivalently, by the operator product expansion at short distances and the renormalisation group (Lepage and Brodsky 1980, Duncan and Mueller 1980, Brodsky *et al.* 1980b, Peskin 1979, Efremov and Radyushkin 1980b, Brodsky and Lepage 1981).

The QCD prediction for the leading helicity-zero deuteron form factor then has the form (Brodsky and Chertok 1976a, 1976b):

$$F_d(q^2) = \left[\frac{\alpha_s(Q^2)}{Q^2} \right]^5 \sum_{m,n} d_{mn} \left(\ln \frac{Q^2}{\Lambda^2} \right)^{-\gamma_n^d - \gamma_m^d} \left[1 + O\left(\alpha_s(Q^2), \frac{m}{Q} \right) \right] \,, \tag{20}$$

where the main dependence $[\alpha_s(Q^2)/Q^2]^5$ comes from the hard-gluon exchange amplitude T_H. The anomalous dimensions γ_n^d are calculated from the evolution equations for $\psi_d(x_i, Q)$.

The evolution equation for six quark systems in which the constituents have the light-front longitudinal momentum fractions x_i ($i = 1, 2, \ldots, 6$) can be obtained from a generalisation of the proton (three quark) case (Lepage and Brodsky 1980, Duncan and Mueller 1980, Brodsky *et al.* 1980b, Peskin 1979, Efremov and Radyushkin 1980b, Brodsky and Lepage 1981). A nontrivial extension is the calculation of the colour factor, C_d, of six quark systems. Since in leading order only pair-wise interactions, with transverse momentum Q, occur between quarks, the evolution equation for the six-quark system becomes $\{[dy] = \delta(1 - \sum_{i=1}^{6} y_i) \prod_{i=1}^{6} dy_i, C_F = (n_c^2 - 1)/2n_c = (4/3), \beta = 11 - (2/3)n_f$, and n_f is the effective number of flavours)

$$\prod_{k=1}^{6} x_k \left[\frac{\partial}{\partial \xi} + \frac{3C_F}{\beta}\right] \widetilde{\Phi}(x_i, Q) = -\frac{C_d}{\beta} \int_0^1 [dy] V(x_i, y_i) \widetilde{\Phi}(y_i, Q) , \qquad (21)$$

where the factor 3 in the square bracket comes from the renormalisation of the six quark field. In Equation 21 we have defined $\Phi(x_i, Q) = \prod_{k=1}^{6} x_k \widetilde{\Phi}(x_i, Q)$. The evolution is in the variable

$$\xi(Q^2) = \frac{\beta}{4\pi} \int_{Q_0^2}^{Q^2} \frac{dk^2}{k^2} \alpha_s(k^2) \sim \ell n \left(\frac{\ell n \frac{Q^2}{\Lambda^2}}{\ell n \frac{Q_0^2}{\Lambda^2}}\right) . \qquad (22)$$

By summing over interactions between quark pairs $\{i, j\}$ due to exchange of a single gluon, $V(x_i, y_i) = V(y_i, x_i)$ is given by

$$V(x_i, y_i) = 2 \prod_{k=1}^{6} x_k \sum_{i \neq j}^{6} \theta(y_i - x_i) \prod_{\ell \neq i, j}^{6} \delta(x_\ell - y_\ell) \frac{y_j}{x_j} \left(\frac{\delta_{h_i \bar{h}_j}}{x_i + x_j} + \frac{\Delta}{y_i - x_i}\right) , \qquad (23)$$

where $\delta_{h_i \bar{h}_j} = 1(0)$ when the constituents' $\{i, j\}$ helicities are antiparallel (parallel). The infrared singularity at $x_i = y_i$ is cancelled by the factor $\Delta \widetilde{\Phi}(y_i, Q) = \widetilde{\Phi}(y_i, Q) - \widetilde{\Phi}(x_i, Q)$ since the deuteron is a colour singlet.

The six-quark bound states have five independent colour singlet components ($3 \times 3 \times 3 \times 3 \times 3 \times 3 \supset \underline{1} + \underline{1} + \underline{1} + \underline{1} + \underline{1}$). It can be shown in general that the colour factor C_d is given by

$$C_d = \frac{1}{5} S_{ijk\ell mn}{}^\alpha \left(\frac{1}{2}\lambda_a\right)_{i'}{}^i \left(\frac{1}{2}\lambda_a\right)_{j'}{}^j S_\alpha^{i'j'k\ell mn} , \qquad (24)$$

where $\lambda_a(a = 1, 2, \ldots, 8)$ are Gell-Mann matrices in $SU(3)^c$ group and $s_{ijk\ell mn}^\alpha(\alpha = 1, 2, \ldots, 5)$ are the five independent colour singlet representations. We shall focus on results for the leading contribution to the distribution amplitude and form factor at large Q. Since the leading eigensolution to the evolution Equation 21 turns out to be completely symmetric in its orbital dependence, the dominant asymptotic deuteron wavefunction is fixed by overall antisymmetry to have spin-isospin symmetry $\{3\}_{TS}$ which is dual to its colour symmetry $[222]_c$. Thus the coefficient for each c (and TS) component has equal weights:

$$\phi_{6_q}([222]_c \otimes \{33\}_{TS}) = \frac{1}{\sqrt{5}} \sum_{\alpha=1}^{5} (-1)^\alpha [222]_c{}^\alpha \{33\}_{TS}^\alpha . \qquad (25)$$

Since the evolution potential is diagonal in isospin and spin, C_d is computed by the trace of the colour representation. The colour factor is $-2/3$ for the colour antisymmetric pair $\{i, j\}$ and $+1/3$ for the colour symmetric pair $\{i, j\}$. Since three colour antisymmetric pairs $\{i, j\}$ and two colour symmetric pairs $\{i, j\}$ exist in this state, the colour factor is

$$C_d = \frac{1}{5}\left(-\frac{2}{3} \times 3 + \frac{1}{3} \times 2\right) = \frac{C_F}{5}. \tag{26}$$

To solve the evolution Equation 21, we factorise the Q^2 dependence of $\widetilde{\Phi}(x_i, Q)$ as

$$\widetilde{\Phi}(x_i, Q) = \tilde{\Phi}(x_i)\, e^{-\gamma\xi} = \tilde{\Phi}(x_i)\left[\ell n\frac{Q^2}{\Lambda^2}\right]^{-\gamma}, \tag{27}$$

where the eigenvalues of γ will provide the anomalous dimensions γ_n. The leading anomalous dimension γ_0 [corresponding to the eigenfunction $\tilde{\Phi}(x_i) = 1$] is

$$\gamma_0 = \frac{3C_F}{\beta} + \frac{C_d}{\beta}\sum_{i \neq j}^{6}\delta_{h_i \bar{h}_j}, \tag{28}$$

so that the asymptotically dominant result for the helicity zero deuteron is given by $\gamma_0 = (6/5)(C_F/\beta)$.

In order to have logarithmic evolution of the deuteron distribution amplitude, the six-quark valence light-front wavefunction must fall, nominally, as $\psi_{qqqqqq/d}(x_i, k_{\perp i}) \simeq [\frac{1}{k_\perp^2}]^5$. This is also the prediction of conformal invariance and the AdS/CFT correspondence. More generally, consistency with the operator product expansion for the moments of the distribution amplitude requires the power law fall off $\psi_n(x_i, k_{\perp i}) \simeq [\frac{1}{k_\perp^2}]^{n-1}$ for all n-parton LFWFs with $L_z = 0$.

At high Q^2 the deuteron form factor is sensitive to wavefunction configurations where all six quarks overlap within an impact separation $b_{\perp i} < \mathcal{O}(1/Q)$. Since the deuteron form factor contains the probability amplitudes for the proton and neutron to scatter from $p/2$ to $p/2 + q/2$, it is natural to define the reduced deuteron form factor (Brodsky and Chertok 1976b, Brodsky *et al.* 1983a):

$$f_d(Q^2) \equiv \frac{F_d(Q^2)}{F_{1N}\left(\frac{Q^2}{4}\right) F_{1N}\left(\frac{Q^2}{4}\right)}. \tag{29}$$

The effect of nucleon compositeness is removed from the reduced form factor. Since the leading anomalous dimensions of the nucleon distribution amplitude is $C_F/2\beta$, the QCD prediction for the asymptotic Q^2 behaviour of $f_d(Q^2)$ is

$$f_d(Q^2) \sim \frac{\alpha_s(Q^2)}{Q^2}\left(\ell n\frac{Q^2}{\Lambda^2}\right)^{(2/5)\,C_F/\beta}, \tag{30}$$

where $(2/5)(C_F/\beta) = -(8/145)$ for $n_f = 2$.

QCD thus predicts essentially the same scaling law for the reduced deuteron form factor as a meson form factor. This scaling is consistent with experiment for $Q^2 > 1\ \text{GeV}^2$.

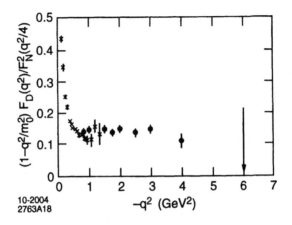

10-2004
2763A18

Figure 9. *Reduced Deuteron Form Factor showing the scaling predicted by perturbative QCD and conformal scaling. The data show two regimes: a fast-falling behaviour at small Q^2 characteristic of normal nuclear binding, and a hard scattering regime with monopole fall-off controlled by the scale $m_0^2 = 0.28$ GeV^2. The latter contribution is attributable to non-nucleonic hidden-colour components of the deuteron's six-quark Fock state. (From Brodsky and Chertok (1976b).)*

In fact as seen in Figure 9, the deuteron reduced form factor contains two components: (1) a fast-falling component characteristic of nuclear binding with probability 85%, and (2) a hard contribution falling as a monopole with a scale of order 0.5 GeV with probability 15%. The normalisation of the deuteron form factor observed at large Q^2 (Arnold *et al.* 1975), as well as the presence of two mass scales in the scaling behaviour of the reduced deuteron form factor (Brodsky and Chertok 1976b) thus suggests sizable hidden-colour Fock state contributions such as $|(uud)_{8_C}(ddu)_{8_C}\rangle$ with probability of order 15% in the deuteron wavefunction (Farrar *et al.* 1995).

In general, one would expect corrections from the leading twist QCD predictions from higher-twist effects (*eg* mass and $k_\perp$ smearing) and higher-order contributions in $\alpha_s(Q^2)$, as well as nonleading anomalous dimensions. However, the agreement of the data with simple $Q^2 f_d(Q^2) \sim$ constant behaviour for $Q^2 > 1/2$ GeV^2 implies that, unless there are fortuitous cancellations, all of the scale-breaking effects are small, and the present QCD perturbation calculations are viable and applicable even in the nuclear physics domain. The lack of deviation from the QCD parameterisation suggests that the parameter Λ in Equation 30 is small. Alternatively, this can be taken as evidence for fixed point behaviour of the QCD coupling in the infrared. A comparison with a standard definition such as $\Lambda_{\overline{MS}}$ would require a calculation of next-to-leading effects. A more definitive check of QCD can be made by calculating the normalisation of $f_d(Q^2)$ from a perturbative calculation of T_H and the evolution of the deuteron wave function to short distances. It is also important to confirm experimentally that the $h = h' = 0$ form factor is indeed dominant.

Note that the deuteron wavefunction which contributes to the asymptotic limit of the form factor is the totally antisymmetric wavefunction corresponding to the orbital Young symmetry given by [6] and isospin (T)+ spin (S) Young symmetry given by $\{33\}$. The deuteron state with this symmetry is related to the NN, $\Delta\Delta$, and hidden-colour (CC) physical bases, for both the $(TS) = (01)$ and (10) cases, by the formula

$$\psi_{[6]\{33\}} = \left(\frac{1}{9}\right)^{1/2} \psi_{NN} + \left(\frac{4}{45}\right)^{1/2} \psi_{\Delta\Delta} + \left(\frac{4}{5}\right)^{1/2} \psi_{CC}. \tag{31}$$

Thus the physical deuteron state, which is mostly ψ_{NN} at large distance, must evolve to the $\psi_{[6]\{33\}}$ state when the six-quark transverse separations $b_\perp^i \le O(1/Q) \to 0$. Since this state is 80% hidden colour, the deuteron wave functions cannot be described by the nucleonic degrees of freedom in this domain. The fact that the six-quark colour-singlet state inevitably evolves in QCD to a dominantly hidden-colour configuration at small transverse separation also has implications for the form of the nucleon-nucleon potential, which can be considered as one component in a coupled-channel system. As the two nucleons approach each other, the system must do work in order to change the six-quark state to a dominantly hidden-colour configuration; *ie*, QCD requires that the nucleon-nucleon potential must be repulsive at short distances (Harvey 1981).

Thus a rigorous prediction of QCD is the "hidden colour" of nuclear wavefunctions at short distances. QCD predicts that nuclear wavefunctions contain "hidden colour" (Matveev and Sorba 1977, Brodsky *et al.* 1983a) components: colour configurations not dual to the usual nucleonic degrees of freedom. In general, the six-quark wavefunction of a deuteron is a mixture of five different colour-singlet states. The dominant colour configuration at large distances corresponds to the usual proton-neutron bound state where transverse momenta are of order $\vec{k}^2 \sim 2M_d\epsilon_{BE}$. However, at small impact space separation, all five Fock colour-singlet components eventually acquire equal weight, *ie*, the deuteron wavefunction evolves to 80% hidden colour.

10.1 Hadron helicity conservation

The distribution amplitudes are $L_z = 0$ projections of the LF wavefunction, and the sum of the spin projections of the valence quarks must equal the J_z of the parent hadron. Higher orbital angular momentum components lead to power-law suppressed exclusive amplitudes (Lepage and Brodsky 1980, Ji *et al.* 2003). Since quark masses can be neglected at leading twist in T_H, one has quark helicity conservation, and thus, finally, hadron-helicity conservation: the sum of initial hadron helicities equals the sum of final helicities. In particular, since the hadron-helicity violating Pauli form factor is computed from states with $\Delta L_z = \pm 1$, PQCD predicts $F_2(Q^2)/F_1(Q^2) \sim 1/Q^2$ [modulo logarithms]. A detailed analysis shows that the asymptotic fall-off takes the form $F_2(Q^2)/F_1(Q^2) \sim \log^2 Q^2/Q^2$ (Belitsky *et al.* 2003). One can also construct other models (Brodsky *et al.* 2004) incorporating the leading-twist perturbative QCD prediction which are consistent with the Jefferson Laboratory polarisation transfer data (Jones *et al.* 2000) for the ratio of proton Pauli and Dirac form factors. This analysis can also be extended to study the spin structure of scattering amplitudes at large transverse momentum and other processes which are dependent on the scaling and orbital angular momentum structure of light-front wavefunctions. Recently Chen *et al.* (2004) have shown that the interfering two-photon exchange contribution to elastic electron-proton scattering, including inelastic intermediate states, can account for the discrepancy between Rosenbluth and Jefferson Laboratory spin transfer polarisation data (Jones *et al.* 2000).

10.2 Timelike form factors

A crucial prediction of models for proton form factors is the relative phase of the time-like form factors, since this can be measured from the proton single spin symmetries in $e^+e^- \to p\bar{p}$ or $p\bar{p} \to \ell\bar{\ell}$. Brodsky *et al.* (2004d) have shown that measurements of the proton's polarisation strongly discriminate between the analytic forms of models which fit the proton form factors in the spacelike region. In particular, the single-spin asymmetry normal to the scattering plane measures the relative phase difference between the timelike G_E and G_M form factors. The dependence on proton polarisation in the time-like region is expected to be large in most models, of the order of several tens of percent. The continuation of the spacelike form factors to the timelike domain $t = s > 4M_p^2$ is very sensitive to the analytic form of the form factors; in particular it is very sensitive to the form of the PQCD predictions including the corrections to conformal scaling. The forward-backward $\ell^+\ell^-$ asymmetry can measure the interference of one-photon and two-photon contributions to $\bar{p}p \to \ell^+\ell^-$.

11 Complications from final-state interactions

Although it has been more than 35 years since the discovery of Bjorken scaling (Bjorken 1969) in electroproduction (Bloom *et al.* 1969), there are still many issues in deep-inelastic lepton scattering and Drell-Yan reactions which are only now being understood from a fundamental basis in QCD.

It is usually assumed—following the parton model—that the leading-twist structure functions measured in deep inelastic lepton-proton scattering are simply the probability distributions for finding quarks and gluons in the target nucleon. In fact, gluon exchange between the fast, outgoing quarks and the target spectators effects the leading-twist structure functions in a profound way, leading to diffractive leptoproduction processes, shadowing of nuclear structure functions, and target spin asymmetries.

As I shall discuss in this section, the final-state interactions from gluon exchange between the outgoing quark and the target spectator system lead to single-spin asymmetries in semi-inclusive deep inelastic lepton-proton scattering at leading twist in perturbative QCD; *ie*, the rescattering corrections of the struck quark with the target spectators are not power-law suppressed at large photon virtuality Q^2 at fixed x_{bj} (Brodsky *et al.* 2002a). The final-state interaction from gluon exchange occurring immediately after the interaction of the current also produces a leading-twist diffractive component to deep inelastic scattering $\ell p \to \ell'p'X$ corresponding to colour-singlet exchange with the target system; this in turn produces shadowing and anti-shadowing of the nuclear structure functions (Brodsky *et al.* 2002a, Brodsky and Lu 1990). In addition, one can show that the pomeron structure function derived from diffractive DIS has the same form as the quark contribution of the gluon structure function (Brodsky *et al.* 2004c). The final-state interactions occur at a short light-front time $\Delta\tau \simeq 1/\nu$ after the virtual photon interacts with the struck quark, producing a nontrivial phase. Here $\nu = p \cdot q/M$ is the laboratory energy of the virtual photon. Thus none of the above phenomena is contained in the target light-front wavefunctions computed in isolation. In particular, the shadowing of nuclear structure functions is due to destructive interference effects from leading-twist diffraction of the virtual photon, physics not included in the nuclear light-front wavefunctions. Thus the structure functions measured in deep inelastic lepton scattering are affected by final-state rescattering, modifying their connection to light-front probability distributions. As an alternative formalism, one can augment the light-front wavefunctions with a gauge link corresponding to an external field created by the virtual photon $q\bar{q}$ pair current (Belitsky *et al.* 2003, Collins and Metz 2004). Such a gauge link is process dependent (Collins 2002), so the resulting augmented LFWFs are not universal (Brodsky *et al.* 2002a, Belitsky *et al.* 2003, Collins 2003). Such rescattering corrections are not contained in the target light-front wavefunctions computed in isolation.

Single-spin asymmetries in hadronic reactions provide a remarkable window to QCD mechanisms at the amplitude level. In general, single-spin asymmetries measure the correlation of the spin projection of a hadron with a production or scattering plane (Sivers 1991). Such correlations are odd under time reversal, and thus they can arise in a time-reversal invariant theory only when there is a phase difference between different spin amplitudes. Specifically, a non-zero correlation of the proton spin normal to a production plane measures the phase difference between two amplitudes coupling the proton target with $J_p^z = \pm\frac{1}{2}$ to the same final-state. The calculation requires the overlap of target light-front wavefunctions with different orbital angular momentum: $\Delta L^z = 1$; thus a single-spin asymmetry (SSA) provides a direct measure of orbital angular momentum in the QCD bound state.

The shadowing and anti-shadowing of nuclear structure functions in the Gribov-Glauber picture is due to the destructive and constructive coherence, respectively, of amplitudes arising from the multiple-scattering of quarks in the nucleus. The effective quark-nucleon scattering amplitude includes pomeron and odderon contributions from multi-gluon exchange as well as Reggeon quark exchange contributions (Brodsky and Lu 1990). The multi-scattering nuclear processes from pomeron, odderon and pseudoscalar Reggeon exchange leads to shadowing and anti-shadowing of the electromagnetic nuclear structure functions in agreement with measurements. An important conclusion is that anti-shadowing is nonuniversal—different for quarks and antiquarks and different for strange quarks versus light quarks. This picture thus leads to substantially different nuclear effects for charged and neutral currents, particularly in anti-neutrino reactions, thus affecting the extraction of the weak-mixing angle $\sin^2 \theta_W$ and the constant ρ_o which are determined from the ratios of charged and neutral current contributions in deep inelastic neutrino and anti-neutrino scattering. In recent work, Brodsky *et al.* (2004e) have shown that a substantial part of the difference between the standard model prediction and the anomalous NuTeV result (Zeller *et al.* 2002) for $\sin^2 \theta_W$ could be due to the different behaviour of nuclear anti-shadowing for charged and neutral currents. Detailed measurements of the nuclear dependence of charged, neutral and electromagnetic DIS processes are needed to establish the distinctive phenomenology of shadowing and anti-shadowing and to make the NuTeV results definitive.

11.1 The paradox of diffractive deep inelastic scattering

A remarkable feature of deep inelastic lepton-proton scattering at HERA is that approximately 10% events are diffractive (Abramowicz 2000, Adloff *et al.* 1997, Breitweg *et al.* 1999): the target proton remains intact and there is a large rapidity gap between the proton and the other hadrons in the final state. These diffractive deep inelastic scattering (DDIS) events can be understood most simply from the perspective of the colour-dipole model (Raufeisen 2000): the $q\bar{q}$ Fock state of the high-energy virtual photon diffractively dissociates into a diffractive dijet system. The colour-singlet exchange of multiple gluons between the colour dipole of the $q\bar{q}$ and the quarks of the target proton leads to the diffractive final state. The same hard pomeron exchange also controls diffractive vector meson electroproduction at large photon virtuality (Brodsky *et al.* 1994b). One can show by analyticity and crossing symmetry that amplitudes with $C = +$ hard-pomeron exchange have a nearly imaginary phase.

This observation presents a paradox: deep inelastic scattering is usually discussed in terms of the parton model. If one chooses the conventional parton model frame where the photon light-front momentum is negative $q+ = q^0 + q^z < 0$, then the virtual photon cannot produce a virtual $q\bar{q}$ pair. Instead, the virtual photon always interacts with a quark constituent with light-front momentum fraction $x = \frac{k^+}{p^+} = x_{bj}$. If one chooses light-front gauge $A^+ = 0$, then the gauge link associated with the struck quark (the Wilson line) becomes unity. Thus the struck "current" quark experiences no final-state interactions. The light-front wavefunctions $\psi_n(x_i, k_{\perp i})$ of the proton which determine the quark probability distributions $q(x, Q)$ are real since the proton is stable. Thus it appears impossible to generate the required imaginary phase, let alone the large rapidity gaps associated with DDIS.

This paradox was resolved by Brodsky *et al.* (2002). It is helpful to consider the case where the virtual photon interacts with a strange quark – the $s\bar{s}$ pair is assumed to be produced in the target by gluon splitting. In the case of Feynman gauge, the struck s quark continues to interact in the final state via gluon exchange as described by the Wilson line. The final-state interactions occur at a light-front time $\Delta\tau \simeq 1/\nu$ after the virtual photon interacts with the struck quark. When one integrates over the nearly-on-shell intermediate state, the amplitude acquires an imaginary part. Thus the rescattering of the quark produces a separated colour-singlet $s\bar{s}$ and an imaginary phase.

In contrast, in the case of the light-front gauge $A^+ = n \cdot A = 0$, one must consider the final state interactions of the (unstruck) $\bar{s}$ quark. Light-front gauge is singular: in particular, the gluon propagator

$$d_{LC}^{\mu\nu}(k) = \frac{i}{k^2 + i\varepsilon}\left[-g^{\mu\nu} + \frac{n^\mu k^\nu + k^\mu n^\nu}{n \cdot k}\right] \qquad (32)$$

has a pole at $k^+ = 0$ which requires an analytic prescription. In final-state scattering involving nearly on-shell intermediate states, the exchanged momentum k^+ is of $O(1/\nu)$ in the target rest frame, which enhances the second term in the propagator. This enhancement allows rescattering to contribute at leading twist even in LC gauge. Thus the rescattering contribution survives in the Bjorken limit because of the singular behaviour of the propagator of the exchanged gluon at small k^+ in $A^+ = 0$ gauge. The net result is gauge invariant and identical to the colour dipole model calculation.

The calculation of the rescattering effects on DIS in Feynman and light-front gauge through three loops is given in detail for a simple Abelian model in Brodsky *et al.* (2002). Figure 10 illustrates two Light Cone Perturbation Theory (LCPTh) diagrams which contribute to the forward $\gamma^*T \rightarrow \gamma^*T$ amplitude, where the target T is taken to be a single quark. In the aligned jet kinematics the virtual photon fluctuates into a $q\bar{q}$ pair with

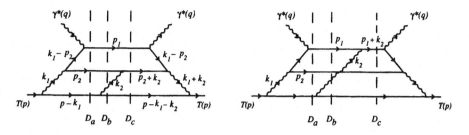

Figure 10. *Two types of final state interactions. (Left) Scattering of the antiquark (p_2 line), which in the aligned jet kinematics is part of the target dynamics. (Right) Scattering of the current quark (p_1 line). For each light-front time-ordered diagram, the potentially on-shell intermediate states—corresponding to the zeroes of the denominators D_a, D_b, D_c—are denoted by dashed lines.*

limited transverse momentum, and the (struck) quark takes nearly all the longitudinal momentum of the photon. The initial q and q̄ momenta are denoted p_1 and $p_2 - k_1$, respectively. The result is most easily expressed in eikonal form in terms of transverse distances r_T, R_T conjugate to p_{2T}, k_T. The DIS cross section can be expressed as

$$Q^4 \frac{d\sigma}{dQ^2 dx_B} = \frac{\alpha_{em}}{16\pi^2} \frac{1-y}{y^2} \frac{1}{2M\nu} \int \frac{dp_2^-}{p_2^-} d^2\vec{r}_T d^2\vec{R}_T |\bar{M}|^2 , \tag{33}$$

where

$$|\bar{M}(p_2^-, \vec{r}_T, \vec{R}_T)| = \left| \frac{\sin\left[g^2 W(\vec{r}_T, \vec{R}_T)/2\right]}{g^2 W(\vec{r}_T, \vec{R}_T)/2} \tilde{A}(p_2^-, \vec{r}_T, \vec{R}_T) \right| \tag{34}$$

is the resummed result. The Born amplitude is

$$\tilde{A}(p_2^-, \vec{r}_T, \vec{R}_T) = 2eg^2 MQp_2^- V(m_{PhysLett} r_T)W(\vec{r}_T, \vec{R}_T) , \tag{35}$$

where $m_{PhysLett}^2 = p_2^- M x_B + m^2$ and

$$V(m r_T) \equiv \int \frac{d^2\vec{p}_T}{(2\pi)^2} \frac{e^{i\vec{r}_T \cdot \vec{p}_T}}{p_T^2 + m^2} = \frac{1}{2\pi} K_0(m r_T) . \tag{36}$$

The rescattering effect of the dipole of the qq̄ is controlled by

$$W(\vec{r}_T, \vec{R}_T) \equiv \int \frac{d^2\vec{k}_T}{(2\pi)^2} \frac{1 - e^{i\vec{r}_T \cdot \vec{k}_T}}{k_T^2} e^{i\vec{R}_T \cdot \vec{k}_T} = \frac{1}{2\pi} \log\left(\frac{|\vec{R}_T + \vec{r}_T|}{R_T}\right) . \tag{37}$$

The fact that the coefficient of $\tilde{A}$ in Equation 34 is less than unity for all $\vec{r}_T, \vec{R}_T$ shows that the rescattering corrections reduce the cross section in analogy to nuclear shadowing.

A new understanding of the role of final-state interactions in deep inelastic scattering has thus emerged. The final-state interactions from gluon exchange occurring immediately after the interaction of the current produce a leading-twist diffractive component to deep inelastic scattering $\ell p \rightarrow \ell' p' X$ due to the colour-singlet exchange with the target system. This rescattering is described in the Feynman gauge by the path-ordered exponential (Wilson line) in the expression for the parton distribution function of the target. The multiple scattering of the struck parton via instantaneous interactions in the target generates dominantly imaginary diffractive amplitudes, giving rise to an effective "hard pomeron" exchange. The presence of a rapidity gap between the target and diffractive system requires that the target remnant emerges in a colour-singlet state; this is made possible in any gauge by the soft rescattering of the final-state $s - \bar{s}$ system.

11.2 Diffractive deep inelastic reactions and rescattering

Brodsky *et al.* (2004c) have recently discussed some further aspects of the QCD dynamics of diffractive deep inelastic scattering. We show that the quark structure function of the effective hard pomeron has the same form as the quark contribution of the gluon structure function. The hard pomeron is not an intrinsic part of the proton; rather it must be considered as a dynamical effect of the lepton-proton interaction.

Our QCD-based picture also applies to diffraction in hadron-initiated processes. The rescattering is different in virtual photon- and hadron-induced processes due to the different colour environment, which accounts for the observed non-universality of diffractive parton distributions. In the hadronic case the colour flow at tree level can involve colour-octet as well as colour-triplet separation. Multiple scattering of the quarks and gluons can set up a variety of different colour singlet domains. This framework also provides a theoretical basis for the phenomenologically successful Soft Colour Interaction (SCI) model which includes rescattering effects and thus generates a variety of final states with rapidity gaps.

11.3 Origin of nuclear shadowing and anti-shadowing

The physics of nuclear shadowing in deep inelastic scattering can be most easily understood in the laboratory frame using the Glauber-Gribov picture (Glauber 1955, Gribov 1969a, Gribov 1969b). The virtual photon, W, or Z^0 produces a quark-antiquark colour-dipole pair which can interact diffractively or inelastically on the nucleons in the nucleus. The destructive interference of diffractive amplitudes from pomeron exchange on the upstream nucleons then causes shadowing of the virtual photon interactions on the back-face nucleons (Stodolsky 1967, Brodsky and Pumplin 1969, Brodsky and Lu 1990, Ioffe 1969, Frankfurt and Stikman 1989, Kopeliovich *et al.* 1998, Kharzeev and Raufeisen 2002). The Bjorken-scaling diffractive interactions on the nucleons in a nucleus thus leads to the shadowing (depletion at small x_{bj}) of the nuclear structure functions.

As emphasised by Ioffe (1969), the coherence between processes which occur on different nucleons at separation L_A requires small Bjorken $x_B : 1/Mx_B = 2\nu/Q^2 \geq L_A$. The coherence between different quark processes is also the basis of saturation phenomena in DIS and other hard QCD reactions at small x_B (Mueller and Shoshi 2004), and coherent multiple parton scattering has been used in the analysis of $p + A$ collisions in terms of the perturbative QCD factorisation approach (Qiu and Vitev 2004). An example of the interference of one- and two-step processes in deep inelastic lepton-nucleus scattering illustrated in Figure 11.

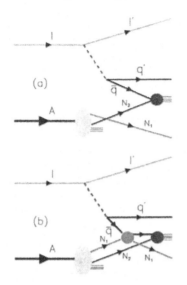

Figure 11. *The one-step and two-step processes in DIS on a nucleus. If the scattering on nucleon N_1 is via pomeron exchange, the one-step and two-step amplitudes are opposite in phase, thus diminishing the $\bar{q}$ flux reaching N_2. This causes shadowing of the charged and neutral current nuclear structure functions.*

An important aspect of the shadowing phenomenon is that the diffractive contribution $\gamma^* N \rightarrow X N'$ to deep inelastic scattering (DDIS), where the nucleon N_1 in Figure 11 remains intact, is a constant fraction of the total DIS rate, confirming that it is a leading-twist contribution. The Bjorken scaling of DDIS has been observed at HERA (Adloff *et al.* 1997, Martin *et al.* 2004, Ruspa 2004). As shown in Brodsky *et al.* (2002), the leading-twist contribution to DDIS arises in QCD in the usual parton model frame when one includes the nearly instantaneous gluon exchange final-state interactions of the struck quark with the target spectators. The same final state interactions also lead to leading-twist single-spin asymmetries in semi-inclusive DIS (Brodsky *et al.* 2002a). Thus the shadowing of nuclear structure functions is also a leading-twist effect.

It was shown in Brodsky and Lu (1990) that if one allows for Reggeon exchanges which leave a nucleon intact, then one can obtain *constructive* interference among the multi-scattering amplitudes in the nucleus. A Bjorken-scaling contribution to DDIS from Reggeon exchange has in fact also been observed at HERA (Adloff *et al.* 1997, Ruspa 2004). The strength and energy dependence of the $C = +$ Reggeon $t-$channel exchange contributions to virtual Compton scattering is constrained by the behaviour (Kuti and Weisskopf 1971): $F_2(x) \sim x^{1-\alpha_R}$ of the non-singlet electromagnetic structure functions at small x. The phase of the Reggeon exchange amplitude is determined by its signature factor. Because of this phase structure (Brodsky and Lu 1990), one obtains constructive interference and *anti-shadowing* of the nuclear structure functions in the range $0.1 < x < 0.2$ – a pronounced excess of the nuclear cross section with respect to nucleon additivity (Arneodo 1994).

In the case where the diffractive amplitude on N_1 is imaginary, the two-step process has the phase $i \times i = -1$ relative to the one-step amplitude, producing destructive interference. (The second factor of i arises from integration over the quasi-real intermediate state.) In the case where the diffractive amplitude on N_1 is due to $C = +$ Reggeon exchange with intercept $\alpha_R(0) = 1/2$, for example, the phase of the two-step amplitude is $\frac{1}{\sqrt{2}}(1-i) \times i = \frac{1}{\sqrt{2}}(i+1)$ relative to the one-step amplitude, thus producing constructive interference and anti-shadowing.

The effective quark-nucleon scattering amplitude includes pomeron and Odderon contributions from multi-gluon exchange as well as Reggeon quark-exchange contributions (Brodsky and Lu 1990). The coherence of these multiscattering nuclear processes leads to shadowing and anti-shadowing of the electromagnetic nuclear structure functions in agreement with measurements. The Reggeon contributions to the quark scattering amplitudes depend specifically on the quark flavour; for example the isovector Regge trajectories couple differently to u and d quarks. The s and $\bar{s}$ couple to yet different Reggeons. This implies distinct anti-shadowing effects for each quark and antiquark component of the nuclear structure function. Brodsky et al. (2004e) have shown that this picture leads to substantially different anti-shadowing for charged and neutral current reactions.

Figures 12–13 illustrate the individual quark q and anti-quark $\bar{q}$ contributions to the ratio of the iron to nucleon structure functions $R = F_2^A/F_2^{No}$ in a model calculation where the Reggeon contributions are constrained by the Kuti-Weisskopf (1971) behaviour of the nucleon structure functions at small x_{bj}. Because the strange quark distribution is much smaller than u and d quark distributions, the strange quark contribution to the ratio is very close to 1 although s^A/s^{No} may significantly deviate from 1.

Our analysis leads to substantially different nuclear anti-shadowing for charged and neutral current reactions; in fact, the neutrino and antineutrino DIS cross sections are each modified in different ways due to the various allowed Regge exchanges. The non-universality of nuclear effects will modify the extraction of the weak-mixing angle $\sin^2 \theta_W$, particularly because of the strong nuclear effects for the F_3 structure function. The shadowing and anti-shadowing of the strange quark structure function in the nucleus can also be considerably different than that of the light quarks. We thus find that part of the anomalous NuTeV result (McFarland et al. 2003) for $\sin^2 \theta_W$ could be due to the non-universality of nuclear anti-shadowing for charged and neutral currents. Our picture also implies non-universality for the nuclear modifications of spin-dependent structure functions.

Thus the anti-shadowing of nuclear structure functions depends in detail on quark flavour. Careful measurements of the nuclear dependence of charged, neutral, and electromagnetic DIS processes are needed to establish the distinctive phenomenology of shadowing and anti-shadowing and to make the NuTeV results definitive. It is also important to map out the shadowing and anti-shadowing of each quark component of the nuclear structure functions to illuminate the underlying QCD mechanisms. Such studies can be carried out in semi-inclusive deep inelastic scattering for the electromagnetic current at Hermes and at Jefferson Laboratory by tagging the flavour of the current quark or by using pion and kaon-induced Drell-Yan reactions. A new determination of $\sin^2 \theta_W$ is also expected from the neutrino scattering experiment NOMAD at CERN (Petti et al. 2004). A systematic program of measurements of the nuclear effects in charged and neutral current reactions could also be carried out in high energy electron-nucleus colliders such as HERA and eRHIC, or by using high intensity neutrino beams (Geer 2003).

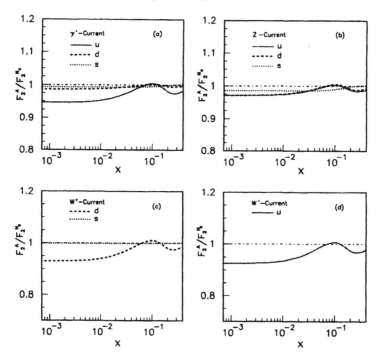

Figure 12. *The quark contributions to the ratios of structure functions at $Q^2 = 1 \text{ GeV}^2$. The solid, dashed and dotted curves correspond to the u, d and s quark contributions, respectively. This corresponds in our model to the nuclear dependence of the $\sigma(\bar{u} - A)$, $\sigma(\bar{d} - A)$, $\sigma(\bar{s} - A)$ cross sections, respectively. In order to stress the individual contribution of quarks, the numerator of the ratios $F_2^A / F_2^{N_0}$ shown in this figure is obtained from the denominator by a replacement q^{N_0} into q^A for only the considered quark. As a result, the effect of anti-shadowing appears diminished.*

11.4 Single-spin asymmetries (SSA) from final-state interactions

Spin correlations provide a remarkably sensitive window to hadronic structure and basic mechanisms in QCD. Among the most interesting polarisation effects are single-spin azimuthal asymmetries in semi-inclusive deep inelastic scattering, representing the correlation of the spin of the proton target and the virtual photon to hadron production plane: $\vec{S}_p \cdot \vec{q} \times \vec{p}_H$ (Avakian *et al.* 2002). Such asymmetries are time-reversal odd, but they can arise in QCD through phase differences in different spin amplitudes.

Until recently, the traditional explanation of pion electroproduction single-spin asymmetries in semi-inclusive deep inelastic scattering is that they are proportional to the transversity distribution of the quarks in the hadron h_1 (Jaffe 1996, Boer 2002a, Boer 2002b) convoluted with the transverse momentum dependent fragmentation (Collins) function $H_1^\perp$, the distribution for a transversely polarised quark to fragment into an unpolarised hadron with non-zero transverse momentum (Collins 1993, Barone *et al.* 2002, Ma *et al.* 2002, Goldstein and Gamberg 2002, Gamberg *et al.* 2003).

An alternative physical mechanism for the azimuthal asymmetries also exists (Brod-

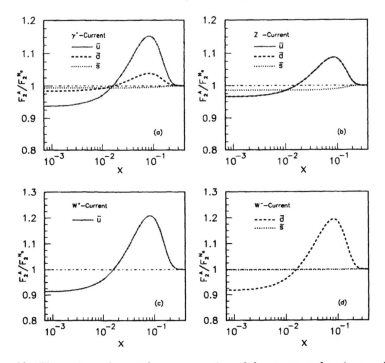

Figure 13. *The anti-quark contributions to ratios of the structure functions at $Q^2 = 1$ GeV^2. The solid, dashed and dotted curves correspond to $\bar{u}$, $\bar{d}$ and $\bar{s}$ quark contributions, respectively. This corresponds in our model to the nuclear dependence of the $\sigma(u-A)$, $\sigma(d-A)$, $\sigma(s-A)$ cross sections, respectively. In order to stress the individual contribution of quarks, the numerator of the ratios $F_2^A/F_2^{N_0}$ shown in this figure is obtained from the denominator by a replacement $\bar{q}^{N_0}$ into $\bar{q}^A$ for only the considered anti-quark.*

sky *et al.* 2002a, Collins 2002, Ji and Yuan 2002). The same QCD final-state interactions (gluon exchange) between the struck quark and the proton spectators which leads to diffractive events also can produce single-spin asymmetries (the Sivers effect) in semi-inclusive deep inelastic lepton scattering which survive in the Bjorken limit. This is illustrated in Figure 14. In contrast to the SSAs arising from transversity and the Collins fragmentation function, the fragmentation of the quark into hadrons is not necessary; one predicts a correlation with the production plane of the quark jet itself $\vec{S}_p \cdot \vec{q} \times \vec{p}_q$.

The final-state interaction mechanism provides an appealing physical explanation within QCD of single-spin asymmetries. Remarkably, the same matrix element which determines the spin-orbit correlation $\vec{S} \cdot \vec{L}$ also produces the anomalous magnetic moment of the proton, the Pauli form factor, and the generalised parton distribution E which is measured in deeply virtual Compton scattering. Physically, the final-state interaction phase arises as the infrared-finite difference of QCD Coulomb phases for hadron wave functions with differing orbital angular momentum. An elegant discussion of the Sivers effect including its sign has been given by Burkardt (2005).

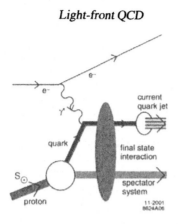

Figure 14. *The origin of the Sivers effect in semi-inclusive deep inelastic scattering.*

The final-state interaction effects can also be identified with the gauge link which is present in the gauge-invariant definition of parton distributions (Collins 2002). Even when the light-front gauge is chosen, a transverse gauge link is required. Thus in any gauge the parton amplitudes need to be augmented by an additional eikonal factor incorporating the final-state interaction and its phase (Ji and Yuan 2002, Belitsky *et al.* 2003). The net effect is that it is possible to define transverse momentum dependent parton distribution functions which contain the effect of the QCD final-state interactions.

A related analysis also predicts that the initial-state interactions from gluon exchange between the incoming quark and the target spectator system lead to leading-twist single-spin asymmetries in the Drell-Yan process $H_1 H_2^{\uparrow} \to \ell^+ \ell^- X$ (Collins 2002, Brodsky *et al.* 2002b). Initial-state interactions also lead to a $\cos 2\phi$ planar correlation in unpolarised Drell-Yan reactions (Boer *et al.* 2003).

11.5 Calculations of single-spin asymmetries in QCD

Brodsky *et al.* (2002) have calculated the single-spin Sivers asymmetry in semi-inclusive electroproduction $\gamma^* p^{\uparrow} \to HX$ induced by final-state interactions in a model of a spin-1/2 proton of mass M with charged spin-1/2 and spin-0 constituents of mass m and λ, respectively, as in the QCD-motivated quark-diquark model of a nucleon. The basic electroproduction reaction is then $\gamma^* p \to q(qq)_0$. In fact, the asymmetry comes from the interference of two amplitudes which have different proton spin, but couple to the same final quark spin state, and therefore it involves the interference of tree and one-loop diagrams with a final-state interaction. In this simple model the azimuthal target single-spin asymmetry $A_{UT}^{\sin \phi}$ is given by

$$
A_{UT}^{\sin \phi} = C_F \alpha_s(\mu^2) \frac{\left(\Delta M + m \right) r_{\perp}}{\left[\left(\Delta M + m \right)^2 + \vec{r}_{\perp}^2 \right]}
$$
$$
\times \left[\vec{r}_{\perp}^2 + \Delta(1 - \Delta)(-M^2 + \frac{m^2}{\Delta} + \frac{\lambda^2}{1 - \Delta}) \right]
$$

$$\times \quad \frac{1}{\vec{r}_\perp^2} \ln \frac{\vec{r}_\perp^2 + \Delta(1-\Delta)(-M^2 + \frac{m^2}{\Delta} + \frac{\lambda^2}{1-\Delta})}{\Delta(1-\Delta)(-M^2 + \frac{m^2}{\Delta} + \frac{\lambda^2}{1-\Delta})} . \qquad (38)$$

Here $r_\perp$ is the magnitude of the transverse momentum of the current quark jet relative to the virtual photon direction, and $\Delta = x_{Bj}$ is the usual Bjorken variable. To obtain Equation 38 from Equation 21 of Brodsky *et al.* (2002), we used the correspondence $|e_1 e_2|/4\pi \to C_F \alpha_s(\mu^2)$ and the fact that the sign of the charges e_1 and e_2 of the quark and diquark are opposite since they constitute a bound state. The result can be tested in jet production using an observable such as thrust to define the momentum $q + r$ of the struck quark.

The predictions of our model for the asymmetry $A_{UT}^{\sin\phi}$ of the $\vec{S}_p \cdot \vec{q} \times \vec{p}_q$ correlation based on Equation 38 are shown in Figure 15. As representative parameters we take $\alpha_s = 0.3$, $M = 0.94$ GeV for the proton mass, $m = 0.3$ GeV for the fermion constituent and $\lambda = 0.8$ GeV for the spin-0 spectator. The single-spin asymmetry $A_{UT}^{\sin\phi}$ is shown as a function of Δ and $r_\perp$ (GeV). The asymmetry measured at HERMES (Airapetian *et al.* 2000) $A_{UL}^{\sin\phi} = K A_{UT}^{\sin\phi}$ contains a kinematic factor $K = \frac{Q}{\nu}\sqrt{1-y} = \sqrt{\frac{2Mx}{E}}\sqrt{\frac{1-y}{y}}$ because the proton is polarised along the incident electron direction. The resulting prediction for $A_{UL}^{\sin\phi}$ is shown in Figure 15(b). Note that $\vec{r} = \vec{p}_q - \vec{q}$ is the momentum of the current quark jet relative to the photon momentum. The asymmetry as a function of the pion momentum $\vec{p}_\pi$ requires a convolution with the quark fragmentation function.

Since the same matrix element controls the Pauli form factor, the contribution of each quark current to the SSA is proportional to the contribution $\kappa_{q/p}$ of that quark to the proton target's anomalous magnetic moment $\kappa_p = \sum_q e_q \kappa_{q/p}$ (Brodsky *et al.* 2002a, Burkardt 2005). Avakian *et al.* (2002) have shown that the data from HERMES and Jefferson Laboratory could be accounted for by the above analysis. The HERMES collaboration has recently measured the SSA in pion electroproduction using transverse target polarisation (Airapetian *et al.* 2005). The Sivers and Collins effects can be separated using planar correlations; both contributions are observed to contribute, with values not in disagreement with theory expectations.

It should be emphasised that the Sivers effect occurs even for jet production; unlike transversity, hadronisation is not required. There is no Sivers effect in charged current reactions since the W only couples to left-handed quarks (Brodsky *et al.* 2003c).

The corresponding single spin asymmetry for the Drell-Yan processes, such as $\pi p^{\leftrightarrow}$ (or $pp^{\leftrightarrow}$) $\to \gamma^* X \to \ell^+ \ell^- X$, is due to initial-state interactions. The simplest way to get the result is applying crossing symmetry to the Semi-Inclusive Deep Inelastic Scattering (SIDIS) processes. The result that the SSA in the Drell-Yan process is the same as that obtained in SIDIS, with the appropriate identification of variables, but with the opposite sign (Collins 2002, Brodsky *et al.* 2002b).

We can also consider the SSA of $e^+ e^-$ annihilation processes such as $e^+ e^- \to \gamma^* \to \pi \Lambda^{\leftrightarrow} X$. The Λ reveals its polarisation via its decay $\Lambda \to p\pi^-$. The spin of the Λ is normal to the decay plane. Thus we can look for a SSA through the T-odd correlation $\epsilon_{\mu\nu\rho\sigma} S_\Lambda^\mu p_\Lambda^\nu q_{\gamma^*}^\rho \cdot p_\pi^\sigma$. This is related by crossing to SIDIS on a Λ target.

Measurements from Jefferson Laboratory (Avakian *et al.* 2004) also show significant beam single spin asymmetries in deep inelastic scattering. Afanasev and Carlson (2004) have recently shown that this asymmetry is due to the interference of longitudinal

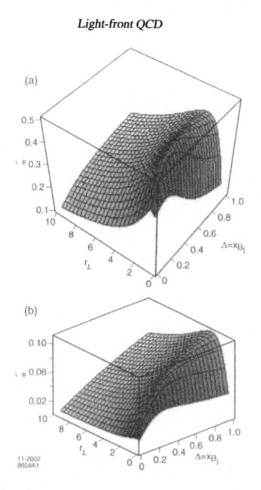

Figure 15. *Model predictions for the target single-spin asymmetry $A_{UT}^{\sin\phi}$ for charged and neutral current deep inelastic scattering with gluon exchange in the final state. $r_\perp$ is the transverse momentum of the outgoing quark relative to the photon or vector boson direction. $\Delta = x_{bj}$ is the light-front momentum fraction of the struck quark. The model parameters are given in the text. (a) The target polarisation transverse to the incident lepton direction. (b) The asymmetry $A_{UL}^{\sin\phi} = K A_{UT}^{\sin\phi}$ includes a kinematic factor $K = \frac{Q}{\nu}\sqrt{1-y}$ for the case where the target nucleon is polarised along the incident lepton direction. We have taken $K = 0.26\sqrt{x}$, corresponding to the kinematics of the HERMES experiment (Airapetian et al. 2000) with $E_{lab} = 27.6$ GeV and $y = 0.5$.*

and transverse photoabsorption amplitudes which have different phases induced by the final-state interaction between the struck quark and the target spectators just as in the calculations of Brodsky *et al.* (2002). Their results are consistent with the experimentally observed magnitude of this effect. Thus similar FSI mechanisms involving quark orbital angular momentum appear to be responsible for both target and beam single-spin asymmetries.

12 New directions for QCD

As I have emphasised in these lectures, the light-front wavefunctions of hadrons are the central elements of QCD phenomenology, describing bound states in terms of their fundamental quark and gluon degrees of freedom at the amplitude level. Given the light-front wavefunctions one can compute quark and gluon distributions, distribution amplitudes, generalised parton distributions, form factors, and matrix elements of local currents such as semileptonic B decays. The diffractive dissociation of hadrons on nucleons or nuclei into jets or leading hadrons can provide new measures of the LFWFs of the projectile as well as tests of colour transparency, hidden colour, and intrinsic charm. The advent of the 12 GeV upgrade of the Jefferson Laboratory electron accelerator and the new 15 GeV antiproton storage ring HESR at GSI will open up important new tests of these properties of QCD in hadronic and nuclear reactions.

Although we are still far from solving QCD explicitly, a number of properties of the light-front wavefunctions of the hadrons are known from both phenomenology and the basic properties of QCD. For example, the endpoint behaviour of light-front wavefunctions and structure functions can be determined from perturbative arguments and Regge arguments. There are also correspondence principles. For example, for heavy quarks in the non-relativistic limit, the light-front formalism reduces to conventional many-body Schrödinger theory. On the other hand, one can also build effective three-quark models which encode the static properties of relativistic baryons.

It is thus imperative to compute the light-front wavefunctions from first principles in QCD. Lattice gauge theory can provide moments of the distribution amplitudes by evaluating vacuum-to-hadron matrix elements of local operators (DelDebbio *et al.* 2000). The transverse lattice is also providing new non-perturbative information (Dalley 2004, Burkardt and Dalley 2002). The DLCQ method is also a first-principles method for solving non-perturbative QCD; at finite harmonic resolution K the DLCQ Hamiltonian acts in physical Minkowski space as a finite-dimensional Hermitian matrix in Fock space. The DLCQ Heisenberg equation is Lorentz-frame independent and has the advantage of providing not only the spectrum of hadrons, but also the complete set of LFWFs for each hadron eigenstate. An important feature the light-front formalism is that J_z is conserved; thus one simplify the DLCQ method by projecting the full Fock space on states with specific angular momentum. As shown in Brodsky *et al.* (2003b), the Karmanov-Smirnov operator uniquely specifies the form of the angular dependence of the light-front wavefunctions, allowing one to transform the light-front Hamiltonian equations to differential equations acting on scalar forms. A complementary method would be to construct the T-matrix for asymptotic $q\bar{q}$ or qqq or gluonium states using the light-front analog of the Lippmann-Schwinger method. This allows one to focus on states with the specific global quantum numbers and spin of a given hadron. The zeros of the resulting resolvent then provides the hadron spectrum and the respective light-front Fock state projections.

In principle, the complete spectrum and bound-state wavefunctions of a quantum field theory can be determined by finding the eigenvalues and eigensolutions of its light-cone Hamiltonian.

The DLCQ method has a number of attractive features for solving 3+1 quantum field theories non-perturbatively because of the ability to truncate the Fock state to low particle number sectors. One of the challenges in obtaining non-perturbative solutions for gauge theories such as QCD using light-cone Hamiltonian methods is to renormalise the theory while preserving Lorentz symmetries and gauge invariance. For example, the truncation of the light-cone Fock space leads to uncompensated ultraviolet divergences. Recently we presented two methods for consistently regularising light-cone-quantised gauge theories in Feynman and light-cone gauges (Brodsky *et al.* 2004f): (1) the introduction of a spectrum of Pauli–Villars fields which produces a finite theory while preserving Lorentz invariance; (2) the augmentation of the gauge-theory Lagrangian with higher derivatives. Finite-mass Pauli–Villars regulators can also be used to compensate for neglected higher Fock states. As a test case, we have applied these regularisation procedures to an approximate non-perturbative computation of the anomalous magnetic moment of the electron in QED as a first attempt to meet Feynman's famous challenge.

12.1   Testing hidden colour

In traditional nuclear physics, the deuteron is a bound state of a proton and a neutron where the binding force arise from the exchange of a pion and other mesonic states. However, as I have reviewed, QCD provides a new perspective (Brodsky and Chertok 1976b, Matveev and Sorba 1977): 6 quarks in the fundamental 3_C representation of $SU(3)$ colour can combine into 5 different colour-singlet combinations, only one of which corresponds to a proton and neutron In fact, if the deuteron wavefunction is a proton-neutron bound state at large distances, then as their separation becomes smaller, the QCD evolution resulting from coloured gluon exchange introduce 4 other "hidden colour" states into the deuteron wavefunction (Brodsky *et al.* 1983a). As I have discussed, the normalisation of the deuteron form factor observed at large Q^2 (Arnold *et al.* 1975), as well as the presence of two mass scales in the scaling behaviour of the reduced deuteron form factor (Brodsky and Chertok 1976b) thus suggests sizable hidden-colour Fock state contributions such as $|(uud)_{8_C}(ddu)_{8_C}\rangle$ with probability of order 15% in the deuteron wavefunction (Farrar *et al.* 1995).

The hidden colour states of the deuteron can be materialised at the hadron level as $\Delta^{++}(uuu)\Delta^-(ddd)$ and other novel quantum fluctuations of the deuteron. These dual hadron components become more and more important as one probes the deuteron at short distances, such as in exclusive reactions at large momentum transfer. For example, the ratio

$$\frac{\frac{d\sigma}{dt}(\gamma d \to \Delta^{++}\Delta^-)}{\frac{d\sigma}{dt}(\gamma d \to np)}$$

should increase dramatically with increasing transverse momentum p_T. Similarly the Coulomb dissociation of the deuteron into various exclusive channels

$$ed \to e' + pn, pp\pi^-, \Delta\Delta, \cdots$$

should have a changing composition as the final-state hadrons are probed at high transverse momentum, reflecting the onset of hidden colour degrees of freedom.

12.2   Perspectives on QCD from AdS/CFT

An outstanding consequence of Maldacena's duality (Maldacena 1998) between 10-dimensional string theory on $AdS_5 \times S^5$ and conformally invariant Yang-Mills theories (Gubser *et al.* 1998, Witten 1998) is the potential to describe processes for physical QCD which are valid at strong coupling and do not rely on perturbation theory. As shown by Polchinski and Strassler (2001), dimensional counting rules (Brodsky and Farrar 1973) for the leading power-law fall-off of hard exclusive scattering can be derived from a gauge theory with a mass gap dual to supergravity in warped spacetimes. The modified theory generates the hard behaviour expected from QCD, instead of the soft behaviour characteristic of strings. Other examples are the description of form factors at large transverse momentum (Polchinski and Susskind 2001) and deep inelastic scattering (Polchinski and Strassler 2003). The discussion of scaling laws in warped backgrounds has also been addressed in Boschi-Filho and Braga (2003), Brower and Tan (2002), and in Andreev (2002).

The AdS/CFT correspondence has now provided important new information on the short-distance structure of hadronic LFWFs; one obtains conformal constraints which are not dependent on perturbation theory. The large $k_\perp$ fall-off of the valence LFWFs is also rigorously determined by consistency with the evolution equations for the hadron distribution amplitudes (Lepage and Brodsky 1979a). Similarly, one can also use the structure of the evolution equations to constrain the $x \to 1$ endpoint behaviour of the LFWFs. One can use these strong constraints on the large $k_\perp$ and $x \to 1$ behaviour to model the LFWFs. Such forms can also be used as the initial approximations to the wavefunctions needed for variational methods which minimise the expectation value of the light-front Hamiltonian. The derivation is carried out in terms of the lowest dimensions of interpolating fields near the boundary of AdS, treating the boundary values of the string states $\Psi(x,r)$ as a product of quantised operators which create n-partonic states out of the vacuum (Brodsky and de Teramond 2004). The AdS/CFT derivation validates QCD perturbative results and confirm the dominance of the quark interchange mechanism (Gunion *et al.* 1972, 1973) for exclusive QCD processes at large N_C. The predicted orbital dependence coincides with the fall-off of light-front Fock wavefunctions derived in perturbative QCD (Ji *et al.* 2003). Since all of the Fock states of the LFWF beyond the valence state are a manifestation of quantum fluctuations, it is natural to match, quanta to quanta, the additional dimensions with the metric fluctuations of the bulk geometry about the fixed AdS background. For example, the quantum numbers of each baryon, including intrinsic spin and orbital angular momentum, are determined by matching the dimensions of the string modes $\Psi(x,r)$, with the lowest dimension of the baryonic interpolating operators in the conformal limit.

The AdS/CFT correspondence also provides a novel way to compute the hadronic spectrum. The essential assumption is to require the hadron wavefunctions to vanish at the fifth-dimensional coordinate $r_0 = \Lambda_{QCD}$. As an example, Figure 16 shows the orbital spectrum of the nucleon states and in Figure 17 the Δ orbital resonances recently computed by de Teramond and Brodsky (2004). The values of $\mathcal{M}^2$ are computed as a function of orbital angular momentum L. The nucleon states with intrinsic spin $S = \frac{1}{2}$ lie on a curve below the nucleons with $S = \frac{3}{2}$. We have chosen our boundary conditions by imposing the condition $\Psi^+(x, z_o) = 0$ on the positive chirality modes for $S = \frac{1}{2}$ nucleons, and $\Psi_\mu^-(x, z_o) = 0$ on the chirality minus strings for $S = \frac{3}{2}$. In contrast to the nucleons, all of the known Δ orbital states with $S = \frac{1}{2}$ and $S = \frac{3}{2}$ lie on the same trajectory. The boundary conditions in this case are imposed on the chirality minus string modes. The numerical solution corresponding to the roots of Bessel functions

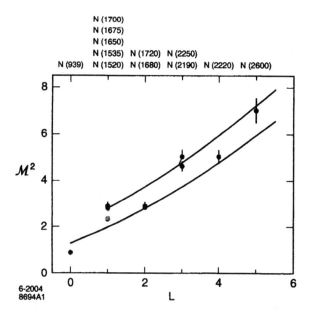

Figure 16. *Nucleon orbital spectrum for a value of* Λ_{QCD} = 0.22 GeV. *The lower curve corresponds to nucleon states dual to spin-$\frac{1}{2}$ string modes in the bulk. The upper curve corresponds to nucleon states dual to string-$\frac{3}{2}$ modes.*

give the nonlinear trajectories indicated in the figures. All the curves correspond to the value Λ_{QCD} = 0.22 GeV, which is the only actual parameter aside from the choice of the boundary conditions. The results for each trajectory show a clustering of states with the same orbital L, consistent with strongly suppressed spin-orbit forces; this is a severe problem for QCD models using one-gluon exchange. The results also indicate a parity degeneracy between states in the parallel trajectories shown in Figure 16, as seen by displacing the upper curve by one unit of L to the right. Nucleon states with $S = \frac{3}{2}$ and Δ resonances fall on the same trajectory (Klempt 2004).

Since only one parameter, the QCD scale Λ_{QCD}, is used, the agreement of the model with the pattern of the physical light baryon spectrum is remarkable. This agreement possibly reflects the fact that our analysis is based on a conformal template, which is a good initial approximation to QCD (Brodsky 2004). We have chosen a special colour representation to construct a three-quark baryon, and the results are effectively independent of N_C. The gauge/string correspondence appears to be a powerful organising principle to classify and compute the mass eigenvalues of baryon resonances.

Acknowledgments

I wish to thank Guenther Rosner, Stanislav Belostotski, David Ireland, Douglas MacGregor and their colleagues at Glasgow University and Edinburgh University for organising the 2004 Scottish Universities Summer School in Physics at St. Andrews. These lec-

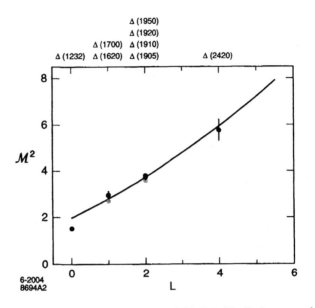

Figure 17. *Delta orbital spectrum for* $\Lambda_{QCD} = 0.22$ *GeV. The Delta states dual to spin-$\frac{1}{2}$ and spin-$\frac{3}{2}$ string modes in the bulk lie on the same trajectory.*

tures are based on collaborations with Carl Carlson, Guy de Teramond, Markus Diehl, Rikard Enberg, John Hiller, Kent Hornbostel, Paul Hoyer, Dae Sung Hwang, Gunnar Ingelman, Chueng Ji, Volodya Karmanov, Peter Lepage, Gary McCartor, Chris Pauli, Joerg Raufeisen, Johan Rathsman, and Dave Robertson.

References

Abramowicz H, 2000, in *Proc 19th Int Symp on Photon and Lepton Interactions at High Energy LP99*, ed J A Jaros and M E Peskin, *Int J Mod Phys A* **15S1** 495; *eConf C* **990809** 495.

Ackerstaff K *et al.* [OPAL Collaboration], 1999, *Eur Phys J C* **7** 571.

Aclander J L S *et al.* , 2004, *Phys Rev C* **70** 015208.

Adloff C *et al.* [H1 Collaboration], 1997, *Z Phys C* **76** 613.

Afanasev A and Carlson C E, 2004, hep-ph/0308163.

Airapetian A *et al.* [HERMES Collaboration], 2000, *Phys Rev Lett* **84** 4047.

Airapetian A *et al.* [HERMES Collaboration], 2005, *Phys Rev Lett* **94** 012002.

Aitala E M *et al.* [E791 Collaboration], 2001a, *Phys Rev Lett* **86** 4773.

Aitala E M *et al.* [E791 Collaboration], 2001b, *Phys Rev Lett* **86** 4768.

Andersson B *et al.* , 1983, *Phys Rep* **97** 31.

Andreev O, 2002, *Phys Rev D* **67** 046001.

Anjos J C *et al.* , 2001, *Phys Lett B* **523** 29.

Antonuccio F *et al.* , 1997, *Phys LettB* **412** 104.

Arneodo M, 1994, *Phys Rep* **240** 301.

Arnold R G *et al.* , 1975, *Phys Rev Lett* **35** 776.

Ashery D, 2002, *Comments Nucl Part Phys* **2** A235.

Avakian H *et al.* [CLAS Collaboration], 2002, *Workshop on Testing QCD through Spin Observables in Nuclear Targets,* Charlottesville, Virginia.

Avakian H *et al.* [CLAS Collaboration], 2004, *Phys Rev D* **69** 112004.

Badalian A M and Veselov A I, 2004, hep-ph/0407082.

Badier J *et al.* [NA3 Collaboration], 1981, *Phys Lett B* **104** 335.

Bakker *et al.* , 2003, *Few-Body Systems* **33** 27.

Baldicchi M and Prosperi G M, 2002, *Phys Rev D* **66** 074008.

Ball P *et al.* , 1995, *Nucl Phys B* **452** 563.

Ball P and Braun V M, 1999, *Nucl Phys B* **543** 201.

Banks T and Zaks A, 1982, *Nucl Phys B* **196** 189.

Bardeen W A, 1980, *Phys Rev D* **21** 1037.

Bari G *et al.* , 1991, *Nuovo Cim A* **104** 1787.

Barone V *et al.* , 2002, *Phys Rep* **359** 1.

Belitsky A V, 2003, *Phys Rev Lett* **91** 092003.

Belitsky A V *et al.* , 2003, *Nucl Phys B* **656** 165.

Beneke M and Braun V M, 1995, *Phys Lett B* **348** 513.

Beneke M *et al.* , 2000, *Nucl Phys B* **591** 313.

Bertsch G *et al.* , 1981, *Phys Rev Lett* **47** 297.

Bjorken J D, 1969, *Phys Rev* **179** 1547.

Bloom E D *et al.* , 1969, *Phys Rev Lett* **23** 930.

Bochna C *et al.* [E89-012 Collaboration], 1998, *Phys Rev Lett* **81** 4576.

Bodwin G T, 1984, *Phys Rev D* **31** 2616; 1985, Erratum *Phys Rev D* **34** 3932.

Boer D, 2002a, *Nuc Phys Proc Suppl* **105** 76.

Boer D, 2002b, *Nucl Phys B* **711** 21.

Boer D *et al.* , 2003, *Phys Rev D* **67** 054003.

Borisov A B *et al.* [HERMES Collaboration], 2002, *Nucl Phys B* **711** 269.

Boschi-Filho H and Braga N R, 2003, *Phys Lett B* **560** 232.

Braun V M *et al.* , 1999, *Nucl Phys B* **553** 355.

Braun V M *et al.* , 2003, *Prog Part Nucl Phys* **51** 311.

Breitweg J *et al.* [ZEUS Collaboration], 1999, *Eur Phys J C* **6** 43.

Brodsky S J and Pumplin J, 1969, *Phys Rev* **182** 1794.

Brodsky S J and Farrar G R, 1973, *Phys Rev Lett* **31** 1153.

Brodsky S J and Farrar G R, 1975, *Phys Rev D* **11** 1309.

Brodsky S J and Chertok B T, 1976a, *Phys Rev Lett* **37** 269.

Brodsky S J and Chertok B T, 1976b, *Phys Rev D* **14** 3003.

Brodsky S J and Lepage G P, 1979, in *Proc Workshop on Current Topics in High Energy Physics,* Cal Tech, Pasadena, California, USA, SLAC-PUB-2294.

Brodsky S J and Drell S D, 1980, *Phys Rev D* **22** 2236.

Brodsky S J *et al.* , 1980a, *Phys Lett B* **93** 451.

Brodsky S J *et al.* , 1980b, *Phys Lett B* **91** 239.

Brodsky S J and Lepage G P, 1981a, *Phys Rev D* **24** 2848.

Brodsky S J and Lepage G P, 1981b, *Phys Rev D* **24** 1808.

Brodsky S J *et al.* , 1983a, *Phys Rev D* **28** 228.

Brodsky S J *et al.* , 1983b, *Phys Rev Lett* **51** 83.

Brodsky S J *et al.* , 1984, *Proc of 1984 Summer Study on the SSC, Snowmass, CO, USA, Jun 1984.*

Brodsky S J *et al.* , 1986a, *Phys Rev D* **33** 1881.

Brodsky S J *et al.* , 1986b, *Phys Lett B* **167** 347.

Brodsky S J and de Teramond G F, 1988, *Phys Rev Lett* **60** 1924.

Brodsky S J and Mueller A H, 1988, *Phys Lett B* **206** 685.

Brodsky S J and Lepage G P, 1989, in *Perturbative Quantum Chromodynamics*, ed A H Mueller, (World Scientific, Singapore), 93.

Brodsky S J and Lu H J, 1990, *Phys Rev Lett* **64** 1342.

Brodsky S J et al. , 1994a, *Nucl Phys B* **441** 197.

Brodsky S J et al. , 1994b, *Phys Rev D* **50** 3134.

Brodsky S J and Lu H J, 1995, *Phys Rev D* **51** 3652.

Brodsky S J and Ma B Q, 1996, *Phys Lett B* **381** 317.

Brodsky S J et al. , 1996, *Phys Lett B* **372** 133.

Brodsky S J and Karliner M, 1997, *Phys Rev Lett* **78** 4682.

Brodsky S J and Huet P, 1998, *Phys Lett B* **417** 145.

Brodsky S J et al. , 1998a, *Phys Rep* **301** 299.

Brodsky S J et al. , 1998b, *Phys Rev D* **57** 245.

Brodsky S J and Hwang D S, 1999, *Nucl Phys B* **543** 239.

Brodsky S J, 2001a, SLAC-PUB-8649 in *At the frontier of particle physics, Handbook of QCD: Boris Ioffe Festschrift*, ed M Shifman, **2*** 1343-1444.

Brodsky S J, 2001b, in *Proc e^+e^- Physics at Intermediate Energies Conf*, ed D Bettoni, eConf **C010430**, W01.

Brodsky S J, 2001c, in *Ise-Shima 2001, B physics and CP violation*, 229-234.

Brodsky S J et al. , 2001a, *Nucl Phys B* **593** 311.

Brodsky S J et al. , 2001b, *Nucl Phys B* **596** 99.

Brodsky S J et al. , 2001c, *Phys Rev D* **63** 094017.

Brodsky S J, 2002a, *Phys Rev D* **65** 114025.

Brodsky S J, 2002b, in *Newport News 2002, Exclusive processes at high momentum transfer* 1.

Brodsky S J and Gardner S, 2002, *Phys Rev D* **65** 054016.

Brodsky S J et al. , 2002a, *Phys Lett B* **530** 99.

Brodsky S J et al. , 2002b, *Nucl Phys B* **642** 344.

Brodsky S J et al. , 2003c, *Phys Lett B* **553** 223.

Brodsky S J, 2003a, in *Nagoya 2002, Strong coupling gauge theories and effective field theories*. 1.

Brodsky S J, 2003b, *Invited talk at Int Conf on colour Confinement and Hadrons in Quantum Chromodynamics - Confinement 2003, Wako, Japan*.

Brodsky S J et al. , 2003a, *Annals Phys* **305** 266.

Brodsky S J et al. , 2003b, *Phys Rev D* **67** 055008.

Brodsky S J, 2004, hep-ph/0408069.

Brodsky S J and de Teramond G F, 2004, *Phys Lett B* **582** 211.

Brodsky S J, Goldhaber A S and Karliner M, 2004a, *work in progress*.

Brodsky S J et al. , 2004b, *Phys Rev D* **69** 076001.

Brodsky S J et al. , 2004c, hep-ph/0409119.

Brodsky S J et al. , 2004d, *Phys Rev D* **69,** 054022.

Brodsky S J et al. , 2004e, *Phys Rev D* **70,** 116003.

Brodsky S J et al. , 2004f, *Nucl Phys B* **703** 333.

Brower R C and Tan C I, 2002, *Nucl Phys B* **662** 393.

Burkardt M, 1989, *Nucl Phys B* **504** 762.

Burkardt M, 1993, *Phys Rev D* **47** 4628.

Burkardt M and Dalley S, 2002, *Prog Part Nucl Phys* **48** 317.

Burkardt M, 2005, *Nucl PhysProc Suppl* **141** 86.

Carbonell J et al. , 1998, *Phys Rep* **300** 215.

Chen Y C et al. , 2004, *Phys Rev Lett* **93** 122301.

Chua C K et al. , 2002, *Phys Rev D* **66** 054004.

Collins J C et al. , 1989, *Adv Ser Direct High Energy Phys* **5** 1.

Collins J C, 1993, *Nucl Phys B* **396** 161.

Collins J C, 2002, *Phys Lett B* **536** 43.

Collins J C, 2003, *Act Phys Polon B* **34** 3103.

Collins J C and Metz A, 2004, *Phys Rev Lett* **93** 252001.

Cornwall J M, 1982, *Phys Rev D* **26** 1453.

Crewther R J, 1972, *Phys Rev Lett* **28** 1421.

Dalley S and Klebanov I R, 1993, *Phys Rev D* **47** 2517.

Dalley S, 2002, *Nucl Phys B (Proc Suppl)* **108** 145.

Dalley S, 2004, hep-ph/0409139.

Date G D *et al.* , 1987, *Nucl Phys B* **283** 365.

DeGrand T, 2004, *Int J Mod Phys A* **19** 1337.

DelDebbio L *et al.* [UKQCD collaboration], 2000, *Nucl Phys Proc Suppl* **83** 235.

Demeterfi K *et al.* , 1994, *Nucl Phys B* **418** 15.

de Teramond G F and Brodsky S J, 2004, hep-th/0409074.

Dirac P A M, 1949, *Rev Mod Phys* **21** 392.

Dokshitzer Y L *et al.* , 1996, *Nucl Phys B* **469** 93.

Drell S D and Yan T M, 1970, *Phys Rev Lett* **24** 181.

Duncan A and Mueller A H, 1980, *Phys Rev D* **21** 1636.

Efremov A V and Radyushkin A V, 1980a, *Theor Math Phys* **42** 97; *Teor Mat Fiz* **42** 147.

Efremov A V and Radyushkin A V, 1980b, *Phys Lett B* **4** 245.

Eller T *et al.* , 1987, *Phys Rev D* **35** 1493.

Farrar G R *et al.* , 1995, *Phys Rev Lett* **74** 650.

Frankfurt L L and Strikman M I, 1989, *Nucl Phys B* **316** 340.

Frankfurt L L *et al.* , 2000, *Found Phys* **30** 533.

Franz M *et al.* , 2000, *Phys Rev D* **62** 074024.

Fritzsch H *et al.* , 1973, *Phys Lett B* **47,** 365.

Furui S and Nakajima H, 2004a, in *Proc 10th Int Conf on Hadron Spectroscopy,* Aschaffenburg, Germany, 2003, *AIP Conf Proc* **717** 685.

Furui S and Nakajima H, 2004b, hep-lat/0403021.

Gamberg L P *et al.* , 2003, *Phys Rev D* **67** 071504.

Geer S, 2003, *J Phys G: Nucl Part Phys* **29** 1485.

Glauber R J, 1955, *Phys Rev* **100** 242.

Goldstein G R and Gamberg L, 2002, *Amsterdam 2002*, ICHEP, 452.

Greensite J, 2003, *Prog Part Nucl Phys* **51** 1.

Gribov V N, 1969a, *Sov Phys JETP* **29** 483; *Zh Eksp Teor Fiz* **56** 892.

Gribov V N, 1969b, *Sov Phys JETP* **30** 709; *Zh Eksp Teor Fiz* **57** 1306.

Gronberg J *et al.* [CLEO Collaboration], 1998, *Phys Rev D* **57** 33.

Gross D J and Wilczek F, 1973, *Phys Rev Lett* **30** 1343.

Gross D J *et al.* , 1998, *Phys Rev D* **57** 6420.

Grunberg G, 2001, *J High Energy Phys* **0108** 019.

Gubser S S *et al.* , 1998, *Phys Lett B* **428** 105.

Gunion J F *et al.* , 1972, *Phys Lett B* **39** 649.

Gunion J F *et al.* , 1973, *Phys Rev D* **8** 287.

Halzen F, 2004, *Nucl PhysProc Suppl* **136** 93; 2005, *Nucl Phys A* **752** 3.

Hamer C J, 1982, *Nucl Phys B* **195** 503.

Harada M *et al.* , 2004, *Phys Rev D* **70** 045015.

Harindranath A and Vary J P, 1987, *Phys Rev D* **36** 1141.

Harindranath A and Vary J P, 1988, *Phys Rev D* **37** 3010.

Harris B W *et al.* , 1996, *Nucl Phys B* **461** 181.

Harvey M, 1981, *Nucl Phys B* **352** 326.

Holt R J, 1990, *Phys Rev C* **41** 2400.

Hornbostel K *et al.* , 1990, *Phys Rev D* **41** 3814.

Howe D M and Maxwell C J, 2002, *Phys Lett B* **541** 129.

Howe D M and Maxwell C J, 2004, *Phys Rev D* **70** 014002.

Hoyer P *et al.* , 1990, *Phys Lett B* **246** 217.

Ioffe B L, 1969, *Phys Lett B* **30** 123.

Jaffe R L and Manohar A, 1990, *Nucl Phys B* **337** 509.

Jaffe R L, 1996, hep-ph/9602236.

Ji X D and Yuan F, 2002, *Phys Lett B* **543** 66.

Ji X D, 2003, *Nucl PhysProc Suppl* **119** 41.

Ji X D *et al.* , 2003, *Phys Rev Lett* **90** 241601.

Jones M K *et al.* [Jefferson Lab Hall A Collaboration], 2000, *Phys Rev Lett* **84** 1398.

Kaluza M and Pauli H C, 1992, *Phys Rev D* **45** 2968.

Karmanov V A and Smirnov A V, 1992, *Nucl Phys B* **546** 691.

Keum Y Y *et al.* , 2001, *Phys Rev D* **63** 054008.

Kharzeev D E and Raufeisen J, 2002, *Campos do Jordao 2002, New states of matter in hadronic interactions* 27, and references therein.

Klempt E, 2000, *Zuoz 2000, Phenomenology of guage interactions* 61.

Klempt E, 2004, hep-ph/0404270.

Kopeliovich B Z *et al.* , 1998, *Phys Lett B* **440** 151.

Krautgartner M *et al.* , 1992, *Phys Rev D* **45** 3755.

Kretzer S, 2004, hep-ph/0408287.

Kuti J and Weisskopf V F, 1971, *Phys Rev D* **4** 3418.

Landshoff P V, 1974, *Phys Rev D* **10** 1024.

Le Diberder F and Pich A, 1992, *Phys Lett B* **289** 165.

Lepage G P and Brodsky S J, 1979a, *Phys Lett B* **87** 359.

Lepage G P and Brodsky S J, 1979b, *Phys Rev Lett* **43** 545.

Lepage G P and Brodsky S J, 1980, *Phys Rev D* **22** 2157.

Ma B Q *et al.* , 2002, *Phys Rev D* **66** 094001.

Maldacena J M, 1998, *Adv Theor Math Phys* **2**, 231; 1999, *Int J Theor Phys* **38** 1113.

Maris P and Roberts C D, 2003, *Int J Mod Phys E* **12** 297.

Martin A D *et al.* , 2004, *Eur Phys J C* **37** 285.

Mattingly A C and Stevenson P M, 1994, *Phys Rev D* **49** 437.

Matveev V A *et al.* , 1973, *Lett Nuovo Cim* **7** 719.

Matveev V A and Sorba P, 1977, *Lett Nuovo Cim* **20** 435.

McCartor G, 1991, *Z Phys C* **52** 611.

McCartor G, 1994, *Z Phys C* **64** 349.

McFarland K S *et al.* , 2003, *Int J Mod Phys A* **18** 3841.

Melic B *et al.* , 2002, *Phys Rev D* **65** 053020.

Mueller A H and Shoshi A I, 2004, *Nucl Phys B* **692** 175.

Muller D, 1995, *Phys Rev D* **51** 3855.

Munger C T *et al.* , 1994, *Phys Rev D* **49** 3228.

Nikolaev N N *et al.* , 2001, *Phys Rev D* **63** 014020.

Ocherashvili A *et al.* [SELEX Collaboration], 2004, hep-ex/0406033.

Olness F *et al.* , 2004, *Eur Phys J C* **40** 145.

Pauli H C and Brodsky S J, 1985a, *Phys Rev D* **32** 1993.

Pauli H C and Brodsky S J, 1985b, *Phys Rev D* **32** 2001.

Pauli H C, 2004, hep-ph/0312300.

Parisi G, 1972, *Phys Lett B* **39** 643.

Peskin M, 1979, *Phys Lett B* **88** 128.

Petti R *et al.* [NOMAD collaboration], 2004, presented at ICHEP.

Polchinski J and Strassler M J, 2001, *Phys Rev Lett* **88** 031601.

Polchinski J and Susskind L, 2001, *Bled 2000/2001, What comes beyond the standard model* **1** 105.

Polchinski J and Strassler M J, 2003, *J High Energy Phys* **0305** 012.

Politzer H D, 1973, *Phys Rev Lett* **30** 1346.

Portheault B, 2004, hep-ph/0406226.

Qiu J W and Vitev I, 2004, hep-ph/0405068.

Rajagopal K and Wilczek F, 2000, *Festschrift B L Ioffe, At the frontier of particle physics*, ed M Schifman, **3** 2061.

Rathsman J, 2001, in *Proc 5th Int Symp on Radiative Corrections (RADCOR 2000)*, ed H E Haber.

Raufeisen J, 2000, PhD thesis, U Heidelberg.

Raufeisen J and Brodsky S J, 2004, *Phys Rev D* **70** 085017.

Rey S J and Yee J T, 2001, *Eur Phys J C* **22** 379.

Rischke D H, 2003, *Prog Part Nucl Phys* **52** 197.

Rossi P *et al.* [CLAS Collaboration], 2005, *Phys Rev Lett* **94** 012301.

Ruspa M, 2004, *Acta Phys Polon B* **35** 473.

Schwarz D J, 2003, *Annalen Phys* **12** 220.

Sivers D W, 1991, *Phys Rev D* **43** 261.

Stodolsky L, 1967, *Phys Rev Lett* **18** 135.

Szczepaniak A *et al.* , 1990, *Phys Lett B* **243** 287.

't Hooft G, 1974, *Nucl Phys B* **75** 461.

van Iersel M and Bakker B L G, 2004, hep-ph/0407318.

Vogt R and Brodsky S J, 1995, *Phys Lett B* **349** 569.

von Smekal L *et al.* , 1997, *Phys Rev Lett* **79** 3591.

Weinstein M, 2004, hep-th/0410113.

Wilson K G, 1976, *Phys Rep* **23** 331.

Witten E, 1998, *Adv Theor Math Phys* **2** 253.

Zeller G P *et al.* [NuTeV Collaboration], 2002, *Phys Rev Lett* **88** 091802; 2003, Erratum *Phys Rev Lett* **90** 239902.

Zhan H *et al.* , 2004, *Phys Rev C* **69** 034302.

Zwanziger D, 2004, *Phys Rev D* **69** 016002.

States beyond $q\bar{q}$ in QCD: diquarks, tetraquarks, pentaquarks and no quarks

Frank E Close

Department of Theoretical Physics, University of Oxford,
Keble Road, Oxford, OX1 3NP, United Kingdom

1 Introduction

These lectures summarise evidence for states beyond those expected in the simple constituent quark model. I focus on the scalar glueball and its mixing with states in the $q\bar{q}$ nonet, and also on correlations in Strong QCD that may form diquarks and seed $qq\bar{q}\bar{q}$ states. Wavefunctions of pentaquarks with careful attention to phase conventions are given and their implications for the flavour dependence of decays described. Some models of the pentaquark candidate $\Theta(1540)$ are critically discussed.

2 Chromostatics

Quarks carry any of three colours. Technically these form the basis representation of SU(3) but for this illustrative introduction we can think of them as three varieties of "positive" charge; antiquarks then have the negative colours. Now use familiar rules of electrostatics suitably generalised: "like charges (colours) repel; opposite attract". This forms a $q\bar{q}$ system: a meson, which is the simplest "chromostatic" analogue of electrostatic positronium.

The existence of **three** colours enable quarks to attract one another making baryons, such as the proton. Identical (positive) colours repel while unlike pairs attract if they form an antisymmetric combination under exchange:

$$Repel: RR; BB; GG; (RB + BR)/\sqrt{2}; (RG + GR)/\sqrt{2}; (BG + GB)/\sqrt{2};$$

$$Attract : (RB - BR)/\sqrt{2}; (BG - GB)/\sqrt{2}; (GR - RG)/\sqrt{2}.$$

These rules of attraction lead to a three body attraction, which in normalised form is

$$[(RB - BR)G + (BG - GB)R + (GR - RG)B]/\sqrt{6}.$$

In the language of SU(3) this is the (colour) singlet state. It is manifestly antisymmetric under the exchange of any pair of colours. The repulsion and attraction for pairs formed respectively the **6** and **$\bar{3}$** representations of colour.

As electric charges within atoms lead to molecules, so qualitatively the colour charges within baryons lead to nuclei. One of the questions that is currently attracting a lot of interest is whether there are analogous "colour molecular" systems involving mesons (two quarks and two antiquarks) or meson and baryon (four quarks and an antiquark, colloquially known as a pentaquark).

The relativistic quantum field theory of electric charge, QED, leads to the existence of photons and prescribes their interactions. Analogously for colour charge we have QCD. This implies the existence of (coloured) gluons. Being coloured, the gluons can mutually interact. This leads to a different evolution of the coupling strength, α_s, in QCD to that in QED. At low energies or long range the interaction becomes strong; perturbative techniques fail and for the solution to the theory from first principles we hope for eventual practical successes from Lattice QCD.

While Lattice QCD can compute many things, as described in the lectures by (Davies 2004), it is still far from confronting decays, mixings and other dynamical questions. Thus we have to resort to models, preferably those that have some overlap with insights developed from Lattice QCD.

In particular the coloured gluons can mutually interact and in principle form bound states of pure glue: glueballs. Lattice QCD (Bali *et al.* 1993, Morningstar and Peardon 1997, Weingarten 1999) and models (Barnes *et al.* 1982) predict that the lightest glueballs are respectively $0^{++} \sim 1.6$ GeV, with 2^{++}, 0^{-+} being the next lightest states at ~ 2GeV.

One of the main unknowns in the standard model is how the gluonic degrees of freedom behave in Strong QCD. The lightest scalar glueball in particular is predicted to lie in the same mass region as the scalar $q\bar{q}$ states. So we can anticipate mixing. To understand what we can expect, first we need to review what we know about the simplest quarkonium spectroscopy.

3   States beyond $q\bar{q}$ in QCD

The QCD effective potential between massive $Q\bar{Q}$ has a short range $1/r$ behaviour, like the familiar Coulomb potential, but behaves different at large r. Lattice simulations imply that it has a linear behaviour in r, and the resulting form

$$V(r) = -\frac{4}{3}\frac{\alpha_s}{r} + \kappa r$$

where $\kappa \sim 1\ GeV/fm$ is the tension of the effective tube of colour flux that this potential suggests. This has led to flux-tube models of mesons. For a review see (Isgur and Paton

1985). Such models make predictions for the excitation not just of the $Q\overline{Q}$ but also the flux-tube, or gluonic degrees of freedom. When these gluonic degrees of freedom are excited in the presence of pre-existing $Q\overline{Q}$ we have "hybrid" mesons. If the gluonic degrees of freedom are excited alone, then we have a "glueball". In general there will be mixing among $Q\overline{Q}$, hybrids and glueballs of the same J^{PC}. One of the challenges for phenomenology is to decode from the spectroscopy what these mixings are.

Before getting into such detail there is even the question of whether such manifestly non-relativistic ideas should be applicable to the light flavours of quark. Empirically Nature seems to have been kind and there is a clear sight of the underlying $Q\overline{Q}$ potential even for light flavours. However, this is not a general truth and my first theme will be to assess when and where the simple model works, and where it does not. The purpose of hadron spectroscopy in practice is to determine the best approximation(s), and should not to be to attempt to shoe-horn a $Q\overline{Q}$ picture onto everything. By defining the realms of applicability of the quark model, of chiral symmetry, or other approximations we may eventually come to a more complete phenomenological description of Strong QCD.

If you take the $b\overline{b}$ or $c\overline{c}$ states from the Data Tables (Particle Data Group 2004) and plot their masses by analogy with atomic notation they tend to be ordered as $^1S_0$; $^3S_1$ then $^3P_{0,1,2}$; $^1P_1$ and $^3D_{1,2,3}$ and so on. Not all states are yet identified and there are also radial excitations seen (eg $1^3S_1\psi(3095)$; $2^3S_1\psi(3685)$; for the $b\overline{b}$ states one even sees clearly up to 4^3S_1). The relative ordering and mass gaps between these states are in accord with a linear potential, which supports the results of Lattice QCD.

Notice also that there is a systematic tendency for $m(^1S_0) < m(^3S_1)$ for any combination of flavours. This mass gap is small for heavy flavours and large for light flavours. It is in accord with expectations from QCD where there is a chromomagnetic perturbation (Zeldovitch and Sakharov 1967, de Rujula *et al.* 1975), analogous to the hyperfine splitting in QED. The operator that causes this is the contact interaction between the colour-magnetic dipoles of the Q and $\overline{Q}$. Its form is

$$\delta H = -\sum_{i>j} \vec{\lambda}_i \cdot \vec{\lambda}_j \vec{S}_i \cdot \vec{S}_j V_{ij}/m_i m_j$$

where m_i is the effective mass of quark i, $\vec{S}$ the spin operator and V_{ij} a hyperfine interaction with the flavour dependence made explicit in the inverse mass dependence, being larger for light flavours than for heavy, as realised by data. There are also analogues of the spin-orbit and tensor interactions, such as separate the $^3P_{0,1,2}$ states, but their details are more subtle, not fully understood quantitatively and are not the theme of these lectures. My starting point is to focus on the scalar mesons and their $^3P_{1,2}$ partners.

First, note that in this heavy flavour sector there are clearly established scalar mesons $c\overline{c}$ and $b\overline{b}$. They behave as canonical $^3P_0$ states which partner $^3P_{1,2}$ siblings. Their production, eg in radiative transitions from 2^3S_1 states where the relative rates (Particle Data Group 2004) are consistent with being dominantly E1, and their decays, into 1^3S_1 or light hadrons, are all in accord with this. There is nothing to suggest that there is anything "exotic" about such scalar mesons.

These states are readily identified because they are below threshold for decays into $D\overline{D}$ or $B\overline{B}$ pairs. Once one is above threshold the $Q\overline{Q}$ states become harder to identify, their Fock states being contaminated by the di-meson continuum. For light flavours, with the exception of the lightest 0^{-+}, all candidates are above threshold for strong decays. There appears to be an empirical rule (Close 2004) that the $q\bar{q}$ seed shows through so long as the $q\bar{q}$ are in an orbital angular momentum state which is lower than that of the partial waves for hadron decays: $L_q < L_H$. For example $\rho \rightarrow \pi\pi$ has $L_H = 1$ whereas $\rho =^3 S_1$ has $L_q = 0$; the $q\bar{q}$ content dominates the spectroscopy. Contrast this with 0^{++} where $q\bar{q}$ has $L_q = 1$ whereas $\pi\pi$ has $L_H = 0$; the $q\bar{q}$ content of the scalar states are hidden. For 1^{++} the $a_1 \rightarrow \rho\pi$ in S-wave obscures the $L_q = 1$ but other members of the nonet have helped to identify the states. For 2^{++} the hadron decays are D-wave and the $L_q = 1$ states are clear.

This suggests that the effective Fock state description of hadrons has S-wave combinations as the leading term, followed by P-wave and higher waves. For 1^{--} this is achieved with $q\bar{q}$; with 0^{++} it is achieved by 0^-0^- dimesons or 0^+0^+ diquark-antidiquark; for 1^{++} it can be done with 0^-1^- and so on. The major decide seems to be whether these channels are open or closed. For the strange axial mesons the $K^*\pi$ S-wave is open and so this hadronic loop can couple to both $^1P_1$ and $^3P_1$ states; the $K_1(1270)$ and $K_1(1400)$ do appear to be strongly mixed. Why this rule should work (*eg* why the imaginary part and not the real part is so critical) is an unanswered question. So bearing this caveat in mind, let's move on.

Empirically for light flavours there are clearly identified $^3P_{1,2}$ nonets which call for analogous $^3P_0$ siblings. If we list the light flavoured mesons from the Particle Data Group (2004) we find that there are clear nonets that fit with the $q\bar{q}$ classification so long as the above rule of "the lowest partial wave dominates the Fock state" is used as a guide.

The light scalars empirically stand out as singular. Part of the problem is the above rule: the hadron decays are in S-wave. The question is whether, nonetheless, we can disentangle the spectroscopy.

The interpretation of the nature of the lightest scalar mesons has been controversial for over thirty years. There is still no general agreement on where are the $q\bar{q}$ states, whether there is necessarily a glueball among the light scalars, and whether some of the too numerous scalars are multiquark, $K\overline{K}$ or other meson-meson bound states. These are fundamental questions of great importance in particle physics. The mesons with vacuum quantum numbers are known to be crucial for a full understanding of the symmetry breaking mechanisms in QCD, and presumably also for confinement.

Theory and data are now converging that QCD forces are at work but with different dynamics dominating below and above 1 GeV/c^2 mass. It is as if the $q\bar{q}$ states are sitting in an S-wave di-meson continuum, which leads to mixing and repulsive shifts of the eigenvalues. This picture fits qualitatively with the experimental proliferation of light scalar mesons as two nonets, one just below the 1 GeV region (a meson-meson nonet) and another one near 1.5 GeV (a $q\bar{q}$ nonet), with evidence for glueball degrees of freedom.

I will describe this as a working hypothesis and attempt to define experiments that can refute or support it. Let's therefore consider the scalar mesons above and then below 1 GeV.

4 No quarks?

Lattice QCD predictions for the mass of the lightest (scalar) glueball are now mature. In the quenched approximation the mass is ~ 1.6 GeV (Bali 1993, Morningstar 1997, Weingarten 1999). Flux tube models imply that if there is a $q\bar{q}$ nonet nearby, with the same J^{PC} as the glueball, then $G - q\bar{q}$ mixing will dominate the decay (Amsler and Close 1995). This is found more generally (Anisovich 1998) and recent studies on coarse-grained lattices appear to confirm that there is indeed significant mixing between G and $q\bar{q}$ together with associated mass shifts, at least for the scalar sector (McNeile and Michael 2001).

Furthermore the maturity of the $q\bar{q}$ spectrum tells us that we anticipate the $0^{++}q\bar{q}$ nonet to occur in the 1.2 to 1.6 GeV region. There are candidates: $a_0(\sim 1400)$; $f_0(1370)$; $K(1430)$; $f_0(1500)$ and $f_0(1710)$.

One immediately notes that if all these states are real there is an excess of isocalars, precisely as would be expected if the glueball predicted by the lattice is mixing in this region. Any such states will have widths and so will mix with a scalar glueball in the same mass range. It turns out that such mixing will lead to three physical isoscalar states with rather characteristic flavour content (Close and Teper 1996, Close and Kirk 2001a, Close and Kirk 2000). Specifically; two will have the $n\bar{n} \equiv (u\bar{u} + d\bar{d})/\sqrt{2}$ and $s\bar{s}$ in phase ("singlet tendency"), their mixings with the glueball having opposite relative phases; the third state will have the $n\bar{n}$ and $s\bar{s}$ out of phase ("octet tendency") with the glueball tending to decouple in the limit of infinite mixing.

$$\begin{pmatrix} Meson & G & s & n \\ 1710: & + & + & + \\ 1500: & - & + & - \\ 1370: & - & + & + \end{pmatrix}$$

There are now clear sightings of prominent scalar resonances $f_0(1500)$ and $f_0(1710)$ and, probably also, $f_0(1370)$. Confirming the resonant status of the latter is one of the critical pieces needed to clinch the proof. The production and decays of these states are in remarkable agreement with this flavour scenario (Close and Kirk 2000).

$$\begin{pmatrix} Meson & G & s & n \\ 1710: & 0.4 & 0.9 & 0.1 \\ 1500: & -0.6 & 0.3 & -0.7 \\ 1370: & -0.7 & 0.15 & 0.7 \end{pmatrix}$$

The fact that mass mixing and also meson decays are consistent with this set of relative phases is interesting. The numerical values should not be taken seriously; the errors on them are probably considerable, but the relative phases and separation of "large, medium, small" is probably reliable. As example, a recent model calculation by Giacosa *et al.* (2005) finds a similar pattern (their glueball phase is defined opposite to here and so I have changed it to conform with that used here).

$$\begin{pmatrix} Meson & G & s & n \\ 1710: & 0.3 & 0.95 & 0.1 \\ 1500: & -0.8 & 0.3 & -0.6 \\ 1370: & -0.6 & 0.1 & 0.8 \end{pmatrix}$$

The question now is what further experimental tests we can do to test this.

There are two examples in ψ decays that may add to our knowledge. First $\psi \to M_1 M_2$ where $M_{1,2}$ are mesons. If M_1 is an ideal flavour combination, such as $\phi = s\bar{s}$ then the folklore is that M_2 will be produced via its $s\bar{s}$ content as this leads to a flavour connected diagram. Similarly $M_1 \equiv \omega$ selects out $n\bar{n}$ for the M_2. The test of this hypothesis has been when $M_2 \equiv 2^{++}$; this nonet consists of ideal states $a_2(1320)$; $f_2(1270) \equiv n\bar{n}$; $f_2(1525) \equiv s\bar{s}$ and therefore is rather clean and confirms the dominance of the "hairpin" diagram. However, the case $M_2 = 0^{++}$ has no simple solution. Indeed, some channels which ought to have been dominant appear to be all, but absent (Jin 2004).

For example: $f_0(1370)$ has strong affinity for $\pi\pi$ and hence $n\bar{n}$ in its wavefunction yet is not seen in $\psi \to \omega\pi\pi$. This being anomalous does not require one to suppose that there is $n\bar{n}$ alone; multiquark components containing non-strange flavours ought to be enough. One explanation could be that some other contribution leads to destructive interference. The G component in the $f_0(1370)$ wavefunction is a natural candidate for this and it has even been predicted (Close and Zhao 2004a) that the strength of $b(\psi \to \phi G) \sim 10^{-3}$; if this also applies to $b(\psi \to \omega G) \sim 10^{-3}$ then the destructive interference becomes plausible. The test will be to see if the relative phases of G and flavoured components are in line with the observed pattern of suppressed and observed decays $\psi \to \omega/\phi f_0$. This will become clear soon as over 10 billion ψ decays will become available from BES in Beijing and CLEO-c at Cornell within a few years.

These huge data sets will provide another route into weighing the flavour and glue content of the 0^{++} and indeed other $C = +$ mesons. The suggestion here is to use the ideal mixing of the vector mesons as a flavour filter by studying the radiative decays $0^{++} \to \gamma(\rho; \omega; \phi)$. The transitions to ρ and ϕ respectively weigh the $n\bar{n}$ and $s\bar{s}$ content of the initial states. The problem has been that radiative decays have small branching ratios, $\sim 1\%$, and there has not been a reliable source of a large number of scalar mesons with which we can study the radiative decays. This is now about to change. For every billion ψ, $\psi \to \gamma M$ to individual $C = +$ mesons are typically of order 10^{-3}, giving a million events per meson. Even with a pessimistic branching ratio of $\sim 10^{-3}$ for the subsequent radiative decay $M \to \gamma V (= \rho; \phi)$, given the promised N-billion ψ decays, we may hope for the order of N-thousand events per channel and the ability to "weigh" the flavour contents. This can be applied to any $C = +$ meson produced in $\psi \to \gamma M$.

5 Diquarks and tetraquarks

As pointed out by Jaffe (1977) and Jaffe and Low (1979), there is a strong QCD attraction among qq and $\bar{q}\bar{q}$ in S-wave, 0^{++}, whereby a low lying nonet of scalars may be expected. As far as the quantum numbers are concerned these states will be like two 0^{-+} $q\bar{q}$ mesons in S-wave. In the latter spirit, Weinstein and Isgur (1982, 1983) had noticed that they could motivate an attraction among such mesons, to the extent that the $f_0(980)$ and $a_0(980)$ could be interpreted as $K\overline{K}$ molecules.

The relationship between these is being debated (Tornqvist 1995, Achasov and Shestakov 1997, Boglione and Pennington 1997, Achasov *et al.* 1979, Krehl *et al.* 1997), but while the details remain to be settled, there is a rather compelling message of the data as follows (Close and Tornqvist 2002). Below 1 GeV the phenomena point clearly towards an S-wave attraction among two quarks and two antiquarks (either as $(qq)^{\overline{3}}(\bar{q}\bar{q})^3$, or $(q\bar{q})^1(q\bar{q})^1$ where superscripts denote their colour state), while above 1 GeV it is the P-wave $q\bar{q}$ that is manifested. There is a critical distinction between them: the "ideal" flavour pattern of a $q\bar{q}$ nonet on the one hand, and of a $qq\bar{q}\bar{q}$ or meson-meson nonet on the other, are radically different; in effect they are flavoured inversions of one another. Thus whereas the former has a single $s\bar{s}$ heaviest, with strange in the middle and and I=0; I=1 set lightest ("ϕ; K; ω, ρ-like"), the latter has the I=0; I=1 set heaviest ($K\overline{K}$; $\pi\eta$ or $s\bar{s}(u\bar{u} \pm d\bar{d})$) with strange in the middle and an isolated I=0 lightest ($\pi\pi$ or $u\bar{u}d\bar{d}$) (Jaffe 1977, Jaffe amd Low 1979, Weinstein and Isgur 1982, Weinstein and Isgur 1983).

The phenomenology of the 0^{++} sector appears to exhibit both of these patterns with ~ 1 GeV being the critical threshold (Close and Tornqvist 2002). Below 1 GeV the inverted structure of the four quark dynamics in S-wave is revealed with $f_0(980)$; $a_0(980)$; κ and σ as the labels. One can debate whether these are truly resonant or instead are the effects of attractive long-range t−channel dynamics between the colour singlet 0^{-+} $K\overline{K}$; $K\pi$; $\pi\pi$, but the systematics of the underlying dynamics seems clear.

As concerns the region below 1 GeV, the debate centres on whether the phenomena are truly resonant or driven by attractive t-channel exchanges, and if the former, whether they are molecules or $qq\bar{q}\bar{q}$. The phenomena are consistent with a strong attraction of QCD in the scalar S-wave nonet channels. The difference between molecules and compact $qq\bar{q}\bar{q}$ will be revealed in the tendency for the former to decay into a single dominant channel – the molecular constituents – while the latter will feed a range of channels driven by the flavour spin Clebsch-Gordans. For the light scalars it has its analogue in the production characteristics.

The picture that is now emerging from both phenomenology (Close and Kirk 2000, Close and Kirk 2001b, Aloisio *et al.* 2001) and theory (Jaffe 2004) is that both components are present. As concerns the theory (Jaffe 2004), think for example of the two component picture as two channels. One, the quarkish channel (QQ) is somehow associated with the $(qq)_{\overline{3}}(\bar{q}\bar{q})_3$ coupling of a two quark-two antiquark system, and is where the attraction comes from. The other, the meson-meson channel (MM) could be completely passive (*eg* no potential at all). There is some off-diagonal potential which flips that system from the QQ channel to MM. The way the object appears to experiment depends on the strength of the attraction in the QQ channel and the strength of the off-diagonal potential. The nearness of the f_0 and a_0 to $K\overline{K}$ threshold suggests that the QQ component cannot be too dominant, but the fact that there is an attraction at all means that the QQ component cannot be negligible. So in this line of argument, a_0 and f_0 must be superpositions of four-quark states and $K\overline{K}$ molecules.

A question is whether these correlations may seed a dynamical breakdown of the simple $q\bar{q}$ model. The relative strengths of the colour potentials are (minus sign being attractive)

$$q_1 q_2 : \overline{3_c} = -4/3; 6_c = +2/3, \tag{1}$$

$$q_1 \bar{q}_2 : 1_c = -8/3; 8 = +1/3. \tag{2}$$

There is also the spin dependent correlation

$$S \cdot S \rightarrow +1/2\,(S=1); -3/2\,(S=0). \tag{3}$$

Thus there are attractive correlations among pairs $\overline{3_c}$; $S = 0$ and 6_c; $S = 1$. Each of these is symmetric in colour-spin and so, for relative S-wave, the Pauli principle implies they are flavour antisymmetric. Thus they consist of different flavours: (ud); (us) : (ds) and their extensions to heavy flavours.

In the lectures by (Weise 2004) we saw how the breaking of chiral symmetry in a Nambu Jona-Lasinio model can lead to an effective "constituent" mass and also set the scale of the diquark mass. From the gap equation, the current quark mass m_0 gains a constituent mass M through interactions with the self consistent field

$$M = m_0 - 2g\langle\bar{\psi}\psi\rangle.$$

Using QCD sum rules to set the scale $\langle\bar{\psi}\psi\rangle = -(250~MeV)^3$ fixes $M_{(u,d)} \sim 0.4~GeV$, $M_s \sim 0.55~GeV$ and a constituent quark size $\sim \frac{1}{3}$fm. The Bethe-Salpeter equation in the diquark channels leads to (Vogl and Weise 1991)

$$M_{[ud]} = 0.25 - 0.3~GeV; M_{[us]} = 0.5~GeV. \tag{4}$$

If this is realised in Nature then it implies that adding a light flavour in the attractive state costs nothing in mass, or even that it is energetically favoured for quarks to correlate like this.

This immediately raises the question whether baryons are quasi-two body systems whose colour structure parallels that of mesons. In S-wave the Δ has any qq in $S = 1$; we denote this symmetric combination (qq). Relative to the $N - \Delta$ average, this implies an upward mass shift of 150 MeV associated with (qq). The nucleon has its quark pairs in a 50:50 mixture of $[qq]$ (S=0) and (qq). The $-3:1$ relative energy shift (Equation 3) would suggest that the $[ud]$ is shifted down by 450 MeV relative to their individual values. However, one needs to take care here; in the Δ there are three possible combinations of $q(qq)$ and so in the symmetric limit the shift is only 50 MeV per (qq), and correspondingly only -150 MeV down for $[ud]$. The $[ud]$ is in a relative S-wave to the remaining q, which is itself a colour source and so some polarisation of the simple diquark $[ud]$ and (qq) may be expected.

Any clear signature for diquark correlations may only show up in higher partial waves. Surprisingly this does appear to ensue and with remarkable consistency with the quantitative masses from the NJL model above.

First I summarise the baryon representations for L=1. Denote by ρ, λ the antisymmetric =[..] and symmetric =(..) combinations for spin (χ) and flavour (ϕ). The spatial oscillators are ψ^ρ and ψ^λ for L=1 excitation. The former corresponds to exciting the diquark constituents to L=1 whereas the latter excites the relative coordinate of the diquark-quark state. The SU(6) combinations associated with the ψ^λ oscillator are (Close 1979, Isgur and Karl 1978)

$$
\begin{aligned}
^2 8 &: \phi^\rho \chi^\rho - \phi^\lambda \chi^\lambda = N(1520;\ 1535), \\
^2 10 &: \phi^S \chi^\lambda = \Delta(1620;\ 1700), \\
^4 8 &: \phi^\lambda \chi^S = N(1650;\ 1700;\ 1675), \\
^2 1 &: \phi^A \chi^\rho = \Lambda(1405;\ 1520).
\end{aligned}
\tag{5}
$$

There is the well-known (Isgur and Karl 1978) separation of masses driven by the different weights of the [..] and (..) spin states. The (..) states in the second line are all more massive than the 50:50 mix in line one and the pure [..] in the strange states of the third line.

Now focus on the maximally stretched states $\Lambda(1520) : J^P = 3/2^-$ state and compare with the L=1 strange meson $K(1420) J^P = 2^+$. This suggests the relative masses $[ud] - u \sim 90$ MeV. Comparison with the $J^P = 2^+ : f_2(1525)$ $s\bar{s}$ state suggests that when $L = 1$

$$
M_{[ud]} \sim M_s;\ L = 1.
\tag{6}
$$

I chose these comparisons as the other states are affected by the S-wave mass shifts or mixings as discussed earlier. The Λ channel with $1/2^-$ couples to KN or $\pi\Sigma$ in S-wave; the $\Lambda(1405)$ is anomalously light and at 30 MeV below KN threshold is suspected of being a KN bound state. The 0^+, 1^+ strange mesons also are disturbed by S-waves. The 0^+ involves the $\kappa(800)$ phenomenon and more general coupling to $K\pi$; the 1^+ states occur in $^1P_1$ and $^3P_1$, couple in S-wave to $K^*\pi$, and mix. So it is only the $\Lambda(1520)$ and $K_2(1430)$; $f_2(1525)$ that merit comparison within the strict quark Fock states as qqq or $q[qq]$ versus $q\bar{q}$. Had we ignored this we would have obtained a lighter mass for the $[ud]$; that in Equation 6 is in accord with what follows from the $N - \Delta$ mass gap in S-wave.

Now look at the L=2 states. Baryons form a 56-plet for which there are candidates

$$
\begin{aligned}
^2 8 &: \phi^\rho \chi^\rho + \phi^\lambda \chi^\lambda = N(P_{13} 1720;\ F_{15} 1680), \\
^4 10 &: \phi^S \chi^S = \Delta(P_{31}(1910);\ P_{33} 1920;\ F_{35} 1905;\ F_{37} 1950).
\end{aligned}
\tag{7}
$$

The mass gap is similar to that of the ground state L=0 $N - \Delta$. As there is no contact interaction between the constituents that are in relative L=2, the mass gap can be associated with that of the intrinsic diquarks $(ud) - [ud]$. Compare with the L=2 $q\bar{q}$ states. The $^3D_{1,3}$ states are known: $\rho_1(1700)$; $\rho_3(1690)$. The decay $\rho_1 \to \rho + 0^+$ in relative S-wave may distort the mass of the ρ_1. The comparison with the ρ_3 suggests that

$$
M_{[ud]} \sim M_u;\ L = 2
\tag{8}
$$

This result, (Equation 8), is in accord with the prediction of Equation 4. It is not in practice possible to draw more solid conclusions going to higher L-states as the opening of decay channels in S-waves becomes more prevalent in all but the highest spin states. Nonetheless, there does appear to be some suggestion, as we have seen above, that the energy cost in forming a diquark in the [..] configuration is minimal. Similar results have also been noted by Wilczek (2004).

These hints suggest that we might be entitled to take the results of Equation 4 seriously and consider their implications. The mass of [ud] being similar to that of a single q implies that the lightest mesons are either $q\bar{q}$ or [qq][qq] in S-wave. The masses of the latter are then

$$m[ud][ud] \sim 500 - 600MeV \equiv \sigma(500), \tag{9}$$

$$m[ud][qs] \sim 800MeV \equiv \kappa(800), \tag{10}$$

$$m[qs][qs] \sim 1000MeV \equiv f_0(980); \ a_0(980). \tag{11}$$

This is similar to the results of Jaffe, earlier, though the approach differs. We have generated the actual mass scales by comparison with other phenomenology and also using an unusually large gap for the $[us] - [ud]$ diquark masses. This diquark motivated approach can be generalised to heavy flavours and to more complicated systems. As regards the latter we can imagine two diquarks (in a relative P-wave due to Bose symmetry (Jaffe and Wilczek 2003) forming an overall colour singlet with an extra antiquark (forming a pentaquark) or with another diquark (forming a state with the same overall quantum numbers as the deuteron).

With four flavours we can form [..] combinations for either diquark or antidiquark and hence three manifestly exotic combinations for which there is no possibility of annihilation into $c\bar{q} + gluons$: $[cu][\bar{s}\bar{d}]$; $[cd][\bar{s}\bar{u}]$; $[cs][\bar{u}\bar{d}]$.

The first two transform like a normal $D_s(c\bar{s})$ but with I=1; they are partnered by another exotic superposition $[cu][\bar{s}\bar{u}] \pm [cd][\bar{s}\bar{d}]$ where the relative minus phase protects against annihilation. The third state is also exotic having both positive charm and strangeness (Lipkin 1977).

If these states are significantly heavier than the thresholds for S-wave meson pairs accessible by rearrangement, such as $D_s\pi$ for the first two states and DK for the third state, then the collapse into these channels, analogous to the σ and κ for light flavours, and dissolving of the $[cq][\bar{q}\bar{q}]$ configurations will occur.

The $[cs][\bar{u}\bar{d}]$ state is especially interesting if the [ud] diquark (and antidiquark) are light as its state could be near or even below the DK threshold at 2.36 GeV. Let us be pessimistic to generate as large a mass as possible for this tetraquark configuration. The [cs] mass involves a colour and spin attraction; it is unlikely to be heavier than that of $c\bar{s}$ in $^3S_1$, namely 2010 MeV. Thus if Equation 4 is a guide it is possible that this tetraquark could lie below the DK threshold. In this case the decay would involve the weak interaction, via $s \to ue^-\nu$, with subsequent annihilation of $u\bar{u} \to \pi$: hence $[cs][\bar{u}\bar{d}] \to D\pi^0e^-\nu$. Observation of such a metastable state would be a radical discovery.

6 Heavy tetraquarks

There are hints that tetraquark (diquark - antidiquark) systems occur involving heavy flavours, though realised as loosely bound combinations of colour singlet mesons. In the charmonium sector we have $X(3871.8)$ seen as a narrow state decaying into $\psi\pi\pi$. Being narrow yet above $D\overline{D}$ threshold suggests that it is forbidden to decay into $D\overline{D}$. This could occur if it is a hybrid charmonium, though the mass scale for these is expected to be somewhat above 4 GeV. More directly, it would be forbidden if its J^{PC} were any of $0^{-+}, 2^{--}, 2^{-+}, 3^{--}$ among others; I have isolated these as they could *a priori* be among the missing states of charmonium. Unfortunately, or tantalisingly, each of these runs into problems with other data. Their masses are wrong, or their electromagnetic widths or angular distributions do not fit with those expected for such charmonium states (Olsen *et al.* 2005).

The conclusion is either that the quark model description of charmonium has been exposed as a fraud, or that the state is not simply charmonium. The latter is suspected to be the case in part driven by the remarkable coincidence between its mass and that of the threshold for $D^0 D^{0*}$ which agree to better than one part in 10,000. (Close and Page 2004, Tornqvist 2004) suggest that it is a molecular or tetraquark bound state of these mesons in S-wave; thus 1^{++}.

Indirect support for this comes from the fact that such a state is not an isospin eigenstate and so the decays would violate isospin radically. The decay to $\psi\pi\pi$ would have the $\pi\pi$ forming a ρ and not a σ; the very limited data are consistent with this. This 1^{++} assignment is supported by the observation of $\psi\omega$ decay (Olsen 2004); the ratio of $\psi\rho : \psi\omega$ will become a sharp test of isospin violation and the molecular interpretation. Further tests include verifying that there is no $\psi\pi^0\pi^0$, which would be forbidden for the ρ but allowed for σ. Also the hadronic decays into *eg* $K\overline{K}\pi$ will be dominated by neutral $K^0\overline{K}^0\pi$ relative to $K^+K^-\pi$.

7 Pentaquarks and exotic baryons

This year we have seen several hadrons announced that do not fit easily with the simple valence picture of $q\bar{q}$ or qqq mesons and baryons. With hindsight one might wonder why it took so long. This simple picture exploits degrees of freedom that transform like the fields of L_{QCD} but are not identical to them. As we have discussed above, two quarks attract one another in $\overline{\mathbf{3}}_{\mathbf{c}}$ with about the strength of $q\bar{q}$ coupled to colour singlet and so should play a significant role in in generating the colour degrees of freedom in Strong QCD. Then for light flavours the lectures by Weise (2004), an old paper (Vogl and Weise 1991) and possibly phenomenology suggested that the effective mass of the antisymmetric $[ud]$ pair, "scalar diquark", could be comparable to that of a single $q \equiv u, d$. If so it is energetically as easy to make colour singlets from $[qq][\overline{q}\overline{q}]$ as from $q\bar{q}$, and we have seen that this is realised empirically.

So if we take all this as evidence for such strong correlations, why stop here? The idea that baryons may emerge naturally as excitations of a quasi-two centred system has been resurrected (Wilczek 2004). In turn this raises questions about other energetically favoured examples of such correlations. Two $[ud][ud]$ couple attractively to $\mathbf{3}_{\mathbf{c}}$ (probably forced into $L = 1$ by Bose symmetry (Jaffe and Wilczek 2003) and so need a further $\overline{3_c}$ to saturate. One way would be to add a third $[ud]$ (and so the diquark mass cannot be too low if we are not to end up with a state more stable than the deuteron!) or a $\bar{q}$. If the latter is $\bar{s}$ we have a manifestly exotic strange baryon with the quantum numbers of the Θ.

But why stop at the diquark? A $[ud\bar{s}]$ combination also is strongly attractive and with different flavours does not suffer annihilation via gluons. This enables one to construct the Θ quantum numbers with a quasi-two centred system, $[ud\bar{s}][ud]$ with $L = 1$ needed to keep would-be repulsive correlations apart (Karliner and Lipkin 2004a). Completing the simple quasi-two centred states are attractive combinations such as $[ud\bar{s}](\bar{s})$. The phenomenology of these includes flavour 10 and $\overline{10}$ mesons, which might relate to exotic mesons with $J^{PC} = 1^{-+}$ at 1.4–1.6 GeV (Burns and Close 2004), but the detailed similarities and differences with the $[qq][\overline{qq}]$ remains to be investigated.

Most attention has focussed on the Θ and its implications for correlations as above. I shall not review this literature due to limitations of space and because much of the empirical situation is covered in the lectures of Burkert (2004), but I shall raise some questions that remain to be answered.

If $[ud]$ diquarks form that are lightweight, then energetically one should consider the possibility of $[ud][ud]\bar{q}$ states. To the extent that the $[ud]$ is a tight scalar diquark, the pair may follow Bose statistics and hence be in a P-wave (Jaffe and Wilczek 2003). If the $\bar{q} \equiv \bar{s}$ then one has a manifestly exotic combination: a baryon with positive strangeness. There will be two other exotic combinations, $[ds][ds]\bar{u}$ and $[us][us]\bar{d}$, which are strangeness -2 Ξ states with electric charge -2 or $+1$, inaccessible to normal Ξ states. These ideas have come into focus this year due to the possible observation of states with such correlations. The experimental situation is reviewed elsewhere in this school; I shall briefly set the scene and then make a critical summary of some models that have arisen.

The original conception of the constituent quark model, and of our modern picture, was based on the observation that hadrons exist with (apparently) unlimited amounts of spin, but with only very restricted amounts of electric charge and strangeness. In particular, all baryons seen hitherto in 60 years of research carry either no strangeness (like the proton and neutron) or **negative** amounts (like the Λ, Σ and Ω^-).

During 2003 evidence emerged from a range of experiments that a metastable particle known as the theta baryon may exist (Close 2004, Burkert 2004). Described most simply: it is like a heavier version of the proton but which possesses **positive** strangeness in addition to its positive electrical charge and it is denoted as Θ^+. This makes it utterly novel. As the absence of "positive strangeness baryons" in part is what helped establish the quark model in the first place, the claims are indeed radical.

QCD allows more complicated clusters of quarks or antiquarks and there is good evidence for this. For example, when the proton is viewed at high resolution, as in inelastic electron scattering, its wavefunction is seen to contain configurations where its three "valence" quarks are accompanied by further quarks and antiquarks in its "sea". The three quark configuration is thus the simplest required to produce its overall positive charge and zero strangeness. The question thus arises whether there are baryons for which the minimal configuration cannot be satisfied by three quarks. A baryon with a positive amount of strangeness would be an example; in this case the positive strangeness could only be produced by the presence of a strange-antiquark $\bar{s}$, the overall baryon number requiring four further quarks to accompany it. Thus we would have three quarks accompanied by an additional quark and antiquark, making what is known as a "pentaquark".

The existence of such a state is not of itself necessarily radical; it is the light mass and, most dramatically, its narrow width that tantalise. The challenge is to explain why such a "pentaquark" has such unexpected metastability: whereas conventional hadrons that decay by the action of strong forces have widths of order of hundreds of MeV, that of the Θ^+ is less than 10 MeV, perhaps no more than 1 MeV if consistency is to be maintained with phase shift analyses of extensive data on the interactions of kaons and nucleons.

There may be a sense of *deja-vu* here in that strange particles were so called, because of their strange behaviour: they were produced readily in strong interactions, but had metastability due to their decays being controlled by the weak interaction. Thus one suggestion has been that the Θ is but one of a family of particles, each with positive strangeness but with electrical charges that span the range from +3 to –1; such a family with a mass around 1550 MeV would be too light to decay by the strong interactions as all isospin conserving pathways would be forbidden by energy conservation, leaving the weak interaction to cause their decays.

However, to date there is no sight of states such as Θ^{++} that are isospin partners of the Θ^+. The most immediate concern must be to establish not simply the spin and parity of the Θ, or other examples like it, but to verify that it indeed exists and is not some artefact. A programme of photoproduction at Jefferson Laboratory may begin to answer some of these questions.

Whether or not it turns out to be real, the stimulus to theory has already reinvigorated interest in the chiral soliton and Skyrme models (which even predicted that such a state should exist, at such a mass) and the pentaquark dynamics of the quark model. The chiral soliton model and the quark model are both rooted in QCD though their relation has been obscure.

When the chiral soliton model is extended to incorporate strangeness, the lightest baryon families consist of the well established **8** with $J^P = 1/2^+$ (which includes the nucleons) and a **10** with $J^P = 3/2^+$ (which includes the Δ and Ω^-) and a further family of ten (technically transforming like a $\overline{\mathbf{10}}$, in the group structure of SU(3)) with spin-parity $1/2^+$. This is the family that can not be formed from three quarks and requires the pentaquark as a minimum. By identifying the members of this $\overline{\mathbf{10}}$ with nucleon quantum numbers with the $P_{11}(1710)$ [which is itself surprising as such a state is forbidden by U-spin to be photoexcited from a proton, in contradiction to data (Particle Data Group 2004)], (Diakanov *et al.* 1997) predicted a mass of $\sim$ 1540 MeV for the strangeness +1 Θ.

7.1 Symmetries, wavefunctions and selection rules for pentaquarks

The hadron decays of non-exotic members of the $\overline{\mathbf{10}}$, in particular those of $\Xi^{0,-}$ are especially sensitive to the interquark dynamics in pentaquark models. A specific example has been discussed in Jaffe and Wilczek (2004) but I shall show here that there is a more extensive set of relations and selection rules that arise in pentaquark models and which can discriminate among various dynamic and mixing schemes (Close and Dudek 2004).

In particular the relative strengths of decays $\Xi^- \to \Xi^-\pi^0 : \Xi^0\pi^-$ test $\overline{\mathbf{10}}$ - $\mathbf{8}$ mixing (Jaffe and Wilczek 2003); $\Xi^- \to \Lambda K^- : \Sigma^0 K^-$ have selection rules that test the decay dynamics that have been hypothesised (Close 2004, Jennings and Maltman 2004, Carlson *et al.* 2004) to suppress the pentaquark widths; and $\Xi \to \Xi^*\pi$ is predicted to vanish for both $\overline{\mathbf{10}}$ and $\mathbf{8}_5$ initial pentaquark states in such dynamics. The electromagnetic mass splittings of the Ξ states also contain important information.

7.1.1 Pentaquark wavefunctions, mixing and decays

In pentaquark models where the $(qqqq)$ is in $\overline{\mathbf{6}}_F$, then $\overline{\mathbf{6}} \otimes \mathbf{3} = \overline{\mathbf{10}} \oplus \mathbf{8}_5$ leads to an $\mathbf{8}_5$ that is degenerate with the $\overline{\mathbf{10}}_5$ before mixing; chiral soliton models can accommodate an $\mathbf{8}$ (as a radial excitation of the ground state nucleon octet) though degeneracy is accidental. A challenge will be to decode the mixings between $\overline{\mathbf{10}}_5$, this $\mathbf{8}_5$ and possible contamination with excited $\mathbf{8}_3$ in experiment. This is our main focus in what follows.

Models of low mass pentaquarks with positive parity are based on the strong correlation among quarks, which form antisymmetric flavour pairs, in $\overline{\mathbf{3}}$ of $SU(3)_F$. In order to study the decays and mixings of these states it is important to have a well defined convention for their wavefunctions (Jaffe 2004, Karliner and Lipkin 2004a). We define the $\overline{\mathbf{3}}_F = (\mathbf{3}_F \otimes \mathbf{3}_F)$ basis states as

$$A \equiv (ud) \equiv (ud - du)/\sqrt{2} \sim \overline{s}$$
$$B \equiv (ds) \equiv (ds - sd)/\sqrt{2} \sim \overline{u}$$
$$C \equiv (su) \equiv (su - us)/\sqrt{2} \sim \overline{d} \tag{12}$$

for which $U_-A = -C$; $V_-B = -A$; $I_-C = -B$. The $\Theta^+ \equiv AAA \equiv (ud)(ud)\overline{s}$ and all other members of the $\overline{\mathbf{10}}$ follow by operating on this state sequentially by U_- and I_- until all states have been achieved. For reference they are listed in table 1. We shall always understand the first two labels to refer to the diquarks and the rightmost to refer to the antiquark in Jaffe and Wilczek (2003), and for Karliner and Lipkin (2004a) the latter pair of labels is understood to be in the triquark. The flavour correlations in the two models are thus identical.

In addition to the three manifestly exotic combinations AAA, BBB, CCC the non-exotic states can also form an octet. In the specific dynamics advocated in Jaffe and Wilczek (2003), the quark pairs are strongly correlated into scalar pairs with colour $\overline{\mathbf{3}}$. These scalar "diquarks" are then forced to satisfy Bose symmetry, which leads naturally to the following correlations. Their colour degree of freedom is antisymmetric $\overline{\mathbf{3}} \otimes \overline{\mathbf{3}} \to \mathbf{3}$; their relative $L = 1$ provides an antisymmetric spatial state; their spin coupling is trivially symmetric; and Bose symmetry is completed by their flavour pairings being symmetric. This leads naturally to the positive parity $\overline{\mathbf{10}}$. For the $\mathbf{8}_5$ it leads to the mixed symmetric $\mathbf{8}^\lambda$ states of table 1; in this extreme dynamics there are no mixed antisymmetric $\mathbf{8}^\rho$ analogues (*eg* $p \equiv (AC - CA)A/\sqrt{2}$). This $\mathbf{8}^\lambda$ decays to $\mathbf{8} \otimes \mathbf{8}$ with $F/D = 1/3$ as will become apparent later. Similar properties occur for the Karliner and Lipkin (2004a) correlation where the assumption that the triquark is in a $\overline{\mathbf{6}}_F$ implies that the pentaquark system form $\overline{\mathbf{10}} \oplus \mathbf{8}$ with the same symmetry type as in table 1.

Thus the selection rules that we obtain are common to all these pentaquark models and a consequence of the assumed decay dynamics. The proposal of references (Close

	$\overline{10}$	8_5
Θ^+	AAA	
p	$-(ACA + CAA + AAC)/\sqrt{3}$	$-(ACA + CAA - 2AAC)/\sqrt{6}$
n	$(ABA + BAA + AAB)/\sqrt{3}$	$(ABA + BAA - 2AAB)/\sqrt{6}$
Σ^+	$(CAC + ACC + CCA)/\sqrt{3}$	$(CAC + ACC - 2CCA)/\sqrt{6}$
Σ^0	$-(ABC + BAC + ACB$ $+CAB + BCA + CBA)/\sqrt{6}$	$-(ABC + BAC + ACB + CAB$ $-2BCA - 2CBA)/\sqrt{12}$
Λ^0		$-(ABC - ACB + BAC - CAB)/2$
Σ^-	$(BAB + ABB + BBA)/\sqrt{3}$	$(BAB + ABB - 2BBA)/\sqrt{6}$
Ξ^+	$-CCC$	
Ξ^0	$(CBC + BCC + CCB)/\sqrt{3}$	$(CBC + BCC - 2CBB)/\sqrt{6}$
Ξ^-	$-(CBB + BCB + BBC)/\sqrt{3}$	$-(CBB + BCB - 2BBC)/\sqrt{6}$
Ξ^{--}	BBB	

Table 1. *Pentaquark wavefunctions where ABC are defined in the text. Note that consistency requires the meson octet to be defined with each $q\bar{q}$ positive except for $\pi^+ = -u\bar{d}$; $\overline{K}^0 = -s\bar{d}$ and then $\pi^0 = (u\bar{u} - d\bar{d})/\sqrt{2}$. In this convention $\eta_8 = (2s\bar{s} - u\bar{u} - d\bar{d})/\sqrt{6}$. For Jaffe and Wilczek (2003) A_1, A_2 refer to diquarks and A_3 to the antiquark; for Karliner and Lipkin (2004a) A_1 is diquark and $[A_2 A_3]$ is the triquark in $\overline{6}_F$. The $\overline{6} \otimes \overline{3}$ gives the $\overline{10}$ and 8 as listed above.*

2004, Jennings and Maltman 2004, Carlson *et al.* 2004) is that such pentaquarks can naturally have narrow widths due to the mismatch between the colour-flavour-spin state in an initial pentaquark and the meson-baryon colour singlet states into which they decay. For a simple attractive square well potential of range 1 fm the width of a P-wave resonance 100 MeV above KN threshold is of order 200 MeV (Jaffe and Wilczek 2003, Jennings and Maltman 2004). However, this has not yet taken into account any price for recoupling colour and flavour-spin to overlap the $(ud)(ud)\bar{s}$ onto colour singlets uud and $d\bar{s}$ say for the KN.

If decays are assumed to arise by "fall-apart" (Close 2004, Jennings and Maltman 2004, Carlson *et al.* 2004, Buccella and Sorba 2004) without need for gluon exchange to trigger the decay (even though gluon exchange may be important in determining the eigenstates), then in amplitude, starting with the Jaffe-Wilczek configuration, the colour recoupling costs $\frac{1}{\sqrt{3}}$. It is further implicitly assumed that the fall-apart decay to a specific channel occurs only when the flavour-spin correlation in the initial wavefunction matches that of the said channel. In such a case the flavour-spin correlation to any particular channel (*eg* K^+n) costs a further $\frac{1}{2\sqrt{2}}$, hence a total suppression in rate of $\frac{1}{24}$. This was originally noted in Jennings and Maltman (2004).

The minimal assumption then is that a diquark must cleave such that one quark enters the baryon and the other enters the meson. *While this is necessary, implicitly it is assumed also to be sufficient*: any components in the wavefunction that are not kinematically allowed to decay are assumed to be absolutely forbidden. Selection rules that we obtain here assume this and therefore are implicitly a test of this decay dynamics. There is also a penalty for the spatial overlaps. If once organised into colour singlets, the constituents then simply fall apart in a P-wave with no momentum transfer, only the $L_z = 0$ part of the wavefunction contributes. This implies a further suppression from the $L = 1 \otimes S = 1/2 \to J = 1/2; 3/2$ coupling. Thus a total suppression of $1/72$ for the $1/2^+$ and $1/36$ for $3/2^+$ may be expected (Close and Dudek 2004).

The general conclusion is that if such dynamics govern the decays, then in such models a width of $O(1-10)$ MeV for $\Theta \to KN$ may be reasonable. The above dynamics also implies that $g^2(\Theta NK^*)/g^2(\Theta NK) = 3$ (this is also implicit in Carlson *et al.* (2004)). Although the NK^* decay mode is kinematically inaccessible this relation may eventually be tested in photoproduction experiments (Close and Zhao 2004b). Analogously this implies that $g^2(\Theta_c ND^*)/g^2(\Theta_c ND) = 3$. The Θ_c is predicted in Jaffe and Wilczek (2003) to lie below strong decay threshold but spin-orbit effects (Close and Dudek 2004) could elevate its mass such that it is even above D^*N threshold (see *eg* Karliner and Lipkin (2004a)). Thus if $m(\Theta_c) > 2.95$ GeV, as may be the case if hints of $\Theta_c \sim 3.1$ GeV (Burkert 2004) are confirmed, an enhanced intrinsic coupling to D^*N could be searched for.

With the wavefunctions given in table 1 we can immediately account for the relative strengths of final states by carefully exploiting the symmetries of the wavefunctions. For example, the $\Theta \equiv (ud - du)(ud - du)\bar{s}/2 \to [(ud - du)u][d\bar{s}]/2 - [(ud - du)d][u\bar{s}]/2$ which maps onto $\Theta \to pK^0/\sqrt{2} - nK^+/\sqrt{2}$.

These amplitudes for decays into meson (M) and baryon (B) also depend on the flavour-spin symmetry of the baryon. If we make this explicit (ϕ, χ referring to the flavour and spin wavefunctions respectively and ρ, λ denoting their mixed symmetry properties under interchange (Close 1979) we have

$$(\mathbf{\bar{3}_F}, S = 0)(\mathbf{\bar{3}_F}, S = 0) \to M + B(\phi^\rho \chi^\rho).$$

The same colour-orbital configuration for tetraquarks ($qqqq$) in overall spin $S = 0$ can be realised with diquarks in $\mathbf{6_F}, S = 1$. The pattern of decays from this configuration mirrors those above except that the baryon's flavour-spin symmetry is swapped

$$(\mathbf{6_F}, S = 1)(\mathbf{6_F}, S = 1) \to M + B(\phi^\lambda \chi^\lambda).$$

Thus if one imposed overall antisymmetry on the tetraquark wavefunction one encounters for the flavour-spin part of the wavefunction

$$|(\mathbf{\bar{3}_F}, S = 0)(\mathbf{\bar{3}_F}, S = 0)\rangle \pm |(\mathbf{6_F}, S = 1)(\mathbf{6_F}, S = 1)\rangle.$$

Noting that there is an $L = 1$ within the ($qqqq$) system, the above wavefunctions imply that the ($+$) phase decays to $M + B(\mathbf{56})$ in a P-wave and the ($-$) phase decays to $M + B(\mathbf{70}(L = 1))$ in an S-wave. The latter would naïvely be kinematically forbidden and as such lead to a suppressed width if the Θ were in this representation, which is in the $\mathbf{105}$ dimensional mixed symmetry representation of flavour-spin (Buccella and Sorba 2004). However, one needs also to confront the kinematically allowed decays to $M + B(\mathbf{56})$ from the ($+$) phase state (in the symmetric $\mathbf{126}$ representation and discussed in Carlson *et al.* (2004). In practice decays shared by the $(\mathbf{\bar{3}}, S = 0)(\mathbf{\bar{3}}, S = 0)$ and $(\mathbf{6}, S = 1)(\mathbf{6}, S = 1)$ states lead to mixing. If this is stronger than the mass gap between these two states, then one would obtain the above two configurations, leading to the possibility of the Θ as a narrow state in **105** partnered by a (yet unobserved) broad partner in **126** (see also Karliner and Lipkin (2004b)). By contrast, if the mixing is small on the scale of the mass gap, the wavefunction of the light eigenstate in this spin-zero tetraquark sector will be dominated by the $(\overline{3}, S = 0)(\overline{3}, S = 0)$ configuration, which is the Jaffe-Wilczek model (Jaffe and Wilczek 2003). As noted in Jennings and Maltman (2004), mixing with the spin-one tetraquark sector as manifested in the Karliner-Lipkin correlation can lead to a lower eigenstate. The discussion in the rest of our present paper does not depend on this dynamical question.

Within the assumption that decays are driven by the fall-apart dynamics, the flavour patterns follow for all of these configurations.

It is especially instructive to apply our study of the fall-apart to the Ξ_5 states. In what follows we assume that only the ground state baryon + 0^- meson channels are kinematically accessible. If other channels such as 1^- mesons could be accessed these would cause the intrinsic suppression to be less dramatic.

7.1.2 Decays of Ξ_5 states

Starting with the wavefunction (see table 1)

$$|\Xi^-(\overline{10})\rangle = -\frac{1}{2\sqrt{3}}\left([(ds - sd)(su - us) + (su - us)(ds - sd)]\overline{u} + (ds - sd)^2\overline{d}\right)$$

we can rewrite this in flavour space in the form $(qqq)(q\bar{q})$.

$$\begin{aligned}|\Xi^-(\overline{10})\rangle &= -\frac{1}{2\sqrt{3}}([(ds - sd)s](u\overline{u} - d\overline{d}) + [(su - us)d + (sd - ds)u](s\overline{u}) \\ &\quad -[(su - us)s](d\overline{u}) + [(ds - sd)d](s\overline{d}))\end{aligned}$$

which maps onto the following ground state hadrons

$$\Xi^-(\overline{10}) \rightarrow -\frac{1}{\sqrt{6}}\left(\sqrt{2}\Xi^-\pi^0 + \Xi^0\pi^- - \sqrt{2}K^-\Sigma^0 - \overline{K}^0\Sigma^-\right).$$

These agree in relative magnitudes and phases with the standard de Swart results (de Swart 1963, Particle Data Group 2004); they agree in relative magnitudes with Oh *et al.* (2004) but their phases differ from ours. Oh *et al.* (2004), Jaffe and Wilczek (2004) do not discuss the 8 decays as these depend in general on an undetermined F/D. However with the pentaquark wavefunctions, as specified as in table 1, the octet from $\overline{6}_F \otimes \overline{3}_F$ that is orthogonal to the $\overline{10}$ is

$$|\Xi_5^-(8)\rangle = -\frac{1}{2\sqrt{3}}\left([(ds - sd)(su - us) + (su - us)(ds - sd)]\overline{u} - \sqrt{2}(ds - sd)^2\overline{d}\right)$$

and for the assumed decay dynamics employed in Close (2004), Jennings and Maltman (2004), and Carlson *et al.* (2004), the particle decomposition is

$$\begin{aligned}\Xi_5^-(8) \rightarrow &-\frac{1}{\sqrt{24}}(\sqrt{6}\Xi^-\eta_1 - \sqrt{3}\Xi^-\eta_8 - \Xi^-\pi^0 + \\ &\sqrt{2}\Xi^0\pi^- + 2\sqrt{2}\Sigma^-\overline{K}^0 - 2\Sigma^0 K^- + 0\Lambda K^-)\end{aligned}$$

which corresponds to $8 \to 8 \otimes 8$ with $F/D=1/3$ [or $g_1 = \sqrt{5}g_2$ in the de Swart convention (de Swart 1963, Particle Data Group 2004)]. With this one can therefore deduce the branching ratios for N, Σ, Λ states in 8_5 immediately from existing tables (de Swart 1963, Particle Data Group 2004) and we do not discuss them further here.

For the Ξ_5 we see immediately distinctions between the two states.

(i) Isospin ($I = 3/2$ versus $I = 1/2$) is responsible for the distinctive ratios

$$\frac{\Gamma(\Xi_5^- \to \Xi^- \pi^0)}{\Gamma(\Xi_5^- \to \Xi^0 \pi^-)} = \begin{cases} 1/2 & \mathbf{8} \\ 2 & \overline{\mathbf{10}} \end{cases}$$

and analogous for the ΣK modes.

(ii) There is a selection rule that ΛK^- modes vanish. For the $\overline{\mathbf{10}}$ this is a trivial consequence of isospin; for the 8_5 it is a result of the pentaquark wavefunction, in particular that the $qqqq$ flavour wavefunction of the pair of diquarks is symmetric in flavour, *ie* $\overline{\mathbf{6}}_F = \overline{\mathbf{3}}_F \otimes \overline{\mathbf{3}}_F$, leading to $F/D=1/3$.

A pedagogic explanation of the selection rule is as follows. The Ξ_5 state wavefunctions contain two pieces of generic structure $(dssu)\overline{u}$ and $(dssd)\overline{d}$. The $I = 3/2$ and $I = 1/2$ states differ in the relative proportions of these two. However, only the first component $(dssu)\overline{u}$ contains the $\overline{u}$ required for the K^- and this is common to both the $\Xi(I = 3/2)$ and $\Xi(I = 1/2)$. Thus as the $\Xi(I = 3/2) \to K\Lambda$ is trivially forbidden by isospin, the $\Xi(I = 1/2) \to K\Lambda$ must be also unless there is cross-talk between the two components in the wavefunction. This would happen if annihilation $(dssu)\overline{u} \to (dss) \to (dssd)\overline{d}$ occurs. Thus observation of ΛK^- could arise if there are admixtures of 8_3 in the wavefunction.

Rescattering from kinematically forbidden channels, such as $\Xi\eta$ can feed both $K\Sigma$ and $K\Lambda$, though this is not expected to be a large effect if experience with light hadrons is relevant (such as the small width of the $f_1(1285)$ not being affected by rescattering from the kinematically closed KK^* channel, and the predicted $\pi_2 \to b_1\pi \sim 0$ (Barnes 2003) not being affected by rescattering from the allowed channels $\pi f_2; \pi\rho$). Whether this carries over to pentaquarks may be tested qualitatively in models by comparing the relative suppression of Θ, Ξ^{--} and Ξ^- states; if there is no rescattering and the $\Xi\eta$ channels are closed in the initial pentaquark wavefunction, its width will be further suppressed from $1/24 \to \sim 3/115$ and $K\Lambda \sim 0$. In this case the width of Ξ^- (after phase space effects have been removed) will be less than that of Ξ^{--}. A dominance of $K\Lambda > K\Sigma$ can arise if there are pentaquark configurations having $F = D$. In this latter case the $\Sigma^0 K^-$ would be forbidden but ΛK^- allowed. The $\Lambda K : \Sigma K$ ratio in general can be used to constrain the F/D ratio and begin to discriminate between various dynamical schemes.

(iii) Decays to $\Xi^*\pi$ and $\Sigma^* K$ for Σ^*, Ξ^* in the $\mathbf{10}$ are forbidden (even if allowed by phase space). For the $\overline{\mathbf{10}}$ this is a result of $\overline{\mathbf{10}} \neq \mathbf{8} \otimes \mathbf{10}$ as noted in Jaffe and Wilczek (2004) who also discuss $SU(3)_F$ breaking as a potential source of violation of this zero. However, this selection rule may be stronger in the pentaquark models of Jaffe and Wilczek (2003) and Karliner and Lipkin (2004a) due to the diquarks having antisymmetric flavour $(\overline{\mathbf{3}})$ and spin zero, both of which prevent simple overlap of flavour-spin with the $\mathbf{10}, S = 3/2$ baryon decuplet resonances. Thus although $SU(3)_F$ allows $\mathbf{8} \to \mathbf{10} \otimes \mathbf{8}$ to occur, for the 8_5 states of table 1 it is again forbidden as a consequence of the antisymmetric flavour content of the wavefunction, at least within the models of suppressed decay widths considered here. While we discussed this for the Jaffe-Wilczek wavefunction, Karliner and Lipkin have one of their quark pairs strongly correlated into a vector spin state within a triquark (*eg* $ud\bar{s}$) so the flavour antisymmetries and explicit scalar diquark in the residual wavefunctions suggest that this dynamics also would be challenged to accommodate a violation of this selection rule.

The $I = 3/2$ states will all be narrow. They are degenerate up to electromagnetic mass shifts. Across the $I = 3/2$ multiplet the mass split is $\Xi^{--} - \Xi^{+} = (d-u) + \langle e^2/R \rangle$ (Close 1979) where the Coulomb contribution in known hadrons is $\sim 2 - 9$ MeV, hence a spread of 3-10 MeV is expected. For the non-exotic states $m(\Xi_{8_5}) > m(\Xi_{\overline{10}})$, with

$$m(\Xi_{8_5}^0) - m(\Xi_{\overline{10}}^0) = \frac{1}{2} \left[m(\Xi_{8_5}^-) - m(\Xi_{\overline{10}}^-) \right] = \frac{1}{3}[m(d) - m(u)] \sim 1.5 - 2.5 \text{ MeV}$$

and hence degenerate to within better than 5 MeV. If the coupling to $\Xi^*\pi$ vanishes for the 8_5 as well as the $\overline{10}$, then mixing by the common $\Xi\pi$ decay channels will be destructive. If the widths are truly narrow the mass eigenstates become $\Xi_u \equiv (ds)(su)\bar{u}$ and $\Xi_d \equiv (ds)(ds)\bar{d}$ separated by ~ 10 MeV. The heavier state $b.r.(\Xi_d \rightarrow \Xi^-\pi^0) = 2 \times b.r.(\Xi_d \rightarrow \Sigma^-\overline{K}^0)$ (apart from phase effects) but it does not decay to either $\Xi^0\pi^-$ nor $\Sigma^0 K^-$. In contrast the lighter state $\Xi_u \rightarrow \Xi^0\pi^- : \Xi^-\pi^0 = 2$, as for the pure **8** (but with opposite relative phase), while it does not decay to $\Sigma^-\overline{K}^0$.

Violation of these relations would imply either mixing with excited 8_3 states be due to pentaquark components in the wavefunction beyond those above, or because the width suppression is realised by some dynamics other than implicit in Close (2004), Jennings and Maltman (2004) or Carlson *et al.* (2004). In the former case one would expect the 8_3 components to decay without suppression and dominate the systematics of the widths. In this case there will be narrow $\Xi^{-,0}$ with $I=3/2$ partnering the exotic $\Xi^{+,--}$ and broad $I=1/2$ states that are akin to normal excited Ξ states. By contrast, were the $\Xi\pi$ charge ratios to show mixing between the two Ξ_5 states with two narrow states such as Ξ_d and Ξ_u, then observation of any ΛK or $\Xi^*\pi$ would require components in the pentaquark wavefunction with different symmetries to those above.

7.1.3 Decays of p_5 and n_5 states

These follow immediately from $SU(3)$ tables with $F/D=1/3$. In general there will be mixing between these as suggested by Jaffe and Wilczek. For the extreme $p_5(s\bar{s})$ and $p_5(d\bar{d})$ we have

$$p_5(s\bar{s}) \rightarrow \frac{1}{2} \left(\frac{1}{\sqrt{2}} \Sigma^0 K^+ + \sqrt{\frac{3}{2}} \Lambda K^+ - \Sigma^+ K^0 + p\eta_s \right)$$

where $\eta_s \equiv \eta_1/\sqrt{3} + 2\eta_8/\sqrt{6}$; and while phase space only admits trivial $p_5(d\bar{d})$ decays to $N\pi$.

It is immediately apparent that the decays of $P_{11}(1440; 1710)$ do not fit well with this scheme. First, there is a dominance of non-strange hadrons in the heavier $P_{11}(1710)$ with prominent $\Delta\pi$ in the decays of both $P_{11}(1440; 1710)$. This mode is not possible for the p_5 states in $\overline{10}$ nor in 8_5 unless overwritten by rescattering or mixing with 8_3.

It is clear that $P_{11}(1440)$ is partnered by $P_{33}(1660)$ as in a traditional **56** multiplet of SU(6) qqq states. There is no obvious sign of pentaquarks here. A possibility is that the states are linear combinations of p_3 and p_5; the p_5 could even dominate the wavefunction but its $O(1\text{ MeV})$ width is swamped by the $O(100\text{ MeV})$ width of the unsuppressed p_3 component. The p_5 decays listed above would then show up as rare decays at the $O(1\%)$ level.

7.2 Mass gaps

In the original chiral soliton model (Diakanov *et al.* 1997), the mass gap between the Θ and the exotic Ξ has to be **larger** than that in the conventional ten, spanned by the $\Delta(1236)$ and $\Omega^-(1672)$. Indeed, Diakanov *et al.* (1997) predicted this gap in the $\overline{10}$ to be over 500 MeV leading to a mass for the Ξ exceeding 2 GeV. In the pentaquark picture, by contrast, this looks surprising as at first sight it appears that one need only pay the price for one extra strange mass throughout the ten-bar. This implies a relatively light mass for the $\Xi \sim 1700$ MeV with the possibility that these states also could be relatively stable.

However, the $\overline{10}$ requires strong correlations among the flavours with the $[ud]$ diquark being tightly bound. So in $[ud][ud]\bar{s}$ versus $[us][us]\bar{d}$ the mass gap

$$m(\Xi) - m(\Theta) = 2\Delta m([us] - [ud]) - \Delta m(\bar{s} - \bar{d}).$$

If $\Delta m([us] - [ud]) \sim 2\Delta m(\bar{s} - \bar{d})$ as Vogl and Weise (1991), and Equation 4 suggests, then $m(\Xi) - m(\Theta) \sim 3\Delta m(s - d)$. So there is not necessarily any inconsistency so long as strong correlations are present.

Further differences emerge for the first excited states. In the pentaquark models these have $J^P = 3/2^+$ arising from the spin-orbit forces that split the $L = 1 \otimes S = 1/2 \rightarrow J = 1/2 \oplus 3/2$ and will again be $\overline{10}$. Such a multiplet does not occur in the chiral soliton models. These allow $J^P = 3/2^+$ but in higher dimensions such as **27, 35**, in which case the excited partner of the Θ can occur only in a range of charge states. In pentaquark models, such configurations also are possible, but it is only in such models that the excited $J^P = 3/2^+\overline{10}$ also occurs.

In general if narrow width pentaquarks exist with positive parity, this implies there are strong correlations at work in the strong QCD sector. Two particular models that build on this are those of Jaffe and Wilczek (2003) and Karliner and Lipkin (2004a), exploiting the ideas already described here that in QCD there are strong attractions between distinct flavours in net spin zero. The stability of scalar diquarks is an open question; their effective boson nature and consistency with hadron spectroscopy also need better understanding and it has not been demonstrated that they maintain such strong boson identities that Bose symmetry forces a relative L=1 in the ways required. But first we need to establish whether this state is real before getting in too deep. I shall now review various features.

8 Properties of Θ

8.1 Mass

The original prediction (Diakanov *et al.* 1997) assumed that the 1710 N^* is in the $\overline{10}$ and used this to set the scale of mass. However $\gamma p \to p^*(\overline{10})$ is forbidden by U-spin which argues against this. The mass gap of 180 MeV per unit of strangeness is also suspect in a quark model interpretation as it leads to a 540 MeV spread across the $\Theta - \Xi$ multiplet even though there is only one extra strange mass in going from ($udud\bar{s}$) to ($usus\bar{d}$) and so a much smaller gap would be anticipated (Jaffe and Wilczek 2003). Beware the naïve application of Gell Mann-Okubo mass formulae which do not distinguish between $|S|$ and S as one goes from $\Theta(S = +1)$ to $\Xi(S = -2)$.

If the Θ should prove to be real, then no simple mapping from chiral soliton onto a pentaquark description seems feasible. The relation between these is more profound.

Nonetheless a narrow state of mass $\sim$1540 MeV has been claimed. But when one compares the masses reported in K^+n versus K^0p there appears to be a tantalising trend towards a difference (Close and Zhao 2004c). Is this a hint of an explanation (see later) or that we are being fooled by poor statistics?

No models successfully predict the mass; in all cases it is fitted relative to some other assumed measure. The original chiral soliton normalised to the 1710, as we already discussed. Jaffe and Wilczek (2003) assume that the Roper 1440 is the $udud\bar{d}$ (but this state is partnered by $\Delta(1660)$ which along with its electromagnetic and other properties, is in accord with it being a radial qqq excitation of the nucleon). Karliner and Lipkin (2004a) noted the kinematic similarity between reduced masses in their diquark-triquark model and the $c\bar{s}$ system. They adopted a 200 MeV orbital excitation energy from the $1^- - 0^+(2317)$ mass gap to realise a 1540 MeV mass for the Θ. However, if one makes a spin averaged mass for the $L = 0, 1$ levels, notwithstanding the questions about the low mass of the 2317, one gets nearer to a 450–480 MeV energy gap; this would lead to a Θ nearer 1800 MeV. In summary, all models appear to normalise to some feature and do not naturally explain the low mass of an orbitally excited pentaquark.

8.2 Width

The chiral soliton model Lagrangian contains three terms with arbitrary strengths, A, B, C. Linear combinations of these can be related to the observable transition $\Delta N\pi$ and the F/D ratio for the $NN\pi$ vertex. The ΘNK vertex is then given by $g(\overline{10}) = 1 - B - C$. We thus have one unknown $g(\Theta NK)$ described by another unknown, C. Ellis *et al.* (2004) show the coupling is relatively insensitive to F/D and that it is C that controls $g(\Theta NK)$. In the non relativistic quark model it is argued (Diakonov *et al.* 1997, Ellis *et al.* 2004) that $F/D = 2/3$ and the absence of $s\bar{s}$ in the nucleon lead to $B = 1/5; C = 4/5$. This has the remarkable implication that $g(\Theta NK) = 0$. If the Θ phenomenon survives then a deeper understanding of this result and its implications would be welcome. It would also raise the challenge of how the Θ is strongly produced.

Phenomenologically it has been suggested that the $\Gamma(\Lambda(1520) \to KN) \sim 7$ MeV is a measure for narrow widths. However this is D-wave and phase space limited: the

P-wave $\Lambda(1660)$ width is ~ 100 MeV. Furthermore in these cases one has to create a $q\bar{q}$ pair to initiate the decay; for the pentaquark one has $qqqq\bar{q}$ and the challenge is to stop its decay. There are no indications in conventional spectroscopy that underpin the narrow width of ~ 1 MeV for the Θ.

Colour spin and flavour mismatches between the pentaquark and NK wavefunctions have been proposed to suppress the natural width by factors of 24 (Jennings and Maltman 2004, Close and Dudek 2004) or even more (Carlson *et al.* 2004). However it is easy to overcome these: soft gluon exchange defeats the colour; spin flip costs little and flavour rearrangement can occur. Further there is colour singlet $q\bar{q}$ in relative S-wave within the correlated models of Jaffe-Wilczek and Karliner-Lipkin (Hogassen and Sorba 2004, Dudek 2004) and their dissociation into NK seems hard to prevent.

Melikhov *et al.* (2004) have suggested that the overlaps of spatial wavefunctions between pentaquark and nucleon may lead to a suppression. However it has not been demonstrated that such is generated dynamically. Dudek (2004) has shown that such an effect can arise but this involves taking a non-relativistic picture rather literally and the self-consistency of the picture remains to be tested. There is also the question of how a colour $\bar{3}$ diquark is attracted into a tighter (smaller?) configuration than a colour singlet meson.

We almost have a paradox here. The small width implies a feeble coupling to KN. Yet something must couple to it very strongly if its production rate is comparable (Jin 2004) to those of conventional hadrons. This is an enigma which we must confront.

8.3 Production

Several experiments place rather strong limits on the hadroproduction of the Θ (Jin 2004). Some are not yet restrictive, *eg* the limit in $\psi \rightarrow \Theta\bar{\Theta}$ which is phase space limited or that in ψ' decay where one can claim that there is a big price to pay for creating ten q and $\bar{q}$. So it is possible to wriggle. However on balance it seems to me that the limits in high statistics hadroproduction are impressive. The onus is on supporters to explain them away or find a loophole.

An example of such a loophole (Lipkin 2004, Lipkin and Karliner 2004) asks why signals are in photoproduction but not in hadroproduction. The photon contains $s\bar{s}$ and so may be able to feed the $\bar{s}$ needed to make $\Theta(udud\bar{s})$ in a way not so readily accessible in hadroproduction. Further appeal is made to a CLAS observation that suggests that a narrow N^* at ~ 2.4 GeV may be the source of $\Theta + K$. While such a dynamics can be tested by searching for other decay modes, forced by SU(3) (Lipkin 2004), there remain problems with this. CLAS see this (statistically insignificant) N^* in π exchange and so the photon does not appear to be essential: why is this object (and its progeny, the Θ) not also made in hadroprodcution if it is made by πN? Second; while a 2.4 GeV N^* may be produced in the 3–5 GeV CLAS experiment, it is kinematically inaccessible in the original SpRING8 experiment and in the earlier CLAS γd. So the source of Θ in this latter pair would still remain to be explained.

Photoproduction has also been suggested as a source of kinematic peaks that fake a Θ (Dzierba *et al.* 2004). $\gamma N \rightarrow a_2/\rho_3 N$ followed by the $K\overline{K}$ decays of these mesons in D/F waves give a forward-backward peaking in the c.m. along the direction of the recoil nucleon. If there were no charge exchange the K^+ and K^- would be equally likely to follow the nucleon and so a kinematically generated peak would be as likely in K^+n as in K^-n. The experimental absence of such peaks in K^-n has been cited as support for the reality of the peak in K^+n. However it is not necessarily so simple. Charge exchange introduces a charge asymmetry and it is claimed to be possible to choose phases such that a narrow peak can arise in K^+n (after feeding through Monte Carlo) whereas a broad structure would arise in K^-n. It has been suggested in the discussions that the different Q-values could cause a mass shift in the kinematic peak in K^+n versus K^0p, in accord with the trend of the data noted by Close and Zhao (2004c). Whether this kinematic effect is responsible may be settled when higher statistics data and significant Dalitz plots become available.

9 A low-lying $J^P = 3/2^+$ $\overline{10}$ multiplet

An essential difference between the pentaquark and chiral soliton (Skyrme) models appears to be in their implications for the first excited state of the Θ. In $qqqq\bar{q}$ with positive parity $1/2^+$ there is necessarily angular momentum present, which implies a family of siblings but with $J^P = 3/2^+$ (Close and Dudek 2004). The spin-orbit forces among the quarks and antiquark lead to a mass gap between any member of the $J^P = 1/2^+$ and its $J^P = 3/2^+$ counterpart, which was calculated in Close and Dudek (2004) to be significantly less than m_π and possibly only $O(10 - 50)$ MeV in the models of Jaffe and Wilczek (2003) and Karliner and Lipkin (2004a). Similar remarks hold for all the members of the $\overline{10}$, such as $\Xi^{+,--}$, and their non-exotic analogues that can also occur in 8_5, such as $\Xi^{0,-}$. Such a $\overline{10}$ family of $J^P = 3/2^+$ states does **not** occur in the present formulation of chiral soliton models, nor can it if the Wess-Zumino constraint selects allowed multiplets (Praszalowicz 2003).

In the Skyrme model there are exotic states with $J^P = 3/2^+$ or higher but these are in **27** and **35** multiplets of $SU(3)_F$. Such states are also expected in pentaquark models (eg isotensor resonance with states ranging from $uuuu\bar{s}$ with charge +3 to $dddd\bar{s}$ with charge -1) (Bijker 2004). The essential difference then is that in the chiral soliton Skyrme models any spin 3/2 partner of the Θ will exist in a variety of charge states with $I = 1, 2$ whereas the unique feature of the pentaquark models (Jaffe and Wilczek 2003, Karliner and Lipkin 2004a) is that the first excited state is an **isoscalar** analogue of the Θ. (There may be versions of pentaquark models where this state is higher in mass but that it is isoscalar is universal in any quark model description).

9.1 Λ_5 state with $J^P = 3/2^+$

There is one further potentially narrow state in pentaquark models, which has little opportunity for mixing with qqq states. This is the Λ_5 state that is the $J^P = 3/2^+$ spin-orbit partner of Λ_5 in 8_5.

First note that $\overline{10}$ contains Σ_5 but has no Λ_5. The 8_5 contains a Λ_5, and there will be no mixing with $\overline{10}$ so long as isospin is good. If there were no mixing with $\Lambda(qqq)$ excited states, this Λ_5 would be narrow, with width identical to that of the Θ apart from phase space factors.

The Λ_5 wavefunction shows that it has only one strange mass quark and hence is similar to the Θ in this regard. Jaffe and Wilczek (2003) estimate ~ 1600 MeV for such a state (the excess ~ 60 MeV relative to the Θ arising because the mass of a $(us)\bar{d}$ set is larger than $(ud)\bar{s}$ due to the relatively smaller downward mass shift in the (us) diquark). Scaling the spin-orbit splitting from Close and Dudek (2004) and allowing for the relative masses of the $\bar{s}/\bar{d}$ and $m(us)/m(ud)$ gives 40-70 MeV for the Λ_5 mass gap of $3/2^+ - 1/2^+$ and hence 1600-1700 as a conservative estimate for the mass range for the partner $\Lambda_5(3/2^+)$.

Perusal of the data (Particle Data Group 2004) shows that, for the $1/2^+$, mixing with qqq states is likely (given the existence of a candidate $\mathbf{56}, 0^+$ multiplet containing $P_{11}(1440)$, $\Lambda(1600)$, $\Sigma(1660)$, $\Xi(?)$). However there is no $3/2^+$ multiplet with a $\Lambda(1600 - 1700)$ seen, nor is one expected in standard qqq models. The first such is the set containing $P_{13}(1720)$, $\Lambda(1890)\cdots$. Thus there is a significant gap between $\Lambda(1890)$ and our predicted $\Lambda_5(3/2^+)$.

The branching ratios for either the spin 1/2 or 3/2 states can be determined from the breakdown

$$\Lambda_5 \rightarrow -\frac{1}{2\sqrt{2}} \left(pK^- - n\overline{K}^0 + \Sigma^-\pi^+ - \Sigma^+\pi^- - \Sigma^0\pi^0 - \sqrt{3}\Lambda\eta_{n\bar{n}} \right).$$

Decays to $\Sigma^*\pi$ should be suppressed, even if they are kinematically accessible. The production rate of the spin 1/2 state in $\gamma p \rightarrow K^+\Lambda_5$ should be similar to that of $\gamma n \rightarrow K^-\Theta$ (perhaps a factor of four smaller if K exchange drives the production and $g(KN\Theta) = 2g(KN\Lambda_5)$). If the arguments about $L \otimes S$ coupling and fall-apart dynamics are correct, then we can expect the spin 3/2 state to be enhanced by a factor of two relative to the spin 1/2 counterpart. A search in $\gamma p \rightarrow K^+\Lambda_5$ therefore seems appropriate.

If $\overline{\mathbf{10}}$-$\mathbf{8}_5$ mixing is ideal, then also charged $\Sigma_d^\pm$ states will occur which for $J^P = 3/2^+$ should be unmixed. For $J^P = 1/2^+$, the amplitudes $g(\Theta^+ K^+ n) = \sqrt{2}g(\Sigma_5^- K^- n)$ and so the relative photoproduction cross sections should scale as $\sigma(\gamma n \rightarrow K^-\Theta^+) \sim 2 \times \sigma(\gamma n \rightarrow K^+\Sigma_5^-)$ (Close and Zhao 2004c). If the Σ_5 is mixed into the $\Sigma(1660)$ then the latter state should be photoproduced at least at the above rate and so may be a test for consistency.

10   A final problem

It will be important to establish the existence of other members of the $\overline{\mathbf{10}}$ containing the Θ. Close and Zhao (2004c) have noted that the relative photoproduction strengths of Θ and the related Σ_5^+ may be predicted even though the scale of each individually is highly model dependent. The implication is that the production rates should be similar, in particular that for a pentaquark Σ_5 one expects $\sigma(\gamma p \rightarrow \Sigma_5^+ K^0) \sim 0.2 - 0.5\sigma(\gamma p \rightarrow \Theta^+\overline{K^0})$. As either of these can decay into $K_s p$, the absence of any Σ_5^+ signal (even after mixing with known Σ^*) accompanying the claimed Θ in the HERMES data for example raises questions. Thus $\gamma p \rightarrow pK_S^0 K_S^0$ should be a source of information about such states.

However, there are enough miracles required to explain the various weird aspects of this state, and there are apparent inconsistencies, such as the absence of other states, differing conclusions on the width (is it $\sim 10 - 20$ MeV or ~ 1 MeV?) that one has to keep clearly in mind the possibility that it simply does not exist. Time, and most important, statistics will tell.

Acknowledgments

I am indebted to Jo Dudek and Qiang Zhao, with whom much of the content of these lectures was first developed.

References

Achasov NN *et al.* , 1979, *Phys Lett B* **88** 367.
Achasov NN and Shestakov GN, 1997, *Phys Rev D* **56** 212.
Aloisio A *et al.* (KLOE Collaboration), 2001, *Proc Lepton Photon 2001*, hep-ex/0107024.
Amsler C and Close F E, 1995, *Phys Lett B* **353** 385.
Anisovich VV, 1998, *Phys Uspekhi* **41** 419.
Bali G *et al.* , 1993, *Phys Lett B* **309** 378.
Barnes T *et al.* , 1982, *Nucl Phys B* **198** 380.
Barnes T, 2003, rapporteur talk at *Hadron03*.
Bijker R *et al.* , 2004, *Eur Phys J A* **22** 319.
Boglione M and Pennington M R, 1997, *Phys Rev Lett* **79** 1998.
Buccella F and Sorba P, 2004, *Mod Phys Lett A* **19** 1547.
Burkert V D, 2004, *contribution to these proceedings*.
Burns T and Close F E, 2004, *in preparation*.
Carlson C E *et al.* , 2004, *Phys Rev D* **70** 037501.
Close F E, 1979, *Introduction to Quarks and Partons* (Academic Press).
Close F E and Teper M J, 1996, *On the lightest Scalar Glueball* Rutherford Appleton Lab report
    no. RAL-96-040; Oxford University report no. OUTP-96-35P.
Close F E and Kirk A, 2000, *Phys Lett B* **483** 345.
Close F E and Kirk A, 2001a, *Eur Phys J C* **21** 531.
Close F E and Kirk A, 2001b, *Phys Lett B* **515** 13.
Close F E and Tornqvist N, 2002, *J Phys G: Nucl Part Phys* **28** R249.
Close F E, 2004, in *Proc Hadron03*, *AIP Conf Proc* **717** 919.
Close F E and Page P R, 2004, *Phys Lett B* **578** 119-123.
Close F E and Dudek J J, 2004, *Phys Lett B* **586** 75.
Close F E and Zhao Q, 2004a, *Phys Lett B* **586** 332.
Close F E and Zhao Q, 2004b, *Phys Lett B* **590** 176.
Close F E and Zhao Q, 2004c, hep-ph/0404075.
Davies C T H, 2004, *Lectures given to SUSSP58 Hadron Physics Summer School, St Andrews*.
de Rujula A *et al.* , 1975, *Phys Rev D* **12** 147.
de Swart JJ, 1963, *Rev Mod Phys* **35** 916.
Diakanov D *et al.* , 1997, *Z Phys A* **359** 305.
Dudek J J, 2004, hep-ph/0403235.
Dzierba A *et al.* , 2004, *Phys Rev D* **69** 051901.
Ellis J *et al.* , 2004, *J High Energy Phys* **0405** 002.
Giacosa F *et al.* , 2005, *Phys Rev C* **71** 025202.
Hogassen H and Sorba P, 2004, *Mod Phys Lett A* **19** 2403.
Isgur N and Karl G, 1978, *Phys Rev D* **18** 4187.

Isgur N and Paton J, 1985, *Phys Rev D* **31** 2910.

Jaffe R L, 1977, *Phys Rev D* **15** 281.

Jaffe R L and Low F E, 1979, *Phys Rev D* **19** 2105.

Jaffe R L, 2004, *priv com.*

Jaffe R L and Wilczek F, 2003, *Phys Rev Lett* **91** 232003.

Jaffe R L and Wilczek F, 2004, *Phys Rev D* **69** 114017.

Jaffe R, Karliner M and Lipkin H J, 2004, unpublished discussions.

Jennings B and Maltman K, 2004, *Phys Rev D* **69** 094020.

Jin S, 2004, *Rapporteur talk on Hadron Spectroscopy (experiment), in Proc ICHEP04.*

Karliner M and Lipkin H J, 2004a, hep-ph/0307243.

Karliner M and Lipkin H J, 2004b, *Phys Lett B* **586** 303.

Krehl O *et al.* , 1997, *Phys Lett B* **390** 23.

Lipkin H J, 1977, *Phys Lett B* **70** 113.

Lipkin H J, 2004, *comments in parallel session at ICHEP04.*

Lipkin H J and Karliner M, 2004, *Phys Lett B* **597** 309.

McNeile C and Michael C, 2001, *Phys Rev D* **63** 114503.

Melikhov D *et al.* , 2004, *Phys Lett B* **594** 265.

Morningstar C J and Peardon M, 1997, *Phys Rev D* **56** 4043.

Oh Y *et al.* , 2004, *Phys Rev D* **69** 094009.

Olsen S L *et al.* [Belle collaboration], 2004, parallel session at ICHEP04; ICHEP04 8-0685.

Olsen S L *et al.* [Belle collaboration], 2005, *Int J Mod Phys A* **20** 240.

Particle Data Group, 2004, *Phys Lett B* **592** 1.

Praszalowicz M, 2003, *Phys Lett B* **575** 234.

Tornqvist N A, 1995, *Z Phys C* **68** 647.

Tornqvist N A, 2004, *Phys Lett B* **590** 209.

Vogl U and Weise W, 1991, *Prog Part Nucl Phys* **27** 195.

Weingarten D, 1999, *Nucl Phys Proc Suppl* **73** 249.

Weinstein J and Isgur N, 1982, *Phys Rev Lett* **48** 659.

Weinstein J and Isgur N, 1983, *Phys Rev D* **27** 588.

Weise W, 2004, *contribution to these proceedings.*

Wilzcek F, 2004, hep-ph/0409168.

Zeldovitch Y and Sakharov A, 1967, *Sov J Nucl Phys* **4** 283.

The QCD vacuum and its hadronic excitations

Wolfram Weise

Physik-Department, Technische Universität München,
D-85747 Garching, Germany

1 Introduction: QCD - its phases and structures

These lectures deal with the complex structure of the ground state, or vacuum, of Quantum Chromo Dynamics (QCD) and its low-energy excitations, the hadrons. QCD, the theory of quarks, gluons and their interactions, is a self-contained part of the Standard Model of elementary particles. It is a consistent quantum field theory with a simple and elegant underlying Lagrangian, based entirely on the invariance under a local gauge group, $SU(3)_{colour}$. Out of this Lagrangian emerges an enormously rich variety of physical phenomena, structures and phases. Exploring and understanding these phenomena is undoubtedly one of the most exciting challenges in modern science.

Figure 1 shows a schematic phase diagram of QCD for first orientation. At high temperatures, above a critical temperature T_c of about 0.2 GeV, the elementary quark and gluon degrees of freedom are released from their confinement in hadrons. Correlations between these basic constituents are expected still to persist up to quite high temperatures, above which matter presumably exists in the form of a quark-gluon plasma.

At temperatures below T_c and at low baryon density, matter exists in aggregates of quarks and gluons with their colour charges combined to form neutral (colour-singlet) objects. This is the domain of low-energy QCD, the physics of the hadronic phase in which mesons, baryons and nuclei reside. In this phase the QCD vacuum has undergone a qualitative change to a ground state characterised by strong condensates of quark-antiquark pairs and gluons. In another sector, at very high baryon chemical potential (ie at large quark densities and Fermi momenta), it is expected that Cooper pairing of quarks sets in and induces transitions to a complex pattern of superconducting and superfluid phases.

Major parts of these lectures will focus on concepts and strategies used to investigate the hadronic sector of the QCD phase diagram. Basic QCD symmetries and the

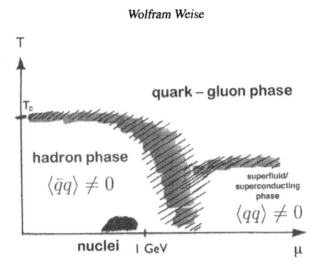

Figure 1. *Illustration of the QCD phase diagram, in the temperature T and baryon chemical potential μ plane.*

corresponding conserved currents are used as a guiding principle to construct effective Lagrangians which represent QCD at low energies and momenta. A rapidly advancing approach to deal with non-perturbative QCD is Lattice Gauge Field Theory. Considerable progress is being made solving QCD on a discretised, Euclidean space-time lattice using powerful computers. Both applications of effective field theory and some selected examples of lattice QCD results will be discussed in these notes. Introductory materials have in part been updated from a previous lecture series (Weise 2003).

2   Basics

2.1   QCD primer

The elementary spin-1/2 particles of QCD, the quarks, come in six species, or flavours, grouped in a field $\psi(x) = (u(x), d(x), s(x), c(x), b(x), t(x))^T$. Each of the $u(x)$, $d(x)$, ... is a four-component Dirac spinor field. Quarks experience all three fundamental interactions of the Standard Model: weak, electromagnetic and strong. Their strong interactions involve $N_c = 3$ "colour" charges for each quark. These interactions are mediated by the gluons, the gauge bosons of the underlying gauge group of QCD, $SU(3)_{colour}$.

The QCD Lagrangian (Fritzsch *et al.* 1973) is

$$\mathcal{L}_{QCD} = \overline{\psi}\left(i\gamma_\mu \mathcal{D}^\mu - m\right)\psi - \frac{1}{4}G^j_{\mu\nu}G^{\mu\nu}_j \,, \qquad (1)$$

with the gauge-covariant derivative

$$\mathcal{D}^\mu = \partial_\mu - ig\sum_{j=1}^{8}\frac{\lambda_j}{2}\mathcal{A}^j_\mu(x) \qquad (2)$$

and the gluon field tensor

$$G^i_{\mu\nu}(x) = \partial_\mu \mathcal{A}^i_\nu(x) - \partial_\nu \mathcal{A}^i_\mu(x) + g f_{ijk} \mathcal{A}^j_\mu(x) \mathcal{A}^k_\nu(x). \tag{3}$$

The λ_j are Gell-Mann matrices and the f_{ijk} are the antisymmetric structure constants of the SU(3) Lie algebra. The non-linear three- and four-point couplings of the gluon fields $\mathcal{A}^j_\mu$ with each other are at the origin of the very special phenomena encountered in QCD and strong interaction physics.

Apart from the quark masses collected in the mass matrix $m = diag(m_u, m_d, m_s, ...)$ in Equation 1, there is no primary scale in $\mathcal{L}_{QCD}$. Renormalisation of the quantum field theory introduces such a scale, to the effect that the "bare" coupling constant g that enters Equations 2 and 3 turns into a "running" coupling, $g(\mu)$, depending on the scale μ at which it is probed. It is common to introduce the QCD coupling strength as $\alpha_s(\mu) = g^2(\mu)/4\pi$, by analogy with the fine-structure constant in QED. Leading-order perturbative QCD gives

$$\alpha_s(\mu) = \frac{4\pi}{\beta_0 \ln(\mu^2/\Lambda^2)}, \tag{4}$$

with $\beta_0 = 11 - 2N_f/3$ where N_f is the number of quark flavours. The QCD scale parameter Λ is determined empirically ($\Lambda \simeq 0.2$ GeV for $N_f = 4$). The fact that α_s decreases with increasing μ leads to a property known as "asymptotic freedom", the domain $\mu \gg 1$ GeV in which QCD can indeed be treated as a perturbative theory of quarks and gluons. The theoretical discovery of asymptotic freedom was honoured with the 2004 Nobel Prize in Physics (Gross and Wilczek 1973, Politzer 1973). At the scale of the Z-boson mass, $\alpha_s(M_Z) \simeq 0.12$ (Eidelmann *et al.* [Particle Data Group] 2004). So while α_s is small at large μ, it is of order 1 at $\mu < 1$ GeV. At low energies and momenta, an expansion in powers of α_s is therefore no longer justified: we are entering the region commonly referred to as non-perturbative QCD.

The quarks are classified as "light" or "heavy" depending on their entries in the mass matrix m of Equation 1. These masses are "running" as well: they depend on the scale μ at which they are determined. The masses of the lightest (u and d) quarks,

$$m_{u,d} < 10 \, MeV \tag{5}$$

(estimated at a renormalisation scale $\mu \simeq 1$ GeV) are very small compared to typical hadron masses of order 1 GeV, such as those of the ρ meson or the nucleon. The strange quark mass,

$$m_s \simeq (100 - 150) \, MeV \tag{6}$$

is an order of magnitude larger than $m_{u,d}$ but still counted as "small" on hadronic scales. The charm quark mass $m_c \simeq (1.1 - 1.4)$ GeV takes an intermediate position while the b and t quarks ($m_b \simeq (4.1-4.4)$ GeV, $m_t = (174\pm5)$ GeV) fall into the "heavy" category. These different quark masses set a hierarchy of scales, each of which is governed by distinct physics phenomena.

2.2 Concepts and strategies

There exist two limiting situations in which QCD is accessible with "controlled" approximations. At momentum scales exceeding several GeV (corresponding to short distances, r < 0.1 fm), QCD is a theory of weakly interacting quarks and gluons (Perturbative QCD). At low momentum scales considerably smaller than 1 GeV (corresponding to long distances, r > 1 fm), QCD is characterised by confinement and a non-trivial vacuum (ground state) with strong condensates of quarks and gluons. Confinement implies the spontaneous breaking of a symmetry which is exact in the limit of massless quarks: chiral symmetry. Spontaneous chiral symmetry breaking in turn implies the existence of pseudoscalar Goldstone bosons. For two flavours ($N_f = 2$) they are identified with the isotriplet of pions (π^+, π^0, π^-). For $N_f = 3$, with inclusion of the strange quark, this is generalised to the pseudoscalar meson octet. Low-energy QCD is thus realised as an Effective Field Theory (EFT) in which these Goldstone bosons are the active, light degrees of freedom.

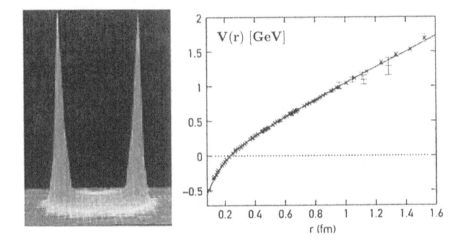

Figure 2. *Confinement from Lattice QCD (Bali 2001). Left: Gluonic flux tube extending between static colour sources (infinitely heavy quarks). Right: Static quark-antiquark potential in physical units.*

Much of the interesting physics lies between these extreme short and long distance limits. This is where large-scale computer simulations of QCD on discretised Euclidean space-time volumes (lattices) are progressing steadily (Davies 2004). Figure 2 presents a key result of lattice QCD (Bali 2001): the gluonic flux tube that connects two infinitely heavy colour sources (quark and antiquark) fixed at lattice sites with a given distance. The resulting quark-antiquark potential is well approximated by the form

$$V(r) = -\frac{4\alpha_s}{3r} + \sigma r . \tag{7}$$

Coulomb-like (perturbative) one-gluon exchange with $\alpha_s \simeq 0.3$ is seen at short distance. At long range, the potential shows the linear rise with increasing distance between the static colour charges characteristic of confinement and parameterised by a string tension $\sigma \simeq 1$ GeV/fm. Confining potentials of the form (7) have been used successfully in accurate decriptions of bottomonium spectroscopy, solving the Schrödinger equation for $b\bar{b}$ bound states with inclusion of spin-dependent interactions (Bali *et al.* 1997).

Confinement has a relatively simple interpretation for heavy quarks and the "string" of (static) gluonic field strength that holds them together, expressed in terms of a static potential. When light quarks are involved, the situation is different. Colour singlet quark-antiquark pairs pop out of the vacuum as the gluon fields propagate over larger distances. Light quarks are fast movers: they do not act as static sources. The potential picture is not applicable. The common features of the confinement phenomenon can nevertheless be phrased as follows: non-linear gluon dynamics in QCD does not permit the propagation of coloured objects over distances of more than a fraction of a Fermi. Beyond the one-Fermi scale, the only remaining relevant degrees of freedom are colour-singlet composites (quasiparticles) of quarks, antiquarks and gluons.

2.3 Scales at work: the hierarchy of quark masses

The quark masses are the only parameters that set primary scales in QCD. Their classi-fication into sectors of "light" and "heavy" quarks determines the very different physics phenomena associated with them. The heavy quarks (*ie* the t, b and, within limits, the c quarks) offer a natural "small parameter" in terms of their reciprocal masses. Non-relativistic approximations (expansions of observables in powers of $m_{t,b,c}^{-1}$) tend to work increasingly well with increasing quark mass. This is the area sometimes referred to under the name of Non-Relativistic QCD.

The sector of the light quarks (*ie* the u, d quarks and, to some extent, the s quarks) is governed by quite different principles and rules. Evidently, the quark masses them-selves are now "small parameters", to be compared with a characteristic "large" scale of dynamical origin. This large scale is related to the order parameter of spontaneously bro-ken chiral symmetry, commonly identified with $4\pi f_\pi \sim 1$ GeV, where $f_\pi \simeq 0.09$ GeV is the pion decay constant to be specified later. The square of f_π is, in turn, proportional to the chiral (or quark) condensate $\langle \bar{q}q \rangle$, a key quantity featuring the non-trivial structure of the QCD vacuum. The low-mass hadrons are quasiparticle excitations of this condensed ground state. There is a characteristic mass gap of about 1 GeV which separates the QCD vacuum from almost all of its excitations, with the exception of the pseudoscalar meson octet of pions, kaons and the eta meson. This mass gap is again comparable to $4\pi f_\pi$, the scale associated with spontaneous chiral symmetry breaking in QCD.

Low-energy QCD is the physics of systems of light quarks at energy and momentum scales smaller than the 1 GeV mass gap observed in the hadron spectrum. The rele-vant expansion parameter in this domain is $Q/4\pi f_\pi$, where Q stands generically for low energy, momentum or pion mass.

In this context it is instructive to explore the spectroscopic pattern of pseudoscalar and vector mesons, starting from heavy-light quark-antiquark pairs in $^1S_0$ and $^3S_1$ states and following those states downward in the mass of the quark. This is illustrated in Figure 3 where we show the masses of mesons composed of b, c, s or u quarks with an anti-d-quark attached. Bare quark masses are subtracted from the meson masses in this plot in order to directly demonstrate the evolution from perturbative hyperfine splitting in the heavy systems to the non-perturbative mass gap in the light ones. In the $\overline{B}$ and $\overline{B}^*$ mesons, the $\bar{d}$ quark is tightly bound to the heavy b quark at small average distance, within the range where perturbative QCD is applicable. The spin-spin interaction is well approximated by perturbative one-gluon exchange, resulting in a small hyperfine

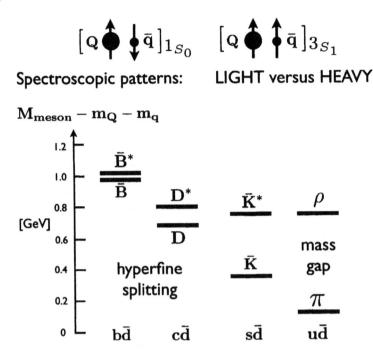

Figure 3. *Evolution of the splitting between spin singlet (lower) and triplet (upper) quark-antiquark states (the pseudoscalar ($J^\pi = 0^-$) and vector ($J^\pi = 1^-$) mesons) when varying the mass of one of the quarks. The bare quark masses are subtracted from the physical meson masses for convenience of demonstration.*

splitting. Moving downward in mass to the D and D^* systems, with the b quark replaced by a c quark, the hyperfine splitting increases but remains perturbative in magnitude. As this pattern evolves further into the light-quark sector, it undergoes a qualitative change via the large mass difference of $\overline{K}$ and $\overline{K}^*$ to the non-perturbative mass gap in the $\pi - \rho$ system, reflecting the Goldstone boson nature of the pion.

3 Low-energy QCD

In the hadronic, low-energy phase of QCD, the active degrees of freedom are not elementary quarks and gluons but mesons and baryons. To illustrate the situation, consider the canonical partition function expressed in terms of the QCD Hamiltonian H:

$$\mathcal{Z} = Tr \exp(-H/T) = \sum_n \langle n|e^{-E_n/T}|n\rangle$$

with $(H - E_n)|n\rangle = 0$. Confinement implies that the eigenstates $|n\rangle$ of H are (colour-singlet) hadrons at $T < T_c$, below the critical temperature for deconfinement. The low-temperature physics is then determined by the states of lowest mass in the spectrum $\{E_n\}$. The very lowest states are the pseudoscalar mesons. All other states are separated by the mass gap mentioned earlier. This separation of scales is the key for approaching low-energy QCD with the systematic theoretical tool of Chiral Effective Field Theory. The guiding principle for this is the approximate chiral symmetry of QCD.

3.1 Chiral symmetry

Consider QCD in the limit of massless quarks, setting $m = 0$ in Equation 1. In this limit, the QCD Lagrangian has a global symmetry related to the conserved right- or left-handedness (chirality) of zero mass spin 1/2 particles. We concentrate here first on the $N_f = 2$ sector of the two lightest (*u*- and *d*-) quarks with $\psi(x) = (u(x), d(x))^T$. Introducing right- and left-handed quark fields,

$$\psi_{R,L} = \frac{1}{2}(1 \pm \gamma_5)\psi, \tag{8}$$

we observe that separate global unitary transformations

$$\psi_R \to \exp[i\theta_R^a \frac{\tau_a}{2}]\psi_R, \qquad \psi_L \to \exp[i\theta_L^a \frac{\tau_a}{2}]\psi_L, \tag{9}$$

with τ_a ($a = 1, 2, 3$) the generators of the $SU(2)$ flavour group, leave $\mathcal{L}_{QCD}$ invariant in the limit $m \to 0$: the right- and left-handed components of the massless quark fields do not mix. This is the chiral $SU(2)_R \times SU(2)_L$ symmetry of QCD. It implies six conserved Noether currents, $J_{R,a}^\mu = \bar{\psi}_R \gamma^\mu \frac{\tau_a}{2} \psi_R$ and $J_{L,a}^\mu = \bar{\psi}_L \gamma^\mu \frac{\tau_a}{2} \psi_L$, with $\partial_\mu J_R^\mu = \partial_\mu J_L^\mu = 0$. It is common to introduce the vector current

$$V_a^\mu = J_{R,a}^\mu + J_{L,a}^\mu = \bar{\psi}\gamma^\mu \frac{\tau_a}{2}\psi, \tag{10}$$

and the axial current,

$$A_a^\mu(x) = J_{R,a} - J_{L,a} = \bar{\psi}\gamma^\mu \gamma_5 \frac{\tau_a}{2}\psi. \tag{11}$$

Their corresponding charges,

$$Q_a^V = \int d^3x \, \psi^\dagger(x)\frac{\tau_a}{2}\psi(x), \qquad Q_a^A = \int d^3x \, \psi^\dagger(x)\gamma_5\frac{\tau_a}{2}\psi(x), \tag{12}$$

are, likewise, generators of $SU(2) \times SU(2)$.

To the extent that the strange quark mass m_s can also be considered "small", it makes sense to generalise the chiral symmetry to $N_f = 3$. The three Pauli matrices τ_a are then replaced by the eight Gell-Mann matrices λ_a of $SU(3)$.

3.2 Spontaneous symmetry breaking

There is evidence from hadron spectroscopy that the chiral $SU(2) \times SU(2)$ symmetry of the QCD Lagrangian (1) with $m = 0$ is spontaneously broken: for dynamical reasons of non-perturbative origin, the ground state (vacuum) of QCD is symmetric only under the subgroup $SU(2)_V$ generated by the vector charges Q^V. This is the well-known isospin symmetry or, correspondingly, the "eightfold way" when extended to $N_f = 3$.

If the ground state of QCD were symmetric under chiral $SU(2) \times SU(2)$, both vector and axial charge operators (Equation 12) would annihilate the vacuum: $Q_a^V|0\rangle = Q_a^A|0\rangle = 0$. This is the Wigner-Weyl realisation of chiral symmetry with a "trivial" vacuum. It would imply the systematic appearance of parity doublets in the hadron spectrum.

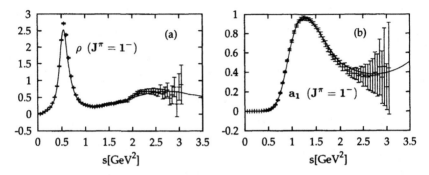

Figure 4. *Spectral distributions $\eta(s)$ of vector (a) and axial vector (b) mesons from τ decays (Barate et al. 1998, Ackerstaff et al. 1999). The curves are results of a QCD sum rule analysis (Marco and Weise 2000).*

Consider for example the correlation functions of vector and axial vector currents:

$$\Pi_V^{\mu\nu}(q) = i \int d^4x \, e^{iq\cdot x} \langle 0|\mathcal{T}[V^\mu(x)V^\nu(0)]|0\rangle\,, \tag{13}$$

$$\Pi_A^{\mu\nu}(q) = i \int d^4x \, e^{iq\cdot x} \langle 0|\mathcal{T}[A^\mu(x)A^\nu(0)]|0\rangle\,, \tag{14}$$

where $\mathcal{T}$ denotes time ordering and $\Pi^{\mu\nu}(q) = (q^\mu q^\nu - q^2 g^{\mu\nu})\Pi(q^2)$. If chiral symmetry were in its (trivial) Wigner-Weyl realisation, these two correlation functions should be identical: $\Pi_V = \Pi_A$. Consequently, their spectral distributions

$$\eta_{V,A}(s) = 4\pi \, Im\, \Pi_{V,A}(s = q^2) \tag{15}$$

which include the vector ($J^\pi = 1^-$) and axial vector ($J^\pi = 1^+$) mesonic excitations, should also be identical. This degeneracy is not observed in nature. The ρ meson mass ($m_\rho \simeq 0.77$ GeV) is well separated from that of the a_1 meson ($m_{a_1} \simeq 1.23$ GeV), as can be seen (Figure 4) in the resonance spectra measured in τ decays to the relevant channels. Likewise, the light pseudoscalar ($J^\pi = 0^-$) mesons have masses much lower than the lightest scalar ($J^\pi = 0^+$) mesons.

One must conclude that $Q_a^A|0\rangle \neq 0$, that is chiral symmetry is spontaneously broken down to isospin: $SU(2)_R \times SU(2)_L \to SU(2)_V$. This is the Nambu-Goldstone realisation of chiral symmetry.

A spontaneously broken global symmetry implies the existence of (massless) Goldstone bosons. If $Q_a^A|0\rangle \neq 0$, there must be a physical state generated by the axial charge, $|\phi_a\rangle = Q_a^A|0\rangle$, which is energetically degenerate with the vacuum. Let H_0 be the QCD Hamiltonian (with massless quarks) which commutes with the axial charge. Setting the ground state energy equal to zero for convenience, we have $H_0|\phi_a\rangle = Q_a^A H_0|0\rangle = 0$. Evidently $|\phi_a\rangle$ represents three massless pseudoscalar bosons (for $N_f = 2$). They are identified with the isotriplet of pions.

When the spontaneous symmetry breaking scenario is extended to chiral $SU(3)_R \times SU(3)_L \rightarrow SU(3)_V$, the corresponding eight Goldstone bosons are identified with the members of the lightest pseudoscalar octet: pions, kaons, antikaons and the η meson. The η' meson, on the other hand, falls out of this scheme. The large η' mass reflects the axial $U(1)_A$ anomaly in QCD. Without this anomaly, QCD with $N_f = 3$ massless quarks would actually have a chiral $U(3)_R \times U(3)_L$ symmetry, and its spontaneous breakdown would lead to nine rather than eight pseudoscalar Goldstone bosons. The axial anomaly removes the $U(1)_A$ symmetry, keeping $SU(3)_R \times SU(3)_L \times U(1)_V$ intact which is then spontaneously broken down to $SU(3)_V \times U(1)_V$. The remaining $SU(3)$ flavour symmetry is accompanied by the conserved baryon number which generates $U(1)_V$.

3.3 The chiral condensate

Spontaneous chiral symmetry breaking goes in parallel with a qualitative re-arrangement of the vacuum, an entirely non-perturbative phenomenon. The ground state is now populated by scalar quark-antiquark pairs. The corresponding ground state expectation value $\langle 0|\bar{\psi}\psi|0\rangle$ is called the chiral condensate (also referred to under the name "quark condensate"). We frequently use the notation

$$\langle \bar{\psi}\psi \rangle = \langle \bar{u}u \rangle + \langle \bar{d}d \rangle \tag{16}$$

(for two flavours, with $\langle \bar{s}s \rangle$ added when going to $N_f = 3$). The precise definition of the chiral condensate is:

$$\langle \bar{\psi}\psi \rangle = -i Tr \lim_{y \to x^+} S_F(x, y) \tag{17}$$

with the full quark propagator, $S_F(x, y) = -i\langle 0|\mathcal{T}\psi(x)\bar{\psi}(y)|0\rangle$. We recall Wick's theorem which states that the time-ordered product $\mathcal{T}\psi(x)\bar{\psi}(y)$ reduces to the normal product $: \psi(x)\bar{\psi}(y) :$ plus the contraction of the two field operators. When considering the perturbative quark propagator, $S_F^{(0)}(x, y)$, the time-ordered product is taken with respect to a trivial vacuum for which the expectation value of $: \bar{\psi}\psi :$ vanishes. Long-range, non-perturbative physics is then at the origin of a non-vanishing $\langle : \bar{\psi}\psi : \rangle$ in Equation 17. (In order to ensure that no perturbative pieces are left in $\langle \bar{\psi}\psi \rangle$ for the case of non-zero quark masses, one should actually make the replacement $S_F \to S_F - S_F^{(0)}$ in Equation 17.)

Let us establish the connection between spontaneous chiral symmetry breaking and the non-vanishing chiral condensate in a more formal way. Introduce the pseudoscalar quantity $P_a(x) = \bar{\psi}(x)\gamma_5 \tau_a \psi(x)$ and derive the (equal-time) commutator relation

$$[Q_a^A, P_b] = -\delta_{ab} \bar{\psi}\psi \tag{18}$$

which involves the axial charge Q_a^A of Equation 12. Taking the ground state expectation value on both sides of Equation 18, we see that $Q_a^A|0\rangle \neq 0$ is indeed consistent with $\langle \bar{\psi}\psi \rangle \neq 0$.

At this point a brief digression is useful in order to demonstrate spontaneous chiral symmetry breaking and its restoration at high temperatures as seen in computer simulations of QCD thermodynamics. In lattice QCD the path integral defining the partition function $\mathcal{Z}$ is regularised on a four-dimensional space-time grid with lattice constant a. Given N_σ lattice points along each space direction and N_τ points along the (Euclidean) time axis with $\tau = it$, volume and temperature are specified as $V = (N_\sigma a)^3$ and $T = (N_\tau a)^{-1}$. The chiral condensate $\langle \bar{\psi}\psi \rangle_T$ at finite temperature is derived starting from the pressure

$$P = T\frac{\partial}{\partial V}\ln \mathcal{Z} \tag{19}$$

by taking the derivative with respect to the quark mass:

$$\langle \bar{\psi}\psi \rangle_T \sim \frac{\partial P(T,V)}{\partial m_q}. \tag{20}$$

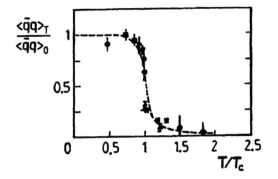

Figure 5. *Temperature dependence of the chiral (quark) condensate from lattice QCD (Boyd et al. 1995).*

An example of a lattice QCD result for $\langle \bar{\psi}\psi \rangle_T$ is given in Figure 5. The magnitude of the quark condensate decreases from its $T = 0$ value to zero beyond a critical temperature T_c. At $T > T_c$ chiral symmetry is restored in its Wigner-Weyl realisation. At $T < T_c$ the symmetry is spontaneously broken in the low-temperature (Nambu-Goldstone) phase. When extrapolated to zero quark masses, one finds a chiral 1st-order transition with critical temperatures depending on the number of flavours (Karsch *et al.* 2001): $T_c = (173\pm8)$ MeV for $N_f = 2$ and $T_c = (154\pm8)$ MeV for $N_f = 3$. The physically relevant situation, with chiral symmetry explicitly broken by finite quark masses $m_{u,d} \sim 5$ MeV and $m_s \sim 120$ MeV, is expected to be not a phase transition but a crossover with a rapid but continuous change of $\langle \bar{\psi}\psi \rangle_T$, as in Figure 5.

Spontaneous symmetry breaking is a widespread phenomenon, and analogues exist in various other areas of physics. Figure 5 is in fact reminiscent of the picture for the magnetisation of a ferromagnet. The basic Hamiltonian of this spin system is invariant under rotations in space. However, its low temperature phase is characterised by a non-vanishing magnetisation which points into a definite direction in space. Rotational symmetry is spontaneously broken. This is a non-perturbative, collective phenomenon. Many spins cooperate to form the macroscopically magnetised material. Slow variations in the direction of the magnetisation produce a collective low-frequency, long wavelength motion of the spins. This spin wave (magnon) is the Goldstone boson of spontaneously broken rotational symmetry. In our case of spontaneous *chiral* symmetry breaking, the analogue of the non-vanishing magnetisation is the chiral condensate; the analogue of the spin wave is the pion.

3.4 Partially conserved axial currents (PCAC)

Let $|\pi_a(p)\rangle$ be the state vectors of the Goldstone bosons associated with the spontaneous breakdown of chiral symmetry. Their four-momenta are denoted $p^\mu = (E_p, \vec{p})$, and we choose the standard normalisation $\langle \pi_a(p)|\pi_b(p')\rangle = 2E_p \delta_{ab}(2\pi)^3 \delta^3(\vec{p} - \vec{p}')$. Goldstone's theorem, briefly sketched in section 3.2, also implies non-vanishing matrix elements of the axial current (Equation 11) which connect $|\pi_a(p)\rangle$ with the vacuum:

$$\langle 0|A_a^\mu(x)|\pi_b(p)\rangle = ip^\mu f \, \delta_{ab} \, e^{-ip\cdot x}. \tag{21}$$

The constant f is called the pion decay constant (taken here in the chiral limit, *ie* for vanishing quark mass). Its physical value (Eidelmann *et al.* [Particle Data Group] 2004)

$$f_\pi = (92.4 \pm 0.3) \, MeV \tag{22}$$

is determined from the decay $\pi^+ \rightarrow \mu^+ \nu_\mu + \mu^+ \nu_\mu \gamma$. The difference between f and f_π is a correction linear in the quark mass m_q.

Non-zero quark masses $m_{u,d}$ shift the mass of the Goldstone boson from zero to the observed value of the physical pion mass, m_π. The relationship between m_π and the u and d quark masses is derived as follows. We start by observing that the divergence of the axial current (Equation 11) is

$$\partial_\mu A_a^\mu = i\bar{\psi}\{m, \frac{\tau_a}{2}\}\gamma_5\psi, \tag{23}$$

where m is the quark mass matrix and $\{,\}$ denotes the anti-commutator. This is the microscopic basis for PCAC, the Partially Conserved Axial Current (exactly conserved in the limit $m \rightarrow 0$) which plays a key role in the weak interactions of hadrons and the low-energy dynamics involving pions. Consider for example the $a = 1$ component of the axial current:

$$\partial_\mu A_1^\mu = (m_u + m_d)\bar{\psi}i\gamma_5\frac{\tau_1}{2}\psi, \tag{24}$$

and combine this with Equation 18 to obtain

$$\langle 0|[Q_1^A, \partial_\mu A_1^\mu]|0\rangle = -\frac{i}{2}(m_u + m_d)\langle \bar{u}u + \bar{d}d\rangle. \tag{25}$$

Now insert a complete set of (pseudoscalar) states in the commutator on the left. Assume, in the spirit of PCAC, that this spectrum of states is saturated by the pion. Then use Equation 21 to evaluate $\langle 0|Q^A|\pi\rangle$ and $\langle 0|\partial_\mu A^\mu|\pi\rangle$ at time $t = 0$, with $E_p = m_\pi$ at $\vec{p} = 0$, and arrive at the Gell-Mann, Oakes, Renner (GOR) relation (Gell-Mann *et al.* 1968):

$$m_\pi^2 f_\pi^2 = -(m_u + m_d)\langle \bar{q}q\rangle + \mathcal{O}(m_{u,d}^2). \tag{26}$$

We have set $\langle \bar{q}q\rangle \equiv \langle \bar{u}u\rangle \simeq \langle \bar{d}d\rangle$ making use of isospin symmetry which is valid to a good approximation. Neglecting terms of order $m_{u,d}^2$ (identifying $f = f_\pi = 92.4$

MeV to this order) and inserting $m_u + m_d \simeq 13$ MeV (Pich and Prades 2000) at a renormalisation scale of order 1 GeV, one obtains

$$\langle \bar{q}q \rangle \simeq -(0.23 \pm 0.03\,GeV)^3 \simeq -1.6\,fm^{-3}. \tag{27}$$

This condensate (or correspondingly, the pion decay constant f_π) is a measure of spontaneous chiral symmetry breaking. The non-zero pion mass, on the other hand, reflects the explicit symmetry breaking by the small quark masses, with $m_\pi^2 \sim m_q$. It is important to note that m_q and $\langle \bar{q}q \rangle$ are both scale dependent quantities. Only their product $m_q \langle \bar{q}q \rangle$ is scale independent, *ie* invariant under the renormalisation group.

3.5   The mass gap and spontaneous chiral symmetry breaking

The appearance of the mass gap $\Gamma \sim 1$ GeV in the hadron spectrum is thought to be closely linked to the presence of the chiral condensate $\langle \bar{\psi}\psi \rangle$ in the QCD ground state. For example, Ioffe's formula (Ioffe 1981), based on QCD sum rules, connects the nucleon mass M_N directly with $\langle \bar{\psi}\psi \rangle$ in leading order:

$$M_N = -\frac{4\pi^2}{\Lambda_B^2} \langle \bar{\psi}\psi \rangle + \dots \, , \tag{28}$$

where $\Lambda_B \sim 1$ GeV is an auxiliary scale (the Borel scale) which separates "short" and "long"-distance physics in the QCD sum rule analysis. While this formula is not very accurate and needs to be improved by including higher order condensates, it nevertheless demonstrates that spontaneous chiral symmetry breaking plays an essential role in giving the nucleon its mass.

The condensate $\langle \bar{\psi}\psi \rangle$ is encoded in the pion decay constant f_π through the GOR relation (Equation 26). In the chiral limit ($m_q \to 0$), this f_π is the only quantity which can serve to define a mass scale ("transmuted" from the QCD scale Λ in Equation 4 through non-perturbative dynamics). It is common to introduce $4\pi f_\pi \sim 1$ GeV as the scale characteristic of spontaneous chiral symmetry breaking. This scale is then roughly identified with the spectral gap Γ.

As another typical example of how the chiral gap translates into hadron masses, consider the ρ and a_1 mesons. Finite-energy sum rules for vector and axial vector current-current correlation functions, when combined with the algebra of these currents as implied by the chiral $SU(2) \times SU(2)$ group, do in fact connect the ρ and a_1 masses directly with the chiral gap (Weinberg 1967, Klingl and Weise 1999, Goltermann and Peris 2000, Marco and Weise 2000):

$$m_{a_1} = \sqrt{2}\,m_\rho = 4\pi f_\pi, \tag{29}$$

at least in leading order (that is, in the large N_c limit, and ignoring decay widths as well as perturbative QCD corrections).

The relations (Equations 28, 29), while not accurate at a quantitative level, give important hints. Systems characterised by an energy gap usually exhibit qualitative changes when exposed to variations of thermodynamic conditions. A typical example is the temperature dependence of the gap in a superconductor. Exploring systematic changes of hadronic spectral functions in a dense and hot medium is therefore a key issue in nuclear and hadron physics (Metag 2004).

4 Effective field theory

4.1 Framework and rules

The scale set by the mass gap $\Gamma \sim 4\pi f_\pi$ offers a natural separation between "light" and "heavy" (or, correspondingly, "fast" and "slow") degrees of freedom. The basic idea of an effective field theory is to introduce the active light particles as collective degrees of freedom, while the heavy particles are frozen and treated as (almost) static sources. The dynamics are described by an effective Lagrangian which incorporates all relevant symmetries of the underlying fundamental theory. We now summarise the necessary steps, first for the pure meson sector (baryon number $B = 0$) and later for the $B = 1$ sector. We work mostly with $N_f = 2$ in this section and turn to the $N_f = 3$ case later.

a) The elementary quarks and gluons of QCD are replaced by Goldstone bosons. They are represented by a 2×2 matrix field $U(x) \in SU(2)$ which collects the three isospin components $\pi_a(x)$ of the Goldstone pion. A convenient choice of coordinates is

$$U(x) = \exp[i\tau_a \phi_a(x)] \, , \tag{30}$$

with $\phi_a = \pi_a/f$ where the pion decay constant f in the chiral limit provides a suitable normalisation. (Other choices, such as $U = \sqrt{1 - \pi_a^2/f^2} + i\tau_a\pi_a/f$, are also common. Results obtained from the effective theory must be independent of the coordinates used.)

b) Goldstone bosons interact weakly at low energy. In fact, if $|\pi\rangle = Q^A|0\rangle$ is a massless state with $H|\pi\rangle = 0$, then a state $|\pi^n\rangle = (Q^A)^n|0\rangle$ with n Goldstone bosons is also massless since the axial charges Q^A all commute with the full Hamiltonian H. Interactions between Goldstone bosons must therefore vanish at zero momentum and in the chiral limit.

c) The QCD Lagrangian (1) is replaced by an effective Lagrangian which involves the field $U(x)$ and its derivatives:

$$\mathcal{L}_{QCD} \to \mathcal{L}_{eff}(U, \partial U, \partial^2 U, ...). \tag{31}$$

Goldstone bosons can only interact when they carry momentum, so the low-energy expansion of (31) is an ordering in powers of $\partial_\mu U$. Lorentz invariance permits only even numbers of derivatives. We write

$$\mathcal{L}_{eff} = \mathcal{L}^{(2)} + \mathcal{L}^{(4)} + ... \tag{32}$$

and omit an irrelevant constant term. The leading term, called non-linear sigma model, involves two derivatives:

$$\mathcal{L}^{(2)} = \frac{f^2}{4} Tr[\partial_\mu U^\dagger \partial^\mu U]. \tag{33}$$

At fourth order, the terms permitted by symmetries are (apart from an extra contribution from the QCD anomaly, not included here):

$$\mathcal{L}^{(4)} = \frac{l_1}{4}(Tr[\partial_\mu U^\dagger \partial^\mu U])^2 + \frac{l_2}{4}Tr[\partial_\mu U^\dagger \partial_\nu U]Tr[\partial^\mu U^\dagger \partial^\nu U], \tag{34}$$

and so forth. The constants l_1, l_2 (following canonical notations of Gasser and Leutwyler 1984) must be determined by experiment. To the extent that the effective Lagrangian includes all terms dictated by the symmetries, the chiral effective field theory is the low-energy equivalent of the original QCD Lagrangian (Weinberg 1979, Leutwyler 1994).

d) The symmetry breaking mass term is small, so that it can be handled perturbatively, together with the power series in momentum. The leading contribution introduces a term linear in the quark mass matrix m:

$$\mathcal{L}^{(2)} = \frac{f^2}{4} Tr[\partial_\mu U^\dagger \partial^\mu U] + \frac{f^2}{2} B_0 \, Tr[m(U + U^\dagger)]. \tag{35}$$

The fourth order term $\mathcal{L}^{(4)}$ also receives symmetry breaking contributions with additional constants l_i.

When expanding $\mathcal{L}^{(2)}$ to terms quadratic in the pion field, one finds

$$\mathcal{L}^{(2)} = (m_u + m_d) f^2 B_0 + \frac{1}{2} \partial_\mu \pi_a \, \partial^\mu \pi_a - \frac{1}{2}(m_u + m_d) B_0 \, \pi_a^2 + 0(\pi^4). \tag{36}$$

At this point we can immediately verify the GOR relation (Equation 26) in the effective theory. The first (constant) term in Equation 36 corresponds to the shift of the vacuum energy density by the non-zero quark masses. Identifying this with the vacuum expectation value of the corresponding piece in the QCD Lagrangian (1), we find $-(m_u \langle \bar{u}u \rangle + m_d \langle \bar{d}d \rangle) = (m_u + m_d) f^2 B_0$ and therefore $\langle \bar{u}u \rangle = \langle \bar{d}d \rangle = -f^2 B_0$ in the chiral limit, $m_{u,d} \to 0$. The pion mass term in Equation 36 is evidently identified as $m_\pi^2 = (m_u + m_d) B_0$. Inserting B_0, we have the GOR relation.

Symmetry breaking terms entering at fourth order are of the form

$$\delta \mathcal{L}^{(4)} = \frac{l_3}{4} B_0^2 \left(Tr[m(U + U^\dagger)] \right)^2 + \frac{l_4}{4} B_0 \, Tr[\partial_\mu U^\dagger \partial^\mu U] \, Tr[m(U + U^\dagger)] + ... \tag{37}$$

e) Given the effective Lagrangian, the framework for systematic perturbative calculations of the S-matrix involving Goldstone bosons, named Chiral Perturbation Theory (ChPT), is then defined by the following rules: Collect all Feynman diagrams generated by $\mathcal{L}_{eff}$. Classify all terms according to powers of a variable Q which stands generically for three-momentum or energy of the Goldstone bosons, or for the pion mass m_π. The small expansion parameter is $Q/4\pi f_\pi$. Loops are subject to dimensional regularisation and renormalisation.

4.2 Pion-pion scattering

When using the GOR relation to leading order in the quark mass, it is tacitly assumed that the chiral condensate is large in magnitude and plays the role of an order parameter for spontaneous chiral symmetry breaking. This basic scenario has been challenged (Knecht et al. 1996) and needs to be confirmed. It can in fact be tested by a detailed quantitative analysis of pion-pion scattering (Leutwyler 2002), the process most extensively studied using ChPT.

Consider s-wave $\pi\pi$ scattering at very low energy. The symmetry of the $\pi\pi$ wavefunction with $l = 0$ requires isospin to be even, $I = 0, 2$. The corresponding scattering lengths, to leading chiral order, are

$$a_0^{I=0} = \frac{7\pi}{2} \zeta \quad , \quad a_0^{I=2} = -\pi \zeta \quad , \tag{38}$$

where $\zeta = (m_\pi/4\pi f_\pi)^2 \approx 1.45 \cdot 10^{-2}$ is the "small parameter". Note that the $\pi\pi$ interaction properly vanishes in the chiral limit, $m_\pi \to 0$. The next-to-leading order (NLO) introduces one-loop iterations of the leading $\mathcal{L}^{(2)}$ part of the effective Lagrangian as well as pieces generated by $\mathcal{L}^{(4)}$. At that level the renormalised coupling constant $\bar{l}_3$ enters, which also determines the correction to the leading-order GOR relation:

$$m_\pi^2 = \overset{o}{m}_\pi^2 - \frac{1}{2}\frac{\bar{l}_3}{(4\pi f)^2}\overset{o}{m}_\pi^4 + \mathcal{O}(\overset{o}{m}_\pi^6) \, , \tag{39}$$

where

$$\overset{o}{m}_\pi^2 = -\frac{m_u + m_d}{f^2}\langle \bar{q}q \rangle \tag{40}$$

involves the pion mass to leading order in the quark mass and f is the pion decay constant in the chiral limit. An accurate determination of the $I = 0$ s-wave $\pi\pi$ scattering length therefore provides a constraint for $\bar{l}_3$ which in turn sets a limit for the NLO correction to the GOR relation.

Such an investigation has recently been performed (Colangelo *et al.* 2001), based on low-energy $\pi\pi$ phase shifts extracted from the detailed final state analysis of the $K \to \pi\pi + e\nu$ decay. The result is that the correction to the leading order prediction (Equation 38) is indeed very small. When translated into a statement about the non-leading term entering Equation 39, it implies that the difference between m_π^2 and the leading GOR expression (Equation 40) is less than 5 percent. Hence the "strong condensate" scenario of spontaneous chiral symmetry breaking in QCD appears to be confirmed. One should note, however, that this conclusion is drawn at the level of QCD with only $N_f = 2$ flavours. Additional corrections may still arise when strange quarks are taken into account. It is nontheless interesting to note that the leading-order relationship $m_\pi^2 \sim m_q$ is also observed in lattice QCD (Aoki *et al.* 2003) up to surprisingly large quark masses.

5 The nucleon in the QCD vacuum

As the lowest-mass excitation of the QCD vacuum with one unit of baryon number, the nucleon is of fundamental importance to our understanding of the strong interaction (Thomas and Weise 2001). The prominent role played by the pion as a Goldstone boson of spontaneously broken chiral symmetry has its impact on the low-energy structure and dynamics of nucleons as well. Decades of research in nuclear physics have established the pion as the mediator of the long-range force between nucleons. When probing the individual nucleon itself with long-wavelength electroweak fields, a substantial part of the response comes from the pion cloud, the "soft" surface of the nucleon.

The calculational framework for this, baryon chiral perturbation theory, has been applied quite successfully to a variety of low-energy processes (such as threshold pion photo- and electroproduction and Compton scattering on the nucleon) for which increasingly accurate experimental data have become available in recent years. A detailed review can be found in Bernard *et al.* (1995). An introductory survey is given in Thomas and Weise (2001).

5.1   Effective Lagrangian including baryons

Let us move to the sector with baryon number $B = 1$ and concentrate on the physics of the pion-nucleon system, restricting ourselves first to the case of $N_f = 2$ flavours. (The generalisation to $N_f = 3$ will follow later).

The nucleon is represented by a Dirac spinor field $\Psi_N(x) = (p, n)^T$ organised as an isospin-1/2 doublet of proton and neutron. The free field Lagrangian

$$\mathcal{L}_0^N = \overline{\Psi}_N(i\gamma_\mu\partial^\mu - M_0)\Psi_N \tag{41}$$

includes the nucleon mass in the chiral limit, M_0. Note that nucleons, unlike pions, are supposed to have at least part of their large mass because of the strong mean field provided by the quark condensate $\langle\bar{\psi}\psi\rangle$. Such a relationship is explicit, for example, in the Ioffe formula (Equation 28).

We can now construct the low-energy effective Lagrangian for pions interacting with a nucleon. The pure meson Lagrangian $\mathcal{L}_{eff}$ is replaced by $\mathcal{L}_{eff}(U, \partial^\mu U, \Psi_N, ...)$ which also includes the mucleon field. The additional term involving the nucleon, denoted by $\mathcal{L}_{eff}^N$, is expanded again in powers of derivatives (external momenta) and quark masses:

$$\mathcal{L}_{eff}^N = \mathcal{L}_{\pi N}^{(1)} + \mathcal{L}_{\pi N}^{(2)} ... \tag{42}$$

Let us discuss the leading term, $\mathcal{L}_{\pi N}^{(1)}$. The modifications, compared to the free nucleon Lagrangian, are twofold. First, there is a replacement of the ∂^μ term by a chiral covariant derivative which introduces vector current couplings between the pions and the nucleon. Secondly, there is an axial vector coupling. This structure of the πN effective Lagrangian is again dictated by chiral symmetry. We have

$$\mathcal{L}_{\pi N}^{(1)} = \overline{\Psi}_N[i\gamma_\mu(\partial^\mu - i\mathcal{V}^\mu) + \gamma_\mu\gamma_5 \mathcal{A}^\mu - M_0]\Psi_N , \tag{43}$$

with vector and axial vector quantities involving the Goldstone boson (pion) fields in the form $\xi = \sqrt{U}$:

$$\mathcal{V}^\mu = \frac{i}{2}(\xi^\dagger\partial^\mu\xi + \xi\partial^\mu\xi^\dagger) = -\frac{1}{4f^2}\varepsilon_{abc}\tau_a\,\pi_b\,\partial^\mu\pi_c + \cdots , \tag{44}$$

$$\mathcal{A}^\mu = \frac{i}{2}(\xi^\dagger\partial^\mu\xi - \xi\partial^\mu\xi^\dagger) = -\frac{1}{2f^2}\tau_a\,\partial^\mu\pi_a + \cdots , \tag{45}$$

where the last steps result when expanding $\mathcal{V}^\mu$ and $\mathcal{A}^\mu$ to leading order in the pion fields.

So far, the only parameters that enter are the nucleon mass, M_0, and the pion decay constant, f, both taken in the chiral limit and ultimately connected with a single scale characteristic of non-perturbative QCD and spontaneous chiral symmetry breaking.

When adding electroweak interactions to this scheme, one observes an additional feature which has its microscopic origin in the substructure of the nucleon, not resolved at the level of the low-energy effective theory. The analysis of neutron beta decay ($n \to pe\bar{\nu}$) reveals that the $\gamma_\mu\gamma_5$ term in Equation 43 is to be multiplied by the axial vector coupling constant g_A, with the empirical value (Eidelmann *et al.* [Particle Data Group] 2004):

$$g_A = 1.270 \pm 0.003 . \tag{46}$$

At next-to-leading order ($\mathcal{L}_{\pi N}^{(2)}$), the symmetry breaking quark mass term enters. It has the effect of shifting the nucleon mass from its value in the chiral limit to the physical one:

$$M_N = M_0 + \sigma_N \ . \tag{47}$$

The sigma term

$$\sigma_N = m_q \frac{\partial M_N}{\partial m_q} = \langle N | m_q (\bar{u}u + \bar{d}d) | N \rangle \tag{48}$$

measures the contribution of the non-vanishing quark mass, $m_q = \frac{1}{2}(m_u + m_d)$, to the nucleon mass M_N. Its empirical value is in the range $\sigma_N \simeq (45 - 55)$ MeV and has been deduced by a sophisticated extrapolation of low-energy pion-nucleon data using dispersion relation techniques (Gasser *et al.* 1991a, Gasser *et al.* 1991b, Sainio 2002).

Up to this point, the πN effective Lagrangian, expanded to second order in the pion field, has the form

$$\mathcal{L}_{eff}^N = \overline{\Psi}_N (i\gamma_\mu \partial^\mu - M_0) \Psi_N - \frac{g_A}{2f_\pi} \overline{\Psi}_N \gamma_\mu \gamma_5 \tau \Psi_N \cdot \partial^\mu \pi \tag{49}$$
$$- \frac{1}{4f_\pi^2} \overline{\Psi}_N \gamma_\mu \tau \Psi_N \cdot \pi \times \partial^\mu \pi - \sigma_N \overline{\Psi}_N \Psi_N \left(1 - \frac{\pi^2}{2f_\pi^2} \right) + \dots \ ,$$

where we have not shown a series of additional terms of order $(\partial^\mu \pi)^2$ included in the complete $\mathcal{L}_{\pi N}^{(2)}$. These terms come with further constants that need to be fitted to experimental data.

5.2 Scalar form factor of the nucleon

While pions are well established constituents seen in the electromagnetic structure of the nucleon, its scalar-isoscalar meson cloud is less familiar. On the other hand, the scalar field of the nucleon is at the origin of the intermediate range nucleon-nucleon force, the source of attraction that binds nuclei. Let us therefore have a closer look, guided by chiral effective field theory.

Consider the nucleon form factor, at squared momentum transfer $q^2 = (p - p')^2$, related to the scalar-isoscalar quark density, $G_S(q^2) = \langle N(p') | \bar{u}u + \bar{d}d | N(p) \rangle$. In fact, a better quantity to work with is the form factor $\sigma_N(q^2) = m_q G_S(q^2)$ associated with the scale invariant object $m_q(\bar{u}u + \bar{d}d)$. Assume that this form factor can be written as a subtracted dispersion relation:

$$\sigma_N(q^2 = -Q^2) = \sigma_N - \frac{Q^2}{\pi} \int_{4m_\pi^2}^{\infty} dt \frac{\eta_S(t)}{t(t + Q^2)} \ , \tag{50}$$

where the sigma term σ_N introduced previously enters as a subtraction constant. We are interested in spacelike momentum transfers with $Q^2 = -q^2 \geq 0$. The dispersion integral in Equation 50 starts out at the two-pion threshold. It involves the spectral function $\eta_S(t)$ which includes all $J^\pi = 0^+, I = 0$ excitations coupled to the nucleon: a continuum of even numbers of pions added to and interacting with the nucleon core.

Chiral perturbation theory at next-to-next-to-leading order (NNLO) in two-loop approximation has been applied in a recent calculation of the spectral function $\eta_S(t)$ (Kaiser

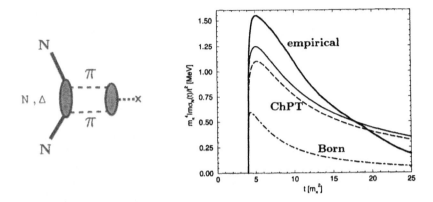

Figure 6. *Left: illustration of the scalar form factor of the nucleon. Right: spectral function $\eta(t)/t^2$ of the scalar form factor (Equation 50). The empirical spectral distribution is taken from Höhler (1983). Also shown are two-loop NNLO chiral perturbation theory (ChPT) calculations (Kaiser 2003). The contribution from the nucleon Born term is given separately.*

2003). This calculation includes not only nucleon Born terms and leading $\pi\pi$ interactions but also important effects of intermediate $\Delta(1230)$ isobar excitations in the two-pion dressing of the nucleon core (see Figure 6 left). The result (Figure 6 right) can be compared with the "empirical" scalar-isoscalar spectral function deduced by analytic continuation from πN, $\pi\pi$ and $\overline{N}N \leftrightarrow \pi\pi$ amplitudes (Höhler 1983). Note that there is no such thing as a "sigma meson" in this spectrum which is completely determined by the (chiral) dynamics of the interacting $\pi\pi$ and πN system.

The integral over $\eta_S(t)t^{-2}$ is proportional to the mean squared scalar radius of the nucleon. One finds (Gasser *et al.* 1991a, 1991b)

$$\langle r_S \rangle^{1/2} \simeq 1.3\, fm\,, \tag{51}$$

the largest of all nucleon radii, considerably larger than the proton charge radius of 0.86 fm.

By its magnitude and range, the form factor $G_S(q^2)$ implies that the nucleon, surrounded with its two-pion cloud, is the source of a strong scalar-isoscalar field with a large effective coupling constant $g_S = G_S(q^2 = 0) = \sigma_N/m_q \simeq 10$. When a second nucleon couples to this scalar field, the resulting two-pion exchange NN interaction $V_{2\pi}$ is reminiscent of a Van der Waals force. More than half of the strength of $V_{2\pi}$ is actually governed by the large spin-isospin polarisability of the nucleon related to the transition $N \to \Delta$ in the intermediate state. At long and intermediate distances it behaves as (Kaiser *et al.* 1998)

$$V_{2\pi}(r) \sim \frac{e^{-2m_\pi r}}{r^6} P(m_\pi r)\,, \tag{52}$$

where P is a polynomial in $m_\pi r$. In the chiral limit ($m_\pi \to 0$), this $V_{2\pi}$ approaches the characteristic r^{-6} dependence of a non-relativistic Van der Waals potential.

The two-pion exchange force is the major source of intermediate range attraction that binds nuclei. This is, of course, not a new observation. For example, the important role of the second-order tensor force from iterated pion exchange had been emphasised long ago (Brown 1971), as well as the close connection of the nuclear force with the strong spin-isospin polarisability of the nucleon (Ericson and Figureau 1981). The new element that has entered the discussion more recently is the systematics provided by chiral effective field theory in dealing with these phenomena. In fact, in-medium chiral perturbation theory combined with constraints from QCD sum rules is now considered a promising basis for approaching the nuclear many-body problem and finite nuclei in the context of low-energy QCD (Kaiser *et al.* 2002, Finelli *et al.* 2004, Vretenar and Weise 2004, Weise 2005).

5.3 Nucleon mass and pion cloud

Understanding the nucleon mass is clearly one of the most fundamental issues in nuclear and particle physics (Thomas and Weise 2001). Progress is now being made towards a synthesis of lattice QCD and chiral effective field theory, such that extrapolations of lattice results to actual observables are beginning to be feasible (Leinweber *et al.* 2004, Procura *et al.* 2004). Accurate computations of the nucleon mass on the lattice have become available (Ali Khan *et al.* 2002, 2004; Aoki *et al.* 2003), but so far with u and d quark masses exceeding their commonly accepted small values by typically an order of magnitude. Methods based on chiral effective theory can then be used, within limits, to interpolate between lattice results and physical observables.

The nucleon mass is determined by the expectation value $\langle N|\Theta^\mu_\mu|N\rangle$ of the trace of the QCD energy-momentum tensor (see *eg* Donoghue *et al.* 1992),

$$\Theta^\mu_\mu = \frac{\beta(g)}{2g}G^j_{\mu\nu}G^{\mu\nu}_j + m_u\bar{u}u + m_d\bar{d}d + \dots\,,$$

where $G^{\mu\nu}$ is the gluonic field tensor, $\beta(g) = \frac{\partial g}{\partial \ln \mu}$ is the beta function of QCD, and $m_q\bar{q}q$ with $q = u, d, \dots$ are the quark mass terms (omitting here the anomalous dimension of the mass operator for brevity). Neglecting small contributions from heavy quarks, the nucleon mass taken in the $N_f = 2$ chiral limit, $m_{u,d} \to 0$, is

$$M_0 = \langle N|\frac{\beta}{2g}G^j_{\mu\nu}G^{\mu\nu}_j|N\rangle\,. \tag{53}$$

This relation emphasises the gluonic origin of the bulk part of M_N, the part for which lattice QCD provides an approriate tool to generate the relevant gluon field configurations. At the same time, QCD sum rules connect M_0 to the chiral condensate $\langle \bar{q}q \rangle$ as in Ioffe's formula (Equation 28).

In chiral effective field theory, the quark mass dependence of M_N translates into a dependence on the pion mass, $m^2_\pi \sim m_q$, at leading order. The dressing of the nucleon with its pion cloud, at one-loop order, is illustrated in Figure 7 (left). The systematic chiral expansion of the nucleon mass gives an expression of the form (Procura *et al.* 2004):

$$M_N = M_0 + cm^2_\pi + dm^4_\pi - \frac{3\pi}{2}g^2_A m_\pi \left(\frac{m_\pi}{4\pi f_\pi}\right)^2 \left(1 - \frac{m^2_\pi}{8M^2_0}\right) + \mathcal{O}(m^6_\pi)\,, \tag{54}$$

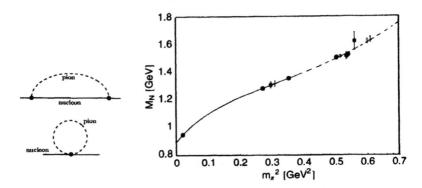

Figure 7. *Left: illustration of pion cloud contributions to the nucleon mass. Right: ChPT interpolation (Procura et al. 2004) between lattice QCD results for the nucleon mass (at $m_\pi > 0.5$ GeV) and the physical point ($m_\pi = 0.14$ GeV). The lattice QCD data are selected from Ali Khan et al. (2003) and Aoki et al. (2003).*

where the coefficients c and d multiplying even powers of the pion mass include low-energy constants constrained by pion-nucleon scattering. The coefficient d also involves a $\log m_\pi$ term. Note that the piece of order m_π^3 (non-analytic in the quark mass) is given model-independently in terms of the known weak decay constants g_A and f_π (strictly speaking: by their values in the chiral limit). The interpolation shown in Figure 7 determines the nucleon mass M_0 in the chiral limit and the sigma term σ_N. One finds $M_0 \simeq 0.89$ GeV and $\sigma_N = (49 \pm 4)$ MeV (Procura *et al.* 2004) in this approach.

Admittedly, the gap between presently available lattice results and the actual physical world is still on the large side, perhaps too large for ChPT expansions to extrapolate reliably up to quark masses $m_q > 70$ MeV. However, the synthesis of lattice QCD and chiral effective field theory offers promising perspectives for future developments once lattice QCD begins to operate with quark masses that reach down to pion masses $m_\pi \sim$ 0.3 GeV.

5.4   Chiral SU(3) dynamics

Chiral perturbation theory as a systematic expansion in small momenta and quark masses is limited to low-energy processes with light quarks. It is an interesting issue to what extent its generalisation including strangeness can be made to work. The $\overline{K}N$ channel is of particular interest in this context, as a testing ground for chiral SU(3) symmetry in QCD and for the role of explicit chiral symmetry breaking by the relatively large strange quark mass. However, any perturbative approach breaks down in the vicinity of resonances. In the K^-p channel, for example, the existence of the $\Lambda(1405)$ resonance just below the K^-p threshold renders SU(3) chiral perturbation theory inapplicable. At this point the combination with non-perturbative coupled-channels techniques has proven useful, by generating the $\Lambda(1405)$ dynamically as an $I = 0$ $\overline{K}N$ quasibound state and as a resonance in the $\pi\Sigma$ channel (Kaiser *et al.* 1995). This theme actually has its early roots in the coupled-channel K-matrix methods developed in the sixties (Dalitz *et al.* 1967). Coupled-channels methods combined with chiral $SU(3)$ dynamics have subsequently been applied to a variety of meson-baryon scattering processes with quite some success (Kaiser *et al.* 1997, Oset and Ramos 1998, Caro Ramon *et al.* 2000, Oller and Meissner 2001, Lutz and Kolomeitsev 2002).

Effective Lagrangian. The starting point is the chiral $SU(3) \times SU(3)$ effective Lagrangian. The chiral field $U(x)$ of Equation 30 is now a 3×3 matrix field which collects the full octet of pseudoscalar Goldstone bosons ($\pi, K, \overline{K}, \eta$). The symmetry breaking mass term in Equation 35 is proportional to the quark mass matrix $diag(m_u, m_d, m_s)$ including the mass of the strange quark. The baryon field in Equations 41 and 43 is generalised in matrix form, Ψ_B, which incorporates the baryon octet ($p, n, \Lambda, \Sigma, \Xi$). The leading meson-baryon term of the effective Lagrangian which replaces Equation 49 becomes:

$$
\begin{aligned}
\mathcal{L}_{eff}^B &= Tr[\overline{\Psi}_B(i\gamma_\mu \mathcal{D}^\mu - M_0)\Psi_B] \\
&+ F\, Tr(\overline{\Psi}_B \gamma_\mu \gamma_5 [\mathcal{A}^\mu, \Psi_B]) + D\, Tr((\overline{\Psi}_B \gamma_\mu \gamma_5 \{\mathcal{A}^\mu, \Psi_B\})
\end{aligned} \tag{55}
$$

with the chiral covariant derivative $\mathcal{D}^\mu \Psi_B = \partial^\mu \Psi_B - i[\mathcal{V}^\mu, \Psi_B]$. The $\mathcal{V}^\mu$ and $\mathcal{A}^\mu$ are the vector and axial vector combinations of octet Goldstone boson fields which generalise those of Equations 44 and 45. The $SU(3)$ axial vector coupling constants $D = 0.80 \pm 0.01$ and $F = 0.47 \pm 0.01$ add up to $D + F = g_A = 1.27$. At next-to-leading order, seven additional constants enter in s-wave channels, three of which are constrained by mass splittings in the baryon octet and the remaining four need to be fixed by comparison with low-energy scattering data.

Coupled channels. Meson-baryon scattering amplitudes based on the $SU(3)$ effective Lagrangian involve a set of coupled channels. For example, The K^-p system in the isospin $I = 0$ sector couples to the $\pi\Sigma$ channel. Consider the T matrix $\mathbf{T}_{\alpha\beta}(p, p'; E)$ connecting meson-baryon channels α and β with four-momenta p, p' in the centre-of-mass frame:

$$
\mathbf{T}_{\alpha\beta}(p, p') = \mathbf{K}_{\alpha\beta}(p, p') + \sum_\gamma \int \frac{d^4q}{(2\pi)^4} \mathbf{K}_{\alpha\gamma}(p, q)\, \mathbf{G}_\gamma(q)\, \mathbf{T}_{\gamma\beta}(q, p') \,, \tag{56}
$$

where $\mathbf{G}$ is the Green's function describing the intermediate meson-baryon loop which is iterated to all orders in integral Equation 56. (Dimensional regularisation with subtraction constants is used in practise). The driving terms $\mathbf{K}$ in each channel are constructed from the chiral $SU(3)$ meson-baryon effective Lagrangian in next-to-leading order. In the kaon-nucleon channels, for example, the leading terms have the form

$$
\mathbf{K}_{K^\pm p} = 2\mathbf{K}_{K^\pm n} = \mp 2M_N \frac{\sqrt{s} - M_N}{f^2} + \cdots \,, \tag{57}
$$

at zero three-momentum, where $\sqrt{s}$ is the invariant CM energy and f is the pseudoscalar meson decay constant ($f \simeq 90$ MeV). Scattering amplitudes are related to the T matrix by $\mathbf{f} = \mathbf{T}/8\pi\sqrt{s}$. Note that $\mathbf{K} > 0$ means attraction, as seen for example in the $K^-p \to K^-p$ channel. Similarly, the coupling from K^-p to $\pi\Sigma$ provides attraction, as well as the diagonal matrix elements in the $\pi\Sigma$ channels. Close to the $\overline{K}N$ threshold, we have

$f(K^-p \to K^-p) \simeq m_K/4\pi f^2$, the analogue of the Tomosawa-Weinberg term in pion-nucleon scattering, but now with the (attractive) strength considerably enhanced by the larger kaon mass m_K.

One should note that when combining chiral effective field theory with the coupled-channels scheme, the "rigorous" chiral counting in powers of small momenta is abandoned in favour of iterating a subclass of loop diagrams to *all* orders. However, the gain in physics insights may well compensate for this sacrifice. Important non-perturbative effects are now included and necessary conditions of unitarity are fulfilled.

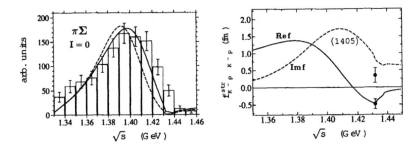

Figure 8. *Left: $\pi\Sigma$ invariant mass spectrum featuring the $\Lambda(1405)$ resonance. Right: real and imaginary parts of the K^-p amplitude. Curves are calculated in the chiral coupled channels approach (Borasoy et al. 2004). The real and imaginary parts (lower and upper data points) of the K^-p scattering length deduced from the DEAR kaonic hydrogen measurements (Beer et al. 2004) are also shown. In both figures $\sqrt{s}$ is the invariant $\overline{K}N$ centre-of-mass energy.*

Low-energy kaon-nucleon interactions. K^-p threshold data have recently been supplemented by new accurate results for the strong interaction shift and width of kaonic hydrogen (Beer *et al.* 2004, Iwasaki *et al.* 1997, Ito *et al.* 1998). These data, together with existing information on K^-p scattering, the $\pi\Sigma$ mass spectrum and measured K^-p threshold decay ratios, set tight constraints on the theory and have therefore revived the interest in this field. Figure 8 shows selected recent results of an improved calculation which combines driving terms from the next-to-leading order chiral $SU(3)$ meson-baryon Lagrangian with coupled-channel equations (Borasoy *et al.* 2004). As in previous calculations of such kind, the $\Lambda(1405)$ is generated dynamically, as an $I = 0$ $\overline{K}N$ quasi-bound state and a resonance in the $\pi\Sigma$ channel. In a quark model picture, this implies that the $\Lambda(1405)$ is not a simple three-quark (q^3) state but has a strong $q^4\bar{q}$ component. The detailed threshold behaviour of the elastic K^-p amplitude in Figure 8 (right) needs to be further examined in view of the improved accuracy of the most recent kaonic hydrogen data from the DEAR experiment.

6   Correlations and quasiparticles

In the preceding sections our interest has been focused primarily on the low-energy, long-wavelength limit of QCD. In this limit the physics is expressed in terms of colour singlet quasiparticle excitations of the QCD vacuum, the lightest hadrons. The spontaneous chiral symmetry breaking scenario in the light-quark sector is the guiding principle for constructing an effective field theory that represents strong interaction physics at momenta small compared to the "chiral gap", $4\pi f_\pi \sim 1$ GeV.

At the other end, on large momentum scales in the multi-GeV range, perturbative QCD operates successfully with quasifree quarks and gluons. Thus there is an obvious and challenging question: how does QCD behave at scales intermediate between those two limits? Do correlations between coloured objects exist in such a way that quasiparticle structures are formed which are not observed asymptotically but may have a meaning at intermediate scales? The phenomenological success of the heuristic constituent quark model might indicate such structures. Pairing interactions between constituent quarks as quasiparticles might form new quasiparticles, such as diquarks. The quasiparticle concept has proved highly successful in condensed matter physics. It makes sense to explore its potential relevance also in strongly interacting many-body systems of quarks and gluon fields. Of course, entering this discussion means that one must at least in part rely on phenomenology rather than controlled expansions and approximation schemes. However, lattice QCD results can again provide useful guidance.

6.1 Gluonic field strength correlations from lattice QCD

The QCD vacuum hosts not only a condensate of scalar quark-antiquark pairs, but also a strong gluon condensate, the vacuum expectation value of $Tr(G_{\mu\nu}G^{\mu\nu}) \sim \mathbf{B}^2 - \mathbf{E}^2$ where $\mathbf{B}$ and $\mathbf{E}$ are the chromomagnetic and chromoelectric fields. We work here in Euclidean space and write

$$\mathcal{G}_0 \equiv \langle \frac{2\alpha_s}{\pi} Tr\, G_{\mu\nu}\, G_{\mu\nu} \rangle \,, \tag{58}$$

with the reduced gluon field tensor $G_{\mu\nu} = G^j_{\mu\nu}\lambda_j/2$. From QCD sum rules and heavy quarkonium spectra one estimates $\mathcal{G}_0 \approx 1.6$ GeV fm^{-3}. The gluon condensate is the local limit $(y \to x)$ of the gluonic field strength correlation function (Di Giacomo *et al.* 2002):

$$\mathcal{D}(x,y) \equiv \langle \frac{2\alpha_s}{\pi} Tr\, G_{\mu\nu}(x)\, \mathcal{S}(x,y)\, G_{\mu\nu}(y)\, \mathcal{S}^\dagger(x,y) \rangle \sim \mathcal{G}_0\, exp(-|x-y|/\lambda_g) \,, \tag{59}$$

where the phase $\mathcal{S}(x,y)$ is a path-ordered exponential of the gauge fields which makes sure that $\mathcal{D}(x,y)$ is gauge invariant. The non-local behaviour of the correlation function is parameterised by a coherence or correlation length λ_g.

Lattice QCD simulations (Di Giacomo and Panagopoulos 1992) for the components of $\mathcal{D}(x,y)$ are shown in Figure 9, with the result that the correlation length λ_g, *ie* the distance over which colour fields propagate in the QCD vacuum, is very small:

$$\lambda_g \leq 0.2\, fm \,. \tag{60}$$

This has interesting consequences. Consider a quark with its (rigorously conserved) colour current

$$\mathbf{J}^i_\mu(x) = \overline{\psi}(x)\gamma_\mu \frac{\lambda^i}{2}\psi(x) \,. \tag{61}$$

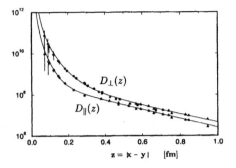

Figure 9. *Lattice QCD results of the gluonic field strength correlation function. Taken from Di Giacomo and Panagopoulos (1991) where also the definitions of $D_\parallel$ and $D_\perp$ can be found.*

This current couples to the gluon field. A second quark interacts with the first one by absorbing and emitting gluons whereby they exchange colour charges. If the distance over which colour can be transported is restricted to the very short range indicated by λ_g, then the quarks experience an interaction which can be approximated by a local coupling between their colour currents:

$$\mathcal{L}_{int} = -G_c \, \mathbf{J}_\mu^i(x) \, \mathbf{J}_i^\mu(x) \, , \tag{62}$$

where $G_c \sim \bar{g}^2 \, \lambda_g^2$ is an effective coupling strength of dimension $length^2$ which encodes the QCD coupling, averaged over the relevant distance scales, in combination with the squared correlation length, λ_g^2.

6.2 A schematic model: Nambu, Jona-Lasinio (NJL)

This is the starting point of a schematic model that dates back to Nambu and Jona-Lasinio (1961) and has been further developed and applied to a variety of problems in hadron physics (Vogl and Weise 1991, Hatsuda and Kunihiro 1994). We adopt the non-linear, local interaction (Equation 62) and write the following Lagrangian for the quark field $\psi(x)$:

$$\mathcal{L} = \overline{\psi}(x)(i\gamma^\mu \partial_\mu - m_0)\psi(x) + \mathcal{L}_{int}(\overline{\psi}, \psi) \, . \tag{63}$$

In essence, by "integrating out" gluon degrees of freedom and absorbing them in the four-fermion interaction $\mathcal{L}_{int}$, the local $SU(N_c)$ gauge symmetry of QCD is now replaced by a global $SU(N_c)$ symmetry of the NJL model. The mass matrix m_0 incorporates small "bare" quark masses. In the limit $m_0 \to 0$, the Lagrangian (Equation 63) evidently has a chiral $SU(N_f) \times SU(N_f)$ symmetry that it shares with the original QCD Lagrangian for N_f massless quark flavours.

A Fierz transform of the colour current-current interaction (Equation 62) produces a set of exchange terms acting in quark-antiquark channels. For the $N_f = 2$ case:

$$\mathcal{L}_{int} \to \frac{G}{2} \left[(\overline{\psi}\psi)^2 + (\overline{\psi}i\gamma_5 \vec{\tau}\psi)^2 \right] + \dots \, , \tag{64}$$

where $\vec{\tau} = (\tau_1, \tau_2, \tau_3)$ are the isospin $SU(2)$ Pauli matrices. For brevity we have not shown a series of terms with combinations of vector and axial vector currents, both in colour singlet and colour octet channels. The constant G is proportional to the colour coupling strength G_c. Their ratio is uniquely determined by N_c and N_f.

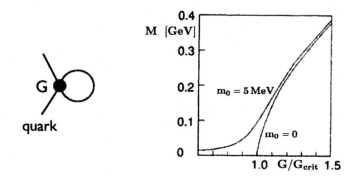

Figure 10. *Left: illustration of the self-consistent gap equation (65). Right: constituent quark mass as function of the coupling strength G for two values of the bare (current) quark mass m_0.*

The vacuum sector and constituent quarks. In the mean field (Hartree) approximation the equation of motion of the NJL model leads to a gap equation

$$M = m_0 - G\langle\bar{\psi}\psi\rangle , \tag{65}$$

illustrated in Figure 10 (left), which links the dynamical generation of a constituent quark mass M to spontaneous chiral symmetry breaking and the appearance of the quark condensate

$$\langle\bar{\psi}\psi\rangle = -Tr \lim_{x\to 0^+}\langle\mathcal{T}\psi(0)\bar{\psi}(x)\rangle = -2iN_f N_c \int \frac{d^4p}{(2\pi)^4} \frac{M\,\theta(\Lambda^2 - \vec{p}^2)}{p^2 - M^2 + i\varepsilon} . \tag{66}$$

Starting from $m_0 = 0$ a non-zero mass develops dynamically, together with a non-vanishing chiral condensate, once G exceeds a critical value as shown in Figure 10 (right). The procedure requires a momentum cutoff $\Lambda \simeq 2M$ beyond which the interaction is "turned off". Note that the strong interactions, by polarising the vacuum and turning it self-consistently into a condensate of quark-antiquark pairs, transmute an initially pointlike quark with its small bare mass m_0 into a massive quasiparticle with a finite size. Such an NJL-type mechanism is thought to be at the origin of the phenomenological constituent quark masses $M \sim M_N/3$.

Whether this concept of constituent quarks as extended objects inside the proton has observable signatures, is presently an interesting and much discussed issue. By detailed examination of moments of structure functions extracted from inelastic electron-proton scattering at moderate Q^2 around 1 GeV^2, at energy transfers above the resonance region but below the deep-inelastic (partonic) domain, Petronzio *et al.* (2003) have argued that constituent quark sizes of about $0.2 - 0.3$ fm may be visible. However, this effect has to be checked against "standard" QCD evolution down to small Q^2.

The meson sector. By solving Bethe-Salpeter equations in the colour singlet quark-antiquark channels, the NJL model has been used to generate the lightest mesons as quark-antiquark excitations of the correlated QCD ground state with its condensate structure. The framework employed here is analogous to the Random Phase Approximation (RPA) method used frequently to decribe collective particle-hole modes in many-body systems.

We briefly report on earlier results obtained with an NJL model based on the interaction (Equation 62) with $N_f = 3$ quark flavours (Klimt *et al.* 1990, Vogl and Weise 1991, Hatsuda and Kunihiro 1994). Such a model has a $U(3)_R \times U(3)_L$ symmetry to start with, but the axial $U(1)_A$ anomaly reduces this symmetry to $SU(3)_R \times SU(3)_L \times U(1)_V$. In QCD, instantons are considered responsible for $U(1)_A$ breaking. In the NJL model, these instanton driven interactions are incorporated in the form of a flavour determinant, $\det[\bar{\psi}_a(1 \pm \gamma_5)\psi_b]$ ('t Hooft 1976). This interaction involves all three flavours u, d, s simultaneously in a genuine three-body term.

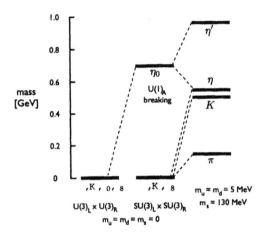

Figure 11. *Symmetry breaking pattern in the pseudoscalar meson nonet calculated in the $N_f = 3$ NJL model (Klimt et al. 1990).*

The symmetry breaking pattern resulting from such a calculation is demonstrated in the pseudoscalar meson spectrum of Figure 11. Starting from massless u, d and s quarks and in the absence of the 't Hooft determinant, the whole pseudoscalar nonet emerges as a set of massless Goldstone bosons of spontaneously broken $U(3) \times U(3)$. Axial $U(1)_A$ breaking removes the singlet η_0 from the Goldstone boson sector. Finite quark masses shift the $J^\pi = 0^-$ nonet into its empirically observed position, including η-η' mixing.

The combination of self-consistent gap and Bethe-Salpeter equations in the NJL model is an approximation which respects the symmetries of the Lagrangian. The GOR relation (Equation 26), based on PCAC, is properly fulfilled, and the pion decay constant derives from the basic quark-antiquark loop integral as follows:

$$f_\pi^2 = -4iN_c \int \frac{d^4p}{(2\pi)^4} \frac{M^2 \, \theta(\Lambda^2 - \vec{p}^2)}{(p^2 - M^2 + i\varepsilon)^2} \,, \tag{67}$$

with the constituent quark mass M determined self-consistently by the gap equation

(Equation 65). In practice, the empirical $f_\pi = 92$ MeV is used to fix the momentum cutoff scale Λ together with the constituent mass $M_u = M_d = 0.36$ GeV and the chiral condensate $\langle \bar{u}u \rangle = \langle \bar{d}d \rangle = -(250 \text{ MeV})^3$.

6.3 Diquarks and pentaquarks

The one-gluon exchange interaction is attractive for colour singlet quark-antiquark pairs and for colour antitriplet diquarks. This gives a first hint that attractive correlations in some of those channels might prevail as the gluon-driven interactions develop non-perturbative strength. It is an interesting feature of the NJL model that it establishes close links between quark-antiquark and diquark channels. As in the many-body theory of Fermion systems where particle-particle and particle-hole interactions are connected by crossing relations, the Fierz exchange transformation of the original colour current-current interaction (Equation 62) has an equivalent representation in diquark channels (Vogl and Weise 1991). For $N_f = 2$ flavours and diquarks interacting in colour antitriplet configurations,

$$\mathcal{L}_{int} \rightarrow \frac{H}{2} \left[(\bar{\psi}_j i\gamma_5 \, \tau_2 \, C \, \bar{\psi}_k^T)(\psi_m^T \, C^{-1} \, i\gamma_5 \, \tau_2 \, \psi_n) + \dots \right] \epsilon^{jkl} \epsilon^{mnl} , \qquad (68)$$

where $C = i\gamma_0\gamma_2$ is the charge conjugation operator. Colour indices $j, k, l, m, \dots$ are explicitly displayed. The term written in Equation 68 applies to scalar diquarks, while a series of additional pieces (representing pseudoscalar, vector and axial vector diquark channels) is not shown explicitly. The coupling strength H is again uniquely determined by the original colour coupling G_c, via Fierz transformation.

When solving the Bethe-Salpeter equation for scalar diquarks (Vogl and Weise 1991) in the isoscalar configuration ($ud - du$), one observes strongly attractive correlations which reduce the mass of the diquark as a quasiparticle cluster of two constituent quarks to $M_{ud} \leq 0.3$ GeV, less than half the sum of the constituent masses. Similarly, strong attraction drives the mass of the ($us - su$) scalar diquark far down to $M_{us} \leq 0.6$ GeV from the unperturbed sum of $M_u + M_s \simeq 0.9$ GeV. The strong pairing in spin singlet diquark configurations is a counterpart of the strong attraction observed in pseudoscalar quark-antiquark channels. Other diquark combinations (pseudoscalar, axial vector, *etc.*) behave differently and stay roughly at the summed mass of their constituent quarks.

QCD at low temperature and very high baryon density is expected to develop Cooper pair condensates and transitions to (colour) superconducting phases. Such phenomena are under lively discussion (Alford *et al.* 1998, Rapp *et al.* 1998). The strongly attractive correlations in scalar diquarks can be considered a precursor of Cooper pairing.

Tightly bound diquark quasiparticles are also at the basis of an interpretation (Jaffe and Wilczek 2003) of the much debated pentaquark state Θ^+, supposedly an "exotic" strangeness $S = +1$ baryon resonance with the quark configuration ($uudd\bar{s}$). Jaffe and Wilczek discuss the Θ^+ as an antidecuplet member of the $\overline{10} + 8$ representation of flavour $SU(3)$, as shown in Figure 12. In order for this scheme to work and produce a Θ^+ at the candidate mass around 1540 MeV (see *eg* Close 2004), a $|[ud]^2\bar{s}\rangle$ structure with two scalar (ud) diquarks is anticipated. The required diquark quasiparticle masses in question are indeed within range of those predicted by the previously mentioned NJL

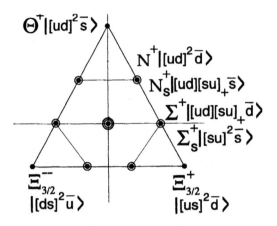

Figure 12. *Quark content of members of the* $(q^4\bar{q})$ *antidecuplet + octet representation of flavour SU(3) (from Jaffe and Wilczek 2003).*

model calculation, although it is so far difficult to underscore these statements at a more detailed quantitative level.

6.4   A testing ground: two-colour QCD

We close these lectures with an outlook into a theoretical laboratory: QCD with $N_c = 2$ colours. This theory does not have the full complexity of QCD. But it already displays many interesting features of non-abelian gauge theory coupled to spin 1/2 particles. In particular, $N_c = 2$ QCD permits complete lattice simulations of the thermodynamics. Unlike the $N_c = 3$ case in which the Fermion determinant at finite chemical potential becomes complex so that standard Monte Carlo techniques do not work, two-colour QCD does not have this problem. The phase diagram is accessible in the full (T, μ) plane, at both finite temperature T and baryon chemical potential μ.

In $N_c = 2$ QCD, diquarks can form colour singlets which are the baryons of the theory. The lightest baryons and the lightest quark-antiquark excitations (pions) have a common mass, m_π, and this spectrum determines the low-energy properties of the theory for small chemical potential. General arguments predict a phase transition to a state with finite baryon density at a critical chemical potential $\mu_c = m_\pi/2$. For $\mu > \mu_c$ Cooper pair condensation of scalar diquarks sets in and the ground state displays superfluidity, associated with spontaneous breaking of the $U(1)_V$ symmetry linked to baryon number.

It is interesting to discuss the thermodynamics of two-colour QCD under the aspect of exploring the relevant correlations and identifying the quasiparticles which determine the dynamically active degrees of freedom of such a system. In particular, quasiparticle models of NJL type in which gluons have been integrated out, can be tested in detail against computer simulations of the full theory on large Euclidean lattices. Consider again the NJL Lagrangian

$$\mathcal{L} = \overline{\psi}(i\gamma^\mu \partial_\mu - m_0)\psi - G_c \sum_{j=1}^{3} \left(\overline{\psi}\gamma^\mu t_j \psi\right)\left(\overline{\psi}\gamma_\mu t_j \psi\right) , \tag{69}$$

with a local colour current-current coupling between quarks. The $SU(2)_{colour}$ generators are denoted t_j, and we work in the two-flavour sector with $\psi(x) = (u(x), d(x))^T$. In quark-antiquark and diquark channels, matrix elements of the interaction term of Equation 69 are again conveniently worked out by taking Fierz transforms, as in Equations 64 and 68. For $N_c = N_f = 2$, the corresponding coupling strengths in those channels are related to the original coupling G_c of the colour currents by $G = H = 3G_c/2$. One notes that there is a symmetry between mesons and diquarks in this model (a realisation of the so-called Pauli-Gürsey symmetry).

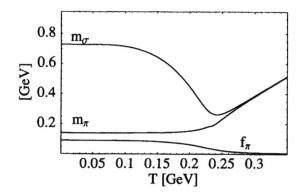

Figure 13. *Temperature dependence of the pion and scalar masses, and of the pion decay constant, at zero baryon chemical potential (Ratti and Weise 2004).*

The partition function is conveniently written by introducing auxiliary boson fields: σ and ϕ_a for scalar-isoscalar and pseudoscalar-isovector quark-antiquark pairs, respectively; Δ and Δ^* for scalar diquarks and antidiquarks. Rephrasing the Lagrangian (Equation 69) in terms of those fields leads to an equivalent Lagrangian

$$\begin{aligned}
\tilde{\mathcal{L}} &= \overline{\psi}(i\gamma^\mu \partial_\mu - m_0 + \sigma + i\gamma_5 \tau \cdot \pi)\psi \\
&+ \frac{1}{2}\Delta^* \psi^T C^{-1}\gamma_5 \tau_2 t_2 \psi + \frac{1}{2}\Delta \overline{\psi}^T \gamma_5 \tau_2 t_2 C \overline{\psi} - \frac{\sigma^2 + \pi^2}{2G} - \frac{|\Delta|^2}{2H} .
\end{aligned} \tag{70}$$

The expectation values $\langle \sigma \rangle$ and $\langle |\Delta| \rangle$ represent chiral (quark) and Cooper pair (diquark) condensates. Fixing the constant G, the bare quark mass m_0 and the momentum space cutoff of the model to reproduce pion properties at $T = \mu = 0$, one can proceed to calculate the thermodynamic potential $\Omega(T, \mu)$ in the mean-field approximation, evaluate the equation of state and derive the phase diagram. Such calculations (Ratti and Weise 2004) can then be compared with lattice QCD results. At zero chemical potential ($\mu = 0$), Figure 13 shows the characteristic pattern of spontaneous chiral symmetry breaking and restoration as a function of temperature. Above a critical temperature $T_c \simeq 0.2$ GeV, the pion and the scalar (σ) become degenerate and jointly move away from the low-energy spectrum as a parity doublet, while the pion decay constant f_π, as order parameter, tends to zero.

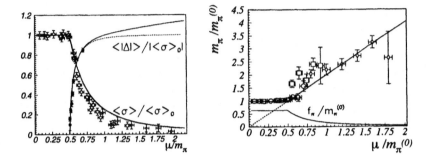

Figure 14. *Chiral and diquark condensates (left) and pion properties (right) at $T = 0$ as function of baryon chemical potential μ in $N_c = 2$ lattice QCD (Hands et al. 2001) and in the NJL model of Equation 69 (Ratti and Weise 2004). Quantities are given in units of their values at $\mu = 0$.*

The condensate structure as function of the chemical potential μ at zero temperature is shown in Figure 14 (left) compared to $N_c = 2$ lattice computations (Hands *et al.* 2001). Above a critical chemical potential $\mu_c = m_\pi/2$, the diquark (Cooper pair) condensate develops while at the same time the chiral condensate decreases correspondingly. The pion mass and decay constant follow this trend as seen in Figure 14 (right).

What is remarkable about these results is that a simple NJL model, with interactions reduced merely to local couplings between the colour currrents of quarks, can draw a realistic picture of the quasiparticle dynamics emerging from $N_c = 2$ lattice QCD. One must note that gluon dynamics linked to the original local colour gauge symmetry has been replaced by a global colour $SU(2)$ symmetry in the model. The symmetry breaking pattern identifies pseudoscalar Goldstone bosons (pions) and scalar diquarks as the thermodynamically active quasiparticles. Colour (triplet) quark-antiquark modes which are the remnants of the gluon degrees of freedom in this model, turn out to be far removed from the low-energy spectrum and "frozen". Of course, the low-energy physics of QCD differs qualitatively between $N_c = 2$ and $N_c = 3$ because of the very different nature of the baryonic quasiparticles in these two theories. However, the present example demonstrates that a synthesis of lattice QCD strategies with effective field theories and quasiparticle approaches promises further insights.

Acknowledgments

Special thanks go to my collaborators Bugra Borasoy, Thomas Hemmert, Norbert Kaiser, Robin Nissler, Massimiliano Procura and Claudia Ratti, whose recent works have contributed substantially to the subjects presented in these lectures. Partial support from BMBF, DFG and GSI is gratefully acknowledged.

References

Ackerstaff K *et al.* [OPAL collaboration], 1999, *Eur Phys J C* **7** *et al.*
Alford M G, *et al.* , 1998, *Phys Lett B* **422** 247.
Ali Khan A *et al.* , 2002, *Phys Rev D* **65** 054505.
Ali Khan A *et al.* , 2004, *Nucl Phys B* **689** 175.
Aoki S *et al.* , 2003, *Phys Rev D* **68** 054502.
Bali G S *et al.* , 1997, *Phys Rev D* **56** 2566.
Bali G S, 2001, *Phys Rep* **343** 1.
Barate M *et al.* [ALEPH collaboration], 1998, *Eur Phys J C* **4** 409.
Bernard V *et al.* , 1995, *Int J Mod Phys E* **4** 193.
Borasoy B *et al.* , 2004, hep-ph/0410305.
Boyd G *et al.* , 1995, *Phys Lett B* **349** 170.
Brown G E, 1971, *Unified Theory of Nuclear Models and Forces*, 3rd ed, North-Holland,
    Amsterdam.
Caro Ramon J *et al.* , 2000, *Nucl Phys B* **672** 249.
Close F E, 2004, *contribution to these proceedings*.
Colangelo G *et al.* , 2001, *Nucl Phys B* **603** 125.
Dalitz R H *et al.* , 1967, *Phys Rev* **153** 1617.
Davies C T H, 2004, *Lectures given to SUSSP58 Hadron Physics Summer School, St Andrews*.
Di Giacomo A and Panagopoulos, 1992, *Phys Lett B* **285** 133.
Di Giacomo A *et al.* , 2002, *Phys Rep* **372** 319.
Donoghue J F *et al.* , 1992, *Dynamics of the Standard Model*, Cambridge University Press.
Eidelman S *et al.* [Particle Data Group], 2004, *Phys Lett B* **592** 1.
Ericson M and Figureau A, 1981, *J Phys G: Nucl Part Phys* **7** 1197.
Fritzsch H *et al.* , 1973, *Phys Lett B* **47** 365.
Finelli P *et al.* , 2004, *Nucl Phys B* **735** 449.
Gasser J and Leutwyler H, 1984, *Ann Phys* **158** 142.
Gasser J *et al.* , 1991a, *Phys Lett B* **253** 252.
Gasser J *et al.* , 1991b, *Phys Lett B* **253** 260.
Gell-Mann M *et al.* , 1968, *Phys Rev* **175** 2195.
Goltermann M and Peris S, 2000, *Phys Rev D* **61** 034018.
Gross D J and Wilczek F, 1973, *Phys Rev Lett* **30** 1343.
Hands S *et al.* , 2001, *Eur Phys J C* **22** 451.
Hatsuda T and Kunihiro T, 1994, *Phys Rep* **247** 221.
Höhler G, 1983, *Pion-Nucleon Scattering*, in: Landolt-Börnstein New Series **I/9b2**,
    Springer, Berlin.
't Hooft G, 1976, *Phys Rev D* **14** 3432.
Jaffe R L and Wilczek F, 2003, *Phys Rev Lett* **91** 232003.
Kaiser N *et al.* , 1995, *Nucl Phys B* **594** 325.
Kaiser N *et al.* , 1997, *Nucl Phys B* **612** 297.
Kaiser N *et al.* , 1998, *Nucl Phys B* **637** 395.
Kaiser N *et al.* , 2002, *Nucl Phys B* **700** 343.
Kaiser N, 2003, *Phys Rev C* **68** 025202.
Karsch F *et al.* , 2000, *Nucl Phys B* **605** 597.
Klimt S *et al.* , 1990, *Nucl Phys B* **516** 429.
Knecht M *et al.* , 1996, *Nucl Phys B* **471** 445.
Klingl F and Weise W, 1999, *Eur Phys J A* **4** 225.
Leinweber D B *et al.* , 2004, *Phys Rev Lett* **92** 242002.
Leutwyler H, 1994, *Ann Phys* **235** 165.
Leutwyler H, 2002, *Nucl PhysB Proc Suppl* **108** 37.

Lutz M F M and Kolomeitsev E, 1999, *Nucl Phys B* **700** 193.

Marco E and Weise W, 2000, *Phys Lett B* **482** 87.

Metag V, 2004, *contribution to these proceedings*.

Nambu Y and Jona-Lasinio G, 1961, *Phys Rev* **122** 345.

Oller J A and Meissner U-G, 2001, *Phys Lett B* **500** 263.

Oset E and Ramos A, 2000, *Nucl Phys B* **635** 99.

Petronzo R *et al.* , 2003, *Phys Rev D* **67** 094004.

Pich A and Prades J, 2000, *Nucl Phys Proc Suppl* **86** 236.

Politzer H D, 1973, *Phys Rev Lett* **30** 1346.

Procura M *et al.* , 2004, *Phys Rev D* **69** 034505.

Rapp R *et al.* , 1998, *Phys Rev Lett* **81** 53.

Ratti C and Weise W, 2004, *Phys Rev D* **70** 054013.

Sainio M E, 2002, *PiN Newslett* **16** 138.

Thomas A W and Weise W, 2001, *The Structure of the Nucleon*, Wiley-VCH, Berlin.

Vogl U and Weise W, 1991, *Prog Part Nucl Phys* **27** 195.

Vretenar D and Weise W, 2004, *Lect Notes Phys* (Springer) **641** 65.

Weinberg S, 1967, *Phys Rev Lett* **18** 507.

Weinberg S, 1979, *Physica A* **96** 327.

Weise W, 2003, in *proc Int School of Physics "Enrico Fermi"*, Course CLIII (Varenna 2002), ed A
    Molinari *et al.* , IOS Press, Amsterdam.

Weise W, 2005, *Nucl Phys B,* **751** 565.

Lattice QCD – A guide for people who want results

Christine T H Davies

Department of Physics and Astronomy, University of Glasgow,
University Avenue, Glasgow G12 8QQ, UK

1 Introduction

Lattice QCD was invented thirty years ago but only in the last few years has it finally fulfilled its promise as a precision tool for calculations in hadron physics. This review will cover the fundamentals of discretising QCD onto a space-time lattice and how to reduce the errors associated with the discretisation. This 'improvement' is the key that has made the enormous computational task of a lattice QCD calculation tractable and enabled us to reach the recent milestone of precision calculations of simple "gold-plated" hadron masses. Accurate decay matrix elements, such as those for leptonic and semileptonic decay of heavy mesons needed by the B factory experimental programme, are now within sight. I will describe what goes into such calculations and what the future prospects and limitations are.

2 Lattice QCD formalism and methods

2.1 The path integral

Lattice QCD is based on the path integral formalism. We can demonstrate this formalism by discussing the solution of the quantum mechanical problem of a particle moving in one dimension (Lepage 1998a). This could be solved using Schrödinger's equation with $H = p^2/2m + V(x)$ and $[x,p] = i$ (in 'particle physics units', $\hbar = c = 1$). We can also solve it using the path integral formulation and this is the basis for lattice QCD.

A key quantity is the transition amplitude between eigenstates of position at, say, time $t = t_i$ and t_f. This is expressed as a functional integral over all possible paths $x(t)$

from $t = t_i$ to t_f weighted by the exponential of the action, S, which is the integral of the Lagrangian, L.

$$\langle x_f(t_f)|x_i(t_i)\rangle = \int \mathcal{D}x(t)e^{iS[x]} \tag{1}$$

$$S[x] \equiv \int_{t_i}^{t_f} dtL(x,\dot{x}) \equiv \int dt[\frac{m\dot{x}(t)^2}{2} - V(x(t))]. \tag{2}$$

The path integral can be evaluated by discretising time into a set of points, $t_j = t_i + ja$ for $j = 0, 1 \ldots N$ and a the lattice spacing $\equiv (t_f - t_i)/N$. The path $x(t)$ then becomes a set of variables, x_j, and Equation 1 above requires us to integrate over each one, *ie* the problem reduces to an ordinary integral but over an $N - 1$-dimensional space. The end points, x_0 and x_N are kept fixed in this example. The most likely path is the classical one, which minimises the action ($m\ddot{x} = V'$), but the path integral allows quantum fluctuations about this, see Figure 1.

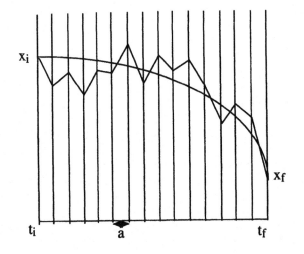

Figure 1. *Discretised classical and possible quantum mechanical paths from x_i to x_f for a particle moving in one dimension.*

The i in front of the action in the exponential gives rise to the problem, both conceptual and numerical, of adding oscillating quantities together. It is simpler to rotate the time axis to Euclidean time, $t \to -it$. Then the path integral becomes

$$\int \mathcal{D}x(t)e^{-S[x]} = A \int_{-\infty}^{\infty} dx_1 dx_2 \ldots dx_{N-1}e^{-S[x]}$$

$$S[x] \equiv \int_{t_i}^{t_f} L(x,\dot{x}) \equiv \sum_{j=0}^{N-1}[\frac{m\dot{x}_j^2}{2} + V(x_j)], \tag{3}$$

where the integrals over the intermediate points in the path, $x(t)$, are now explicit. We will ignore the normalisation of the integral, A. To perform the integral we must discretise the Lagrangian so that it takes values at the discrete time points. Then

$$S = \sum_{j=0}^{N-1} [\frac{m}{2a}(x_{j+1} - x_j)^2 + aV(x_j)]. \tag{4}$$

Clearly the accuracy of our discretisation will depend on a being small. However, as a becomes smaller at fixed physical time length, the number of points, N, increases and so does the computational cost.

For large N an efficient way to perform the integral is by Monte Carlo. A set of possible values for x_j, $j = 1, N$ is called a *configuration*. The configurations with most weight in the integral are those with large e^{-S}. For maximum efficiency we want to generate configurations with probability e^{-S} - this is known as *importance sampling*. A simple method for doing this is the Metropolis algorithm. This starts with an initial configuration (*eg* x_j all zero or chosen randomly). It then passes through the x_j in turn proposing a change to a given x_j of size ϵ, *ie* a random number between $-\epsilon$ and ϵ temporarily added to x_k, say. The change in the action as a result of x_k changing is calculated. Call this ΔS. Note that this calculation only involves the x_j in the neighbourhood of x_k and connected to it through the action. If $\Delta S < 0$ the change is accepted. If $\Delta S > 0$ another random number, uniformly distributed between 0 and 1 is generated. If $e^{-\Delta S} > rand$ then the change is accepted. If not, x_k reverts to its previous value (Lepage 1998a, di Pierro 2000).

In this way a new configuration is generated after each sweep through the lattice. A set of configurations is called an *ensemble*. Calculations of various functions of the x_j on an ensemble then yield the physics results that we are after, such as the quantised energy levels available to the particle.

In wavefunction language we can express

$$\langle x_f|e^{-iH(t_f-t_i)}|x_i\rangle = \sum_n \psi_n^*(x_f)\psi_n(x_i)e^{-iE_n(t_f-t_i)}, \tag{5}$$

by inserting a complete set of eigenstates of the Hamiltonian, H. When rotated to Euclidean time, and taking $t_f - t_i = T$ and $x_i = x_f = x$ we have

$$\langle x|e^{-HT}|x\rangle = \sum_n \psi_n^*(x)\psi_n(x)e^{-E_nT}, \tag{6}$$

where $\psi_n(x) = \langle n|x\rangle$. It now becomes clear that the result will be dominated by the ground state as T becomes large, because all higher states are exponentially suppressed. Integrating over the initial and final x values gives:

$$\int dx\langle x|e^{-HT}|x\rangle \rightarrow e^{-E_0T}, T \rightarrow \infty. \tag{7}$$

This would be the result, up to a normalisation, of integrating by Monte Carlo over all the x_j, setting the initial and final values to be the same, *ie* using periodic boundary conditions. The result looks very similar to a problem in statistical mechanics when formulated in this way, as did Equation 3, and this is not an accident. Indeed we can work at non-zero temperature if we take T finite, but here we will concentrate on the zero temperature case where in principle $T \rightarrow \infty$, and in practice is large.

To investigate excitations above the ground state, we must interrupt the propagation of the ground state by introducing new operators at intermediate times. For example,

$$
\begin{aligned}
\frac{\langle x(T)|x(t_2)x(t_1)|x(0)\rangle}{\langle x(T)|x(0)\rangle} &= \frac{\int \mathcal{D}x \; x(t_2)x(t_1)e^{-S[x]}}{\int \mathcal{D}x \, e^{-S[x]}} \\
&= |\langle E_0|x|E_1\rangle|^2 e^{-(E_1-E_0)(t_2-t_1)}, t_2 - t_1 \to \infty. \quad (8)
\end{aligned}
$$

The state propagating between the insertions of the x operators at t_1 and t_2 cannot be the ground state, since x switches parity. If $t_2 - t_1$ is large enough then the first excited state will dominate, by the same argument as used above for ground state domination. So, if we can evaluate the ratio of path integrals, we can determine the energy splitting $E_1 - E_0$ between the first excited and the ground state. The evaluation is very simply done by taking an ensemble of configurations generated by the Metropolis algorithm above and 'measuring' on each one the value of $x(t_2)x(t_1)$. The ensemble average of this quantity, denoted $<< x(t_2)x(t_1) >>$, is then the ratio above.

The evaluation will suffer from a statistical error depending on how many configurations are in the ensemble. This will improve as the square root of the number of configurations provided that the results on each configuration are statistically independent. Since the configurations were made in a sequence, this will not be strictly true and results on neighbouring configurations in the sequence will be correlated. There are various statistical techniques, such as binning (averaging) neighbouring results and then recalculating the statistical error, that will uncover correlations. If the statistical error grows with the bin size then the results are correlated and the binned results should be used rather than the raw results. It might also be true that results on some number of the initial configurations of an ensemble have to be discarded because these configurations, and measurements on them, are not yet typical of the e^{-S} distribution. For this example, the results can obviously be improved statistically by moving t_1 and t_2 along the time axis, keeping $t_2 - t_1$ fixed, and averaging those results.

To extract $E_1 - E_0$ we should fit the results as a function of $t_2 - t_1$ to the exponential form of Equation 8. Since $t_2 - t_1$ will not be infinite, we can take account of contamination from higher states by fitting to a sum of exponentials in which we constrain the higher $E_n > E_1$. Finally we may also improve the accuracy on the energy by modifying the operator $x(t_1)x(t_2)$ to a product of functions of $x(t_{1,2})$ which has optimum overlap with the first excited state and minimum overlap with higher excited states. This will allow Equation 8 to become true for smaller values of $t_2 - t_1$ and improve the extraction of $E_1 - E_0$ from the fits. Indeed we can fit simultaneously to the results from several different operators and this again will improve the accuracy with which $E_1 - E_0$ can be determined. It is a useful exercise to do the calculation above for, say, the simple harmonic oscillator potential, $V = x^2/2$ and $m = 1$. All the pitfalls of 'measurement' and the improvements above have their mirror in lattice QCD calculations, as we shall see, but this simple example demonstrates them very clearly in a setting where computer time is not an issue.

An important issue is that of the systematic errors, called discretisation errors, introduced because of the non-zero lattice spacing. We can make these explicit by rewriting the finite difference in terms of the continuum (*ie* continuous space-time of the real world) derivative. The exponential of the continuum derivative is the translation operator for moving from one site to the next. So

$$\frac{x_{i+1} - x_i}{a} = \frac{e^{\partial a} - 1}{a} = \partial + \frac{a\partial^2}{2}. \tag{9}$$

This shows that this form of the finite difference has errors linear in a. The actual size of the errors at a given value of a would depend on the effective size of ∂^2 in the quantity being calculated. To halve the errors requires half the lattice spacing at (at least) double the computer cost.

The situation is significantly improved by using an improved discretisation. $\Delta x_i = (x_{i+1} - x_{i-1})/2a$ has errors first at $\mathcal{O}(a^2)$. Further improvement is obtained by correcting for the a^2 error using

$$\Delta x_i - \frac{a^2}{6}(\Delta)^3 x_i \tag{10}$$

where $(\Delta)^3$ is a discretisation of the third order derivative. Errors are then $\mathcal{O}(a^4)$ and this means very much smaller errors at a given value of a, or a huge reduction in computer cost for a given error by being able to work at a larger value of a. The improved discretisation costs a little in computer power to implement but this is completely negligible compared to the computer cost saving of the improvement and it is this that has made lattice QCD calculations tractable.

2.2   Gluon fields in lattice QCD

Now we have all the ingredients for lattice QCD. The position operator as a function of time is replaced by the quark and gluon field operators as a function of four-dimensional space-time. We discretise the space-time into a lattice of points (Figure 2) in order to be able to calculate the Feynman Path Integral numerically, using Monte Carlo methods on a computer. The ground-state which dominates the path integral is the QCD vacuum and we will be interested in excitations of this which correspond to hadrons.

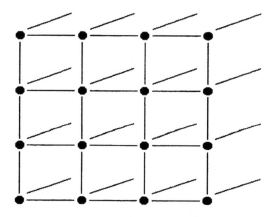

Figure 2. *A 2-dimensional rendition of a 3-dimensional cubic lattice. Lattice QCD calculations use a 4-dimensional grid.*

It is important to remember that lattice QCD is not complicated. It is a straightforward simulation of the theory on a computer, but a numerically intensive one. It is the theory of QCD that is being simulated, not a model. However, we are limited in the things that we can calculate. There are also statistical and systematic errors associated with the calculations and it is important to understand where they come from, so that you can assess the usefulness of a particular calculation for your needs.

Consider the operator $\mathcal{O} = (\overline{\psi}\psi)_y(\overline{\psi}\psi)_x$. This creates a hadron at a point x and destroys it at a point y. This is the QCD generalisation of the $x(t_1)x(t_2)$ operator of the previous section. Then the matrix element in the vacuum of the operator is given in path integral form as:

$$\langle 0|\mathcal{O}|0\rangle = \frac{\displaystyle\int [d\psi]\,[d\overline{\psi}]\,[dA_\mu]\,\mathcal{O}[\psi,\overline{\psi},A]e^{-S}}{\displaystyle\int [d\psi]\,[d\overline{\psi}]\,[dA_\mu]e^{-S_{QCD}}}. \tag{11}$$

The path integral runs over all values of the quark and gluon fields ψ and A at every point in space-time. Discretisation of space-time onto a lattice makes the number of space-time points (and therefore field variables) finite. Continuous space-time (x, t) becomes a grid of labelled points, (x_i, t_i) or $(n_i a, n_t a)$ where a is the lattice spacing (this doesn't have to be the same in all directions but usually is). The fields are then associated only with the sites, $\psi(x, t) \to \psi(n_i, n_t)$. The action must also be discretised, but, as in the previous section, this is straightforward, replacing fields by fields at the lattice sites and the derivatives by finite differences of these fields. The integral over space-time of the Lagrangian becomes a sum over all lattice sites: $(\int d^4x \to \sum_n a^4)$, and the path integral becomes a product of integrals over each of the fields, to be done by Monte Carlo averaging. These integrals will be finite, unlike such integrals in the continuum, because the lattice provides a regularisation of the theory. The lattice spacing provides an ultraviolet cut-off in momentum space since no momenta larger than π/a make sense (since the wavelength is then smaller than a).

To discretise gauge theories such as QCD onto a lattice in fact requires a little additional thought to that described above because of the paramount importance of local gauge invariance. The role of the gluon (gauge) field in QCD is to transport colour from one place to another so that we can rotate our colour basis locally. It should then seem natural for the gluon fields to 'live' on the links connecting lattice points, if the quark fields 'live' on the sites.

The gluon field is also expressed somewhat differently on the lattice to the continuum. The continuum A_μ is an 8-dimensional vector, understood as a product of coefficients A_μ^b times the 8 matrices, T_b, which are generators of the SU(3) gauge group for QCD. On the lattice it is more useful to take the gluon field on each link to be a member of the gauge group itself *ie* a special (determinant = 1) unitary 3×3 matrix. The lattice gluon field is denoted $U_\mu(n_i, n_t)$, where μ denotes the direction of the link, n_i, n_t refer to the lattice point at the beginning of the link, and the colour indices are suppressed. We will often just revert to continuum notation for space-time, as in $U_\mu(x)$. The lattice and continuum fields are then related exponentially,

$$U_\mu = e^{-iagA_\mu} \tag{12}$$

where the a in the exponent makes it dimensionless, and we include the coupling, g, for convenience. If $U_\mu(x)$ is the gluon field connecting the points x and $x+1_\mu$ (see Figure 3), then the gluon field connecting these same points but in the downwards direction must be the inverse of this matrix, $U_\mu^{-1}(x)$. Since the U fields are unitary matrices, satisfying $U^\dagger U = 1$, this is then $U_\mu^\dagger(x)$.

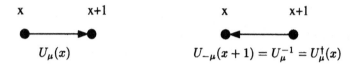

$$\begin{array}{cc} x & x+1 \\ \bullet\!\!-\!\!-\!\!-\!\!-\!\!-\!\!\to\!\bullet & \\ U_\mu(x) & \end{array} \qquad \begin{array}{cc} x & x+1 \\ \bullet\!\leftarrow\!\!-\!\!-\!\!-\!\!-\!\!\bullet & \\ U_{-\mu}(x+1) = U_\mu^{-1} = U_\mu^\dagger(x) & \end{array}$$

Figure 3. *The gluon field on the lattice.*

This form for the gluon field makes it possible to maintain exact local gauge invariance on a lattice. To apply a gauge transformation to a set of gluon fields we must specify an SU(3) gauge transformation matrix at each point. Call this $G(x)$. Then the gluon field $U_\mu(x)$ simply gauge transforms by the (matrix) multiplication of the appropriate G at both ends of its link. The quark field (a 3-dimensional colour vector) transforms by multiplication by G at its site.

$$\begin{aligned} U_\mu^{(g)}(x) &= G(x)U_\mu(x)G^\dagger(x+1_\mu) \\ \psi^{(g)}(x) &= G(x)\psi(x) \\ \overline{\psi}^{(g)}(x) &= \overline{\psi}(x)G^\dagger(x). \end{aligned} \tag{13}$$

To understand how this relates to continuum gauge transformations try the exercise of setting $G(x)$ to a simple U(1) transformation, $e^{-i\alpha(x)}$, and show that Equation 13 is equivalent to the QED-like gauge transformation in the continuum, $A_\mu^g = A_\mu - \partial_\mu\alpha$.

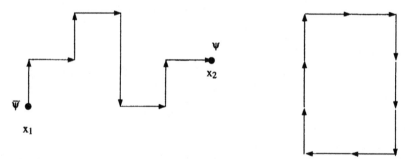

Figure 4. *A string of gluon fields connecting quark and antiquark fields (left) and a closed loop of gluon fields (right).*

Gauge-invariant objects can easily be made on the lattice (see Figure 4) out of closed loops of gluon fields or strings of gluon fields with a quark field at one end and an antiquark field at the other, eg $\overline{\psi}(x_1)U_\mu(x_1)U_\nu(x_1 + 1_\mu)\ldots U_\epsilon(x_2 - 1_\epsilon)\psi(x_2)$. Under a gauge transformation the G matrix at the beginning of one link 'eats' the $G^\dagger$ at the end of the previous link, since $G^\dagger G = 1$. The G matrices at x_1 and x_2 are 'eaten' by those transforming the quark and anti-quark fields, if we sum over quark and antiquark colours.

The same thing happens for any closed loop of Us, provided that we take a trace over colour indices. Then the G at the beginning of the loop and the $G^\dagger$ at the end of the loop, the same point for a closed loop, can 'eat' each other.

The purely gluonic piece of the continuum QCD action is

$$S_{\text{cont}} = \int d^4 x \frac{1}{2} \text{Tr} \, F_{\mu\nu} F^{\mu\nu} \tag{14}$$

where $F_{\mu\nu}$ is the field strength tensor,

$$F_{\mu\nu} = \partial_\mu A_\nu - \partial_\nu A_\mu + ig[A_\mu, A_\nu]. \tag{15}$$

The simplest lattice discretisation of this is the so-called Wilson plaquette action:

$$S_{\text{latt}} = \beta \sum_p \left(1 - \frac{1}{3} \text{Re} \{ \text{Tr} \, U_p \} \right); \quad \beta = \frac{6}{g^2}. \tag{16}$$

In fact the 1 here is irrelevant in the lattice QCD calculation, giving only an overall normalisation that vanishes from the ratio of path integrals, and so it is often dropped from S_{latt}.

Figure 5. *A plaquette on the lattice.*

U_p is the closed 1×1 loop called the plaquette, an SU(3) matrix formed by multiplying 4 gluon links together in a sequence. For the plaquette with corner x in the i, j plane we have (Figure 5):

$$U_{p,ij}(x) = U_i(x) U_j(x + 1_i) U_i^\dagger(x + 1_j) U_j^\dagger(x) \tag{17}$$

Tr in S_{latt} denotes taking the trace of U_p *ie* the sum of the 3 diagonal elements. S_{latt} sums over all plaquettes of all orientations on the lattice. β is a more convenient version for the lattice of the QCD bare coupling constant, g^2. This is the single input parameter for a QCD calculation (whether on the lattice or not) involving only gluon fields. Notice that the lattice spacing is not explicit anywhere, and we do not know its value until *after* the calculation. (This is a difference from the quantum mechanical example of the previous section where we had to choose and input a value for a.) The value of the lattice spacing depends on the bare coupling constant. Typical values of β for current lattice calculations using the Wilson plaquette action are $\beta \approx 6$. This corresponds to $a \approx 0.1$fm. Smaller values of β give coarser lattices, larger ones, finer lattices. This is obvious from the asymptotic freedom of QCD, which tells us that the coupling constant g^2 goes to zero at small distances or, equivalently, high energies. β is the inverse of the bare coupling constant at the scale of the lattice spacing and therefore tends to ∞ as the lattice spacing goes to zero.

That S_{latt} of Equation 16 is a discretisation of S_{cont} is not obvious, and we will not demonstrate it here. It should be clear, however, from Equations 12 and 17 that S_{latt} does contain terms of the form $\partial_\mu A_\nu$ needed for the field strength tensor, $F_{\mu\nu}$.

S_{latt} is gauge-invariant, as will be clear from earlier. Thus lattice QCD calculations do *not* require gauge fixing or any discussion of different gauges or ghost terms, as would be required for continuum calculations using perturbation theory. Our lattice calculation is fully non-perturbative since the Feynman Path Integral includes any number of QCD interactions. In contrast to the real world, however, the calculations are done with a non-zero value of the lattice spacing and a non-infinite volume. In principle we must take $a \to 0$ and $V \to \infty$ by extrapolation. In practice it suffices to demonstrate, with calculations at several values of a and V, that the a and V dependence of our results is small, and understood, and include a systematic error for this in our result.

Before discussing quarks we illustrate here how a lattice calculation is done with only gluon fields, how the lattice spacing is determined, and what the systematic errors are.

A quantity that can be calculated in the pure gluon theory is the expectation value of a closed loop of gluon fields. In fact this can be related to the QCD potential between an infinitely massive quark and antiquark. Although not directly a physical quantity, this is something about which we have some physical understanding. An infinitely massive quark does not move in spatial directions and so simply generates a path which is a string of gluon fields in the time direction. If this is joined to a string generated by an antiquark a distance R in lattice units away then a rectangular $R \times T$ loop of gluon fields is created. This is known as a Wilson loop, see Figure 4 right.

To 'measure' this in pure gluon QCD, we generate configurations of gluon fields with probability $e^{-S_{latt}}$ where S_{latt} is a discretisation of the pure gluon QCD action given by Equation 16. On each of these configurations we calculate the $R \times T$ Wilson loop, averaging over all positions of it on the lattice. We then average over the results on each configuration in the ensemble to obtain a final answer with statistical error, for lots of values of R and T. We have

$$\frac{1}{Z} < 0|\mathcal{O}|0 > = \frac{\int \mathcal{D}U \mathcal{O}[U] e^{-S_{g,QCD}}}{\int \mathcal{D}U e^{-S_{g,QCD}}} = << \mathcal{O} >> = \frac{1}{N_{conf}} \sum_{i=1}^{N_{conf}} O_i \qquad (18)$$

Typically we need many hundreds of configurations in an ensemble for a small statistical error at large R and T.

The ensemble average of O is related to the heavy quark potential by similar arguments to those used for the operator $x(t_1)x(t_1)$ in section 2.1. One end of the Wilson loop creates a set of eigenstates of the Hamiltonian that are based on a massive quark-antiquark pair. These eigenstates are produced with different amplitudes by the Wilson loop operator and have different energies. In this case there is no kinetic energy, so the energies are those of the heavy quark potential. The different eigenstates propagate for time T and, if T is large, the ground state eventually dominates.

$$<< \mathcal{O} >> = C e^{-aV(R)T} + C' e^{-aV'(R)T} + \dots. \qquad (19)$$

By fitting the results as a function of the time length, T, in lattice units, the heavy quark potential in lattice units, $aV(R)$, is obtained. V' is some kind of excitation of the potential which we will not be interested in here. The heavy quark potential at short distances should behave perturbatively and take a Coulomb form. At large distances we expect a 'string' to develop which confines the quark and antiquark and gives a potential which rises linearly with separation. We can therefore fit the lattice potential to the form

$$aV(r = Ra) = -\frac{4}{3}\frac{\alpha_s(r)}{R} + \sigma a^2 R + \tilde{C} \tag{20}$$

where $\tilde{C}$ is a 'self-energy' constant that appears in the lattice calculation. If the results for $aV(R)$ are plotted against R, the slope at large R is the 'string tension', σ, in lattice units, *ie* σa^2. Phenomenological models of the heavy quark potential give values for $\sqrt{\sigma}$ of around 440 MeV. Using this value for σ and the result from the lattice of σa^2, gives a value for a. This is often quoted as a value for a^{-1} in GeV. Note that a^{-1} in GeV $= 0.197/(a \text{ in fm})$.

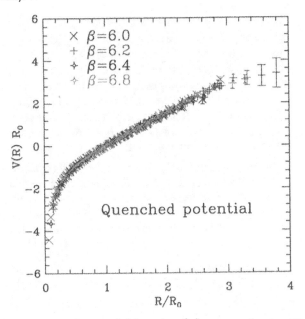

Figure 6. *The heavy quark potential in units of the parameter, r_0, as a function of distance, r, also in units of r_0. The calculations were done in quenched (pure glue) QCD at a variety of different values of the lattice spacing, corresponding to the different values of β quoted. (Bali 2000)*

The results for aV can then be multiplied by a^{-1} to convert them to physical units of GeV and, having removed the constant $\tilde{C}$, the results can be plotted as a function of the physical distance, r in fm. If discretisation errors are small, then results at different values of the lattice spacing should be the same. Figure 6 shows results for the heavy quark potential on relatively fine lattices at different values of β using the Wilson plaquette action for S_g (Bali 2000). $V(r)$ is not in fact given in GeV here, nor is r in fm, but both are given in terms of a parameter called r_0 (Sommer 1994). This is the value of r at which $r^2 \partial V/\partial r = 1.65$ and is a commonly used quantity to determine the lattice spacing (or at least relative lattice spacings), rather than the string tension. r_0 is not a physical parameter, and as such is not available in the Particle Data Tables. We shall see later that there are good hadron masses to use for the determination of a and these can also be used to determine the value of r_0. Meanwhile, $r_0 \approx 0.5$fm. The results at different values of β lie on top of each other and this gives us confidence that the discretisation errors are small.

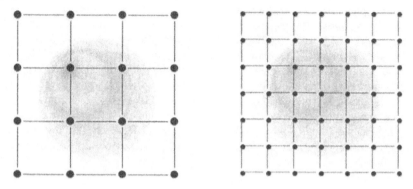

Figure 7. *A physical object on two lattices of different lattice spacing. On the right our updating algorithm takes much longer to register a significant change on the length scale of the object.*

The values of β used correspond to rather fine lattices. The coarsest lattice is at $\beta = 6.0$ and has $a = 0.1$ fm. The finest lattice is at $\beta = 6.8$ and has $a = 0.03$ fm. The finest configurations are very expensive to generate. Yet if we looked closely at the results at $\beta = 6.0$ we would be able to see discretisation errors at the few percent level. If we want accurate results on lattices that are coarse enough for affordable calculation (especially, as we shall see, once we include quarks) then we must improve the discretisation of the action. The cost of lattice calculations grows naively as a^{-4} because of the four-dimensional space-time. In fact it is even worse than this because of critical slowing-down. Physical distances on the lattice grow in lattice units as the lattice spacing gets smaller (see Figure 7). This means that algorithms that update configurations by making local changes to the fields, get slower and slower at making a change on a distance scale of r as a is reduced. Then the cost of the calculation of, say, $V(r)$ actually grows as a^{-6}. It becomes imperative to improve the action, rather than to attempt to beat down the discretisation errors by reducing a.

Improving the gluon action is at first sight straightforward. We expand the Wilson plaquette action in powers of a and notice that it has a^2 errors when compared to the continuum QCD gluon action. We add a higher dimension operator to the action in the form of a 2×1 Wilson loop to cancel this error:

$$S_{g,latt} = \beta \sum (1 - \frac{5c_1}{3}\frac{\mathrm{ReTr}U_p}{3} + \frac{c_2}{12}\frac{\mathrm{ReTr}(U_{2\times1} + U_{1\times2})}{3}). \qquad (21)$$

c_1 and c_2 are chosen to remove the a^2 error and naively would have the value 1. However, in a quantum field theory like QCD, the value of $c_{1,2}$ becomes renormalised by radiative corrections. $S_{g,latt}$ must reproduce $S_{g,cont}$ to a required level of accuracy and so $c_{1,2}$ must absorb the effect of the differences in gluon radiation between the continuum and lattice versions of QCD. This difference arises from gluon radiation with momentum larger than π/a which does not exist on the lattice. Gluon radiation at these high momenta is perturbative and so we can calculate $c_{1,2}$ as a perturbative power series in α_s. Provided a is small enough so that $\alpha_s(\pi/a)$ is small enough, the $c_{1,2}$ will not be very different from 1. There is a complication, however, and that arises from the way in which the lattice gluon field U_μ is related to the usual gluon field A_μ. The exponential relationship in Equation 12 means that the lattice perturbation theory contains vertices with many powers of A_μ. Although these are suppressed by powers of g, they produce rather large contributions and have to be taken into account. In fact they appear as so-called 'tadpole' diagrams which take the same form in many different processes (Lepage and Mackenzie 1993). This allows them to be substantially removed in a universal way by the simple expedient of estimating how far the gluon link field, U_μ, is from the value 1 (the 3×3 unit matrix) that it would take in the continuum, $g^2 \to 0$ limit. We can measure this, for example, from the average plaquette (traced and divided by 3 so that it would be 1 in the continuum limit). This contains 4 U fields so we take the fourth root to determine the 'tadpole-factor', u_0. Taking

$$U_\mu(x) \to \frac{U_\mu(x)}{u_0} \qquad (22)$$

is called 'tadpole-improvement'. If this is done then the calculation of $c_{1,2}$ does indeed give well-behaved perturbative expansions which do not differ substantially from the naive value of 1. The improved gluon action becomes

$$S_{g,latt} = \beta \sum (-\frac{5c_1}{3} \frac{\text{ReTr}U_p}{3u_0^4} + \frac{c_2}{12} \frac{\text{ReTr}(U_{2\times1} + U_{1\times2})}{3u_0^6}), \qquad (23)$$

dropping the constant term in the action. Taking $c_{1,2}$ to be 1 is the tree-level tadpole-improved Symanzik action and this gives results significantly better than the Wilson plaquette action on coarse lattices (Alford *et al.* 1995, Lepage 1996). Note that β here cannot be compared directly to that for the Wilson plaquette action. It only makes sense to compare results between different actions at the same value of the lattice spacing.

Figure 8 compares results for the heavy quark potential, calculated as discussed above, for the two actions on coarse lattices with lattice spacing 0.4fm (Alford *et al.* 1995). The results with the Wilson plaquette action show clear discretisation errors in the fact that $V(r)$ is not a smooth curve, *ie* it does not have rotational invariance. It reflects the fact that distances on the lattice are not in general travelled in straight lines and the result will depend on the actual path used between two points. The curve obtained with the improved gluon action is much smoother. As discretisation errors are removed, the path dependence of the distance measurement is reduced and the rotational invariance of the continuum is restored. Indeed $c_{1,2}$ could be fixed non-perturbatively (*ie* within the lattice simulation) by demanding rotational invariance and it is clear from this that results close to 1 after tadpole-improvement would be obtained.

In section 3 we will describe recent lattice results and these use an improved gluon action to reduce the discretisation errors. In fact the action is improved beyond that used in Figure 8 by including the first α_s terms in c_1 and c_2 and an additional operator that appears at $\mathcal{O}(\alpha_s)$ made of a 3-dimensional parallelogram of U fields (Alford *et al.* 1995, Lepage 1996).

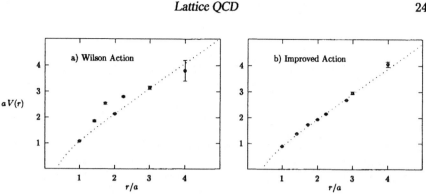

Figure 8. *The heavy quark potential calculated in pure gluon QCD at a lattice spacing, a, of 0.4fm. The left plot uses the Wilson plaquette action, the right plot the tree-level tadpole-improved Symanzik action. The right plot has a much smoother curve than the upper one reflecting the improvement of discretisation errors (Alford et al. 1995).*

2.3 Quark fields in Lattice QCD

Quarks represent a big headache for Lattice QCD. Because quark fields anticommute, they cannot be handled using ordinary numbers on a computer. We therefore have to perform the integral over the quark fields in the path integral by hand. In fact this is easy to do because of the way that the quark fields appear in the QCD action. Including the Dirac piece of the action for quarks, we have

$$Z = \int \mathcal{D}U \, \mathcal{D}\psi \, \mathcal{D}\overline{\psi} \, e^{-S_{QCD}}$$

$$S_{QCD} = S_g + \sum_x \overline{\psi}(\gamma \cdot D + m)\psi = S_g + \overline{\psi} M \psi. \tag{24}$$

The quark field, ψ is a 12-component (3 for colour and 4 for spin) field at every lattice point, so M is a matrix with $12 \times L^3 T$ rows and columns for an $L^3 T$ lattice. D is a covariant derivative that includes a coupling with the gluon field. The integral over quark fields gives

$$Z = \int \mathcal{D}U \det(M) \, e^{-S_{g,QCD}} = \int \mathcal{D}U \, e^{-\tilde{S}_{QCD}} \tag{25}$$

where $\tilde{S} = S_{g,QCD} + \ln \det(M)$, one detM factor per quark flavour.

Typically in lattice QCD we want to calculate the mass of a hadron made of quarks and antiquarks. To do this we must calculate the expectation value of a product of appropriate operators made of quarks and antiquarks. A suitable operator that creates a meson is $\overline{\psi}^{a,\alpha,f_1} \Gamma^{\alpha\beta} \psi^{a,\beta,f_2}$ where Γ is some combination of 4×4 γ matrices that give the right spin-parity (J^P) quantum numbers to the meson, a is a colour index, α and β are spin indices and f_i are flavour indices. We include the conjugate operator to destroy the meson T lattice spacings away in time. Then, suppressing colour and spin,

$$\frac{1}{Z}\langle 0|H^\dagger(T)H(0)|0\rangle = \frac{\int \mathcal{D}U \, \mathcal{D}\psi \, \mathcal{D}\overline{\psi} \, (\sum_{\vec{x}} \overline{\psi}^{f_1} \Gamma \psi^{f_2}(\vec{x}))_T (\overline{\psi}^{f_2} \Gamma \psi^{f_1})_0 e^{-S_{QCD}}}{\int \mathcal{D}U \, \mathcal{D}\psi \, \mathcal{D}\overline{\psi} \, e^{-S_{QCD}}}. \tag{26}$$

On integration, the right-hand-side becomes (for the case where $f_1 \neq f_2$):

$$\frac{\int \mathcal{D}U \, \text{Tr}_{\text{spin,colour},\vec{x}}(M_{f_1(0,T)}^{-1} \Gamma M_{f_2(T,0)}^{-1} \Gamma) \det M \, e^{-S_{g,QCD}}}{\int \mathcal{D}U \, \det M \, e^{-S_{g,QCD}}} \qquad (27)$$

This calculation requires us to generate sets of gluon field configurations with probability $e^{-\tilde{S}_{QCD}}$, calculate the trace over spin and colour of the M^{-1} factors on each configuration and then average these over the ensemble. The calculation is illustrated in Figure 9. It is known as a 2-point function calculations because there are two operators at times 0 and T, represented by the filled ovals. The straight lines at the top and bottom indicate the valence quark propagators (the M^{-1} factors above that connect the creation and annihilation operators for that particular valence quark). As the quark propagates it interacts any number of times with the other quark through the lattice gluon field and this is indicated by the curly lines. Some of these gluon lines may include the effect of a sea quark-antiquark pair (as on the left of the diagram).

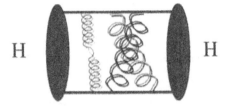

Figure 9. *An illustration of the calculation of a simple 2-point function for a hadron in lattice QCD.*

If we have $f_1 = f_2$ for a flavour-singlet quantity then there are additional 'disconnected' pieces where the M^{-1} factors are generated from each operator separately. This gives, for example, $\text{Tr}(M_{(0,0)}^{-1})\text{Tr}(M_{(T,T)}^{-1})$. These are very difficult to calculate, being very noisy, and accurate results for light flavour-singlet mesons are not yet available. For heavy mesons, $c\bar{c}$ and $b\bar{b}$, this is expected to be a very small effect and is usually ignored.

The hadron mass is determined from the results through the usual multi-exponential form

$$\frac{1}{Z}\langle 0|H^\dagger(T)H(0)|0\rangle = \, << \text{Tr}_{\text{spin,colour},\vec{x}}(M_{f_1}^{-1}\Gamma M_{f_2}^{-1}\Gamma) >> = \sum_n C_n e^{-m_n aT}. \quad (28)$$

n runs over radial excitations of the hadron with a particular set of J^P and flavour quantum numbers, and $m_n a$ are the different hadron masses. In fact the formula above is only correct for a lattice of infinite time extent in lattice units, T'. On the finite lattices we must use there are additional terms from the possibility of quarks 'going backwards' round the lattices. These terms take the form $e^{-m_n a(T'-T)}$ and can readily be taken account of in the fit, although their form does depend on the hadron being studied. C_n in the above equation is related to the square of the matrix element $< 0|H|n >$ by analogy with Equation 8 in section 2.1. The size of C_n then depends on the form of the operator H used to create and destroy the hadron. Any operator with the right quantum numbers can be used and if we make a set of such operators we can calculate a whole matrix of correlators and fit them simultaneously for improved precision. Typically we use operators made of quark and antiquark fields but separated in space according to some kind of "wavefunction" known as a "smearing", eg $\overline{\psi}_x \phi(x - y)\psi_y$. To make a gauge-invariant operator of this kind either requires strings of Us to be inserted between x and y or for the gluon field configuration to be gauge-fixed, typically to Coulomb gauge. It may be possible to choose ϕ so that a particularly good overlap with one of the states of the system (eg the ground state) is obtained, and poor overlap with the others. Then C_0 would be large in the equation above, and the other C_n small, and this would give improved precision for $m_0 a$. The masses of the radial excitations are of interest too and smearings which improve overlap with them have also been studied. With particular operators, H, C_n contains information of physical relevance. For example the matrix element of the temporal axial current, J_{A_0} between the vacuum and a pseudoscalar meson gives its decay constant, related to the rate for its decay purely to leptons via a W boson. The calculation in lattice QCD of matrix elements of this kind is discussed in section 3.3.

The way in which a lattice QCD calculation with quarks is done makes a clear distinction between valence quarks and sea quarks. The valence quarks are those that give the hadron its quantum numbers and these give rise to the M^{-1} factors in Equation 27. The sea quarks are those that are produced as quark-antiquark pairs from energy fluctuations in the vacuum (quark vacuum polarisation). They give rise to the det$M$ factors in Equation 27. The important sea quarks are the light ones, u, d and s. The heavy quarks c and b have no effect as sea quarks and the t quark does not even have to be considered as a valence quark since it does not form bound states before decaying.

Manipulations of the matrix M are computationally costly. Even though it is a sparse matrix (with only a few non-zero entries) it is very large. There are various computational techniques for calculating the M^{-1} factors and the det$M$ factor is included by repeated determination of the calculation of M^{-1}. This makes the inclusion of det$M$ very costly indeed. It becomes increasingly hard as the quark mass becomes smaller because M becomes ill-conditioned. (The eigenvalues of M range between some fixed upper limit and the quark mass, so this range increases as $m_q \to 0$). In the real world the u and d quarks have very small mass and so in the past there has not been sufficient computer power available to include them as sea quarks or, if they have been included, their masses have been much heavier than their real values.

Missing out sea quarks entirely is known as the quenched approximation. It is clearly wrong, but for a long time the presence of other systematic errors and poor statistics obscured this fact. More recently it has become clear that the systematic error in the quenched approximation is around 10-20%. When light quark vacuum polarisation (det$M$) is included the calculation is said to be 'unquenched' or 'dynamical'. The sea quarks are then also called dynamical quarks. We will see in the results section that it is now possible to include realistic quark vacuum polarisation effects and the quenched approximation can be laid aside at last.

Our earlier discussion makes clear that the computational cost of including light quark vacuum polarisation will require a very good (ie highly improved) discretisation of both the gluon and quark actions. We have discussed the gluon action earlier. For the quark action there are several possibilities, or formulations. We will concentrate here on those formulations that have already been used in unquenched simulations.

The naive quark action is a straightforward discretisation of the Dirac action in Euclidean space-time:

$$S_q = \sum_x \overline{\psi}(x)(\gamma \cdot \Delta + ma)\psi(x). \tag{29}$$

The finite difference Δ is

$$\Delta_\mu \psi(x) \equiv \frac{1}{2}(U_\mu(x)\psi(x + 1_\mu) - U_\mu^\dagger(x - 1_\mu)\psi(x - 1_\mu)) \tag{30}$$

showing explicitly how the gluon fields appear, coupled to the quarks, to maintain gauge invariance in the action. QCD with quarks has parameters, in addition to the coupling constant, that are the quark masses. In lattice QCD these appear as quark masses in lattice units, ma. As with the lattice spacing/coupling constant, the quark masses are not known a priori and must be adjusted as a result of a lattice calculation. We therefore iterate until, for example, we get a particular hadron mass correct. Since the quark mass is in lattice units, finer lattices will require smaller values of ma than coarser ones.

The naive quark action has several good properties. It has discretisation errors at $\mathcal{O}(a^2)$. The finite difference piece is anti-hermitian and therefore has purely imaginary eigenvalues that appear as $\pm i\lambda$. With the mass term the eigenvalues of M are then $ma \pm i\lambda$. This is the same as in the continuum and follows from the chiral symmetry of the action, *ie* our ability to rotate independently left- and right-handed projections of the quark field. It means that the bare quark mass is simply multiplicatively renormalised and the renormalised quark mass goes to zero at the same point as the bare quark mass. This is important because we need to work with very small (renormalised) quark masses for the u and d quarks and we want to be able to go in that direction by just taking ma to be smaller and smaller (if we have enough computing power).

Chiral symmetry is spontaneously broken in the real world giving rise to a Goldstone boson, the π, whose mass consequently vanishes at zero quark mass ($m_\pi^2 \propto m_q$). For small quark masses, where m_π is small, we have a well-developed chiral perturbation theory which tells us how hadron masses and properties should depend on the u/d quark masses (or equivalently m_π^2) and we can make use of this to extrapolate down to physical u/d quark masses from the results of our lattice QCD simulation provided that we are able to work at small enough u/d quark masses to be in the regime where chiral perturbation theory works. In general $m_{u/d} < m_s/2$ is necessary for an accurate extrapolation (Arndt 2004). We will discuss this further in section 3.

It would seem that we have everything ready to do the simulations but we must first discuss the infamous doubling problem. The naive quark action in fact describes 16 quarks rather than 1. This is demonstrated most easily in 1-dimension in the absence of gluon fields. It suffices to compare the continuum derivative in momentum space, *ie* p, with the Fourier transform of the finite difference on the lattice, which is $\sin(pa)/a$. These are plotted in Figure 10 as a function of p between $-\pi/a$ and π/a, the limits over which p makes sense on the lattice (the first Brillioun zone). Momenta larger than π/a reappear as negative momenta larger than $-\pi/a$ so these points are periodically connected. Around $p \approx 0$ the sine function goes through zero and mimics a straight line up to a^2 errors as expected. The problem is that this is also true around $p \approx \pi/a$ (with opposite slope). This means that there is a continuuum-like solution of the Dirac equation around $p = \pi/a$ as well as around $p = 0$, *ie* there are 2 quarks instead of 1. In 4-d (and including γ matrix algebra as well) the picture is more complicated but the basic

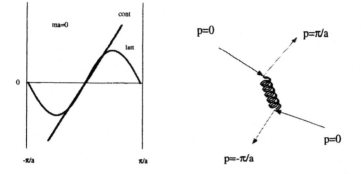

Figure 10. *(Left) The continuum derivative and lattice finite difference compared as a function of momentum within the first Brillouin zone on the lattice. The points $p = \pi/a$ and $-\pi/a$ are joined by lattice periodicity. (Right) Taste-changing interactions for naive quarks. A high-momentum $(p = \pi/a)$ gluon exchange can change one taste of quark into another.*

facts are the same - there are 2^4 quarks instead of 1. The 'doublers' or extra 'tastes' of quark live at the edges of the Brillouin zone, where any component of p is close to π/a.

This was originally thought to be disastrous and huge efforts, which continue today, were made to solve the problem. Wilson introduced a Wilson term into the naive action which has the effect of giving mass to the doublers. The Wilson action is then

$$
\begin{aligned}
S_f^W &= S_f^{naive} - \frac{r}{2} \sum_x \overline{\psi}_x \Box \psi_x \\
\Box \psi_x &= \sum_{\mu=1}^{4} U_\mu(x)\psi_{x+1_\mu} - 2\psi_x + U_\mu^\dagger(x - 1_\mu)\psi_{x-1_\mu}.
\end{aligned} \tag{31}
$$

In the absence of gluon fields the Fourier transform of the extra term is $2r\sin^2(pa/2)$ which has a maximum at $\pm\pi/a$. When folded in to the inverse quark propagator, taking account of the γ matrix structure, this then prevents the inverse propagator from vanishing at the edges of the Brillouin zone, effectively giving the doublers a large mass which increases as $a \to 0$ so that they are completely removed in the continuum limit.

This solution of the doubling problem was used for many years. It does have two big disadvantages. One is that discretisation errors now appear at $\mathcal{O}(a)$ which causes large systematic errors even on relatively fine lattices. The other is that the Wilson term does not respect chiral symmetry and so we lose the simple connection between the bare quark mass parameter and the renormalised quark mass and have to search in ma for the point where m_π vanishes. The eigenvalue spectrum is now complicated as well which creates numerical difficulties. Discretisation errors are ameliorated in the improved form of the action known as the 'clover' action. The $\mathcal{O}(a)$ errors are cancelled by a $\sigma_{\mu\nu}F^{\mu\nu}$ term which is discretised as a set of plaquettes around a central point that looks like a 4-leaf clover. The coefficient of the clover term can be set using tadpole-improvement or non-perturbatively. Again it is close to 1 *after* the tadpole-improvement with a well-behaved perturbative expansion. Work using the clover action continues and it is being used for unquenched simulations (Ishikawa *et al.* 2005). A newer form of the action, called twisted mass QCD, is also being developed. This overcomes several of the numerical problems and is a promising development for the future (Frezzotti 2005).

Another relatively new development is that of a set of actions that maintain chiral symmetry on the lattice but also solve the doubling problem. These require an enormous increase in computer power because they either need a matrix inversion inside a matrix inversion to calculate M^{-1} (overlap quarks) or a calculation in 5-dimensions (domain wall quarks). Calculations using these formalisms have then been restricted to the quenched approximation because of the cost. In future they may be possible to implement including realistic quark vacuum polarisation effects on very powerful supercomputers (Chiu 2004).

The recent progress in lattice QCD calculations has come, however, from returning to the naive discretisation of the quark action. In fact the doubling problem is not a problem if the doublers are simply copies of each other, because we can then simply 'divide by 16' inside our calculation. If M has a 16-fold degeneracy in its eigenvalue spectrum then taking $(\det M)^{1/16}$ gives us the effect of 1 taste for each flavour. The problem is that the different tastes of quarks do interact with each other and change taste and this splits some of the degeneracy. The process by which this happens is a high momentum (π/a) gluon exchange as in Figure 10 which moves the quark from one Brillouin zone boundary to another. These taste-changing interactions are an artefact of using the lattice and they induce large discretisation errors, even though they are formally $\mathcal{O}(a^2)$ (Lepage 1998b). To improve the naive action then requires removing the unphysical taste-changing interactions at leading order as well as the usual a^2 errors from discretising a derivative as a symmetric finite difference, discussed in section 2.1. This markedly improves the discretisation errors, but also significantly reduces the taste-changing interactions, and makes this action a good one for lattice simulations.

The naive quark formalism is fast numerically because it can be converted to the staggered quark formalism in which there is no quark spin. It is a remarkable feature that the quark fields can be rotated so that the naive quark action (or its improved variant) is diagonal in spin space. We have only to simulate one spin component on the lattice and then, if required, we can reconstruct all quantities in terms of naive quarks. If we take

$$\psi(x) \to \Omega(x)\chi(x); \quad \overline{\psi}(x) \to \overline{\chi}(x)\Omega^{\dagger}(x) \tag{32}$$

where

$$\Omega(x) \equiv \prod_{\mu=0}^{3} (\gamma_{\mu})^{x_{\mu}},$$

the unimproved naive action becomes

$$\overline{\psi}(x)(\gamma \cdot \Delta + ma)\psi = \overline{\chi}(x)(\alpha(x) \cdot \Delta + ma)\chi(x)$$

where $\alpha(x)$ is diagonal in spin space, $\alpha_{\mu}(x) = (-1)^{x_0 + \cdots x_{\mu-1}}$. Rewriting the action in terms of a single-spin (but 3-colour) component field χ, we have the unimproved staggered quark action:

$$S_f^S = \sum_x \overline{\chi}_x \{ \frac{1}{2} \sum_{\mu} \alpha_{x,\mu} (U_{\mu}(x)\chi_{x+1_{\mu}} - U_{\mu}^{\dagger}(x - 1_{\mu})\chi_{x-1_{\mu}}) + ma\chi_x \}. \tag{33}$$

The staggered quark action does not solve the doubling problem, but removes an exact 4-fold degeneracy from it. There are then 4 tastes for every quark flavour that we include with M instead of 16. If we improve the action so that the a^2 taste-changing interactions are small then the tastes will be close to being copies of each other. We will expect to see a 4-fold close-to-degeneracy in the eigenvalue spectrum of the improved staggered M, which will become closer and closer as $a \to 0$. This has been demonstrated numerically (Follana *et al.* 2004) and justifies 'dividing by 4' by taking $(\det M)^{1/4}$ in the path integral to include one quark flavour. The improvement of the staggered quark action to remove the leading-order taste-changing interactions of Figure 10 is done by replacing the gluon fields that appear in the finite difference of Equation 33 with a combination of gluon fields that remove the coupling between the quark and a single $p = \pi/a$ gluon. The combination 'smears' out the gluon field in directions perpendicular to its link by including in combination with a single $U_\mu(x)$ paths such as 'the staple', $U_\nu(x)U_\mu(x + 1_\nu)U_\nu^\dagger(x + 1_\mu)$. The simplest improved staggered action with a^2 errors removed uses paths up to those containing 7 U fields in all 3 directions perpendicular to a given link with tadpole-improvement. It is called the asqtad action and has been successfully used in extensive unquenched simulations as described in section 3 on results (Lepage 1998b, Orginos *et al.* 1999).

All quark formulations must give the same physical results in the continuum $a \to 0$ limit and this can be usefully checked in the quenched approximation. Figure 11 shows results for the mass of a vector light meson (a 'ρ') at an arbitrary fixed physical quark mass as a function of lattice spacing (Davies *et al.* 2005). Both axes are given in units of r_1, defined in a similar to way to r_0, but $r_1 \approx 0.35$ fm. The x axis is in terms of the squared lattice spacing since most of the formalisms plotted have discretisation errors at $\mathcal{O}(a^2)$ or better. The Wilson formalism, with errors at $\mathcal{O}(a)$, is clearly much worse than the other formalisms on this plot, with a very steep a dependence, but it does give a consistent continuum limit.

Heavy quarks (b and c) represent a rather different set of issues to those for light quarks in lattice QCD. They have large quark masses, m_Q, and therefore large values of $m_Q a$ at any value of a at which we are able to do QCD simulations. This means that we risk large discretisation errors when handling these quarks if we use a formalism in which the errors are set by the size of ma. If $m_Q a > 1$ then no amount of improvement will give a good discretisation for these quarks. This is particularly true for the b quark which has a mass of around 5 GeV. To reduce $m_Q a$ below 1 requires $a^{-1} > 5$ GeV or a lattice spacing < 0.04 fm which is incredibly expensive to simulate (as well as being wasteful). Luckily physical understanding of these systems comes to our rescue here. The experimental spectrum of hadrons containing b and c quarks shows clearly that the masses of these hadrons are much larger than the differences in mass between the hadrons in different radial and orbital excitations (and these differences are in fact very similar for b and c systems). The differences reflect typical kinetic energies and momenta inside the hadrons of a few hundred MeV to a GeV, *ie* of the same size or somewhat larger than those typically inside light hadrons but $<< m_Q$. The quark mass itself is not important for the dynamics of the bound state but simply provides an overall mass shift (Davies 1998). If we formulate the quark action in a way that simulates accurately nonrelativistic quark momenta and kinetic energies we will be able to accurately determine the properties of the hadrons and it is these scales that will also control the discretisation errors.

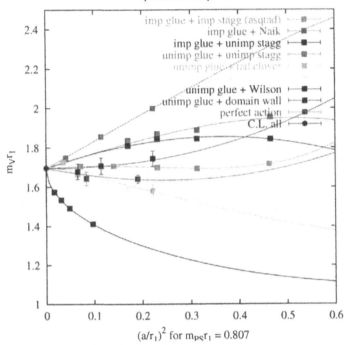

Figure 11. *Results for the vector light meson mass at fixed physical quark mass in quenched lattice QCD using a variety of gluon and quark actions (for the valence quarks). The mass is given in units of r_1 and plotted as a function of a^2, also in units of r_1. The lines represent fits to the lattice results, including appropriate discretisation errors for each case, that require all lines to have a common continuum limit. A good fit is obtained (Davies et al. 2005).*

This means using a nonrelativistic effective field theory and the example we give here is Nonrelativistic QCD, NRQCD. The lattice NRQCD action is a discretised nonrelativistic expansion of the Dirac action, matched to continuum QCD to the desired order in the heavy quark velocity, v_Q and in the strong coupling constant, α_s (Lepage *et al.* 1992). The first few terms of the continuum NRQCD quark Lagrangian are:

$$\mathcal{L}_Q = \overline{\psi}(D_t - \frac{\vec{D}^2}{2m_Q} - c_4 \frac{\vec{\sigma} \cdot \vec{B}}{2m_Q} + \ldots)\psi \tag{34}$$

where we include just the non-relativistic kinetic energy term and the coupling between the quark spin and chromomagnetic field. Higher order terms give the spin-orbit interaction and Darwin term *etc.* ψ is a 2-component spinor here but the antiquark Lagrangian is simply related to the quark one and so the antiquark propagator is easily determined without a separate calculation. On the lattice the covariant time and space derivatives above become finite differences with U fields included. The presence of a single time derivative means that the calculation of the quark propagator, M^{-1}, can be solved as an initial value problem on one pass through the lattice. This is much faster than the iterative methods needed for light quarks. On the lattice m_Q becomes $m_Q a$, the quark mass in lattice units, and this must be determined by getting a heavy hadron mass correct, for example the Υ mass. This is slightly more complicated than in the light quark mass case because the direct mass term is missing from the Lagrangian, so the energy exponent of an Υ correlator at zero momentum will not give the mass (although energy differences do give mass differences). Instead we have to calculate hadron energies at non-zero spatial momentum and extract the mass of the hadron from its kinetic energy.

NRQCD contains, by design, the low momentum physics of QCD for heavy quarks. High-momentum interactions of continuum QCD are missing, and the lattice in any case does not allow for $p > \pi/a$. The effect of these missing high-momentum interactions is included in NRQCD through a renormalisation of the coefficients, c_i, of the higher order terms. Because the effects that we are talking about are high momentum, the calculation can be done in perturbation theory as a powers series in α_s. Again small deviations from 1 are found after tadpole-improvement of the U fields is implemented, provided $m_Q a$ is not large. The c_i contain functions of $m_Q a$, including inverse powers of $m_Q a$, multiplying powers of α_s and so we cannot take a to zero or we lose control of the convergence of the c_i. This is not a problem in practice since, as discussed earlier, discretisation errors can now be controlled in this formalism to high accuracy at values of the lattice spacing that we can simulate and there is no need to take a very small.

For charm quarks the situation is not as clear as for b quarks because $m_c a \approx 1$ on lattices in current use. NRQCD does not necessarily work well and it may be more advantageous to take a relativistic formalism of the c quark and attempt to improve it highly to reduce the discretisation errors in $m_c a$ to a reasonable level. The Fermilab formalism does a mixture of both these things by using a relativistic action, the clover action, but interpreting it non-relativistically where necessary to avoid the largest discretisation errors (El-Khadra *et al.* 1997). This is the method currently being used most extensively for c quarks and results will be described in section 3. Another promising method is to use anisotropic lattices in which the lattice spacing in the time direction is much finer than that in spatial directions. This then allows $m_c a_t$ to be small without requiring a huge amount of computer power to make a fine grid in 4-dimensions. The price of this is a relatively complicated action, however.

3 Results

Recent lattice results which include for the first time realistic quark vacuum polarisation effects, have changed the landscape of lattice calculations. We will concentrate on those results here and the possibilities for the future that they engender.

The new results are based on gluon field configurations generated by the MILC collaboration and analysis by the HPQCD and MILC collaborations (Davies *et al.* 2004). They have worked with a highly improved gluon action and with improved staggered (asqtad) quarks, as described in the previous section. The effect of the quark vacuum polarisation is included in the generation of the gluon configurations by taking the fourth root of the determinant of the improved staggered quark matrix for each flavour of quark included in the sea. Three flavours of quarks are included in the sea, u, d and s. This is believed to be all that is necessary, since a perturbative analysis shows that the effects of c or b quarks in the sea is very small (Nobes 2005). The u and d quarks are taken in fact to have the same mass (as in all current lattice calculations) because this provides for some simplification and should give only a small error when comparson is made to appropriately isospin-averaged experimental quantities. The configurations are then referred to as '2+1' flavour unquenched configurations.

As described above, it is not possible to perform the lattice calculations at the physical u/d quark mass (since we take $m_u = m_d$ this would be $(m_{u,phys} + m_{d,phys})/2$). Instead, many different ensembles of configurations are made with different values of $m_{u/d}$ and extrapolations to find the physical point must be done. These chiral extrapolations will be accurate provided that we are close enough to the physical point and we are able to constrain the functional form of the dependence on $m_{u/d}$ with chiral perturbation theory.

It is convenient, when discussing the u/d quark masses used in the lattice calculations, to give them in terms of the strange quark mass, m_s. This is because there is no problem (with current computers) in doing lattice QCD calculations that include the s quark vacuum polarisation. However, it is also clear that, in the real world, there is a very visible difference between s quarks and u/d quarks. Indeed, if we consider chiral perturbation theory as an expansion in powers of $x_q = (m_{PS}/(4\pi f_\pi))^2$ where m_{PS} is the mass of the pseudoscalar light meson made from the light quarks, then $x_s = 0.33$ and $x_{u/d} = 0.03$. We would therefore not expect low order chiral perturbation theory to work well for light quarks with a mass equal to that of the s quark. This is not a problem since we do not need chiral perturbation theory to reach the s quark mass. However it does mean that our u/d quark mass must be well below m_s to expect to be able to reach the physical point by chiral extrapolation in $m_{u/d}$. However, with a set of ensembles with $m_{u/d} < m_s/2$ and using next-to-leading order chiral perturbation theory in $m_{u/d}$, it should be possible to perform extrapolations with errors at the few % level. This is what has now been done using the configurations generated by the MILC collaboration. The MILC configurations range in $m_{u/d}$ down to $m_{u/d} = m_s/8$ which is within a factor of three of the real world. This is a huge improvement over previous calculations which had only two flavours of sea quarks, meant to represent u and d, but with $m_{u/d} > m_s/2$, and it is this improvement that is responsible for the quality of the new results.

In addition to including the effect of quark vacuum polarisation with light u/d quarks and s quarks, the MILC configurations come in sets with three different values of the lattice spacing. In each set the bare coupling constant must be adjusted as the quark masses are changed to keep the lattice spacing approximately the same (as described earlier the lattice spacing is determined accurately after the simulation, but preliminary results enable this approximate tuning of the lattice spacing to be done). Having a set of configurations which include the effect of sea quarks of different light quark mass but which have the same lattice spacing is important to allow the chiral extrapolations to be done without confusing the (real) effects of changing $m_{u/d}$ with (unphysical) systematic errors from having different values of the lattice spacing. The effect of the discretisation errors can also be gauged accurately (and an extrapolation to the continuum limit be performed if necessary) by having several well-spaced values of the lattice spacing. Again this has not been possible with previous lattice calculations. On the MILC configurations the lattice spacing values are approximately 0.18 fm (supercoarse), 0.12 fm (coarse) and 0.08 fm (fine), chosen to be appropriately spaced in a^2, the expected size of the discretisation errors with the improved action used. Most of the results described here will come from the coarse and fine sets, since the supercoarse set has only been made recently. A superfine set is also now planned, to give a fourth lattice spacing value. In addition another set of ensembles is being made with a different bare sea *s* quark mass. There is no problem in principle with working at the correct *s* mass, but it is hard to tune this accurately until after the calculations have been done. On the coarse ensemble the sea *s* mass is in fact 25% high and it is 10% high on the fine set of ensembles. Results with a different *s* mass will enable a more accurate interpolation to the correct *s* mass.

Whilst singing the praises of the MILC configurations, it should also be said that they have a commendably large volume (2.5 fm on a spatial side) and a very long extent in the time direction (nearly 8 fm). There are several hundred configurations in each ensemble that can be used for calculations. All of these factors lead to small statistical errors, indeed smaller than for a lot of calculations that have been done in the quenched approximation. The MILC configurations are publicly available at www.nersc.gov.

3.1 Results on the spectrum

Figure 12 summarises the new results (Davies *et al.* 2004). It shows the ratio of the lattice calculation to the experimental number for a range of gold-plated quantities from the π decay constant to radial and orbital excitation energies in the Υ system, by way of heavy-light meson masses and baryon masses. The left-hand panel of the figure shows results in the old quenched approximation and the right-hand panel shows the new results including the effect of light quark vacuum polarisation. The clear qualitative difference between the two panels is that the left-hand panel shows clear disagreement with experiment for a number of quantities because the quenched approximation is wrong. The right-hand panel shows instead agreement with experiment for all the quantities within errors of a few percent.

To make Figure 12 the parameters of QCD first had to be fixed. The lattice spacing was determined from the radial excitation energy in the Υ system ($2S - 1S \equiv M_{\Upsilon'} - M_{\Upsilon}$). The u/d quark mass was fixed by finding where the π mass would be correct by extrapolation using chiral perturbation theory. The s quark mass was fixed by determining where the K meson mass was correct. The c quark mass was fixed using the D_s and the b quark mass was fixed by getting the Υ mass correct. All of the hadron masses used to determine the parameters are "gold-plated", *ie* they have very small decay widths and are well below strong decay thresholds. This means that they are well-defined experimentally and theoretically and should be accurately calculable in lattice QCD. Using them to fix parameters will not then introduce unnecessary additional systematic errors into lattice results for other quantities. This is an important issue when lattice QCD is to be used as a precision calculational tool.

Having fixed the parameters, we can then focus on other gold-plated masses and decay constants and Figure 12 shows the predictions for these other quantities. The fact that the right-hand panel demonstrates agreement with experiment for all the quantities shown is an indication that the parameters of QCD are unique. Instead of using the hadron masses of the previous paragraph to fix the parameters we could have used appropriate quantities from Figure 12 and we would have obtained the same answer (and would then have been able to predict m_π, m_K etc). The left-hand panel shows that this is not true in

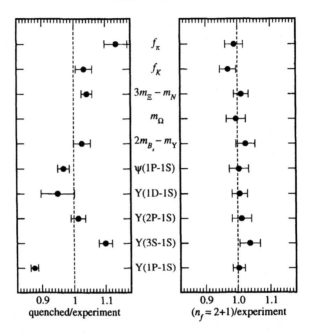

Figure 12. *Lattice QCD divided by experiment for a range of "gold-plated" quantities that cover the full range of QCD physics. The unquenched calculations on the right show agreement with experiment across the board, whereas the quenched approximation on the left gives systematic errors of $\mathcal{O}(10\%)$. (Davies et al. 2004)*

the quenched approximation. The results there show disagreement with experiment and it is clear, for example, that if we had used the orbital excitation energy in the Υ system $(1P - 1S)$ to fix the lattice spacing we would have obtained an answer 10% different. The quenched approximation is then internally inconsistent since the parameters depend on the hadrons used to fix them. The basic reason is that, as the light quark vacuum polarisation is missing, the strong coupling constant runs incorrectly between different momentum scales. Therefore hadrons which are sensitive to different momentum scales cannot simultaneously agree with experiment.

In Figure 13 more details are shown of the lattice results for the radial and orbital energy levels in the Υ system in the old quenched approximation and now using the MILC configurations with their inclusion of a realistic light quark vacuum polarisation (Gray *et al.* 2003). Our physical understanding of the Υ system is very good and there are a lot of gold-plated states below decay threshold so it is a very valuable system for lattice QCD tests and for determining the lattice spacing. We use the standard lattice NRQCD effective theory, described above, for the valence b quarks, accurate through v^4 where v is the velocity of the b quark in its bound state. This means that spin-independent splittings, such as radial and orbital excitations, are simulated through next-to-leading-order and should be accurate to about 1%. The test of QCD using these splittings is then a very accurate one. We do not expect the Υ system to be very sensitive to the masses of the light quarks included in the quark vacuum polarisation, only to their number. The

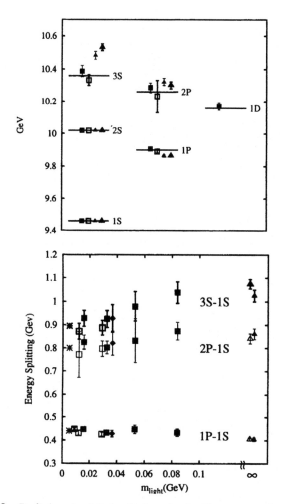

Figure 13. *Radial and orbital splittings in the* Υ *spectrum from lattice QCD in the quenched approximation and including a realistic light quark vacuum polarisation. In these plots the b quark mass was fixed from the* Υ *mass and the lattice spacing from the splitting between the* Υ' *and the* Υ. *Neither of these masses is predicted. (Top) The spectrum of S, P and D levels in the* Υ *system obtained from coarse (filled triangles) and fine (open triangles) quenched lattice calculations and from coarse (filled squares) and fine (open squares) unquenched calculations. Experimental results are shown as lines. (Bottom) Results for different splittings as a function of light u/d quark mass. The leftmost points, at lightest u/d quark mass, are the ones included in the top plot for the unquenched results. (Gray et al. 2003)*

momentum transfer inside an Υ is larger than any of the *u*, *d*, or *s* masses and so we expect these splittings simply to 'count' the presence of the light quarks. This lack of variation with light quark mass is evident in Figure 13.

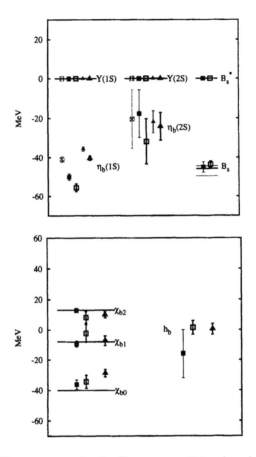

Figure 14. *Fine structure in the* Υ *system. Triangles show quenched results (closed=coarse, open =fine) and squares show unquenched results (crossed =super-coarse, closed =coarse, open =fine). (Top) S-wave splittings between* Υ *and* η_b *and between* Υ' *and* η_b'. *The same NRQCD action is used to determine the splitting between the* B_s^* *and the* B_s *and this is shown on the right. There the experimental situation is summarised by the lines - faint lines show limits for the experimental* B_s *results, the darker line shows the more precise result for the B, which is expected to be very close to the result for the* B_s. *(Bottom) P-wave splittings between the* $^3P_{0,1,2}$ χ_b *states, compared to experiment. The results on the right show the lattice prediction for the unseen* $^1P_1$ h_b *state. (Gray et al. 2003)*

Figure 14 shows the fine structure in the Υ spectrum. Since the fine structure is determined at leading order by spin-dependent terms that appear first at $\mathcal{O}(v^4)$ this has significant systematic errors from missing higher order terms. There are also systematic errors arising from missing radiative corrections to the leading v^4 terms. In fact, encouraging agreement with experiment is obtained and we are able to predict the ground-state hyperfine splitting between the Υ and the unseen η_b to be 60(15) MeV. The results for the fine structure can be improved by improving the NRQCD action and this will be done in the near future. For comparison to the Υ splittings, Figure 14 also shows the splitting between corresponding mesons made of one b quark and one s quark, *ie* the B_s heavy-light system. This is calculated on the lattice using the same NRQCD action of the valence b quarks and the improved staggered action (as used in the quark vacuum polarisation) for the light quark. Once the b quark mass and lattice spacing have been fixed from the Υ system the heavy-light spectrum is entirely predicted by lattice QCD and provides another useful test against experiment.

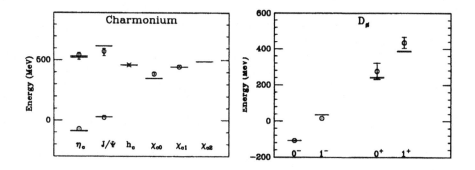

Figure 15. *Spectrum of mesons containing charm quarks from lattice QCD including light quark vacuum polarisation. These results are from the coarse unquenched MILC configurations using the Υ system to determine the lattice spacing and the D_s meson mass to fix the c quark mass. The Fermilab action was used for the c quark. (Left) The charmonium system. (Right) The D_s system. (di Pierro et al. 2004)*

The charm quark is somewhat too light for good results using the NRQCD action above. The results shown in Figure 15 use the Fermilab action described earlier which aims to interpolate between light relativistic quarks and the heavy nonrelativistic limit (di Pierro *et al.* 2004). It should be borne in mind that there are fewer gold-plated states in the $c\bar{c}$ spectrum compared to the $b\bar{b}$ spectrum described above. In fact even the ψ' is rather close to decay threshold to be sure that its mass is not affected by coupling to virtual decay modes which will have a distorted momentum spectrum on a finite lattice volume. The right-hand plot of Figure 15 shows the D_s spectrum. The D_s mass was used to fix the c quark mass, so that is not predicted. The two newly discovered 0^+ and 1^+ states are shown. The simplest explanation for these states is that they are P-wave states, narrow because their masses are too low for the Zweig-allowed decay to D, K. They do decay to D_s, π, however, and are therefore not gold-plated. We expect systematic errors in a lattice QCD calculation of their masses, even when light quark vacuum polarisation is included. The current lattice results, shown in Figure 15, are consistent with this picture and certainly do not require any more exotic explanation of the meson internal structure.

With a good action for b quarks and a good action for c quarks it is possible to predict the mass of the B_c meson. The results are shown in Figure 16, (Allison *et al.* 2005) compared to the very recent experimental results (Acosta *et al.* 2005). The quantity that is calculated on the lattice is the mass difference between the B_c and a combination of masses of other mesons containing a b quark and a c quark, enabling some of the systematic errors to cancel. There are two combinations available on the lattice. The one that gives the most accurate result we believe is to calculate the difference between the B_c mass and that of the average of the Υ and ψ. This is called $\Delta_{\psi\Upsilon}$. It turns out to be very small, although non-zero. The difference between the B_c and the sum of the masses of the D_s and B_s, $\Delta_{D_s B_s}$, is also useful but less accurate. Both splittings are shown in Figure 16 as a function of the light quark mass included in the quark vacuum polarisation on the MILC unquenched configurations. After a study of systematic errors from fixing the quark masses and lattice spacing we are able to give a final result for the B_c mass of 6304(20) MeV (Allison *et al.* 2005). The error is much better than the 100 MeV obtained for previous calculations in the quenched approximation. There 100 MeV arose directly from the impossibility of consistently determining the parameters of the theory in the quenched approximation, a problem that disappears in the new unquenched results. The very recent experimental result from CDF is 6287(5) MeV, showing impressive agreement with the unquenched lattice calculation (Acosta *et al.* 2005).

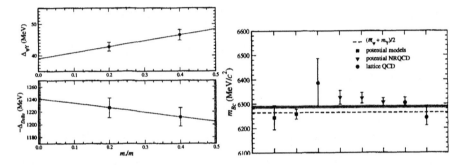

Figure 16. *(Left) The differences between the B_c mass and, above, the average of Υ and ψ masses and, below, the B_s and D_s masses, plotted against the light quark mass included in the quark vacuum polarisation in the MILC unquenched configurations. (Right) Predictions for the mass of the B_c from a variety of methods. In the centre is the old quenched lattice QCD result and on the right the two new lattice QCD predictions. The dashed line shows where the B_c mass would be equal to the average of Υ and ψ masses and the broader coloured line is the new experimental result. (Allison et al. 2005, Acosta et al. 2005)*

Results for light mesons are equally good on the MILC configurations (Aubin *et al.* 2004a, 2004b, Bernard 2001). Figure 17 shows results for the light meson masses and decay constants plotted against light quark mass. The left-hand plot shows the fits to chiral perturbation theory for the squared masses of the π and K, which must then be extrapolated (for the π) or interpolated (for the K) to find the quark masses which correspond to u/d (bearing in mind that the u and d quark masses have been taken to be the same) and s. Once this has been done, predictions for other light meson quantities can be made. Two very clean quantities to calculate are the light meson decay constants, f_π or f_K. The decay constant is obtained from the matrix element of the temporal axial current between the pseudoscalar light meson and the QCD vacuum and it is related, for the charged π and K, to the rate of purely leptonic decay. This rate is well-known experimentally for these mesons and can be calculated accurately on the lattice. This is because, unlike the matrix elements we will discuss below, the axial current needs no renormalisation for improved staggered quarks. The right-hand plot of Figure 17 shows the chiral extrapolations for the π and K decay constants for the fine MILC lattices. Only points for which the u/d quark mass is less than half the s quark mass are used in the fits to chiral perturbation theory. The curves, however, show what happens when these fits are extrapolated to higher u/d quark masses. For f_π it is clear that the heavier u/d quark are not on the same chiral perturbation theory curve as the lighter ones, *ie* using heavier u/d masses would have distorted the chiral fit and given the wrong result in the chiral limit where the u/d quark mass takes the correct value to give the experimental π meson mass. This should be borne in mind when looking at earlier lattice results that have only $m_{u/d} > m_s/2$.

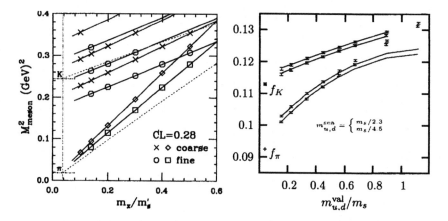

Figure 17. *(Left) Chiral fits and extrapolations for light meson (π and K) meson masses from the MILC configurations. The square of the meson mass is plotted against the light (u/d) quark mass used in the quark vacuum polarisation. This is denoted m_X and given in units of the s quark mass used in the quark vacuum polarisation (denoted here m'_s). Several different values of valence s quark mass were used so there are several different sets of results for the K mass. The lattice results are the points and lines are high-order chiral perturbation theory fits. At small light quark mass these are indistinguishable from the lowest order straight line. (Right) Chiral fits for f_π and f_K, the π and K decay constants. The points are lattice calculations on the fine unquenched MILC configurations, for two different values of the light quark mass in the quark vacuum polarisation ($m_{u/d}^{sea}$). The results are plotted against the valence u/d quark mass in units of the valence s quark mass. Note that only points with $m_{u/d} < m_s/2$ (for both valence and sea) were used in the chiral fits, shown as lines. The fits were then extrapolated back to compare to the lattice results with $m_{u/d}^{valence} > m_s/2$. (Aubin et al. 2004a, 2004b, Lepage and Davies 2004)*

Light baryons are also gold-plated particles and calculating their masses provides a further opportunity for predictions from lattice QCD since all the QCD parameters have been fixed from the meson sector. Baryons are harder to work with on the lattice and the masses are not as precise. The nucleon also requires, for example, a more complicated chiral fit and this has not yet been done. Figure 18 shows the nucleon results on the super-coarse, coarse and fine MILC unquenched configurations (Aubin *et al.* 2004a).

There are signs that the results change slightly with lattice spacing, *ie* there is a visible discretisation error. This needs to be studied further. The line shows the low order chiral perturbation theory result and the lattice results do seem to be heading towards that on the finer lattices. The right-hand plot shows results for the Ω baryon made of three valence s quarks (Toussaint and Davies 2005). This baryon is not very dependent on the u/d quark mass used in the quark vacuum polarisation so chiral extrapolations are much simpler. Its mass is sensitive to the s quark mass, however, so the fact that agreement with the experimental result is obtained with the s quark mass fixed from the K is another strong confirmation that lattice QCD is internally consistent once realistic quark vacuum polarisation is included.

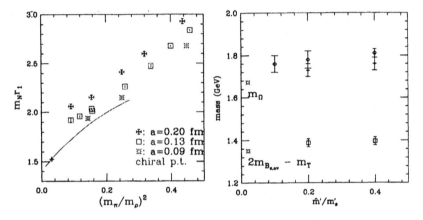

Figure 18. *(Left) The nucleon mass in units of the r_1 parameter from the unquenched MILC configurations at three different values of the lattice spacing and for several values of the light quark mass, denoted on the x axis by the ratio of π to ρ masses made of the light quarks. The line shows continuum chiral perturbation theory for this quantity (Aubin et al. 2004a). The fancy diamond gives the experimental point in these units. (Right) The Ω baryon mass as a function of sea light quark mass m' divided by sea s quark mass. The Ω mass is calculated for a valence s quark mass which is the correct one, fixed from the K (Toussaint and Davies 2005). The burst gives the experimental point. Also on this plot is the splitting between the spin average of the B_s and B_s^* and the Υ, a quantity sensitive to the s quark mass, but not the b quark mass. Again it shows good agreement with experiment denoted by a burst (Aubin et al. 2004c).*

It is important to realise that accurate lattice QCD results are not going to be obtainable in the near future for every hadronic quantity of interest. What these results show is that "gold-plated" quantities should be now be calculable. Unstable hadrons, or even those within 100 MeV or so of Zweig-allowed decay thresholds, have strong coupling to their real or virtual decay channel and this is not correctly simulated on the lattice volumes currently being used, *eg* the smallest non-zero momentum on typical current lattices exceeds 400 MeV. This could significantly distort the decay channel contribution to the hadron mass. Much larger volumes will then be necessary to handle these hadrons. Unfortunately the list of non-gold-plated hadrons is a long one - it includes the ρ, ϕ, D^*, Δ, N^*, glueballs, hybrids *etc*. Some of these may be more accurately calculable than others and qualitative results may also be useful. These points must be borne in mind, however, when making quantitative comparison between lattice QCD and experiment.

3.2 Determination of the parameters of QCD

Lattice QCD calculations are an excellent way to determine the parameters of QCD, masses and coupling constant, because the method is a very direct one.

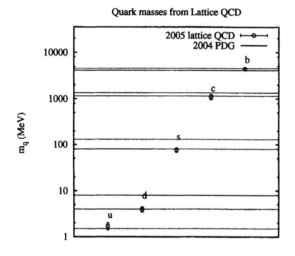

Figure 19. *Quark masses in the $\overline{MS}$ scheme at a relevant scale (2 GeV for u, d and s and their own mass for c and b) as determined from lattice QCD using the unquenched MILC configurations (Aubin et al. 2004c, 2004b, Gray et al. 2003, Nobes et al. 2005). The lines give the range of the current values quoted in the Particle Data Tables (PDG 2004).*

For the quark masses we directly determine the bare quark masses in the lattice QCD Lagrangian required to give the correct answer for a given hadron mass. With the inclusion of light quark vacuum polarisation we have seen above that this can be done both accurately and unambiguously. Masses are more conventionally quoted in the continuum $\overline{MS}$ renormalisation scheme rather than the lattice scheme. To convert from the lattice scheme to $\overline{MS}$ requires the calculation of a finite renormalisation factor to take into account the gluon radiation with momenta larger than π/a that does not exist on the lattice. The renormalisation factor can be calculated in perturbation theory since it involves large momenta, and it then appears as a power series in α_s. This calculation has been done to $\mathcal{O}(\alpha_s)$ for the light quark masses for the improved staggered quark action (Hein et al. 2003). The u/d quark mass was fixed from m_π and the s quark mass from m_K as described above. They were then converted to the $\overline{MS}$ scheme using the renormalisation factor. This gave, quoting the masses at the conventional scale:

$$m_s^{\overline{MS}}(2GeV) = 76(8)MeV,$$
$$m_{u/d}^{\overline{MS}}(2GeV) = 2.8(3)MeV, \tag{35}$$

where $m_{u/d}$ is the average of u and d masses (Aubin *et al.* 2004c). The main source of error from the lattice calculation is from unknown higher order terms in the perturbative renormalisation factor to convert to $\overline{MS}$. A two-loop calculation will be available shortly and this will reduce the error to a few %. This is a huge improvement over previous determinations of the masses from *eg* QCD sum rules. Figure 19 shows the results for all 5 quark masses (u and d are determined separately by matching to different combinations of charged and neutral pions and kaons) obtained on the MILC configurations and using 1-loop matching to convert to the conventional $\overline{MS}$ scheme, compared to the results from the 2004 Particle Data Tables (PDG 2004). It is quite clear that the lattice will take the lead in providing accurate quark masses now.

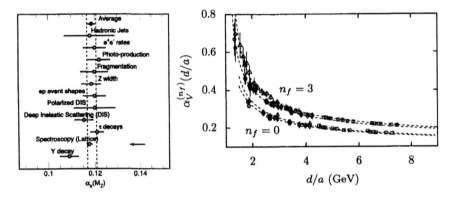

Figure 20. *(Left) A comparison of the new lattice determination of α_s, 0.1170(12) with the results from other determinations in the Particle Data Tables (PDG 2004). (Right) The determination of α_s from the lattice in the V scheme on quenched ($n_f = 0$) and unquenched ($n_f = 3$) gluon configurations at various energy scales, d/a for a variety of Wilson loop operators. The dashed curves show the expected running of α_s in the two cases from QCD. (Mason et al. 2005)*

The strong coupling constant, α_s, is also well-determined on the lattice (Mason *et al.* 2005). The determination of α_s proceeds by the calculation in perturbation theory to high order of some lattice operator (Trottier 2004). Recently calculations up to and including terms in α_s^3 became available for a lot of different Wilson loops and combinations of them. These could then be readily 'measured' (*ie* calculated) on the MILC configurations. From each non-perturbative lattice calculation compared to perturbation theory a value for α_s is extracted at some momentum scale in lattice units. From the determination of the lattice spacing, this scale can then be converted to a physical scale in GeV and α_s evolved to different scales use the QCD β function. Because results at three different values of the lattice spacing are available, it is possible to do a consistency check for the determination of α_s at a given physical scale from three different determinations of a particular Wilson loop at three different lattice scales. This allows estimates of fourth order terms in the perturbation theory, which puts this calculation into a new regime of accuracy for α_s determinations. Altogether 28 different loops and loop combinations were studied. The α_s determined was converted to the $\overline{MS}$ scheme and run to the scale of M_Z since again this is the conventional comparison point. The final answer obtained is 0.1170(12) which compares very well with other determinations quoted in the Particle

Data Tables, see Figure 20.

Another interesting point is that α_s is well able to distinguish between quenched and unquenched configurations. Figure 20 shows corresponding determinations of α_s from MILC quenched and unquenched configurations for different Wilson loops. The relevant momentum for the α_s determination varies from loop to loop and also depends on whether the result comes from the super-coarse, coarse or fine lattices. This enables a comparison of α_s values with the expected curve for the running. The agreement is very good and shows both that the quenched ($n_f=0$) and unquenched ($n_f=3$) α_s figures differ markedly and also run differently, both in agreement with the perturbative expectation (Mason *et al.* 2005).

3.3   Results on matrix elements

A key point where lattice QCD calculations are needed and can make an impact is in the determination of the elements of the Cabibbo-Kobayashi-Maskawa matrix that links the quark flavours under weak decay in the Standard Model of particle physics. When a quark changes flavour inside a hadron with the emission of a W the quark level process is quite simple (see Figure 24). However, this decay unavoidably takes places inside a hadron because the quarks are confined by QCD and the QCD corrections to the decay rate are significant. This is why the decay matrix element need to be calculated in lattice QCD so that all of the gluonic radiation around the decay vertex can be taken into account. A CKM element $V_{f_1 f_2}$ multiplies the vertex but this appears in a simple way in the final theoretical answer for the decay rate since there is only one weak vertex in the process. A comparison of the experimental decay rate and the lattice QCD results times $V_{f_1 f_2}^2$ then gives $V_{f_1 f_2}$.

Decay rates which can be accurately calculated in lattice QCD are those for gold-plated hadrons in which there is at most one (gold-plated) hadron in the final state. This therefore includes leptonic and semi-leptonic decays and the mixing of neutral B and K mesons. Luckily there is a gold-plated decay mode available to extract each element (except V_{tb}) of the CKM matrix which mixes quark flavours under the weak interactions in the Standard Model:

$$
\begin{pmatrix}
\mathbf{V_{ud}} & \mathbf{V_{us}} & \mathbf{V_{ub}} \\
\pi \to l\nu & K \to l\nu & B \to \pi l\nu \\
 & K \to \pi l\nu & \\
\mathbf{V_{cd}} & \mathbf{V_{cs}} & \mathbf{V_{cb}} \\
D \to l\nu & D_s \to l\nu & B \to D l\nu \\
D \to \pi l\nu & D \to K l\nu & \\
\mathbf{V_{td}} & \mathbf{V_{ts}} & \mathbf{V_{tb}} \\
\langle B_d | \overline{B}_d \rangle & \langle B_s | \overline{B}_s \rangle &
\end{pmatrix}
$$

The determination of the CKM elements and tests of the self-consistency of the CKM matrix are the current focus for the search for Beyond the Standard Model physics and lattice calculations of these decay rates will be a key factor in the precision with which this can be done.

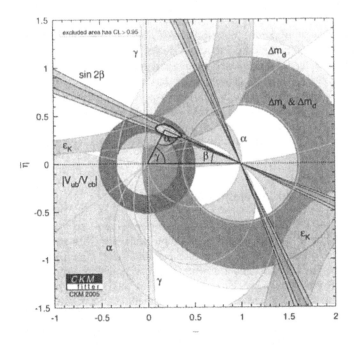

Figure 21. *Current constraints on the vertex of the 'unitarity triangle' made from the Cabibbo-Kobayashi-Maskawa matrix. (CKMfitter 2005)*

Figure 21 shows the current 'unitarity triangle' picture in which limits are placed on various combinations of elements of the CKM matrix and the result is expressed as a search for the vertex of a triangle. The limits that depend on results from B factories and the associated lattice calculations are the dark ring centred on the origin and the two lighter rings to the right in the figure. The light hyperbola comes from kaon physics and associated lattice calculations. The dark ring is fixed from semileptonic decays of B mesons to π mesons or D mesons. The other two rings result from mixing of neutral B or B_s mesons. We will discuss further below the lattice results for these matrix elements. The angles of the unitarity triangle, such as $\sin(2\beta)$, are determined directly by the experiment (light sloping lines) without theoretical input.

Decay matrix elements of this kind can be calculated on the lattice from the amplitudes of the exponentials in the fit functions for hadron masses, Equation 28. For example, the rate at which a B meson decays completely to leptons via a W boson depends on the matrix element of the heavy-light axial vector current, $J_{A_\mu} = \overline{\psi}_b \gamma_\mu \gamma_5 \psi_u$, between a B meson and the (QCD) vacuum. The matrix element of this current is parameterised by the decay constant, f_B, so that

$$\langle 0|J_{A_\mu}|B\rangle = p_\mu f_B \tag{36}$$

(using a relativistic normalisation of the state). For a B meson at rest only J_{A_0} is relevant and the right-hand-side of the equation above becomes $m_B f_B$.

The way in which f_B is calculated on the lattice is illustrated in Figure 22. On the

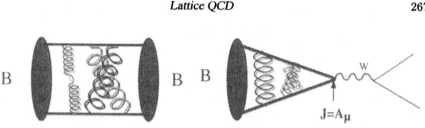

Figure 22. *(Left) The calculation on the lattice of the 2-point correlator for a B meson. (Right) The calculation on the lattice of a 2-point function in which a B meson decays leptonically.*

left-hand side is an illustration of the usual operator (correlator) whose expectation value we calculate to determine the B meson mass or energy. Then, as earlier,

$$<< H^\dagger(T)H(0) >> = \sum_n C_n e^{-E_n aT}. \tag{37}$$

For the B meson C_0 in the fit above is equal to $(\langle 0|H|B\rangle)^2/2E_0$, with a relativistic normalisation of the states. On the right-hand-side of Figure 22 we illustrate the operator in which the B meson is destroyed by the axial vector current (the W decay to leptons is handled analytically and separately from the lattice QCD calculation). Then

$$\frac{1}{Z}\langle 0|H^\dagger(T)J_{A_\mu}(0)|0\rangle = << H^\dagger(T)J_{A_\mu}(0) >> = \sum_n D_n e^{-E_n aT}. \tag{38}$$

We again have a product of M^{-1} factors to average over configurations, since J_{A_μ} contains the same kind of product of $\overline{\psi}$ and ψ fields as H. The same energies appear in this fit (indeed the correlators should be fit simultaneously to ensure that this is true) but the amplitudes are different. Now $D_0 = (\langle 0|H|B\rangle)(\langle B|J_{A_\mu}|0\rangle)/2E_0$. The second factor is the matrix element that we want and we can extract this as $2E_0 D_0/\sqrt{(2E_0 C_0)}$. If the temporal axial current is used for a B meson at rest then $f_B\sqrt{m_B}a^{3/2} = \sqrt{2/C_0}D_0$.

The current J_{A_μ} that we use on the lattice needs to be well-matched to the continuum current. The leading term in the lattice version of the current will be just be the obvious transcription of the continuum current, $A_\mu = \overline{\psi}_b\gamma_\mu\gamma_5\psi_u$. However this current has discretisation errors at $\mathcal{O}(\alpha_s a)$ and these can be improved by adding higher order operators to cancel the errors in the same way as for the action earlier. Again we can use perturbation theory to calculate the coefficients of these correction terms. If we use NRQCD for the b quark in the current then there are also relativistic corrections that can be applied to make the current a more accurate version of the continuum current. In fact the relativistic correction operators and the discretisation correction operators are the same and simply pick up perturbative coefficients that are functions of $m_Q a$ (Morningstar and Shigemitsu 1998).

We also have to renormalise the matrix elements from the lattice to an appropriate continuum renormalisation scheme such as $\overline{MS}$. This can be done in perturbation theory as, again, it takes account of gluons with momenta above the lattice cut-off. Such calculations have only been done through $\mathcal{O}(\alpha_s)$ so far and this means that the final quoted result has errors at $\mathcal{O}(\alpha_s)^2$ (Morningstar and Shigemitsu 1998). Now that the systematic error from working in the quenched approximation has been overcome, this is often the largest source of error and much more work must be done in future to reduce this error. Methods for renormalisation and matching that use direct numerical methods on the lattice (often called nonperturbative) are also being explored by many people. Matrix elements of conserved currents do not need renormalisation and this explains why, for example, the calculations of f_π and f_K described above are so accurate.

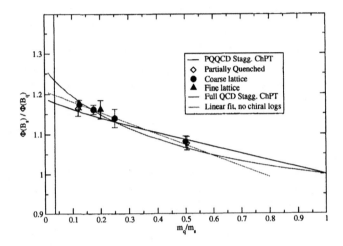

Figure 23. *The ratio of $f_{B_S}\sqrt{m_{B_s}}/f_B\sqrt{m_B}$ from unquenched lattice calculations on the MILC configurations at two values of the lattice spacing (Gray et al. 2005). The lines represent fits using chiral perturbation theory to various combinations of valence and sea light quark masses. The pink curve is the full QCD curve that extrapolates to the physical answer.*

The B meson decay constant is of interest, both because it sets the rate of B leptonic decay but also because it appears in the mixing rate of neutral B mesons. The mixing rate is parameterised by $f_B^2 B_B$ where B_B is the 'bag constant'. $f_B^2 B_B$ is being calculated in lattice QCD but it is harder than the calculation of f_B alone. We believe that the B_B factor is fairly benign with a value around 1. The calculation of f_B then provides a good indicator of the size of mixing effects (for the two lighter rings in Figure 21) until $f_B^2 B_B$ is known accurately.

Figure 23 shows the current results by the HPQCD collaboration, using the MILC unquenched configurations and the NRQCD formalism for the valence b quarks and asqtad improved staggered formalism for the valence light quarks (Gray *et al.* 2005, Wingate 2004). What is plotted is the ratio of $f_{B_s}\sqrt{m_{B_s}}/f_B\sqrt{m_B}$ in which the overall renormalisation constant for the lattice J_{A_0} cancels giving an accurate result. What is interesting here is the approach to the light quark mass limit in which $m_{u/d}$ takes its physical value. Chiral perturbation theory expects fairly strong logarithmic dependence but earlier results worked at such heavy $m_{u/d}$ that they were not in the region in which chiral perturbation theory was valid. It is clear that the new results now have light $m_{u/d}$ and an accurate result for this ratio will be possible.

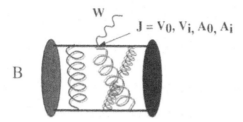

Figure 24. *The calculation of a 3-point function for B → π semileptonic decay on the lattice.*

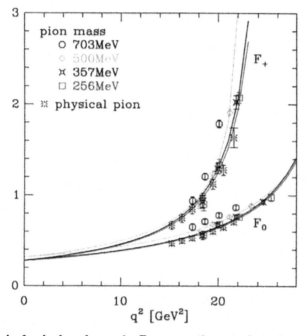

Figure 25. *Results for the form factors for B → π semileptonic decay from unquenched lattice QCD (Shigemitsu et al. 2005). Results are shown for a variety of light quarks, given in terms of the mass of the π made from these quarks. The bursts show the (small) extrapolation to the real π mass.*

The determination of semileptonic decay rates requires the calculation of a 3-point function on the lattice. This is illustrated in Figure 24 for $B \to \pi$. We now have two hadron operators at 0 and T for different hadrons and an intermediate current operator

J_{V_μ} or J_{A_μ} which causes the quark flavour change and the emission of a W. Now

$$\frac{1}{Z}\langle 0|H^\dagger(T)J(t)H'(0)|0\rangle = << H^\dagger(T)J(t)H'(0) >>= \sum_n D_n e^{-E_n a(T-t)} e^{-E'at}$$

(39)

The amplitudes, D_n, are now related to $< 0|H|B >< B|J|\pi >< \pi|H'|0 >$ and the matrix element that we want, *ie* $< B|J|\pi >$ can be extracted by simultaneously fitting the relevant 2-point functions to determine the other factors in D_0. The matrix element is now a function of the 4-momentum transfer between B and π, q^2. It is expressed in terms of two different form factors, $f_+(q^2)$ and $f_0(q^2)$, with different momentum-dependent prefactors. $f_+(q^2)$ is the form factor that translates directly into the rate of semileptonic decay, since f_0 ends up in this rate multiplied by the mass of the lepton into which the W decays, which is very small.

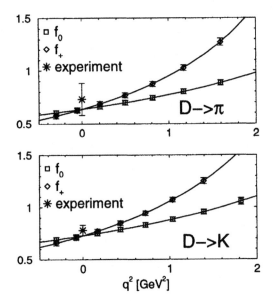

Figure 26. *Results for the form factors for D semileptonic decay to π and K. The experimental points shown at $q^2 = 0$ use the experimental decay rate at that point and the current PDG results for the appropriate CKM elements (Aubin et al. 2005).*

Once again we must match the lattice current to a continuum current and renormalise to obtain results in a continuum renormalisation scheme. This limits the accuracy with which these calculations can now be done and more work is required to improve this situation. Figure 25 and Figure 26 show current results obtained by the HPQCD and FNAL/MILC collaborations for the formfactors for $B \to \pi$ decay and $D \to \pi/K$ decay respectively. The B results use NRQCD b quarks (Shigemitsu *et al.* 2005); the D results use the Fermilab formalism for the c quark (Aubin *et al.* 2005). The usefulness of the D results is to compare to imminent experimental results from the CLEO-c collaboration that will provide a check of lattice methods and systematic errors for confidence in our precision B results (Shipsey 2005).

4   Conclusions

Lattice QCD has come a long way from the original calculations of the 1970s. The original idea that we could solve a simple discretisation of QCD numerically by 'brute force' has been replaced by a more sophisticated approach. Improved discretisation of both the gluon action and the quark action has led to the possibility of performing realistic simulations of QCD on current computers. Indeed this has been done, and I hope that I have conveyed something of the excitement of seeing accurate lattice calculations reproduce well-known experimental numbers for the first time. These first calculations are of the masses of "gold-plated" hadrons, those for which lattice QCD must be able to get the right answer if it is to be trusted at all. Leading on from this we have been able to make the first accurate lattice QCD prediction of the mass of a new meson, the B_c.

It is now important to beat down the sources of systematic error in the lattice calculation of decay matrix elements for B and D physics to obtain results that can be combined with experiment to give an accurate determination of elements of the CKM matrix. The timescale for this programme is the next two years and, as well as lattice QCD calculations, it requires $\mathcal{O}(\alpha_s^2)$ calculations in perturbation theory to renormalise the lattice results to numbers appropriate to the continuous real world. On a longer timescale (say five years) studies of more complicated baryonic matrix elements will be undertaken and progress will be made in understanding to what extent accurate lattice calculations can be done of some of the more interesting, but not gold-plated, particles in the hadron spectrum.

Acknowledgments

It was a pleasure to contribute to this interesting school. I am grateful to a large number of lattice colleagues, but particularly my long-standing collaborators Peter Lepage and Junko Shigemitsu, for numerous useful discussions over many years.

References

There are a number of books on lattice QCD that provide more information on the theoretical background, such as Smit, 2002, *An Introduction to Quantum Fields on a Lattice*, Cambridge University Press. The annual lattice conference provides up-to-date review talks and access to the literature. The 2004 Conference Proceedings is published in *Nucl Phys B Proc Suppl* **140**. Below I provide a few references, concentrating on other summer school lectures where possible.

Acosta D *et al.* [CDF collaboration], 2005, hep-ex/0505076.
Alford M *et al.* , 1995, *Phys LettB* **361** 87.
Allison I *et al.* [HPQCD/FNAL collaborations], 2005, *Phys Rev Lett* **94** 172001.
Arndt D, PhD thesis, 2004, University of Washington, USA.
Aubin C *et al.* [MILC collaboration], 2004a, *Phys Rev D* **70** 094505.

Aubin C *et al.* [MILC collaboration], 2004b, *Phys Rev D* **70** 114501.

Aubin C *et al.* [HPQCD/MILC collaborations], 2004c, *Phys Rev D* **70** 031504.

Aubin C *et al.* [FNAL/MILC/HPQCD collaborations], 2005, *Phys Rev Lett* **94** 011601.

Bali G, 2000, *Phys Rep* **343** 1.

Bernard C [MILC collaboration], 2001, *Phys Rev D* **64** 054506.

Chiu T, 2004, *Nuc Phys B Proc Suppl* **129** 135.

CKMfitter, 2005, http://ckmfitter.in2p3.fr/

Davies C T H, 1998, Springer lecture notes in physics, Eds Gausterer, Lang, hep-ph/9710394.

Davies C T H *et al.* [HPQCD/UKQCD/MILC/FNAL collaborations], 2004, *Phys Rev Lett* **92** 022001.

Davies C T H *et al.* , 2005, *Nuc Phys B Proc Suppl* **140** 261.

di Pierro M, 2000, lectures given at the GSA summer school on Physics on the Frontier, hep-lat/0009001. Includes code for example calculations.

di Pierro M *et al.* [Fermilab collaboration], 2004, *Nucl Phys B Proc Suppl* **129** 328; *Nucl Phys B Proc Suppl* **129** 340.

El-Khadra A *et al.* , 1997, *Phys Rev D* **55** 3933.

Follana E *et al.* [HPQCD/UKQCD collaborations], 2004, *Phys Rev Lett* **93** 241601.

Frezzotti R, 2005, *Nuc Phys B Proc Suppl* **140** 134.

Gray A *et al.* [HPQCD/UKQCD collaborations], 2003, *Nuc Phys B Proc Suppl* **119** 592; and in preparation.

Gray A *et al.* [HPQCD collaboration], 2005, *Nuc Phys B Proc Suppl* **140** 446; and in preparation.

Hein J *et al.* [HPQCD/UKQCD collaborations], 2003, *Nucl Phys B Proc Suppl* **119** 317.

Ishikawa K *et al.* [JLQCD/CP-PACS collaborations], 2005, *Nuc Phys B Proc Suppl* **140** 225.

Lepage P, 1996, Schladming winter school, hep-lat/9607076.

Lepage P, 1998a. For a useful discussion of path integrals in quantum mechanics and their relevance to lattice QCD see *Lattice QCD for novices*, HUGS98, hep-lat/0506036. Includes code for example calculations.

Lepage P, 1998b, *Phys Rev D* **59** 074502.

Lepage P *et al.* , 1992, *Phys Rev D* **46,** 4052.

Lepage P and Mackenzie P, 1993, *Phys Rev D* **48,** 2250.

Lepage P and Davies C T H, 2004, *Int J Mod Phys A* **19** 877.

Mason Q *et al.* [HPQCD/UKQCD collaborations], 2005, *Phys Rev Lett* (in press).

Morningstar C and Shigemitsu J, 1998, *Phys Rev D* **57** 6741.

Nobes M, 2005, hep-lat/0501009.

Nobes M *et al.* , 2005, in preparation.

Orginos K *et al.* , 1999, *Phys Rev D* **60** 054503.

PDG, 2004, http://pdg.lbl.gov/

Shigemitsu J *et al.* [HPQCD collaboration], 2005, *Nuc Phys B Proc Suppl* **140** 464.

Shipsey I, 2005, *Nuc Phys B Proc Suppl* **140** 58.

Sommer R, 1994, *Nuc Phys B* **411** 839.

Toussaint D and Davies C T H, 2005, *Nuc Phys B Proc Suppl* **140** 234.

Trottier H, 2004, *Nuc Phys B Proc Suppl* **129** 142.

Wingate M [HPQCD collaboration], 2004, *Phys Rev Lett* **92** 162001.

Resonances from coupled channel chiral unitarity

Angels Ramos

Departament d'Estructura i Constituents de la Matèria, Universitat de Barcelona
E-08028 Barcelona, Spain

1 Introduction

Establishing the nature of hadronic resonances is one of the primary goals in the field of hadronic physics. The interest lies in understanding whether they behave as genuine three quark states or they are dynamically generated through the iteration of appropriate non-polar terms of the hadron-hadron interaction, not being preexistent states that remain in the large N_c limit where the multiple scattering is suppressed. In the last decade, chiral perturbation theory (χPT) has emerged as a powerful scheme to describe low-energy meson-meson and meson-baryon dynamics. In recent years, the introduction of unitarity constraints has allowed the extension of the chiral description to much higher energies and, in addition, it has lead to the generation of many hadron resonances both in the mesonic and the baryonic sectors.

The $\Lambda(1405)$ resonance is a clear example of a dynamically generated state appearing naturally from the multiple scattering of coupled meson-baryon channels with strangeness $S = -1$ (Jones et al. 1977, Kaiser et al. 1995, Kaiser et al. 1997, Oset and Ramos 1998, Oller and Meissner 2001). Recently, the interest in studying its properties has been revived by the observation in the chiral models that the nominal $\Lambda(1405)$ is in fact built up from two poles of the T-matrix in the complex plane (Oller and Meissner 2001, Jido et al. 2002a, Garcia-Recio et al. 2003) both contributing to the invariant $\pi\Sigma$ mass distribution, as it was the case within the cloudy bag model (Fink et al.). The fact that these two poles have different widths and partial decay widths into $\pi\Sigma$ and $\overline{K}N$ states opens the possibility that they might be experimentally observed in hadronic or electromagnetic reactions.

The unitary chiral dynamical models have been extended by various groups (Oller and Meissner 2001, Garcia-Recio et al. 2003, Nacher et al. 2000, Inoue et al. 2002,

Jido *et al.* 2002b, Oset *et al.* 2002, Ramos *et al.* 2002, Nieves and Ruiz Arriola 2000, Nieves and Ruiz Arriola 2001, Lutz and Kolomeitsev 2002), covering an energy range of about 1.4–1.7 GeV and giving rise to a series of resonant states in all isospin and strangeness sectors. All these observations have finally merged into the classification of the dynamical generated baryon resonances into SU(3) multiplets (Jido *et al.* 2003), as seen also in (García-Recio *et al.* 2003).

In this contribution we present a summary of our latest developments in the field of baryon resonances generated from chiral unitary dynamics.

2   Meson-baryon scattering model

The search for dynamically generated resonances proceeds by first constructing the meson-baryon coupled states from the octet of ground state positive-parity baryons (B) and the octet of pseudoscalar mesons (Φ) for a given strangeness channel. Next, from the lowest order Lagrangian

$$L_1^{(B)} = \langle \overline{B} i \gamma^\mu \frac{1}{4f^2} [(\Phi \partial_\mu \Phi - \partial_\mu \Phi \Phi) B - B(\Phi \partial_\mu \Phi - \partial_\mu \Phi \Phi)] \rangle \tag{1}$$

one derives the driving kernel in s-wave

$$V_{ij} = -C_{ij} \frac{1}{4f^2} (2\sqrt{s} - M_i - M_j) \left(\frac{M_i + E_i}{2M_i} \right)^{1/2} \left(\frac{M_j + E_j}{2M_j} \right)^{1/2}, \tag{2}$$

where the constants C_{ij} are SU(3) coefficients encoded in the chiral Lagrangian and f is the meson decay constant, which we take to have an average value of $f = 1.123 f_\pi$, where $f_\pi = 92.4$ MeV is the pion decay constant. While at lowest order in the chiral expansion all the baryon masses are equal to the chiral mass M_0, the physical masses are used in Equation (2) as done in (Oset and Ramos 1998, Oset *et al.* 2002). We recall that, in addition to the Weinberg-Tomozawa or seagull term of Equation (2), one also has at the same order of the chiral expansion the direct and exchange diagrams considered in (Oller and Meissner 2001). Their contribution increases with energy and represents around 20% of that from the seagull term at $\sqrt{s} \simeq 1.5$ GeV.

The scattering matrix amplitudes between the various meson-baryon states are obtained by solving the coupled channel equation

$$T_{ij} = V_{ij} + V_{il} G_l T_{lj}, \tag{3}$$

where i, j, l are channel indices and the V_{il} and T_{lj} amplitudes are taken on-shell. This is a particular case of the N/D unitarisation method when the unphysical cuts are ignored (Oller and Oset 1999, Oller and Meissner 2000). Under these conditions the diagonal matrix G_l is simply built from the convolution of a meson and a baryon propagator and can be regularised either by a cut-off ($q_{\max}^l$), as in (Oset and Ramos 1998), or alternatively by dimensional regularisation depending on a subtraction constant (a_l) coming from a subtracted dispersion relation (Oller and Meissner 2001, Oset *et al.* 2002).

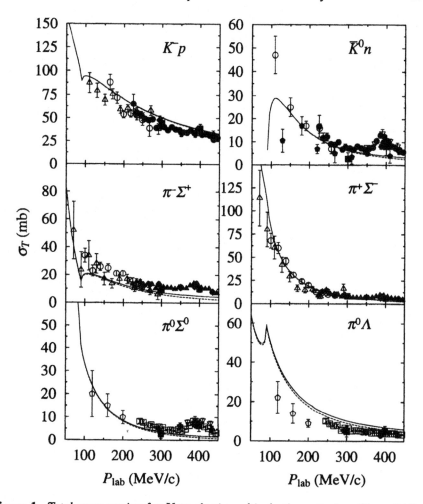

Figure 1. *Total cross section for K^-p elastic and inelastic scattering. The solid line denotes our results including both s-wave and p-wave. The dashed line shows our results without the p-wave amplitudes. The data are taken from (Coborowski et al. 1982, Bangerter et al. 1981, Mast et al. 1975, Mast et al. 1976, Sakitt et al. 1965, Nordin 1961, Berley et al. 1970, Ferro-Luzzi et al. 1962, Watson et al. 1963, Eberhard et al. 1959, Kim 1965).*

3 Strangeness $S = -1$

In the case of K^-p scattering, we consider the complete basis of meson-baryon states, namely K^-p, $\overline{K}^0 n$, $\pi\Lambda$, $\eta\Lambda$, $\eta\Sigma^0$, $\pi^+\Sigma^-$, $\pi^-\Sigma^+$, $\pi^0\Sigma^0$, $K^+\Xi^-$ and $K^0\Xi^0$, thus preserving SU(3) symmetry in the limit of equal baryon and meson masses. Taking a cut-off of 630 MeV, the scattering observables, threshold branching ratios and properties of the

$\Lambda(1405)$ resonance were well reproduced (Oset and Ramos 1998, Jido *et al.* 2002b), as shown in Table 1 and Figures 1 and 2. The inclusion of the $\eta\Lambda$, $\eta\Sigma$ channels was found crucial to obtain a good agreement with experimental data in terms of the lowest order chiral Lagrangian.

Table 1. *Branching ratios at the K^-p threshold. Experimental values are taken from (Tovee et al. 1971, Nowak et al. 1978).*

Ratio	Exp.	Model
$\gamma = \frac{\Gamma(K^-p\to\pi^+\Sigma^-)}{\Gamma(K^-p\to\pi^-\Sigma^+)}$	2.36 ± 0.04	2.32
$R_c = \frac{\Gamma(K^-p\to\text{charged particles})}{\Gamma(K^-p\to\text{all})}$	0.664 ± 0.011	0.627
$R_n = \frac{\Gamma(K^-p\to\pi^0\Lambda)}{\Gamma(K^-p\to\text{all neutral states})}$	0.189 ± 0.015	0.213

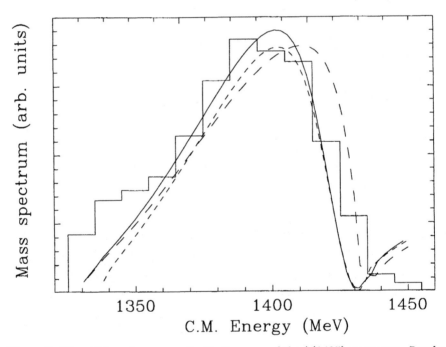

Figure 2. *The $\pi\Sigma$ invariant mass distribution around the $\Lambda(1405)$ resonance. Results in particle basis (solid line), isospin basis (short-dashed line) or omitting the $\eta\Lambda$, $\eta\Sigma^0$ channels. Experimental histogram taken from (Hemingway 1985).*

Our model extrapolated smoothly to high energies (Oset *et al.* 2002) by using the dimensional regularisation scheme. The subtraction constants have a "natural" size (Oller and Meissner 2001) which permits qualifying the generated resonances as being dynam-

Table 2. *Pole positions and couplings to meson-baryon states of the dynamically generated resonances in the $S = -1$ sector (Oset et al. 2002).*

	z_R (MeV)	$\mid g_{\pi\Sigma} \mid^2$	$\mid g_{\overline{K}N} \mid^2$	$\mid g_{\eta\Lambda} \mid^2$	$\mid g_{K\Xi} \mid^2$
$\Lambda(1405)$	1390−i66	8.4	4.5	0.59	0.38
	1426−i16	2.3	7.4	2.0	0.12
$\Lambda(1670)$	1680−i20	0.01	0.61	1.1	12
		$\mid g_{\pi\Lambda} \mid^2$	$\mid g_{\pi\Sigma} \mid^2$	$\mid g_{\overline{K}N} \mid^2$	$\mid g_{\eta\Sigma} \mid^2$	$\mid g_{K\Xi} \mid^2$
$\Sigma(1620)$	1579−i274	4.2	7.2	2.6	3.5	12

ical. While the $I = 0$ components of the $\overline{K}N \to \overline{K}N$ and $\overline{K}N \to \pi\Sigma$ amplitudes displayed a clear signal from the $\Lambda(1670)$ resonance, the $I = 1$ amplitudes are smooth and featureless without any trace of resonant behaviour, in line with the experimental observation. In Table 2 we display the value of the poles of the scattering amplitude in the second Riemann sheet, $z_R = M_R - i\Gamma/2$, together with the corresponding couplings to the various meson-baryon states, obtained from identifying the amplitudes T_{ij} with $g_i g_j/(z - z_R)$ in the limit $z \to z_R$. Two poles define the $\Lambda(1405)$ resonance. The pole at lower energy is wider and couples mostly to $\pi\Sigma$ states, while that at higher energy is narrower and couples mostly to $\overline{K}N$ states. The consequences of this two-pole nature of the $\Lambda(1405)$ are discussed in detail in (Jido *et al.* 2003). We also find poles corresponding to the $\Lambda(1670)$ and $\Sigma(1620)$ resonances. The large coupling of the $\Lambda(1670)$ to $K\Xi$ states allows one to identify this resonance as a "quasibound" $K\Xi$ state. The large width associated to the $\Sigma(1620)$ resonance, rated as 1-star by the Particle Data Group (PDG) (Particle Data Group 2002), explains why there is no trace of this state in the scattering amplitudes.

4 Strangeness $S = -2$

The unitarity chiral meson-baryon approach has also been extended to the $S = -2$ sector (Ramos *et al.* 2002) to investigate the nature of the lowest possible s-wave Ξ states, the $\Xi(1620)$ and $\Xi(1690)$, rated 1- and 3-star, respectively, and quoted with unknown spin and parity by the PDG (Particle Data Group 2002). Allowing the subtraction constants to vary around a natural size of -2, a pole is found at $z_R = 1605 - i66$, the real part showing a strong stability against the change of parameters. The imaginary part would apparently give a too large width of 132 MeV compared to the experimental ones reported to be of 50 MeV or less. However, due to a threshold effect, the actual $\pi\Xi$ invariant mass distribution, displayed in Figure 3, shows a much narrower width and resembles the peaks observed experimentally.

The couplings obtained are $\mid g_{\pi\Xi} \mid^2= 5.9$, $\mid g_{\overline{K}\Lambda} \mid^2= 7.0$, $\mid g_{\overline{K}\Sigma} \mid^2= 0.93$ and $\mid g_{\eta\Xi} \mid^2= 0.23$. The particularly large values for final $\pi\Xi$ and $\overline{K}\Lambda$ states rule out identifying this resonance with the $\Xi(1690)$, which is found to decay predominantly to $\overline{K}\Sigma$ states. Therefore, the dynamically generated $S = -2$ state can be safely identified with the $\Xi(1620)$ resonance and this also enables us to assign the values $J^P = 1/2^-$ to its unmeasured spin and parity. The model of (García-Recio *et al.* 2003) finds this state at

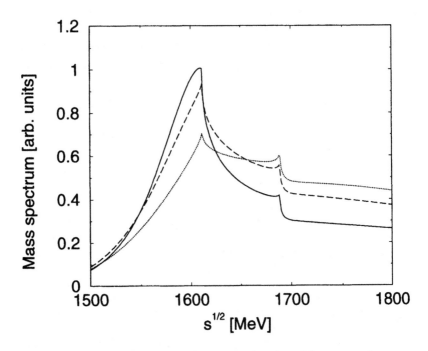

Figure 3. *The $\pi\Xi$ invariant mass distribution as a function of the centre-of-mass energy, for several sets of subtraction constants. Solid line: $a_{\pi\Xi} = -3.1$ and $a_{\overline{K}\Lambda} = -1.0$; Dashed line: $a_{\pi\Xi} = -2.5$ and $a_{\overline{K}\Lambda} = -1.6$; Dotted line: $a_{\pi\Xi} = -2.0$ and $a_{\overline{K}\Lambda} = -2.0$. The value of the two other subtraction constants, $a_{\overline{K}\Sigma}$ and $a_{\eta\Xi}$, is fixed to -2.0 in all curves.*

$z_R = 1565 - i124$, together with another pole at $z_R = 1663 - i2$, identified with the $\Xi(1690)$ because of its strong coupling to $\overline{K}\Sigma$ states.

5 Strangeness $S = 0$

For completeness, we briefly mention here the work done in the $S = 0$ sector (Inoue *et al.* 2002, Nacher *et al.* 2000) where the $N(1535)$ was generated dynamically within the same approach. In order to reproduce the phase shifts and inelasticities, four subtraction constants were adjusted to the data leading to a $N(1535)$ state with a total decay width of $\Gamma \simeq 110$ MeV, divided into $\Gamma_\pi \simeq 43$ MeV and $\Gamma_\eta \simeq 67$ MeV, compatible with present data within errors. The dynamical $N(1535)$ is found to have strong couplings to the $K\Sigma$ and ηN final states.

6 SU(3) multiplets of resonant states

The SU(3) symmetry encoded in the chiral Lagrangian permits classifying all these resonances into SU(3) multiplets. We first recall that the meson-baryon states built from the octet of pseudoscalar mesons and the octet of ground state baryons can be classified into the irreducible representations:

$$8 \otimes 8 = 1 \oplus 8_s \oplus 8_a \oplus 10 \oplus \overline{10} \oplus 27 \tag{4}$$

Taking a common meson mass and a common baryon mass, the lowest-order meson-baryon chiral Lagrangian is exactly SU(3) invariant. If, in addition, all the subtraction constants a_l are equal to a common value, the scattering problem decouples into each of the SU(3) sectors. Using SU(3) Clebsh-Gordan coefficients, the matrix elements of the transition potential V in a basis of SU(3) states are

$$V_{\alpha\beta} \propto -\frac{1}{4f^2} \sum_{i,j} \langle i, \alpha \rangle C_{ij} \langle j, \beta \rangle = \frac{1}{4f^2} \text{diag}(-6, -3, -3, 0, 0, 2), \tag{5}$$

taking the following order for the irreducible representations: $1, 8_s, 8_a, 10, \overline{10}$ and 27. The attraction in the singlet and the two octet channels gives rise to bound states in the unitarised amplitude, with the two octet poles being degenerate (Jido *et al.* 2003). By breaking the SU(3) symmetry gradually, allowing the masses and subtraction constants to evolve to their physical values, the degeneracy is lost and the poles move along trajectories in the complex plane as shown in Figure 4, which collects the behaviour of the $S = -1$ states. As discussed further in (Jido *et al.* 2003), two poles in the $I = 0$ sector appear very close in energy and they will manifest themselves as a single resonance, the $\Lambda(1405)$, in invariant $\pi\Sigma$ mass distributions.

7 The two-pole nature of the $\Lambda(1405)$

The $\Lambda(1405)$ is seen through the invariant mass distributions of $\pi\Sigma$ states given by

$$\frac{d\sigma}{dM_I} = |\sum_i C_i t_{i \to \pi\Sigma}|^2 p_{CM} \tag{6}$$

with i standing for any of the coupled channels ($\overline{K}N, \pi\Sigma, \eta\Lambda, K\Xi$) and C_i being coefficients that determine the strength for the excitation of channel i, which eventually evolves into a $\pi\Sigma$ state through multiple scattering. As the two $\Lambda(1405)$ poles couple differently to $\pi\Sigma$ and $\overline{K}N$ states, the amplitudes $t_{\pi\Sigma \to \pi\Sigma}$, $t_{\overline{K}N \to \pi\Sigma}$ are dominated by one or the other pole, respectively, thus making the invariant mass distribution sensitive to the coefficients $C_{\pi\Sigma}$, $C_{\overline{K}N}$, *ie* to the reaction used to generate the $\Lambda(1405)$.

The radiative production reaction $K^-p \to \gamma\Lambda(1405)$ provides an interesting example. In order to access the subthreshold region, the photon must be radiated from the initial K^-p state, ensuring that the $\Lambda(1405)$ resonance is initiated from K^-p states. This selects the pole that couples more strongly to $\overline{K}N$, which is narrower and appears at a higher energy. The calculated invariant mass $\pi\Sigma$ distribution (Nacher *et al.* 1999b)

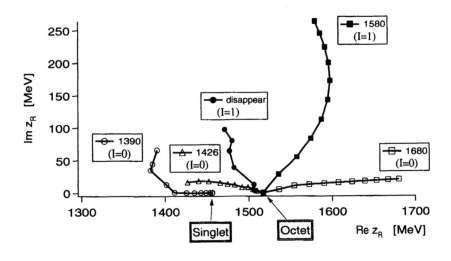

Figure 4. *Trajectories of the poles in the scattering amplitudes obtained by changing the SU(3) breaking parameter x gradually. At the SU(3) symmetric limit (x = 0), only two poles appear, one is for the singlet and the other (two-times degenerate) for the octets. The symbols correspond to the step size $\delta x = 0.1$. The results are from (Jido et al. 2003).*

appears indeed displaced to higher energies (~ 1420 MeV) and it is narrower (35 MeV) than what one obtains from other reactions.

The $\Lambda(1405)$ can also be produced from the reaction (γ, K^+) on protons, recently implemented at LEPS of SPring8/RCNP (LEPS 2003). In this case, the invariant mass distribution of the final meson-baryon state obtained in (Nacher et al. 1999a) shows a width of around 50 MeV. Due to the particular isospin decomposition of the $\pi\Sigma$ states, the $\pi^-\Sigma$ and $\pi^+\Sigma^-$ cross sections differ in the sign of the interference between $I = 0$ and $I = 1$ amplitudes (omitting the negligible $I = 2$ contribution). This difference has been observed in the experiment performed at SPring8/RCNP (LEPS 2003) and provides some information on the $I = 1$ amplitude.

The dynamics that goes into the $\pi^-p \to K^0\pi\Sigma$ reaction, from which the experimental data of the $\Lambda(1405)$ resonance have been extracted (Thomas 1973), has recently been investigated (Hyodo et al. 2003). The contributions from chiral terms and from resonance excitation are dominated by either one or the other pole, and their coherent sum produces a distribution in agreement with the experimental one.

Recently, the production of the $\Lambda(1405)$ through K^* vector meson photoproduction, $\gamma p \to K^*\Lambda(1405) \to \pi K\pi\Sigma$, using linearly polarised photons has been studied (Hyodo et al. 2004). Selecting the events in which the polarisation of the incident photon and that of the produced K^* are perpendicular, the mass distribution of the $\Lambda(1405)$ peaks at 1420 MeV. In this case the process is dominated by t-channel K-meson exchange, hence selecting preferentially the $t_{\overline{K}N \to \overline{K}N}$ amplitude.

8 Summary and Conclusions

By implementing unitarity in the study of meson-baryon scattering using the lowest order chiral Lagrangian, a series of resonant states have been dynamically generated in all strangeness and isospin sectors.

In the SU(3) limit, all these resonances belong to a singlet or to either of the two (degenerate) octets of dynamically generated poles of the SU(3) symmetric scattering amplitude.

In the physical limit, there are two $I = 0$ poles representing the $\Lambda(1405)$, the one at lower energy having a larger imaginary part than the one at higher energy. These poles couple differently to $\pi\Sigma$ and $\overline{K}N$ states and, as a consequence, the properties of the $\Lambda(1405)$ will depend on the particular reaction used to produce it. Various processes that might preferentially select the contribution from one or the other pole have been discussed.

Acknowledgments

The results presented in this lecture were obtained in collaboration with C. Bennhold, D. Jido, U.-G. Meissner, J.A. Oller and E. Oset. I am deeply indebted to them for their valuable contributions.

This work is supported by DGICYT (Spain) projects BFM2000-1326, BFM2002-01868 and FPA2002-03265, by the EU projects EURIDICE HPRN-CT-2002-00311 and HadronPhysics RII3-CT-2004-506078, and by project 2001SGR00064 of the Generalitat de Catalunya.

References

Bangerter R O *et al.* , 1981, *Phys Rev D* **23** 1484.
Berley D *et al.* , 1970, *Phys Rev D* **1** 1996.
Coborowski J *et al.* , 1982, *J Phys G: Nucl Part Phys* **8** 13.
Eberhard P *et al.* , 1959, *Phys Rev Lett* **2** 312.
Ferro-Luzzi M *et al.* , 1962, *Phys Rev Lett* **8** 28.
Fink J *et al.* , 1990, *Phys Rev C* **41** 2720.
Garcia-Recio C *et al.* , 2003, *Phys Rev D* **67** 076009.
García-Recio *et al.* , 2004, *Phys Lett B* **582** 49.
Hemingway R J, 1985, *Nucl Phys B* **253** 742.
Hyodo T *et al.* , 2003, *Phys Rev C* **68** 065203.
Hyodo T *et al.* , 2004, *Phys Lett B* **593** 75.
Inoue T *et al.* , 2002, *Phys Rev C* **65** 035204.
Jido D *et al.* , 2002a, *Phys Rev C* **66** 025203.
Jido D *et al.* , 2002b, *Phys Rev C* **66** 055203.
Jido D *et al.* , 2003, *Nucl Phys A* **725** 181.
Jones M *et al.* , 1977, *Nucl Phys B* **129** 45.
Kaiser N *et al.* , 1995, *Nucl Phys A* **594** 325.
Kaiser N *et al.* , 1997, *Nucl Phys A* **612** 297.

Kim J K, 1965, *Phys Rev Lett* **21** 719.

LEPS collaboration, Ahn J K *et al.* , 2003, *Nucl Phys A* **721** 715c.

Lutz M F M and Elomeitsev E E, 2002, *Nucl Phys A* **700** 193.

Mast T S *et al.* , 1975, *Phys Rev D* **11** 3078.

Mast T S *et al.* , 1976, *Phys Rev D* **14** 13.

Nacher J C *et al.* , 1999a, *Phys Lett B* **455** 55.

Nacher J C *et al.* , 1999b, *Phys Lett B* **461** 299.

Nacher J C, 2000, *Nucl Phys A* **678** 187.

Nieves J and Ruiz Arriola E, 2000, *Nucl Phys A* **679** 57.

Nieves J and Ruiz Arriola E, 2001, *Phys Rev D* **64** 116008.

Nordin P, 1961, *Phys Rev* **123** 2168.

Nowak R J *et al.* , 1978, *Nucl Phys B* **139** 61.

Oset E *et al.* , 2002, *Phys Lett B* **527** 99.

Oset E and Ramos A, 1998, *Nucl Phys A* **635** 99.

Oller J A and Meissner U G, 2000, *Nucl Phys A* **673** 311.

Oller J A and Meissner U G, 2001, *Phys Lett B* **500** 263.

Oller J A and Oset E, 1999, *Phys Rev C* **60** 074023.

Particle Data Group, 2002, *Phys Rev D* **66** 010001.

Ramos A *et al.* , 2002, *Phys Rev Lett* **89** 252001.

Sakitt *et al.* , 1965, *Phys Rev B* **139** 719.

Thomas D W *et al.* , 1973, *Nucl Phys B* **56** 15.

Tovee D N *et al.* , 1971, *Nucl Phys B* **33** 493.

Watson M B *et al.* , 1963, *Phys Rev* **131** 2248.

Chiral extrapolations

Meinulf Göckeler

Institut für Theoretische Physik, Universität Leipzig,
Augustusplatz 10-11, 4109 Leipzig, Germany

1 Introduction

The title 'Chiral extrapolations' of this contribution refers more precisely to chiral extrapolations *of lattice* QCD *data*. We shall deal with low-energy aspects of QCD, which are not accessible to ordinary weak-coupling perturbation theory. Possible alternatives to weak-coupling perturbation theory in the low-energy domain of QCD include the investigation of specific models, Monte Carlo simulations of lattice regularised QCD, and chiral effective field theories (CHEFT), the latter being low-energy theories which incorporate the constraints from (spontaneously broken) chiral symmetry. It is the comparison of CHEFT (or chiral perturbation theory (CHPT)) with results from lattice QCD simulations that will be the subject of the present paper.

However, the reader should be warned that as lattice QCD practitioners we look at CHEFT from an (ab)user's viewpoint. Also a second warning may be in order: This is not a review. The material has been selected according to subjective criteria, and the references are far from being complete. For a review of CHPT see *eg* (Leutwyler 2000) and (Meißner 2000). For reviews on closely related subjects, partly overlapping with the present work, see (Bär 2004) and (Colangelo 2004).

2 Lattice regularisation and Monte Carlo simulation

The basic input for a Monte Carlo simulation of lattice QCD is first of all a lattice action, *ie* a discretised version of Euclidean QCD. Secondly, one has to choose a lattice size (necessarily finite). Thirdly, the bare coupling constant and the quark mass(es) have to be fixed. Then the lattice spacing a and the spatial box size L can (approximately) be given in physical units.

In general it is not possible to choose the simulation parameters such that the results can immediately be identified with experimentally measurable quantities. In particular, three extrapolations are required: the continuum limit $a \to 0$, the thermodynamic limit $L \to \infty$, and the chiral limit, where the masses of the light quarks decrease to their physical values and further down to zero. Unfortunately, in all three limits the simulation costs increase rapidly. It is therefore preferable to appeal to theory in order to relate simulation results obtained for unphysical quark masses, finite volumes, ... to phenomenology. This will then lead to well-justified extrapolation formulae.

3   Chiral effective field theory in the pion sector

The natural starting point for chiral perturbation theory is the pion sector. The very existence of light pions (for the case of two flavours) relies on the spontaneous breakdown of chiral symmetry combined with the weak explicit breaking due to the quark masses. A further consequence is the weakness of the pion-pion interaction at low energies and momenta which makes a perturbative treatment meaningful. This (chiral) perturbation theory is most conveniently set up by means of an effective Lagrangian, *ie* the most general Lagrangian for effective pion (and later on also nucleon, ...) fields which is compatible with chiral symmetry. The effective Lagrangian is constructed out of terms with more and more derivatives as the order of the expansion increases, leading at the same time to an increasing number of effective coupling constants, which are not determined by chiral symmetry. Eventually an expansion of the physical observables in the pion mass M_π and the particle momenta is obtained, where both are considered as quantities of the order of a small parameter p. Here 'small' means $p \ll 4\pi F_\pi \sim 1\,\mathrm{GeV}$.

We shall restrict ourselves to the case of two flavours, $N_f = 2$, with isospin breaking neglected, *ie* we take for the quark masses $m_u = m_d = m_q$. Then the lowest-order expression for M_π is the famous Gell-Mann–Oakes–Renner relation $M_\pi^2 = 2|\langle \bar{q}q \rangle| m_q / F^2$, where $\langle \bar{q}q \rangle$ is the chiral condensate and F denotes the pion decay constant F_π in the chiral limit.

Beyond leading order one finds (see *eg* Colangelo *et al.* 2001)

$$M_\pi^2 = M^2 \left\{ 1 - \frac{1}{2} x \hat{\ell}_3 + \tfrac{17}{8} x^2 \hat{\ell}_M^2 + x^2 k_M + O(x^3) \right\} \tag{1}$$

with

$$\hat{\ell}_M = \frac{1}{51}(28\hat{\ell}_1 + 32\hat{\ell}_2 - 9\hat{\ell}_3 + 49) . \tag{2}$$

Here we have set $M^2 = 2m_q B$ with $B = -\langle \bar{q}q \rangle / F^2$ and $x = M^2/(16\pi^2 F^2)$. The chiral logarithms are hidden in the quantities $\hat{\ell}_i = \ln(\Lambda_i^2/M^2)$, which contain the information on the (renormalised) coupling constants. The term proportional to k_M represents analytic contributions $O(x^2)$, which are expected to be small. Note that k_M and Λ_i depend neither on m_q nor on the renormalisation scale. One can estimate $F = 0.0862\,\mathrm{GeV}$, and phenomenological analyses lead to $\Lambda_1 = 0.12^{+0.04}_{-0.03}\,\mathrm{GeV}$, $\Lambda_2 = 1.20^{+0.06}_{-0.06}\,\mathrm{GeV}$, $\Lambda_3 = 0.59^{+1.40}_{-0.41}\,\mathrm{GeV}$, $\Lambda_4 = 1.25^{+0.15}_{-0.13}\,\mathrm{GeV}$ (see *eg* Colangelo and Dürr (2004)).

4 Comparison with Monte Carlo data in the pion sector

Let us start with some general remarks on the comparison of Monte Carlo data with CHEFT formulae. First of all, one needs results in physical units. A popular way to set the physical scale uses the Sommer parameter r_0, which is a length scale derived from the heavy-quark potential $V(R)$ through the condition $dV(R)/dR|_{R=r_0} = 1.65$. The phenomenological value has been found to be approximately $r_0 = 0.5$ fm, which is the number to be used in the following. Note that this method assumes that the dependence of r_0 on the light quark masses is negligible, an assumption whose validity is not quite clear. Secondly, many quantities, such as *eg* quark masses, have to be renormalised. Finally, we have to deal with lattice artefacts. Ideally, one would eliminate them by an extrapolation to the continuum limit, which is not an easy task, or one could incorporate them in the CHEFT (see the review by Bär (2004)). In the following we shall adopt a simple-minded approach and try to select the data such that cut-off effects are negligible.

Typically there is little structure in the quark-mass dependence of lattice results, and the data are in most cases well described by a linear function of m_q. In other words, there are no obvious chiral logarithms. Of course, this may be due to the relatively large quark masses in the present simulations, where leading order CHPT is unlikely to work. Thus one needs higher-order calculations, and one has to face the question up to which masses CHPT is reliable. Alternatively, one may try to tame the unphysical behaviour of the truncated series at large masses by some cut-off function and in this way arrive at a formula which works in the mass range covered by the simulations. For this approach see *eg* (Leinweber *et al.* 2004), (Young *et al.* 2004), and references therein.

In Figure 1 we compare pion and quark masses obtained by the JLQCD collaboration (Aoki *et al.* 2003) with the quark-mass dependence of the pion mass as predicted by Equation (1). The parameters have been chosen (not fitted) as follows: $F = 0.0862$ GeV, $B = 3.8$ GeV, $\Lambda_1 = 0.12$ GeV, $\Lambda_2 = 1.20$ GeV, $\Lambda_3 = 0.65$ GeV, $\Lambda_4 = 1.25$ GeV. It is a remarkable observation that the Gell-Mann–Oakes–Renner relation $M_\pi^2 = 2Bm_q$ is a rather good approximation for pion masses up to about 0.7 GeV. The above parameters are well compatible with the phenomenological values quoted in the preceding section. Only B is somewhat larger than the value $B = 2.8$ GeV given in (Dürr 2003). Note that B is scale and scheme dependent as are the quark masses, while the product Bm_q is independent of scale and scheme. Here we have employed tadpole improved one-loop perturbation theory with the renormalisation scale $\mu = 2$ GeV in order to convert the bare VWI masses to renormalised quark masses m_q in the $\overline{MS}$ scheme. The use of perturbation theory entails a considerable uncertainty in the renormalised quark masses and hence in B. Indeed, a recent investigation (Göckeler *et al.* 2004) suggests that the non-perturbative mass renormalisation factor (at the bare coupling used by the JLQCD collaboration) is about 2.3 times larger than the perturbative estimate employed here. But it is gratifying to see that chiral perturbation theory with phenomenologically acceptable values of the coupling constants is able to make contact with the low-mass end of the quark mass range that can be reached in present simulations with dynamical quarks. For a more detailed discussion of the quark-mass dependence of M_π in comparison with different Monte Carlo data see (Dürr 2003).

For the pion decay constant F_π (normalised such that the physical value is 92.4 MeV)

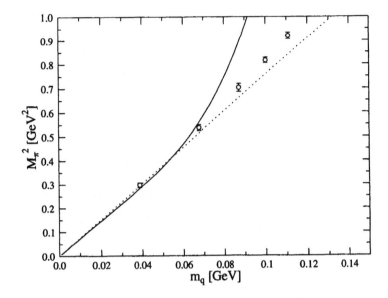

Figure 1. *The square of the pion mass versus the quark mass.* JLQCD *data for* $N_f = 2$
are compared with the Gell-Mann–Oakes–Renner relation (dotted line) and chiral per-
turbation theory (full line).

chiral perturbation theory yields (see *eg* Colangelo *et al.* 2001)

$$F_\pi = F\left\{1 + x\hat{\ell}_4 - \tfrac{5}{4}x^2\hat{\ell}_F^2 + x^2 k_F + O(x^3)\right\} \qquad (3)$$

with

$$\hat{\ell}_F = \frac{1}{30}(14\hat{\ell}_1 + 16\hat{\ell}_2 + 6\hat{\ell}_3 - 6\hat{\ell}_4 + 23) = \ln\frac{\Lambda_F^2}{M^2}. \qquad (4)$$

Again, the analytic contributions $O(x^2)$ (proportional to k_F) are expected to be small.
In Figure 2 we compare this formula to preliminary data from the UKQCD and QCDSF
collaborations. Motivated by our observation that the Gell-Mann–Oakes–Renner relation
works so well, we replace M by M_π. Thus we avoid the problem of renormalising the
quark mass. Choosing $F = 0.0862$ GeV, $k_F = -0.5$, $\Lambda_4 = 1.12$ GeV, $\Lambda_F = 0.67$ GeV
(consistent with phenomenology) we obtain the full curve in Figure 2, which connects
the physical point with the data for the lowest mass but does not describe the data at
larger masses. It is however reassuring that for masses up to the first Monte Carlo points
the chiral expansion seems to be well-convergent: The dotted curve, which corresponds
to $F_\pi = F\{1 + x\hat{\ell}_4\}$, does not deviate dramatically from the full curve in this region.

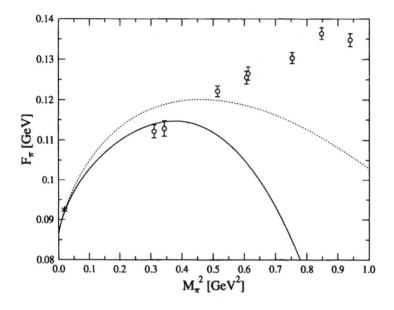

Figure 2. *Preliminary data ($N_f = 2$) for the pion decay constant compared with chiral perturbation theory at different orders. The asterisk indicates the physical point.*

5 Including baryons

The nucleon mass m_N does not vanish even in the chiral limit. Indeed $m_N \gg M_\pi$, and a non-relativistic treatment of the nucleon field seems reasonable. This leads to the so-called heavy-baryon chiral perturbation theory (HBCHPT). A relativistic formulation of chiral perturbation theory in the baryon sector has been given by Becher and Leutwyler (1999). In both cases the physical picture is that of a (heavy) nucleon core surrounded by a cloud of light pions. This is rather different from the situation for the pion, and hence the behaviour of the chiral expansion in the nucleon sector need not be similar to that in the pion sector.

In Becher's and Leutwyler's formulation one obtains for the nucleon mass (Becher and Leutwyler 1999, Procura *et al.* 2004)

$$m_N = m_0 - 4c_1 M_\pi^2 - \frac{3(g_A^0)^2}{32\pi F^2} M_\pi^3 + \left[e_1^r(\lambda) - \frac{3}{64\pi^2 F^2} \left(\frac{(g_A^0)^2}{m_0} - \frac{c_2}{2} \right) \right.$$

$$\left. - \frac{3}{32\pi^2 F^2} \left(\frac{(g_A^0)^2}{m_0} - 8c_1 + c_2 + 4c_3 \right) \ln \frac{M_\pi}{\lambda} \right] M_\pi^4 + \frac{3(g_A^0)^2}{256\pi F^2 m_0^2} M_\pi^5 + O(M_\pi^6).$$

$$(5)$$

Here we have again identified $M_\pi^2 = M^2 = 2Bm_q$, λ is the renormalisation scale, m_0 and g_A^0 are the mass and the axial charge of the nucleon in the chiral limit, c_1, c_2, c_3

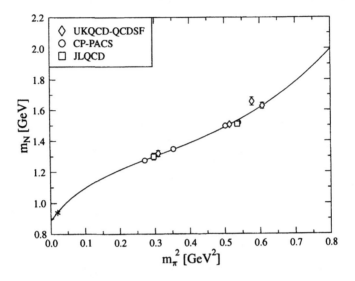

Figure 3. *Nucleon mass data for $N_f = 2$ on (relatively) large and fine lattices. The asterisk indicates the physical point. The curve corresponds to the fit mentioned in the text.*

denote coupling constants from the effective Lagrangian, and $e_1^r(\lambda)$ is a counter-term.

Hadron masses for $N_f = 2$ have been published by the CP-PACS collaboration (Ali Khan *et al.* 2002), the JLQCD collaboration (Aoki *et al.* 2003) as well as the UKQCD and QCDSF collaborations (Allton *et al.* 2002, Ali Khan *et al.* 2004a). We have selected results obtained on (relatively) large and fine lattices to compare with chiral perturbation theory. More precisely, we have considered masses from simulations with $a < 0.15$ fm, $M_\pi < 800$ MeV and $M_\pi L > 5$. These ten data points were fitted with Equation (5) where g_A^0, F, c_2 and c_3 were fixed to phenomenologically reasonable values while m_0, c_1 and $e_1^r(\lambda = 1\,\text{GeV})$ were the fit parameters. For more details see (Ali Khan *et al.* 2004a). The fit curve and the data points are shown in Figure 3. It is first of all remarkable that the masses obtained by different collaborations with different lattice actions and algorithms fall (to rather good accuracy) onto a single curve. Furthermore, the fit parameters are very well compatible with phenomenology, in particular, c_1 is about $-1\,\text{GeV}^{-1}$ and the fit curve comes quite close to the physical point. On the other hand, Equation (5) seems to work up to surprisingly large masses.

6 Axial charge of the nucleon

The quark-mass dependence of the axial charge (or axial-vector coupling constant) g_A of the nucleon has been studied by Hemmert *et al.* (2003) within the framework of the so-called small-scale expansion. This is an extension of HBCHPT which includes explicit $\Delta(1232)$ degrees of freedom.

In the small-scale expansion the expansion parameter is usually called ϵ, and one finds in $O(\epsilon^3)$:

$$
\begin{aligned}
g_A(M_\pi^2) = g_A^0 &- \frac{(g_A^0)^3 M_\pi^2}{16\pi^2 F^2} \\
&+ 4 \left\{ C^{SSE}(\lambda) + \frac{c_A^2}{4\pi^2 F^2} \left[\frac{155}{972} g_1 - \frac{17}{36} g_A^0 \right] + \gamma^{SSE} \ln \frac{M_\pi}{\lambda} \right\} M_\pi^2 \\
&+ \frac{4 c_A^2 g_A^0}{27\pi F^2 \Delta} M_\pi^3 + \frac{8}{27\pi^2 F^2} c_A^2 g_A^0 M_\pi^2 \sqrt{1 - \frac{M_\pi^2}{\Delta^2}} \ln R \\
&+ \frac{c_A^2 \Delta^2}{81\pi^2 F^2} (25 g_1 - 57 g_A^0) \left\{ \ln \left[\frac{2\Delta}{M_\pi} \right] - \sqrt{1 - \frac{M_\pi^2}{\Delta^2}} \ln R \right\} + O(\epsilon^4)
\end{aligned}
\tag{6}
$$

with

$$
\gamma^{SSE} = \frac{1}{16\pi^2 F^2} \left[\frac{50}{81} c_A^2 g_1 - \frac{1}{2} g_A^0 - \frac{2}{9} c_A^2 g_A^0 - (g_A^0)^3 \right], \quad R = \frac{\Delta}{M_\pi} + \sqrt{\frac{\Delta^2}{M_\pi^2} - 1}.
\tag{7}
$$

The new parameters appearing here are g_A^0 (the value of g_A in the chiral limit), Δ (the nucleon Δ mass splitting in the chiral limit), c_A, g_1 (the $N\Delta$ and $\Delta\Delta$ axial coupling constants), and $C^{SSE}(\lambda)$ (a counter-term at the renormalisation scale λ). Hemmert *et al.* (2003) find that for reasonable values of the parameters the formula (6) is able to describe the rather weak mass dependence of the Monte Carlo data as well as the physical point.

7   Chiral effective field theory in a finite volume

Presently, lattice QCD simulations are not only restricted to unphysical quark masses, but also to relatively small (spatial) volumes, usually with periodic boundary conditions. While simulations at the physical quark masses might be possible some day, it will take a bit longer before the ideal case of an infinite volume simulation can be realised. In the meantime we can take advantage of the fact that the chiral effective Lagrangian is volume independent for periodic boundary conditions (Gasser and Leutwyler 1988). So the same Lagrangian governs the quark-mass as well as the volume dependence, and additional information on the coupling constants can be extracted from finite size effects. This description of the finite size effects should work as long as they result from the deformation of the pion cloud in the finite volume, *ie* as long as L is not too small. After all, it is the pion propagation that is predominantly affected by the finite volume, because the pion is the lightest particle in the theory. Treating L^{-1} as a quantity of order p like M_π we arrive at the so-called p expansion (Gasser and Leutwyler 1987). This is to be distinguished from the ϵ expansion, where $L^{-1} = O(\epsilon)$, $M_\pi = O(\epsilon^2)$ with a small parameter ϵ.

In more technical terms, the finite volume (with periodic boundary conditions) discretises the allowed momenta such that the momentum components are restricted to integer multiples of $2\pi/L$. So the loop integrals of CHPT become sums. On the other hand, we can interpret the resulting expressions in the following way. In a finite volume a pion

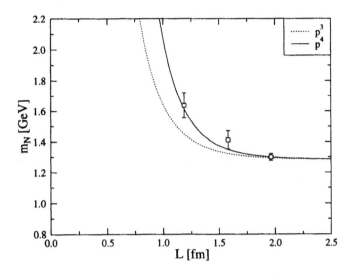

Figure 4. *Volume dependence of the nucleon mass for $M_\pi = 545\,MeV$. The dotted curve shows the contribution of the p^3 term, while the solid curve includes also the p^4 correction.*

emitted from a nucleon can not only be reabsorbed by the same nucleon, but also by one of the periodic images of the original nucleon at a distance which is an integer multiple of L in each of the finite directions. From this point of view the finite size effects arise from pions travelling around the volume once, twice, ... in a given direction, and each crossing of the boundary leads (roughly) to a factor $\exp(-M_\pi L)$ in the final contribution.

As an example, let us consider the nucleon mass in relativistic baryon CHPT. Then the nucleon mass in a spatial box of length L, but for an infinite extent in time direction, can be written as

$$m_N(L) = m_N(\infty) + \Delta_a(L) + \Delta_b(L) + O(p^5), \tag{8}$$

where

$$\Delta_a(L) = \frac{3(g_A^0)^2 m_0 M_\pi^2}{16\pi^2 F^2} \int_0^\infty dx \sum_{\vec{n}\neq\vec{0}} K_0\left(L|\vec{n}|\sqrt{m_0^2 x^2 + M_\pi^2(1-x)}\right) \tag{9}$$

is the $O(p^3)$ contribution to the finite size effect and

$$\Delta_b(L) = \frac{3M_\pi^4}{4\pi^2 F^2} \sum_{\vec{n}\neq\vec{0}} \left[(2c_1 - c_3)\frac{K_1(L|\vec{n}|M_\pi)}{L|\vec{n}|M_\pi} + c_2\frac{K_2(L|\vec{n}|M_\pi)}{(L|\vec{n}|M_\pi)^2}\right] \tag{10}$$

is the additional contribution arising at $O(p^4)$. Here n_i can be interpreted as the number of times the pion crosses the 'boundary' of the box in the i direction. Note that the finite volume does not introduce any new coupling constant. For a comparison with Monte Carlo data we evaluate the finite size corrections using the parameters from the fit discussed in Section 5, choose $m_N(\infty)$ such that $m_N(L)$ on the largest lattice agrees with the Monte Carlo value and take for M_π the value from the largest lattice. This leads to a surprisingly good agreement with the Monte Carlo masses, as is shown for $M_\pi = 545\,\text{MeV}$ in Figure 4 using JLQCD data. Even for larger pion masses the formula reproduces the finite size effects quite well (Ali Khan *et al.* 2004a). This lends further support to the fit shown in Figure 3.

In the case of the pion mass the leading finite volume correction has already been known for some time (Gasser and Leutwyler 1987):

$$M_\pi(L) = M_\pi \left\{ 1 + \frac{M_\pi^2}{(4\pi F_\pi)^2} \sum_{\vec{n} \neq \vec{0}} \frac{K_1(L|\vec{n}|M_\pi)}{L|\vec{n}|M_\pi} + O(M_\pi^4) \right\} . \tag{11}$$

Moreover, neglecting pions which travel around the box more than once ($|\vec{n}| > 1$) Lüscher (1986) has expressed the finite volume correction in terms of the $\pi\pi$ forward scattering amplitude $F(\nu)$, $\nu = (s - u)/(4M_\pi)$:

$$M_\pi(L) - M_\pi = -\frac{3}{16\pi^2 M_\pi L} \int_{-\infty}^{\infty} dy\, F(iy) e^{-\sqrt{M_\pi^2 + y^2}\, L} . \tag{12}$$

The chiral expansion of $F(\nu)$ is known to $O(M_\pi^6)$ and the corresponding coupling constants are reasonably well determined. Using this information one can work out the volume dependence of M_π predicted by Equation (12) (Colangelo and Dürr 2004). The non-leading terms of the chiral expansion give non-negligible contributions, still the finite size effects are considerably underestimated by this approach, at least for $M_\pi > 500\,\text{MeV}$. Probably terms with $|\vec{n}| > 1$ are important for smaller volumes, which is certainly the case in Equation (11). However, Equation (11) alone predicts even smaller finite size effects than Equation (12). So it seems that one would need higher orders in the chiral expansion as well as pions propagating around the volume more than once, and the present understanding of the finite size effects for M_π is unsatisfactory. For another investigation of hadron masses in a finite volume see (Orth *et al.* 2004).

The volume dependence of g_A has recently attracted much interest. First calculations in chiral perturbation theory have been performed by Beane and Savage (2004). However, at the masses used in current simulations these leading-order formulae do not even reproduce the sign of the finite size effects observed in the Monte Carlo data, see eg (Ali Khan *et al.* 2004b), (Sasaki *et al.* 2003). It remains to be seen whether more advanced calculations in CHEFT can solve this discrepancy or whether lower quark masses are required.

8 Conclusions

Only the first steps of an ongoing effort to combine calculations in CHEFT and lattice simulations have been described here. The overall impression is that the range of quark masses that can be used in actual Monte Carlo computations is beginning to overlap with the region of applicability of CHPT. In favourable cases this overlap seems to be so large that meaningful fits are possible. Although fits without phenomenological input are still beyond reach, interesting information on (effective) coupling constants can already be extracted. In most cases the leading chiral logarithm appears to be dominating only for rather small quark masses. Thus CHEFT should be pushed to higher orders while on the lattice side results for lower quark masses are eagerly awaited.

Acknowledgments

The studies reported in this paper have been performed within the QCDSF-UKQCD collaboration. I wish to thank all my colleagues who have contributed to this effort, in particular A Ali Khan, P Hägler, T R Hemmert, R Horsley, A C Irving, H Perlt, D Pleiter, P E L Rakow, A Schäfer, G Schierholz, A Schiller, H Stüben and J M Zanotti.

The numerical calculations have been performed on the Hitachi SR8000 at LRZ (Munich), on the Cray T3E at EPCC (Edinburgh) and on the APEmille at NIC/DESY (Zeuthen). This work has been supported in part by the DFG (Forschergruppe Gitter-Hadronen-Phänomenologie) and by the EU Integrated Infrastructure Initiative 'Hadron Physics' as well as 'Study of Strongly Interactive Matter'.

References

Ali Khan A *et al.* , 2002, *Phys Rev D* **65** 054505; *Phys Rev D* **67** 059901.
Ali Khan A *et al.* , 2004a, *Nucl Phys B* **689** 175.
Ali Khan A *et al.* , 2004b, arXiv:hep-lat/0409161.
Allton C R *et al.* , 2002, *Phys Rev D* **65** 054502.
Aoki S *et al.* , 2003, *Phys Rev D* **68** 054502.
Bär O, 2004, arXiv:hep-ph/0409123.
Beane S R and Savage M J, 2004, arXiv:hep-ph/0404131.
Becher T and Leutwyler H, 1999, *Eur Phys J C* **9** 643.
Colangelo G, 2004, arXiv:hep-ph/0409111.
Colangelo G and Dürr S, 2004, *Eur Phys J C* **33** 543.
Colangelo G, Gasser J, and Leutwyler H, 2001, *Nucl Phys B* **603** 125.
Dürr S, 2003, *Eur Phys J C* **29** 383.
Gasser J and Leutwyler H, 1987, *Phys Lett B* **184** 83.
Gasser J and Leutwyler H, 1988, *Nucl Phys B* **307** 763.
Göckeler M *et al.* , 2004, arXiv:hep-ph/0409312.
Hemmert T R, Procura M, and Weise W, 2003, *Phys Rev D* **68** 075009.
Leinweber D B, Thomas A W, and Young R D, 2004, *Phys Rev Lett* **92** 242002.
Leutwyler H, 2000, arXiv:hep-ph/0008124.
Lüscher M, 1986, *Commun Math Phys* **104** 177.
Meißner U, 2000, arXiv:hep-ph/0007092.
Orth B, Lippert T, and Schilling K, 2004, *Nucl Phys Proc Suppl* **129** 173.
Procura M, Hemmert T R, and Weise W, 2004, *Phys Rev D* **69** 034505.
Sasaki S, Orginos K, Ohta S, and Blum T, 2003, *Phys Rev D* **68** 054509.
Young R D, Leinweber D B, and Thomas A W, 2004, *Nucl Phys Proc Suppl* **128** 227.

Chiral phase transition in hadronic matter: The influence of baryon density

Boris L Ioffe

Institute of Theoretical and Experimental Physics,
B Cheremushkinskaya 25, 117218 Moscow, Russia

1 Introduction

A qualitative analysis of the chiral phase transition in Quantum Chromodynamics (QCD) with two massless quarks and non–zero baryon density is performed. It is assumed that at zero baryonic density, $\rho = 0$, the temperature phase transition is second order and the quark condensate $\eta = \langle 0 \mid \bar{u}u \mid 0 \rangle = \langle 0 \mid \bar{d}d \mid 0 \rangle$ may be taken as the order parameter of phase transition. The baryon masses strongly violate chiral symmetry, $m_B \sim \langle 0 \mid \bar{q}q \mid 0 \rangle^{1/3}$. By supposing, that such specific dependence of baryon masses on the quark condensate takes place up to the phase transition point, it is shown that at finite baryon density ρ the phase transition becomes first order at the temperature $T = T_{\mathrm{ph}}(\rho)$, for $\rho > 0$. At temperatures $T_{\mathrm{cont}}(\rho) > T > T_{\mathrm{ph}}(\rho)$ there is a mixed phase consisting of the (stable) quark phase and the (unstable) hadron phase. At temperature $T = T_{\mathrm{cont}}(\rho)$ the system experiences a continuous transition to the pure, chirally symmetric, phase.

It is well known that chiral symmetry is valid in perturbative QCD with massless quarks. It is also expected that chiral symmetry occurs in full-perturbative and non-perturbative QCD at high temperatures, $(T \gtrsim 200 \text{ MeV})$, if heavy quarks (c, b, t) are ignored. The chiral symmetry is strongly violated, however, in hadronic matter, ie in QCD at $T = 0$ and low density. The order of the phase transition between two phases of QCD, with broken and restored chiral symmetry, with variation of temperature and density is not completely clear now. There are different opinions about this subject; for detailed reviews see (Smilga 2001, Rajagopal and Wilczek 2001) and references therein.

In this talk I discuss the phase transitions in QCD with two massless quarks, u and d. Many lattice calculations (Karsch 1994, Karsch and Laermann 1994, Aoki $et\ al.$

1988, Ali Khan *et al.* 2001) indicate that at zero chemical potential the phase transition is second order. It is shown below that taking account of baryon density drastically changes the situation and the transition becomes first order and, at high density, matter is always in the chirally symmetric phase.

The masses of light u, d, s quarks which enter the QCD Lagrangian, especially the masses of u and d quarks, from which the usual (non-strange) hadrons are built, are very small, $m_u, m_d < 10$ MeV as compared with the characteristic mass scale $M \sim 1$ GeV. Since in QCD the quark interaction proceeds through the exchange of vector gluonic fields, if light quark masses are neglected, the QCD Lagrangian (its light quark part) is chirally symmetric, *ie* not only vector, but also axial currents are conserved and the left and right chiral quark fields are conserved separately. This chiral symmetry is not realized in hadronic matter, in the spectrum of hadrons and their low energy interactions. Indeed, in chirally symmetrical theory the fermion states must be either massless or degenerate in parity. It is evident, that the baryons (particularly, the nucleon) do not possess such properties. This means that the chiral symmetry of the QCD Lagrangian is not realized in the spectrum of physical states and is spontaneously broken. According to the Goldstone theorem, spontaneous breaking of symmetry leads to the appearance of massless particles in the spectrum of physical states – the Goldstone bosons.

Let us first consider the case of zero baryonic density and suppose that the phase transition from chirality violating phase to the chirality conserving one is second order. Second order phase transitions are generally characterized by an order parameter η. The order parameter is a thermal average of some operator which may be chosen in various ways. The physical results are independent of the choice of the order parameter. In QCD the quark condensate, $\eta = |\langle 0|\bar{u}u|0\rangle| = |\langle 0|\bar{d}d|0\rangle| \geq 0$, may be taken as such a parameter. In the phase of broken chiral symmetry (hadronic phase) the quark condensate is non-zero and has the normal hadronic scale, $\langle 0 \mid \bar{q}q \mid 0 \rangle = (250 MeV)^3$; in the phase of restored chiral symmetry it is vanishing.

The quark condensate has the desired properties: as demonstrated in chiral effective theory (Leutwyler 1988, Gerber and Leutwyler 1989) η decreases with increasing temperature and an extrapolation of the curve $\eta(T)$ to higher temperatures indicates that η vanishes at $T = T_c^{(0)} \approx 180$ MeV. Here the superscript "0" indicates that the critical temperature is taken at zero baryon density. The same conclusion follows from lattice calculations (Karsch 1994, Ali Khan *et al* 2001, Karsch 1995), where it was also found that the chiral condensate η decreases with increasing chemical potential (Kogut *et al.* 2001, 2002).

We now apply the general theory of second order phase transitions (Landau and Lifshitz 1977) and consider the thermodynamical potential $\Phi(\eta)$ at a temperature T near $T_c^{(0)}$. Since η is small in this domain, $\Phi(\eta)$ may be expanded in η:

$$\Phi(\eta) = \Phi_0 + \frac{1}{2}A\eta^2 + \frac{1}{4}B\eta^4, \qquad B > 0. \qquad (1)$$

For the moment we neglect possible derivative terms in the potential.

The terms, proportional to η and η^3 vanish for general reasons (Landau and Lifshitz 1977). In QCD with massless quarks the absence of η and η^3 terms can be proved for any perturbative Feynman diagrams. At small $t = T - T_c^{(0)}$ the function $A(t)$ is linear in t: $A(t) = at, a > 0$. If $t < 0$, the thermodynamical potential $\Phi(\eta)$ is minimal at

$\eta \neq 0$, while at $t > 0$ the chiral condensate vanishes, $\eta = 0$. At small t the t-dependence of the coefficient $B(t)$ is insignificant and may be neglected. The minimum, $\bar{\eta}$, of the thermodynamical potential can be found from the condition, $\partial \Phi / \partial \eta = 0$:

$$\bar{\eta} = \begin{cases} \sqrt{-at/B}, & t < 0; \\ 0, & t > 0. \end{cases} \tag{2}$$

It corresponds to the second order phase transition since the potential is quartic in η and – if the derivative terms are included in the expansion – the correlation length becomes infinite at $T = T_c^{(0)}$.

2 Nucleon mass and quark condensate

We show now that the existence of large baryon masses and the appearance of chiral symmetry violating quark condensate are deeply interconnected and furthermore that baryon masses arise due to the quark condensate. We will use the QCD sum rule method invented by Shifman *et al.* (1979) in its applications to baryons (Ioffe 1981, Chung *at al.* 1982, Ioffe 1983, Belyaev and Ioffe 1982, Belyaev and Ioffe 1983). For a review and collection of relevant original papers see Shifman (ed) (1992). The idea of the method is that at virtualities of order $Q^2 \sim 1 \text{ GeV}^2$, the operator product expansion (OPE) may be used in consideration of hadronic vacuum correlators. In OPE the nonperturbative effects reduce to the appearance of vacuum condensates and the lowest dimension condensates play the most important role. The perturbative terms are moderate and do not change the results in any significant way, especially in the case of chiral symmetry violation, where they can appear as corrections only.

Specifically consider the proton mass calculation (Ioffe 1981, Chung *et al.* 1982, Ioffe 1983). We introduce the polarisation operator

$$\Pi(p) = i \int d^4 x e^{ipx} \langle 0|T\xi(x), \bar{\xi}(0)|0\rangle, \tag{3}$$

where $\xi(x)$ is the quark current with proton quantum numbers and p^2 is chosen to be space-like, $p^2 < 0$, $|p^2| \sim 1 \text{ GeV}^2$. The current ξ is the colourless product of three quark fields, $\xi = \varepsilon^{abc} q^a q^b q^c$, $q = u, d$ and the form of the current is specified below. The general structure of $\Pi(p)$ is

$$\Pi(p) = \hat{p} f_1(p) + f_2(p). \tag{4}$$

The first term, proportional to $\hat{p}$, conserves chirality while the second violates chirality.

For each of the functions $f_i(p^2)$, $i = 1, 2$, the OPE can be written as:

$$f_i(p^2) = \sum_n C_n^{(i)}(p^2) \langle 0|O_n^{(i)}|0\rangle, \tag{5}$$

where $\langle 0|O_n^{(i)}|0\rangle$ are vacuum expectation values (vev) of various operators (vacuum condensates), and $C_n^{(i)}$ are coefficient functions calculated in QCD. For the first, chirality

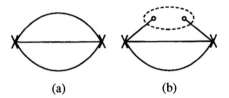

Figure 1. *(a) The bare loop diagram, contributing to the chirality conserving function $f_1(p^2)$: solid lines correspond to quark propagators, crosses represent the interaction with external currents. (b) The diagram corresponding to the chirality violating dimension-3 operator (quark condensate). The dots, surrounded by the circle represent quarks in the condensate phase.*

conserving structure function $f_i(p^2)$, OPE starts with the dimension zero $(d = 0)$ unit operator. Its contribution is described by the diagram in Figure 1(a) and by

$$\hat{p} f_1(p^2) = C_0 \hat{p} p^4 ln[\Lambda_u^2/(-p^2)] + polynomial, \qquad (6)$$

where C_0 is a constant and Λ_u is the ultraviolet cutoff. The OPE for the chirality violating structure $f_2(p^2)$ starts from the $d = 3$ operator, and its contribution is represented by the diagram in Figure 1(b):

$$f_2(p^2) = C_1 p^2 \langle 0|0\bar{q}q|0\rangle ln \frac{\Lambda_u^2}{(-p^2)} + polynomial \qquad (7)$$

Let us, for a moment, restrict ourselves to the first order terms of OPE and neglect higher order terms (as well as perturbative corrections).

On the other hand, the polarisation operator, Equation 3, may be expressed via the characteristics of physical states using dispersion relations.

$$f_i(s) = \frac{1}{\pi} \int \frac{Im f_i(s')}{s' + s} ds' + polynomial, \quad s = -p^2. \qquad (8)$$

The proton contribution to $Im\Pi(p)$ is equal to:

$$Im\Pi(p) = \pi \langle 0|\xi|p\rangle \langle p|\bar{\xi}|0\rangle \delta(p^2 - m^2) = \pi \lambda_N^2 (\hat{p} + m)\delta(p^2 - m^2), \qquad (9)$$

where

$$\langle 0|\xi|p\rangle = \lambda_N v(p), \qquad (10)$$

λ_N is a constant, $v(p)$ is the proton spinor and m is the proton mass. Still restricting ourselves to this rough approximation, we may take the expression calculated in QCD for $\Pi(p)$ (Equations 6, 7) as equal to its phenomenological representation, Equation 9. The best way to get rid of the unknown polynomial, is to apply the Borel(Laplace) transformation to both sides of the equality. This is defined as:

$$\mathcal{B}_{M^2} f(s) = \lim_{\substack{n \to \infty, s \to \infty, \\ s/n = M^2 = Const}} \frac{s^{n+1}}{n!} \left(-\frac{d}{ds}\right)^n f(s) = \frac{1}{\pi} \int\limits_0^\infty ds Im f(s) e^{-s/M^2}, \qquad (11)$$

if $f(s)$ is given by the dispersion relation, Equation 8. Notice, that

$$\mathcal{B}_{M^2} \frac{1}{s^n} = \frac{1}{(n-1)!(M^2)^{n-1}}. \tag{12}$$

Owing to the factor $1/(n-1)!$ in Equation 12 the Borel transformation suppresses the contributions of high order terms in OPE.

We now specify the quark current $\xi(x)$. It is clear from Equation 9 that the proton contribution will dominate in some region of the Borel parameter $M^2 \sim m^2$ only in the case when both of the functions f_1 and f_2, calculated in QCD, are of the same order. This requirement, together with the requirements of absence of derivatives and of renormco-variance, fixes the form of the current in a unique way (for more details see Ioffe 1981, 1983):

$$\xi(x) = \varepsilon^{abc}(u^a C \gamma_\mu u^b)\gamma_\mu \gamma_5 d^c, \tag{13}$$

where C is the charge conjugation matrix. With the current $\xi(x)$ (Equation 13) the calculations of the diagrams in Figure 1(b) can be easily performed. The constants C_0 and C_1 are determined and, after Borel transformation, two equations (sum rules) arise (on the phenomenological sides of the sum rules only the proton state is accounted):

$$M^6 = \tilde{\lambda}_N^2 e^{-m^2/M^2}, \tag{14}$$

$$-2(2\pi)^2 \langle 0|\bar{q}q|0\rangle M^4 = m \tilde{\lambda}_N^2 \, e^{-m^2/M^2}, \tag{15}$$

where $\tilde{\lambda}_N^2 = 32\pi^4 \, \lambda_N^2$.

It can be shown that this rough approximation is valid at $M \approx m$. Using this value of M and dividing Equation 15 by Equation 14 we get a simple formula for proton mass (Ioffe 1981):

$$m = [-2(2\pi)^2 \langle 0|\bar{q}q|0\rangle]^{1/3}. \tag{16}$$

This formula demonstrates the fundamental fact that the appearance of the proton mass is caused by spontaneous violation of chiral invariance: the presence of the quark condensate. (Numerically, Equation 16 gives the experimental value of proton mass with an accuracy better than 10%).

A more refined treatment of the problem of the proton mass calculation was performed: high order terms of OPE were accounted for, as well as excited states in the phenomenological sides of the sum rules and the stability of the Borel mass dependence was checked. In the same way, the hyperons, isobar and some resonance masses were calculated, all in a good agreement with experiment (Ioffe 1981, Belyaev and Ioffe 1983). I will not dwell on these results. The main conclusion is: the origin of baryon masses is in spontaneous violation of chiral invariance – the existence of quark condensate in QCD.

3   The chiral phase transition in the presence of finite baryon density

I would like to consider here the influence of baryon density on the chiral phase transition in hadronic matter. Kogan, Kovner and Tekin (Kogan *et al* 2001) have suggested the idea that baryons may initiate the restoration of chiral symmetry if their density is high – when roughly half of the volume is occupied by baryons. The physical argument in favour of this idea comes from the hypothesis, supported by a calculation in the chiral soliton model of the nucleon (Diakonov and Petrov 2001)), that inside the baryon the chiral condensate has a sign opposite to that in vacuum. This hypothesis is not proved. Furthermore, it is doubtful that the concept of quark condensate inside the nucleon can be formulated in a correct way in quantum theory. But the idea of the strong influence of baryon density on the chiral phase transition looks very attractive. For this reason no assumption on the driving mechanism of the chiral phase transition at zero baryon density will be made here. The problem under consideration is: how the phase transition changes in the presence of baryons. The content of my talk closely follows (Chernodub and Ioffe 2004).

For completeness it must be mentioned that the variation of the quark condensate with temperature is not the only source of baryon effective mass shifts. At low T the effective baryon mass shift also arises from the interaction with pions in the thermal bath (Leutwyler and Smilga 1990, Eletsky and Ioffe 1997). However, this mass shift, which may be called external (unlike the internal, arising from variation of quark condensate), is related to effective mass, *ie* to propagation of baryons in matter and has nothing to do with the properties of matter as a whole or with the phase transition. (A similar phenomenon takes place in case of vector mesons where, because of interaction with pions in the thermal bath, the mixing with axial mesons arises (Eletsky and Ioffe 1993).)

Consider the case of the finite, but small baryon density ρ (by ρ we mean here the sum of baryon and anti-baryon densities). For a moment, consider only one type of baryon, *ie* the nucleon. The temperature of the phase transition T_{ph} is in general dependent on the baryon density, $T_{\mathrm{ph}} = T_{\mathrm{ph}}(\rho)$, with $T_{\mathrm{ph}}(\rho = 0) \equiv T_c^{(0)}$. At $T < T_{\mathrm{ph}}(\rho)$ the term, proportional to $E\rho$, where $E = \sqrt{p^2 + m^2}$ is the baryon energy, must be added to the thermodynamical potential, Equation 1. As shown above, the nucleon mass m (as well as the masses of other baryons) arises due to the spontaneous violation of the chiral symmetry and is approximately proportional to the cubic root of the quark condensate: $m = c\eta^{1/3}$, with $c = (8\pi^2)^{1/3}$ for a nucleon. At small temperatures T the baryon contribution to Φ is strongly suppressed by the Boltzmann factor $e^{-E/T}$ and is negligible. Below we assume that the proportionality $m \sim \eta^{1/3}$ is valid in a broad temperature interval. Arguments in favour of such an assumption are based on the expectation that the baryon masses vanish at $T = T_{\mathrm{ph}}(\rho)$ and on dimensional grounds. Near the phase transition point $E = \sqrt{p^2 + m^2} \approx p + c^2 \eta^{2/3}/(2p)$. As $\eta \to 0$ all baryons accumulate near zero mass and a summation over all baryons gives, instead of Equation 1, the following:

$$\Phi(\eta, \rho) = \Phi_0 + \frac{1}{2}at\,\eta^2 + \frac{1}{4}B\,\eta^4 + C\eta^{2/3}\rho\,, \qquad (17)$$

where $C = \sum_i c_i^2/(2p_i)$. The term $\rho \sum_i p_i$ is absorbed into Φ_0 since it is independent on the chiral condensate η. The typical momenta are of the order of the temperature,

$p_i \sim T$. Thus, Equation 17 is valid in the region $\eta \ll T^3$. In the leading approximation the coefficient C can be considered as independent of temperature at $T \sim T_c^{(0)}$.

Due to the last term in Equation 17 the thermodynamical potential *always* has a local minimum at $\eta = 0$ since the condensate η is never negative. At small negative values of t there is also a local minimum at $\eta > 0$, which is a solution of the equation:

$$\frac{\partial \Phi}{\partial \eta} \equiv (at + B\eta^2)\eta + \frac{2}{3}C\rho\eta^{-1/3} = 0 . \tag{18}$$

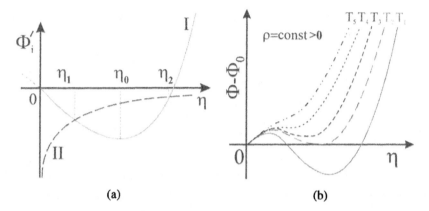

(a)   (b)

Figure 2. *(a) Graphical representation of Equation 18: "I" is the first term and "II" the second term (with the opposite sign) in the rhs of the equation. (b) The thermodynamic potential Equation 17 vs the chiral condensate at a fixed baryon density $\rho > 0$. At low enough temperatures, $T = T_1$, the system resides in the chirally broken (hadron) phase. The first order phase transition to the quark phase takes place at $T_{ph} = T_2 > T_1$. At somewhat higher temperatures, $T_3 > T_{ph}$ the system is in a mixed state. The temperature $T_4 \equiv T_{cont}$ corresponds to a continuous transition to the pure quark phase, in which the thermodynamic potential has the form T_5.*

At small enough baryon density ρ, Equation 18 (see Figure 2(a)) has, in general, two roots, $\eta_1 < \eta_0$ and $\eta_2 > \eta_0$, where $\eta_0 = (-at/3B)^{1/2}$ is the minimum of the first term on the rhs of Equation 18.

The calculation of the second derivative $\partial^2\Phi/\partial\eta^2$ shows that the second root η_2 (if it exists) corresponds to a local minimum of Φ. The point $\eta = \eta_1$ corresponds to a local maximum of the thermodynamical potential since at this point the second derivative is never positive.

The thermodynamical potential $\Phi(\eta, \rho)$ at (fixed) non–zero baryon density ρ has the form plotted in Figure 2(b). At low enough temperatures (curve T_1) the potential has a global minimum at $\eta > 0$ and the system resides in the chirally broken (hadron) phase. As the temperature increases the minima at $\eta = 0$ and at $\eta = \bar\eta_2 > 0$ becomes of equal height (curve $T_2 \equiv T_{ph}$). At this point the first order phase transition to the quark phase takes place. At somewhat higher temperatures, $T = T_3 > T_{ph}$, the $\eta > 0$ minimum of the potential still exists but $\Phi(\eta = 0) < \Phi(\bar\eta_2)$. This is a mixed phase, in which bubbles of the hadron phase may still exist. However, as the temperature increases further, the second minimum disappears (curve $T_4 \equiv T_{\text{cont}}$). This temperature corresponds to a continuous transition to the pure quark phase. At still higher temperatures, within the pure quark phase, the thermodynamic potential has the form T_5.

Let us calculate the temperature of the phase transition, $T_{\text{ph}}(\rho)$, at non–zero baryon density ρ. The transition corresponds to the curve T_2 in Figure 2(b), which is defined by the equation $\Phi(\bar{\eta}_2, \rho) = \Phi(\eta = 0, \rho)$, where $\bar{\eta}_2$ is the second root of Equation 18 as discussed above. The solution is:

$$T_{\text{ph}}(\rho) = T_c^{(0)} - \frac{5}{a}\left(\frac{2C\rho}{3}\right)^{3/5}\left(\frac{B}{4}\right)^{2/5}, \tag{19}$$

and the second minimum of the thermodynamic potential is at $\bar{\eta}_2 = [4a\,(T^{(0)} - T_{\text{ph}}(\rho))/(5\,B)]^{1/2}$.

At a temperature slightly higher than $T_{\text{ph}}(\rho)$ the potential is smallest at $\eta = 0$, but also has an unstable (local) minimum at some $\eta > 0$. The existence of such metastable states is a common feature of first order phase transitions, (eg overheated liquid in the case of a liquid–gas system). With a further increase in the density ρ (at a given temperature) the intersection of the two curves in Figure 2(a) disappears and the two curves only touch one another at one point, $\eta = \bar{\eta}_4$. At this temperature a continuous transition (crossover) takes place. The corresponding potential has the characteristic form denoted as T_4 in Figure 2(b). The temperature $T_4 \equiv T_{\text{cont}}$ is defined by the condition that the first and the second derivatives of Equation 17 vanish:

$$T_{\text{cont}}(\rho) = T_c^{(0)} - \frac{5}{a}\left(\frac{2C\rho}{9}\right)^{3/5}\left(\frac{B}{2}\right)^{2/5}, \tag{20}$$

and the value of the chiral condensate, where the second local minimum of the potential disappears is given by $\bar{\eta}_4 = \left[2a(T_{\text{cont}}(\rho) - T_c^{(0)})/(5\,B)\right]^{1/2}$. At temperatures $T > T_{\text{cont}}(\rho)$ the potential has only one minimum and the matter is in the state with restored chiral symmetry. Thus, in QCD with massless quarks, the type of phase transition corresponding to the restoration of the chiral symmetry strongly depends on the value of the baryonic density ρ. At a fixed temperature, $T < T_c^{(0)}$, the phase transition happens at a certain critical density, ρ_{ph}. According to Equation 19 the critical density has a kind of a "universal" dependence on the temperature, $\rho_{\text{ph}}(T) \propto [T_c^{(0)} - T]^{5/3}$, the power of which does not depend on the parameters of the thermodynamic potential, a and B.

The expected phase diagram is shown qualitatively in Figure 3(a). This diagram does not contain an end-point which was found in lattice simulations of the QCD with a finite chemical potential (Fodor and Katz 2002, 2004). One may expect that this happens because, in our approach, a possible influence of the confinement on the order of the chiral restoration transition was ignored. Intuitively it seems that at low baryon densities such influence is absent. Indeed, the deconfinement phenomenon refers to large quark–antiquark separations while the restoration of chiral symmetry appears due to fluctuations of the gluonic fields in the vicinity of the quark. However, the confinement phenomenon dictates the value of the baryon size which cannot be ignored at high baryon densities, when baryons overlap. If the melting of baryons happens in the hadron phase depicted

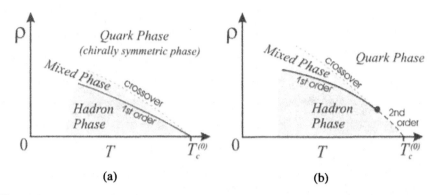

Figure 3. *The qualitative phase diagram at finite baryon density and temperature based on the analysis, (a) without, and (b) with, an indication of the approximate 2nd order transition domain.*

in Figure 3(a), then at high enough density the nature of the transition could be changed. This may give rise in appearance of the end-point observed in (Fodor and Katz 2002, 2004).

The domain where the inequality $|at| \gg C\rho\eta^{2/3}$, $\rho \neq 0$ is fulfilled, has specific features. In this domain the phase transition looks like a smeared second order phase transition: the specific heat has (approximately) a discontinuity at the phase transition point, $\Delta C_p = a^2 T_c / 2B$. This statement follows from general theory (Landau and Lifshitz 1977). At $|at| \gg C\rho\eta^{2/3}$ the last term in Equation 17 may be neglected and we find the entropy is:

$$S = -\frac{\partial \Phi}{\partial T} = S_0 - \frac{1}{2}\frac{\partial A}{\partial T}\eta^2. \tag{21}$$

Above the phase transition, $\eta = 0$ and $S = S_0$, whereas below the phase transition:

$$S = S_0 + \frac{a^2}{2B}(T - T_c). \tag{22}$$

The specific heat $C_p = T(\partial S/\partial T)_p$ in the limit $T \to T_c$, below the phase transition, is:

$$C_p = C_{p0} + \frac{a^2}{2B}T_c. \tag{23}$$

The correlation length increases as $(T - T_c^{(0)})^{-1/2}$ as $T - T_c^{(0)} \to 0$, if we include the derivative terms in the effective thermodynamical potential. The phase diagram with this domain indicated may look as shown in Figure 3(b). Note that the applicability of our considerations is limited to the region $|T - T_c^{(0)}|/T_c^{(0)} \ll 1$ and to low baryon densities.

In real QCD the massive heavy quarks (c, b, t) do not influence this conclusion, since their concentration in the vicinity of $T \approx T_c^{(0)} \sim 200$ MeV is small. However, strange quarks, with masses $m_s \approx 150$ MeV of the same order as the expected $T_c^{(0)}$, may change the situation. This problem deserves further investigation.

Acknowledgments

I am thankful to M N Chernodub, with whom the main part of this work was done. This work was supported in part by INTAS grant 2000-587, RFBR grants 03-02-16209 and by funds from EC to the project "Study of Strongly Interacting Matter" under contract 2004 No. R113-CT-2004-506078.

References

Ali Khan A *et al* [CP-PACS Collaboration], 2001, *Phys Rev D* **63** 034502.
Aoki S *et al* [JLQCD Collaboration] 1998, *Phys Rev D* **57** 3910.
Belyaev V M and Ioffe B L, 1982, *Zh Eksp Teor Fiz* **83** 876.
Belyaev V M and Ioffe B L, 1983, *Zh Eksp Teor Fiz* **84** 1236.
Chernodub M N and Ioffe B L, 2004, *JETP Lett* **79** 606; *Pis'ma Zh Eksp Teor Fiz* **79** 742.
Chung Y, Dosch H G, Kremer M, and Schall D, 1982, *Nucl Phys B* **197** 55.
Diakonov D and Petrov V, 2001, in *Handbook of QCD*, Boris Ioffe Festschrift, ed by M Shifman, World Scientific, **1** 359.
Eletsky V L and Ioffe B L, 1993, *Phys Rev D* **47** 3083.
Eletsky V L and Ioffe B L, 1997, *Phys Lett B* **401** 327.
Fodor Z and Katz S D, 2002, *JHEP* **0203** 014.
Fodor Z and Katz S D, 2004, *JHEP* **0404** 050.
Gerber P and Leutwyler H, 1989, *Nucl Phys B* **321** 387.
Ioffe B L, 1981, *Nucl Phys B* **188** 317, *E* **191** 591.
Ioffe B L, 1983, *Z Phys C* **18** 67.
Karsch F, 1994 *Phys Rev D* **49** 3791.
Karsch F and Laermann E, 1994, *Phys Rev D* **50** 6954.
Karsch F, 1995, *Nucl Phys A* **590** 367C.
Kogan I I, Kovner A and Tekin B, 2001, *Phys Rev D* **63** 116007.
Kogut J B, Sinclair D K, Hands S J and Morrison S E, 2001, *Phys Rev D* **64** 094505.
Kogut J B, Toublan D and Sinclair D K, 2002, *Nucl Phys B* **642** 181.
Landau L and Lifshitz E, 1977, *Statistical Physics*, part 1, Pergamon Press, Oxford, 3r d edition.
Leutwyler H, 1988, *Nucl Phys B* (Proc Suppl) **4** 248.
Leutwyler H and Smilga A V, 1990, *Nucl Phys B* **342** 302.
Rajagopal K and Wilczek F, 2000, The Condensed Matter Physics of QCD, in *Handbook of QCD*, Boris Ioffe Festschrift, ed by M Shifman, World Scientific, Singapore, **3** 2061.
Shifman M A, Vainshtein A I and Zakharov V A, 1979, *Nucl Phys B* **147**, 385.
Shifman M A (ed), 1992, Vacuum Structure and QCD Sum Rules (*Current Physics – Sources and Comments*, **10**), ed by M A Shifman, Amsterdam, North Holland, 1992.
Smilga A V, 2001, Hot and Dense QCD, in *Handbook of QCD*, Boris Ioffe Festschrift, ed by M Shifman, World Scientific, Singapore, **3** 2033.

A covariant quark model

Bernard Metsch

Rheinische Friedrich-Wilhelms-Universität Bonn,
Nussallee 14-16, D-53115 Bonn, Germany

1 Introduction

The ultimate goal of any approach to the structure of hadrons is a unified description of the mass spectra of eg light-flavoured mesons and baryons up to masses of 3 GeV and high angular momenta – accounting for all such features as Regge trajectories, low-lying scalar excitations, parity doublets – together with a reliable calculation of a multitude of electro-weak properties, such as electromagnetic form factors, radiative decays and transitions, semi-leptonic and non-leptonic weak decays.

In principle the most fundamental approach is doubtless lattice gauge theory, which in the present unquenched form is a simulation of QCD itself. In spite of enormous progress in the past decade, unfortunately, as far as the hadronic excitation spectrum is concerned, the results are restricted to the lowest state in each channel only; some results on first excited states, addressing eg the position of the Roper resonance, are emerging now.

Therefore constituent quark model approaches, where it is assumed that the majority of mesonic and baryonic excitations can be understood in terms of $q\bar{q}$– and q^3– bound states of (constituent) quarks, respectively, and supposing that the coupling to strong decay channels can be treated perturbatively, constitute an indispensable tool to correlate a wealth of spectroscopic data and thus provide a reliable framework to judge what is to be considered exotic. However, a relativistic treatment seems to be imperative, since even with effective constituent quark masses of a few hundred MeV quarks inside a hadron are not really slow. Moreover, hadron masses differ appreciably from the sum of its constituents and many observables involve processes at sizable momentum transfers.

A field-theoretical approach which respects relativistic covariance is based on (coupled) Dyson-Schwinger/Bethe-Salpeter equations. Very interesting results on ground state meson properties have been obtained on the basis of a suitable parametrisation of the gluon propagator (eg Maris and Roberts 2003). A more practical approach in the framework of quantum field theory is based on the Salpeter equation (see below), using free-form fermion propagators with constituent masses and instantaneous interaction kernels to model confinement by linearly rising potentials and to explain the remaining spin dependent splittings by instanton effects. In this manner the complete hadronic excitation spectrum can be addressed. This of course is also the usual scope of the more traditional non-relativistic constituent quark models (Capstick and Roberts 2000) which are essentially based on the Schrödinger equation, possibly improved by using relativistic kinematics and some parametrisation of relativistic effects in the quark dynamics. Recently it has been shown, how in the framework of Dirac's relativistic quantum mechanics, especially in its point form, one can obtain a reliable calculation of electromagnetic properties in such a setup (Plessas 2003).

In the framework of quantum field theory composite bound states are described by the homogeneous Bethe-Salpeter equation which involves the full propagators, irreducible interaction kernels and interaction vertices. In the skeleton expansion these in turn fulfil (an infinite set of (inhomogeneous)) integral equations, such as *eg* the Dyson-Schwinger equation for the fermion propagator. In practise this expansion is truncated, followed by an ansatz for some n-point function. The Bethe-Salpeter equation for two-particle bound states or the Dyson-Schwinger equation for the self-energy of lower order are then solved. At this point we refer to results from a renormalisation-group-improved rainbow-ladder approach (DSE) based on an effective gluon propagator with a specific infrared behaviour (Maris and Roberts 2003, Pichowsky 2003).

A simplified ansatz is to assume that the fermion propagator has the free form $S(p) \approx i \left[\gamma^\mu p_\mu - m + i\varepsilon \right]^{-1}$ and accordingly, that all self-energy contributions can be suitably accounted for by introducing a constituent mass m. If in addition all interactions can be taken to be instantaneous (*ie* neglecting retardation) in the form of potentials $V(\vec{p}, \vec{p}\,')$, the Salpeter amplitude can be defined in terms of the Bethe–Salpeter amplitude χ for a $q\bar{q}$-system by $\Phi(\vec{p}) = \int \frac{dp^0}{2\pi} \chi(p^0, \vec{p}) \Big|_{P=(M,\vec{0})}$ which then fulfils the homogeneous Salpeter equation:

$$
\begin{aligned}
\Phi(\vec{p}) &= \Lambda_1^-(\vec{p}) \gamma_0 \frac{\left[\int \frac{d^3 p'}{(2\pi)^3} V(\vec{p}, \vec{p}\,') \Phi(\vec{p}\,') \right]}{M + \omega_1(\vec{p}) + \omega_2(\vec{p})} \gamma_0 \Lambda_2^+(-\vec{p}) \\
&\quad - \Lambda_1^+(\vec{p}) \gamma_0 \frac{\left[\int \frac{d^3 p'}{(2\pi)^3} V(\vec{p}, \vec{p}\,') \Phi(\vec{p}\,') \right]}{M - \omega_1(\vec{p}) - \omega_2(\vec{p})} \gamma_0 \Lambda_2^-(-\vec{p}) ,
\end{aligned}
\tag{1}
$$

where $\Lambda_i^\pm(\vec{p}) := \frac{\omega_i(\vec{p}) \pm H_i(\vec{p})}{2\omega_i(\vec{p})}$ are projectors on positive and negative energy solutions to the usual Dirac one-particle Hamiltonian $H_i(\vec{p}) = \gamma_0 \left((\vec{\gamma} \cdot \vec{p}) + m_i \right)$ and where $\omega_i(\vec{p}) = \sqrt{m_i^2 + |\vec{p}|^2}$. This equation constitutes the basis of virtually all constituent quark models: In the extreme non-relativistic limit the first term on the r.h.s. of this equation is ignored. The remaining component of the Salpeter amplitude can then be interpreted as a Schrödinger wavefunction which fulfils a Schrödinger-type equation with a relativistic kinetic energy and relativistic corrections to the potentials and as such was exploited very successfully both for mesons (Godfrey and Isgur 1985) and for baryons (Capstick and Roberts 2000).

2 Application to meson structure

For the application of the full Salpeter equation to mesons, which accounts for confinement by an instantaneous string-like potential and for the major mass splittings by an effective instanton-induced interaction, see (Koll *et al.* 2000) and (Ricken *et al.* 2000). In spite of the instantaneous approximation involved here, it seems that form factors can be calculated reliably (see Figure 1).

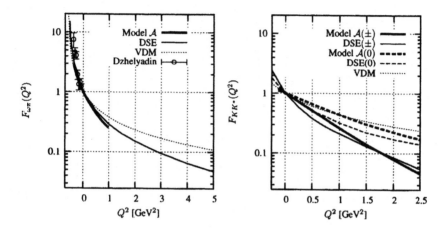

Figure 1. *Comparison of the $\omega\pi\gamma$ – and $K^*K\gamma$– transition form factor results obtained from the Dyson-Schwinger (DSE) approach of (Maris and Roberts 2003) (thin lines) and from the instantaneous Salpeter equation, Model $\mathcal{A}$ of (Koll et al. 2000) (thick lines). In the right panel the solid and dashed lines distinguish the results for the charged and neutral kaons, respectively. Figure adapted from (Maris and Roberts 2003).*

The same model, with a naive extension of the spin-dependent quark dynamics to heavy quark flavours, has also been applied to a calculation of the weak decays of D– and B–mesons, see (Merten *et al.* 2002).

J^π	D			D_s			B			B_s		
	$\mathcal{A}$	exp	$\mathcal{B}$	$\mathcal{A}$	exp	$\mathcal{B}$	$\mathcal{A}$	exp	$\mathcal{B}$	$\mathcal{A}$	exp	$\mathcal{B}$
0^-	1869	1868	1869	1969	1968	1969	5279	5279	5279	5368	5370	5369
	2677	2637	2578	2794		2683	6002		5869	6101		5960
1^-	1993	2009	2034	2049	2112	2116	5325	5325	5346	5369	5417	5425
0^+	2519	2308	2375	2563	2317	2464	5796		5675	5822		5774
1^+	2464	2422	2420	2532	2459	2506	5770		5696	5822		5774
1^+	2464		2420	2532	2536	2506	5770		5696	5822		5774
2^+	2475	2459	2469	2541	2573	2552	5771		5771	5823		5788

Table 1. *Comparison of calculated and experimental heavy flavoured meson masses (in MeV). $\mathcal{A}$ and $\mathcal{B}$ refer to the results from the two model versions discussed in (Merten et al. 2002). Experimental data from (Particle Data Group 2004), except for $D(0^+)$ from (Belle 2003).*

Although most masses could be reproduced reasonably, see Table 1, this does not apply to the recently obtained data for the scalar charmed mesons which are obviously too low to be accounted for by constituent quark models of the type discussed here. Nevertheless the description of the differential decay rates, see Figure 2, and many other observables in semi-leptonic decays and non-leptonic decays in the factorisation approximation are very satisfactory.

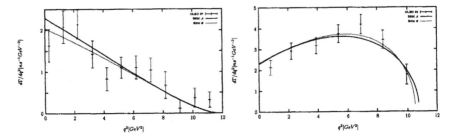

Figure 2. *Comparison of calculated and experimental differential decay rates for the* $B \rightarrow D\ell\nu_\ell$ *(left panel) and* $B \rightarrow D^*\ell\nu_\ell$ *(right panel) semi-leptonic decays. BSM $\mathcal{A}$ (solid) and BSM $\mathcal{B}$ (dotted) refer to the results from the two model versions discussed in (Merten et al. 2002). Experimental data from (CLEO 1997).*

3   Application to baryon structure

With the very same assumptions, *ie* effective constituent quark propagators and instantaneous (three-body) interaction kernels, the Salpeter equation for the Salpeter amplitude Φ_M, which in the baryon rest frame depends on relative internal momenta $\vec{p}_\xi, \vec{p}_\eta$, can likewise be written as an eigenvalue equation

$$(\mathcal{H}\Phi_M)(\vec{p}_\xi, \vec{p}_\eta) = M\,\Phi_M(\vec{p}_\xi, \vec{p}_\eta) = \sum_{i=1}^{3} H_i\,\Phi_M(\vec{p}_\xi, \vec{p}_\eta)$$

$$+\ \left(\Lambda^{+++} + \Lambda^{---}\right)\gamma^0 \otimes \gamma^0 \otimes \gamma^0 \int \frac{d^3 p'_\xi}{(2\pi)^3}\frac{d^3 p'_\eta}{(2\pi)^3}\,V^{(3)}(\vec{p}_\xi, \vec{p}_\eta, \vec{p}'_\xi, \vec{p}'_\eta)\,\Phi_M(\vec{p}'_\xi, \vec{p}'_\eta)$$

$$+\ \left(\Lambda^{+++} - \Lambda^{---}\right)\gamma^0 \otimes \gamma^0 \otimes 1 \int \frac{d^3 p'_\xi}{(2\pi)^3}\,\left[V^{(2)}(\vec{p}_\xi, \vec{p}'_\xi) \otimes 1\right]\,\Phi_M(\vec{p}'_\xi, \vec{p}_\eta)$$

$$+\ \text{cycl. perm. (123)}. \tag{2}$$

Here $\Lambda^{\pm\pm\pm} = \Lambda_1^\pm \otimes \Lambda_2^\pm \otimes \Lambda_3^\pm$ are again projectors on positive and negative energy components and the two-body interactions have been accounted for by an effective instantaneous kernel constructed in lowest order (Löring *et al.* 2001). Although the Salpeter Hamiltonian given above is not positive definite, it can be shown that the negative energy solutions of the eigenvalue equation can be related to positive energy solutions of opposite parity via CPT-symmetry. Hence the number of states in instantaneous approximation is the same as for the non-relativistic constituent quark model but in the relativistic treatment states of opposite parity are coupled. For results on mass spectra, where $V^{(3)}$ was taken to be a three-body linear confinement potential with a suitable spin-dependence and $V^{(2)}$ takes into account the effective instanton-induced interaction, we refer to (Löring *et al.* 2001). Here we merely illustrate the effects of the instanton induced interaction on the positive parity nucleon mass splittings, see Figure 3, where the mass spectrum dependence of the strength of the instanton force is compared to experimental data.

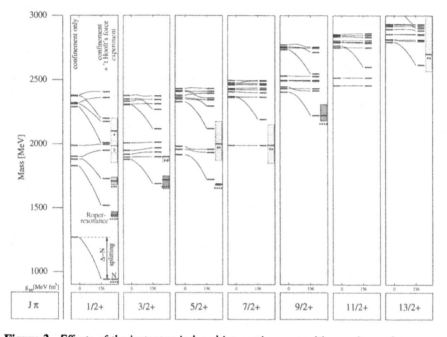

Figure 3. *Effects of the instanton induced interaction on positive parity nucleon resonances. The left of each column represents by bars the spectrum with the confinement potential alone. The curves indicate the dependence of the spectrum on the strength of the 't Hooft coupling g_{nn} up to the value chosen to reproduce the $N - \Delta$ mass splitting. This then also accounts for the most prominent features of the entire mass spectrum, which is given by horizontal bars on the right of each column. Shaded boxes indicate the experimental uncertainty in the mass.*

On the basis of the Salpeter amplitudes the baryonic vertex function (amputated Bethe-Salpeter amplitude) can be reconstructed for any on-shell baryon momentum and thus form factors and various couplings can be calculated covariantly within the Mandelstam formalism; in lowest order (impulse approximation) this involves no additional parameters. The details and results of the calculation of static moments, electroweak form factors and photon couplings of non-strange baryons have been published recently (Merten 2002a). Here we will briefly discuss a novel approach to calculate magnetic moments directly from the Salpeter amplitudes: The baryon magnetic moment can be expressed in momentum space as the expectation value of the 3rd component of an operator $\hat{\mu}$ with respect to the Salpeter amplitudes Φ_M, normalised to $\langle \Phi_M | \Phi_M \rangle = 2M$:

$$\mu = \frac{\langle \Phi_M | \hat{\mu} | \Phi_M \rangle}{2M}, \text{ where } \hat{\mu}^3 = \frac{\omega_1 + \omega_2 + \omega_3}{M} \left(\sum_{\alpha=1}^{3} \frac{\hat{e}_\alpha}{2\omega_\alpha} \left(\hat{\ell}_\alpha^3 + \hat{\Sigma}_\alpha^3 \right) \right) - \hat{\mu}_C^3, \quad (3)$$

where $\omega_\alpha = \sqrt{m_\alpha^2 + p_\alpha^2}$ represents the relativistic energy of quark α with mass m_α and charge $\hat{e}_\alpha$, $\hat{\ell}$ and $\hat{\Sigma}$ represent the single particle angular momentum– and (twice the) spin–operator, respectively, and where

$$\hat{\mu}_C^i = \varepsilon_{ijk} \frac{i}{M} \sum_{\alpha=1}^{3} \frac{\hat{e}_\alpha}{2\omega_\alpha} p_\alpha^k \sum_{\beta=1}^{3} \omega_\beta \frac{\partial}{\partial p_\beta^j} \tag{4}$$

corrects for the relativistic centre of charge motion. We think that this is a remarkable expression, to our knowledge not to be found in the literature. The non-relativistic limit of this expression is of course obvious. The results for some ground state baryons are compared in Table 2 to experimental data and the results obtained by the Graz group (Berger *et al.* 2004).

B	BSE	Exp.	GBE	B	BSE	Exp.	GBE	B	BSE	Exp.	GBE
p	2.77	2.793	2.70	n	-1.71	-1.913	-1.70	Λ	-0.61	-0.613	-0.65
Σ^+	2.51	2.458	2.35	Σ^0	0.75	—	0.72	Σ^-	-1.02	-1.160	-0.92
				Ξ^0	-1.33	-1.250	-1.24	Ξ^-	-0.56	-0.6507	-0.68
Δ^+	2.07	$2.7^{\pm1.5}_{\pm1.3}$	2.08	Δ^{++}	4.14	3.7-7.5	4.17	Ω^-	-1.66	-2.0200	-1.59

Table 2. *Comparison of baryon magnetic moments (in units of μ_N) calculated within the present approach (BSE) with experimental data (Particle Data Group 2004) and the results obtained by the Graz-group on the basis of point form of relativistic quantum mechanics in a Goldstone Boson Exchange (GBE) constituent quark model (Berger et al. 2004).*

Although both approaches differ substantially both in the calculational framework and the quark dynamics adopted, the results are remarkably similar and stress the importance of a relativistically covariant calculation of electromagnetic currents.

The effects of the instanton induced correlations are illustrated by comparing in Figure 4 the electromagnetic nucleon form factors calculated with varying strength of this interaction to experimental data.

Apparently the description of the form factors could be improved by reducing the strength determined from the mass spectra. Also note that the neutron electric form factor would virtually vanish without the instanton induced correlations. See (Merten *et al.* 2002) for a further discussion.

To exemplify the relevance of the relativistic components in the Bethe-Salpeter vertex functions on form factors we present the contributions of these to the $N - \Delta$ magnetic transition form factor in Figure 5: The contribution from the dominant component containing only positive energy spinors $(+++)$ does not account for the form factor at higher momentum transfers which is correctly reproduced if those relativistic components corresponding to contributions with at least one negative energy spinor (such as $+ + -$) are considered. It is also evident that the present description fails at low momentum transfers, clearly indicating the relevance of pionic degrees of freedom not taken into account in the approach. Results on electromagnetic form factors for hyperons which should serve as guidelines for the photon couplings to be used in hadronic models for strangeness photoproduction can be found in (Cauteren *et al.* 2004).

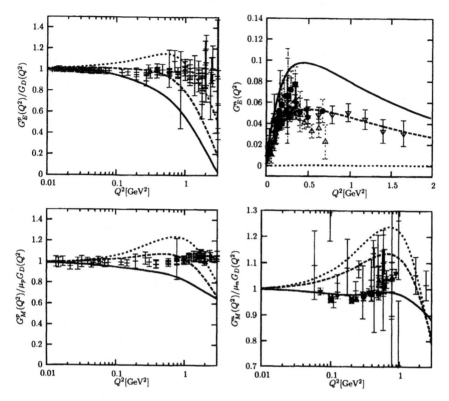

Figure 4. *Effects of the instanton induced interaction on the electromagnetic nucleon form factors. Top left: electric proton form factor normalised to a dipole; top right: electric neutron form factor; bottom left and right: magnetic form factor of the proton and the neutron normalised to a dipole, respectively. The solid, dashed and dotted curves represent the results with full, half and vanishing strength of the instanton induced interaction, respectively. Experimental data are cited in (Merten et al. 2002).*

Some results on strong two-body decays of baryons, calculated in lowest order with the baryonic and mesonic vertex functions are compared in Table 3 to experimental data and to the results obtained in an earlier constituent quark model calculation with the $^3P_0$ model for the strong decay (Capstick and Roberts 1995) and with results obtained from a relativistic meson emission calculation on the basis of the GBE quark model (Plessas 2003).

As for the magnetic moments the results from the latter model are rather similar to those obtained in the present field theoretical approach. Although the calculated partial widths are in general too small to account for the experimental values, appreciable decay widths are found only for experimentally well established resonances. The predicted values for higher lying resonances are smaller by at least an order of magnitude and thus explain why they have not been observed so far in elastic pion-nucleon scattering, in accordance with the conclusion drawn previously in (Maris and Roberts 2003, Capstick

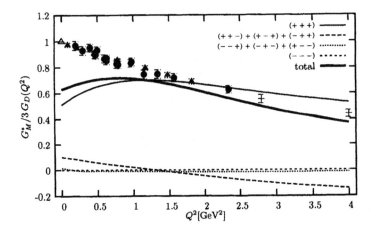

Figure 5. *Comparison of the calculated $N - \Delta$ magnetic transition form factor to the data cited in (Merten et al. 2002). The various thin curves represent the contributions from the various components in the Bethe-Salpeter vertex functions $\Gamma^{\pm\pm\pm} = \Lambda^\pm \otimes \Lambda^\pm \otimes \Lambda^\pm\Gamma$, where Λ^+ and Λ^- project onto positive and negative energy components of spinors (see also text), respectively. The thick solid line represents the full total result.*

Decay	BSE	GBE	$^3P_0$	Exp.	Decay	BSE	$^3P_0$	Exp.
$S_{11}(1535) \to N\pi$	33	93	216	$(68 \pm 15)^{+45}_{-23}$	$\to \Delta\pi$	1	2	< 2
$S_{11}(1650) \to N\pi$	3	29	149	$(109 \pm 26)^{+29}_{-4}$	$\to \Delta\pi$	5	13	$(6 \pm 5)^{+2}_{0}$
$D_{13}(1520) \to N\pi$	38	17	74	$(66 \pm 6)^{+8}_{-5}$	$\to \Delta\pi$	35	35	$(24 \pm 6)^{+3}_{-2}$
$D_{13}(1700) \to N\pi$	0.1	1	34	$(10 \pm 5)^{+5}_{-5}$	$\to \Delta\pi$	88	778	seen
$D_{15}(1675) \to N\pi$	4	6	28	$(68 \pm 7)^{+14}_{-5}$	$\to \Delta\pi$	30	32	$(83 \pm 7)^{+17}_{-6}$
$P_{11}(1440) \to N\pi$	38	30	412	$(228 \pm 18)^{+65}_{-65}$	$\to \Delta\pi$	35	11	$(88 \pm 18)^{+25}_{-25}$
$P_{33}(1232) \to N\pi$	62	34	108	$(119 \pm 0)^{+5}_{-5}$				
$S_{31}(1620) \to N\pi$	4	10	26	$(38 \pm 7)^{+8}_{-8}$	$\to \Delta\pi$	72	18	$(68 \pm 23)^{+14}_{-14}$
$D_{33}(1700) \to N\pi$	2	3	24	$(45 \pm 15)^{+15}_{-15}$	$\to \Delta\pi$	52	262	$(135 \pm 45)^{+45}_{-45}$

Table 3. *Comparison of experimental and calculated (BSE) strong $N\pi$ and $\Delta\pi$ partial decay widths $\Gamma[MeV]$ of baryon resonances. Experimental data are from (Particle Data Group 2004); (GBE) refers to the results from (Plessas 2003); $^3P_0$ refers to the results obtained in (Capstick and Roberts 1995).*

and Roberts 2000).

Some new, representative results for semi-leptonic decays, calculated from the weak baryonic currents in the Mandelstam formalism, are listed in Table 4.

These results show that in the framework of the present relativistically covariant quark model a multitude of electroweak observables of light flavoured systems can be reliably calculated.

As in the meson case we have performed a naive extension of the same quark dynamics within the present framework to the calculation of heavy flavoured systems. Some results for the masses are given in Table 5.

Decay	$\Gamma\,[10^6\mathrm{s}^{-1}]$		g_A/g_V	
	Exp.	Calc.	Exp.	Calc.
$n \to p\,e^-\bar{\nu}_e$			1.2670 ± 0.0035	1.21
$\Lambda \to p\,e^-\bar{\nu}_e$	3.16 ± 0.06	3.10	-0.718 ± 0.015	-0.82
$\Sigma^+ \to \Lambda\,e^+\nu_e$	0.25 ± 0.06	0.20		
$\Sigma^- \to \Lambda\,e^-\bar{\nu}_e$	0.38 ± 0.02	0.34		
$\Sigma^- \to n\,e^-\bar{\nu}_e$	6.9 ± 0.2	4.91	0.340 ± 0.017	0.25
$\Xi^0 \to \Sigma^+\,e^-\bar{\nu}_e$	0.93 ± 0.14	0.91	$1.32^{+0.21}_{-0.17} \pm 0.05$	1.38
$\Xi^- \to \Sigma^0\,e^-\bar{\nu}_e$	0.5 ± 0.1	0.51		
$\Xi^- \to \Lambda\,e^-\bar{\nu}_e$	3.3 ± 0.2	2.30	-0.25 ± 0.05	-0.27
$\Omega^- \to \Xi^0\,e^-\bar{\nu}_e$	68 ± 34	46		
$\Lambda \to p\,\mu^-\bar{\nu}_\mu$	0.60 ± 0.13	0.47		
$\Sigma^- \to n\,\mu^-\bar{\nu}_\mu$	3.04 ± 0.27	1.60		
$\Xi^- \to \Lambda\,\mu^-\bar{\nu}_\mu$	2.1 ± 1.3	1.04		

Table 4. *Decay rates and axial vector couplings of semi-leptonic decays of baryons. Experimental data are from (Particle Data Group 2004).*

Baryon	Calc.	Exp.	Baryon	Calc.	Exp.	Baryon	Calc.	Exp.
$\Lambda_c(\tfrac{1}{2}^+)$	2284	2285	$\Lambda_c(\tfrac{1}{2}^-)$	2605	2593	$\Lambda_c(\tfrac{1}{2}^+)$	2597	2627
$\Sigma_c(\tfrac{1}{2}^+)$	2474	2652	$\Sigma_c(\tfrac{3}{2}^+)$	2561	2517	$\Xi_c(\tfrac{1}{2}^+)$	2477	2469
$\Xi_c(\tfrac{1}{2}^+)$	2608	2576	$\Xi_c(\tfrac{3}{2}^+)$	2661	2645	$\Xi_c(\tfrac{3}{2}^-)$	2776	2815
$\Omega_c(\tfrac{1}{2}^+)$	2714	2698	$\Xi_{cc}(\tfrac{1}{2}^+)$	3561	3519?	$\Xi_{cc}(\tfrac{3}{2}^+)$	3734	
$\Omega_{cc}(\tfrac{1}{2}^+)$	3750		$\Omega_{cc}(\tfrac{1}{2}^-)$	4001		$\Omega_{ccc}(\tfrac{3}{2}^+)$	4787	

Table 5. *Comparison of calculated and experimental charmed and doubly charmed baryon masses (in MeV). Experimental data from (Particle Data Group 2004).*

In addition we quote two preliminary results for the calculation of the decay rates for semi-leptonic decays: $\Gamma[\Lambda_c^+ \to \Lambda\,e^+\,\nu_e] = 1.58\ 10^{11}\ \mathrm{s}^{-1}$ to be compared to the experimental value (Particle Data Group 2004) $(1.02 \pm 0.30)\ 10^{11}\ \mathrm{s}^{-1}$ and $\Gamma[\Lambda_c^+ \to \Lambda\,\mu^+\,\nu_\mu]$: $1.40\ 10^{11}\ \mathrm{s}^{-1}$(calc) to be compared to the experimental value $(0.97 \pm 0.34)\ 10^{11}\ \mathrm{s}^{-1}$. An extensive study of the static and electroweak properties of heavy flavoured baryons is in progress within the framework of I3HP.

4 Conclusion and outlook

Relativistically covariant constituent quark models provide a very useful tool for relating various hadron properties such as mass spectra, electroweak form factors and decay amplitudes and constitute a reliable reference frame for discriminating exotics. This holds in particular for field theoretical frameworks formulated on the basis of Bethe-Salpeter/Dyson-Schwinger equations. It has been shown that especially with instantaneous potentials (full Salpeter equation) accounting for confinement and an instanton induced spin-flavour dependent interaction all the prominent features in the mass spectra upto high masses and spins and subsequently are reproduced and a parameter-free and satisfactory calculation of amplitudes (in lowest order) is obtained for a multitude of observables for light flavoured hadrons. The formal covariance of the setup is absolutely crucial in this respect. The same conclusion has been arrived at in the context of alternative constituent quark models formulated in the framework of Dirac's point-form formulation of relativistic quantum mechanics. The same concept has also been successfully applied to the calculation of weak decays of heavy flavoured mesons and is presently, within the context of the I3HP, extended to the application to heavy flavoured baryons.

Acknowledgments

This work was done in collaboration with Herbert Petry. Calculations have been performed by Tim van Cauteren, Christian Haupt, Matthias Koll, Ulrich Löring, Dirk Merten and Sascha Migura.

References

Abe K *et al.* [Belle collaboration], 2003, hep-ex/0307021.
Athanas M *et al.* [CLEO collaboration], 1997, *Phys Rev Lett* **79** 2208.
Barish B *et al.* [CLEO collaboration], 1995, *Phys Rev D* **51** 1014.
Berger K *et al.* , 2004, nucl-th/0407009.
Capstick S and Roberts W, 2000, *Prog Part Nucl Phys* **45** S241.
Capstick S and Roberts W, 1995, *Phys Rev D***49** 4570.
Cauteren T van *et al.* , 2004, *Eur Phys J A* **20** 283.
Eidelman S *et al.* [Particle Data Group], 2004, *Phys Lett B* **592** 1.
Godfrey S and Isgur N, 1985, *Phys Rev* **32** 189.
Koll M *et al.* , 2000, *Eur Phys J A* **9** 73.
Löring U *et al.* , 2001a, *Eur Phys J A* **10** 309.
Löring U *et al.* , 2001b, *Eur Phys J A* **10** 395; *Eur Phys J A* **10** 447.
Maris M and Roberts C D, 2003, *Int J Mod Phys A* **12** 297.
Merten D *et al.* , 2002a, *Eur Phys J A* **14** 477.
Merten D *et al.* , 2002b, *Eur Phys J A* **13** 477.
Pichowsky M, 2003, *Proc. Workshop on the Physics of Excited Nucleons, NSTAR 2002, Pittsburgh* ed S A Dytman and E S Swanson (New Jersey: World Scientific) p 83.
Plessas W, 2003, *Proc. CFIF Fall Workshop 2002, Nuclear Dynamics, Lisbon Few Body Syst Suppl* **15** ed T M Pena *et al.* (Berlin: Springer) p 139.
Ricken R *et al.* , 2000, *Eur Phys J A* **9** 221.

The lattice calculation of moments of structure functions

Roger Horsley

Department of Physics and Astronomy, The University of Edinburgh,
Edinburgh, EH9 3JZ, UK,

1 Introduction

Much of our knowledge about QCD and the structure of hadrons (mainly nucleons) has been gained from Deep Inelastic Scattering (DIS) experiments such as $eN \rightarrow eX$ or $\nu N \rightarrow \mu^- X$. The (inclusive) cross sections are determined by structure functions F_1 and F_2 when summing over beam and target polarisations (and an additional F_3 when using neutrino beams), and g_1, g_2 when both the beam and target are suitably polarised. Structure functions are functions of the Bjorken variable x and Q^2, the large space-like momentum transfer from the lepton. (Another class of structure functions – the transversity h_1 – can be measured in Drell-Yan processes or certain types of semi-inclusive processes.)

As a direct theoretical computation of structure functions does not seem to be possible, we must turn to the Wilson Operator Product Expansion (OPE) which relates moments of structure functions to (nucleon) matrix elements in a twist (*ie* operator [dimension - spin]) or Taylor expansion in $1/Q^2$. So first defining bilinear quark operators

$$\mathcal{O}_q^{\Gamma;\mu_1\cdots\mu_n} = \overline{q}\Gamma^{\mu_1\cdots\mu_i}i\overset{\leftrightarrow}{D}^{\mu_{i+1}}\cdots i\overset{\leftrightarrow}{D}^{\mu_n}q\,, \tag{1}$$

where q is taken to be either a u or d quark and Γ is an arbitrary Dirac gamma matrix we have for the nucleon matrix elements, the Lorentz decompositions ($s^2 = -m_N^2$)

$$\frac{1}{2}\sum_s \langle \vec{p}, \vec{s}|\widehat{\mathcal{O}}_q^{\gamma;\{\mu_1\cdots\mu_n\}}|\vec{p}, \vec{s}\rangle = 2v_n^{(q)}\left[p^{\mu_1}\cdots p^{\mu_n} - \mathrm{tr}\right]\,,$$

$$\langle \vec{p}, \vec{s}|\widehat{\mathcal{O}}_q^{\gamma\gamma_5;\{\sigma\mu_1\cdots\mu_n\}}|\vec{p}, \vec{s}\rangle = a_n^{(q)}\left[s^{\{\sigma}p^{\mu_1}\cdots p^{\mu_n\}} - \mathrm{tr}\right]\,,$$

$$\langle \vec{p}, \vec{s}|\widehat{\mathcal{O}}_q^{\gamma\gamma_5;[\sigma\{\mu_1]\cdots\mu_n\}}|\vec{p}, \vec{s}\rangle = \frac{nd_n^{(q)}}{n+1}\left[(s^\sigma p^{\{\mu_1} - p^\sigma s^{\{\mu_1\}})p^{\mu_2}\cdots p^{\mu_n\}} - \mathrm{tr}\right]\,,$$

$$\langle \vec{p}, \vec{s} | \widehat{O}_q^{\sigma \gamma_5; \sigma \{\mu_1 \cdots \mu_n\}} | \vec{p}, \vec{s} \rangle = \frac{t_{n-1}^{(q)}}{m_N} \left[(s^\sigma p^{\{\mu_1} - p^\sigma s^{\{\mu_1\}}) p^{\mu_2} \cdots p^{\mu_n\}} - \text{tr} \right], \quad (2)$$

where the symmetrisation/anti-symmetrisation operations on the operator indices also indicates that they are traceless (which gives them a definite spin). v_n, a_n, d_n and t_n can be related to moments of the structure functions. For example we have for v_n and F_2

$$\int_0^1 dx x^{n-2} F_2(x, Q^2) = \frac{1}{3} \sum_{f=u,d,g,\ldots} E_{F_2;n}^{(f)\overline{MS}}(\mu^2/Q^2, g^{\overline{MS}}) v_n^{(f)\overline{MS}}(\mu) + O(1/Q^2),$$

(3)

and similar relations hold between g_1 and a_n; g_2 and a linear combination of a_n and d_n; h_1 and t_n. Although the OPE gives v_n from F_1 (or F_2) for $n = 2, 4, \ldots$; v_n from F_3 for $n = 3, 5, \ldots$; a_n from g_1 for $n = 0, 2, \ldots$; a_n, d_n from g_2 for $n = 2, 4, \ldots$, other matrix elements can be determined form semi-exclusive experiments, for example a_1 by measuring $\pi^\pm$ in the final state.

While the Wilson coefficients, $E^{\overline{MS}}(1, g^{\overline{MS}}(Q))$ are known perturbatively (typically two to three loops) and determine how the moments change with scale, the 'initial condition' ie the matrix element is non-perturbative in nature. The only known way of determining them from QCD in a model independent way is via Lattice Gauge Theory (LGT). In this talk we review our status (QCDSF and UKQCD Collaborations) of some aspects of these determinations including some higher twist results. A more general review may be found in (Göckeler *et al.* 2002a). We shall also restrict ourselves here to forward matrix elements (and so not consider the form factors and the more embracing Generalised Parton Densities or GPDs). Determining moments of structure functions is an active field of research at present, see for example (Dolgov *et al.* 2002, Guagnelli *et al.* 2005, Ohta *et al.* 2005).

2   The lattice approach

The lattice approach involves first Euclideanising the QCD action and then discretising space-time with lattice spacing a. The path integral then becomes a very high dimensional partition function, which is amenable to Monte Carlo methods of statistical physics. This allows ratios of three-point to two-point correlation functions to be defined,

$$R_{\alpha\beta}(t, \tau; \vec{p}) = \frac{\langle N_\alpha(t; \vec{p}) O_q(\tau) \overline{N}_\beta(0; \vec{p}) \rangle}{\langle N(t; \vec{p}) \overline{N}(0; \vec{p}) \rangle} \propto \langle N_\alpha(\vec{p}) | \widehat{O}_q | N_\beta(\vec{p}) \rangle, \quad (4)$$

where N_α is some suitable nucleon wavefunction (with Dirac index α) such as

$$N_\alpha(t; \vec{p}) = \sum_{\vec{x}} e^{-i\vec{p} \cdot \vec{x}} \epsilon^{ijk} u_\alpha^i(\vec{x}, t) [u_\beta^j(\vec{x}, t) (C\gamma_5)_{\beta\gamma} d_\gamma^k(\vec{x}, t)]. \quad (5)$$

The proportionality holds for $0 \ll \tau \ll t \lesssim \frac{1}{2} N_T$ for a lattice of size $N_S^3 \times N_T$. There are two basic types of diagrams to compute in Equation (4): the first is a quark insertion in one of the nucleon quark lines ('quark line connected'), while in the second type the operator interacts only via gluon exchange with the nucleon ('quark line disconnected'). Due to gluon UV fluctuations these latter diagrams are numerically difficult to compute.

However by considering the Non-Singlet, NS, or $\mathcal{O}_u - \mathcal{O}_d$ operators, giving matrix elements such as $v_{n;NS} = v_n^{(u)} - v_n^{(d)}$ in Equation (3) then the $f = s$ and g (gluon) terms cancel. (For higher moments however, one might expect that sea effects anyway are less significant as the integral is more weighted to $x \sim 1$.) Although LQCD is in principle an 'ab initio' calculation there are, of course, several caveats. First our lattice 'box' must be large enough to fit our correlation functions into. A continuum limit $a \to 0$ must be taken. A chiral extrapolation must be made from simulations often at the strange quark mass or larger down to the almost massless u/d quarks, or until we can match to Chiral Perturbation Theory, χPT (the problem there being that the radius of convergence of χPT is not known). Also to save CPU time, the fermion determinant in the action (representing n_f quark flavours) is often discarded - the 'quenched' approximation. Finally in addition to all the above problems the matrix element must be renormalised, in order to be able to compare with the phenomenological $\overline{MS}$ results.

To attempt to address some of these issues we have generated data sets (Bakeyev *et al.* 2004)

1. $O(a)$-improved Wilson fermions ('clover fermions') in the quenched approximation at three couplings $\beta \equiv 6/g^2 = 6.0$, 6.2 and 6.4 (Göckeler *et al* 2004) corresponding to lattice spacings $a^{-1} \sim 2.12$, 2.91 and 3.85 GeV. (This checks lattice discretisation errors, which should be $O(a^2)$.) The pseudoscalar mass, m_{ps}, lies between 580 MeV and 1200 MeV.

2. Unquenched clover fermions at m_{ps} down to ~ 560 MeV in order to see if there are any discernable quenching effects. Various couplings are used, $\beta = 5.20, 5.25, 5.29$ and 5.40 with lattice spacings ranging from $a^{-1} \sim 1.61$ GeV to 2.4 GeV.

3. Wilson fermions at one fixed lattice spacing, $a^{-1} \sim 2.12$ GeV in the quenched approximation at pseudoscalar masses, m_{ps}, down to ~ 310 MeV, $(m_{ps}/m_V \sim 0.4)$ in order to try to match to chiral perturbation theory. This lattice fermion formulation has discretisation errors of $O(a)$.

4. Overlap fermions, in the quenched approximation at one lattice spacing $a^{-1} \sim 2.09$ GeV down to m_{ps} of about 440 MeV. These have a chiral symmetry even with finite lattice spacing and hence have better chiral properties than either Wilson or clover fermions (and also have discretisation errors of $O(a^2)$).

Note that the physical pion mass is about $m_\pi \sim 140$ MeV and we use the force scale $r_0 = 0.5$ fm $\equiv (394.6 \,\text{MeV})^{-1}$ to set the scale. These results cover various patches of (m_{ps}, a, n_f) space. This is however not completely satisfactory. Overlap fermions, although the best formulation of lattice fermions known, are very expensive in CPU time, and are only just beginning to be investigated, eg (Galletly *et al.* 2003, Ohta *et al.* 2005), to which we refer the reader to for more details.

The results obtained from Equation (4) are, of course, bare results and must be renormalised. We shall not discuss this further here, just noting that many one-loop perturbative results are known; but are generally not very satisfactory as lattice perturbative series do not appear to converge very fast. (The convergence can be helped using tadpole-improvement.) A preferred non-perturbative method is also available (Martinelli *et al.* 1995) and the results presented here will have the Zs determined by this method.

All our results are for hadrons containing light (*ie* u/d) quarks. Reaching this limit is extremely costly in CPU time (and except for the overlap formulation, other problems connected with the non-chiral nature of the fermion formulation may arise). Much work has been done recently on chiral perturbation theory and it would be highly desirable to be in a region where these results can be matched to lattice results and then the limit $m_{ps} \rightarrow m_\pi$ can be taken. Although one should take the continuum and chiral limits separately (and preferably in that order) we shall try here a variant procedure of using a simultaneous 'plane' fit containing both limits. This is because at present the unquenched, data set 2 is less complete than the quenched data set 1 and this procedure at least allows for a direct comparison of results. (For set 1 these different fit procedures lead to similar results.) Practically we might thus expect that for a quantity Q of interest

$$Q = F_\chi^Q(r_0 m_{ps}) + d_s^Q(a/r_0)^s . \tag{6}$$

$F_\chi^Q(r_0 m_{ps})$ describes the (chiral) physics and the discretisation errors are $O(a^s)$ where $s = 1$ for Wilson fermions and $s = 2$ for clover fermions. Naively one might expect a Taylor series expansion for F_χ^Q to be sufficient, ie

$$F_\chi^Q(x) = Q(0) + c^Q x^2 + \dots , \tag{7}$$

where $x = r_0 m_{ps}$. Over the last few years expressions for F_χ^Q have been found

$$F_\chi^Q(x) = Q(0) \left(1 - c_\chi^Q x^2 \ln(x/r_0 \Lambda_\chi)^2 \right) + \dots , \tag{8}$$

showing the existence of a chiral logarithm $\sim m_q \ln m_q$ (including the quenched case). For $v_{n;NS}$, $a_{n;NS}$, $t_{n;NS}$ the constant c_χ^Q is known (and positive), see eg (Chen *et al.* 1997). One expects most effect of the chiral logarithm for $t_{1;NS}$ and least for $a_{0;NS}$. The chiral scale, Λ_χ, is usually taken to be $\sim 1\,\text{GeV}$. The range of validity of the expansion, Equation (8) is not known; one might expect that for $m_{ps} > \Lambda_\chi$, pion loops are suppressed, leading to a smooth variation in m_q ie constituent quarks, while for $m_{ps} < \Lambda_\chi$ non-linear behaviour would be seen. Thus building in some of the constituent or heavy quark mass expectations, an equation of the form

$$F_\chi^Q(x) = Q(0) \left(1 - c_\chi^Q x^2 \ln \frac{x^2}{(x^2 + (r_0 \Lambda_\chi)^2)} \right) + c^Q x^2 , \tag{9}$$

has been proposed (Detmold *et al.* 2001).

Present (numerically) investigated matrix elements include $v_2 \equiv \langle x \rangle$ (also part of the momentum sum rule: $\sum_q \langle x \rangle^{(q)} + \langle x \rangle^{(g)} = 1$), $v_3 \equiv \langle x^2 \rangle$, $v_4 \equiv \langle x^3 \rangle$, $a_0 = 2\Delta q$ (also occurring in neutron decay, as well as the Bjorken sum rule, as $\Delta u - \Delta d = g_A$ and connected with quark spin), $a_1 = 2\Delta q^{(2)}$, $a_2 = 2\Delta q^{(3)}$, $t_0 = 2\delta q$, $t_1 = 2\delta q^{(2)}$ and d_2. We shall only discuss v_n, $n = 1, 2, 3$, $a_0/2$, $t_0/2$ and d_2 here.

3 Results of continuum/chiral extrapolations for some twist two operators

We will now show some results, starting by considering $v_n^{\overline{MS}}(2\,\text{GeV})$ for $n = 2, 3, 4$. In Figure 1 we show $v_{2;NS}^{\overline{MS}}$ from data set 1, together with a fit using Equations (6) and (7).

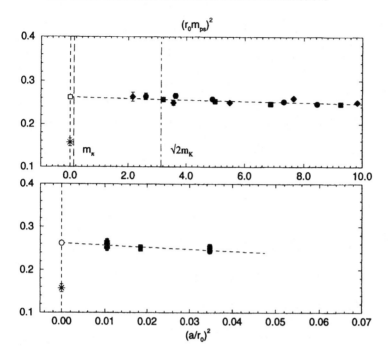

Figure 1. $v_{2;NS}^{\overline{MS}}(2\,GeV)$ versus $(r_0 m_{ps})^2$ (upper plot) and versus $(a/r_0)^2$ (lower plot) using data set 1. Filled circles, squares and diamonds represent the three lattice spacings corresponding to $\beta = 6.0, 6.2, 6.4$. The chiral limit $(r_0 m_{ps})^2 = 0$ is shown as a short-dashed line, while the physical pion mass is denoted by the long-dashed line. Also shown as a dot-dashed line is the mass of a hypothetical $\bar{s}s$ meson calculated as $\sim \sqrt{2}m_K$. The MRST phenomenological value is denoted by a star.

We see that $O(a^2)$ discretisation errors are small and seem to be relatively benign. By this we mean that the only limiting factor with the extrapolation is the amount of data available. We shall (thus) in future assume that this limit is not a problem. This does not seem to be the case with the chiral extrapolation where the data seems to strongly favour a *linear* extrapolation rather than the χPT result in Equation (8). The value found in the chiral limit is about 50% larger than the MRST phenomenological value (Martin *et al.* 2002). (Note however that there are exciting hints that overlap fermions may be closer to the phenomenological value (Galletly *et al.* 2003, Gürtler *et al.* 2004).)

The same situation persists for the higher moments $v_3^{\overline{MS}}$ and $v_4^{\overline{MS}}$. In Figures 2 and 3 we show these moments and compare the results with the MRS phenomenological values. In all cases we find that the moments are too large in comparison with phenomenological result. It is not clear why this is so, again the quarks seem to be acting more like constituent quarks rather than current quarks. Possible causes are quenching and/or a chiral extrapolation from too heavy a quark mass. We first consider possible quenching effects. In Figure 4 we consider $v_{2;NS}^{\overline{MS}}$ again, but this time for $n_f = 2$ flavours using data set 2. No real difference is seen in comparison to the quenched case. Indeed for other matrix elements considered a similar situation prevails.

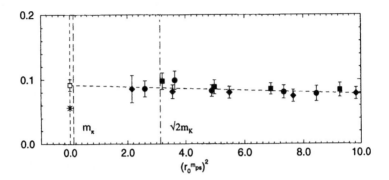

Figure 2. $v_{3;NS}^{\overline{MS}}(2\,GeV)$ *versus* $(r_0 m_{ps})^2$. *Same notation as in Figure 1.*

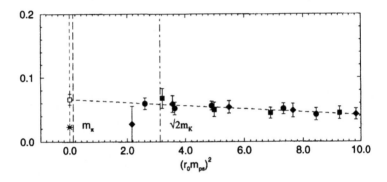

Figure 3. $v_{4;NS}^{\overline{MS}}(2\,GeV)$ *versus* $(r_0 m_{ps})^2$. *Same notation as in Figure 1.*

To try to examine the situation at smaller quark mass, we now turn to the data set 3. Most of the above results have a quark mass at the strange quark mass (or heavier). In this data set we have generated quenched Wilson data at one lattice spacing, at light pion masses down to 310 MeV. In Figure 5 we show $v_{2;NS}^{\overline{MS}}(2\,GeV)$. In comparison with the previous pictures note that the x-scale only runs to $(r_0 m_{ps})^2 \sim 3.0$. Again (except possibly for the lightest pion mass) the data seems rather linear (and constant). Also shown is a forced fit from Equation (9), leaving Λ_χ and $c_\chi^{v_2}$ free but constrained to go through the MRST phenomenological value at $m_{ps} = m_\pi$. Ignoring the lightest quark mass point, this is just possible; however it is very unnatural giving, for example, $\Lambda_\chi \sim 500\,GeV$ which is a very low value.

A similar situation holds for the axial $a_{0;NS}/2 = g_A$ and tensor charge $t_{0;NS}/2$ in Figures 6 and 7.

Again the results in both cases seem very linear. Indeed from Equation (9) due to the negative sign, we must have the lattice data decreasing to the phenomenological value. This is certainly not the case here (although experimentally $t_{0;NS}$ is not known, we expect a similar situation as for g_A; indeed in the non-relativistic limit $t_{0;NS}/2 \to g_A$). Later work including the Δ as well as the N in chiral perturbation theory (Detmold *et al.* 2002, Hemmert *et al.* 2003) reduce the $c_\chi^{g_A}$ coefficient, but this still is a problem. We

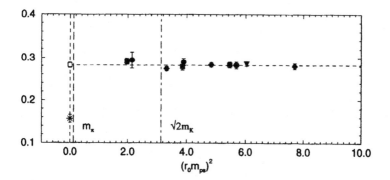

Figure 4. $v_{2;NS}^{\overline{MS}}(2\,GeV)$ *versus* $(r_0 m_{ps})^2$ *for unquenched fermions using data set 2.* $\beta = 5.20$ *results are (filled) circles; 5.25 squares; 5.29, diamonds; 5.40 down triangle. Same notation as in Figure 1.*

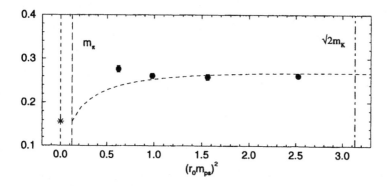

Figure 5. $v_{2;NS}^{\overline{MS}}(2\,GeV)$ *versus* $(r_0 m_{ps})^2$ *for Wilson fermions from data set 3. The 'fit' uses Equation (9). Otherwise the same notation as in Figure 1.*

show only linear fits, giving for g_A a value somewhat lower than the experimental one. (There are two possible caveats: for clover fermions the continuum extrapolation may have significant $O(a^2)$ effects, in distinction to $v_{n;NS}$ and also there may or may not be larger finite volume effects present both in the data and theoretically, see for example the discussion in (Cohen 2001).) But at present the same general picture emerges as for the unpolarised moments.

4 Some higher twist operator results

4.1 Twist three

The prime example is given by d_2, which can be determined from g_2, the first moment of which is a linear combination of a_2 and d_2. The operators for the a_n moments have twist two, but d_n corresponds to twist three and is thus of particular interest. A "straightfor-

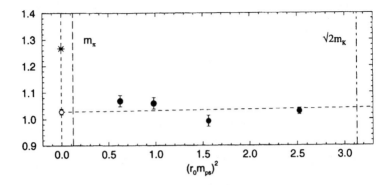

Figure 6. $a_{0;NS}/2 = g_A$ versus $(r_0 m_{ps})^2$ for Wilson fermions (data set 3), with a linear fit. Otherwise the same notation as in Figure 1.

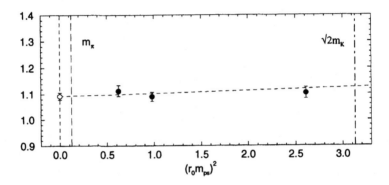

Figure 7. $t_{0;NS}/2$ versus $(r_0 m_{ps})^2$ using data set 3. The same notation as in Figure 6.

ward" lattice computation (Göckeler *et al.* 1996) gave rather large values for d_2^p (where $d_2^{(p)} = Q^{(u)2} d_2^{(u)} + Q^{(d)2} d_2^{(d)}$). A recent experiment (E155 Collaboration 1999) however indicated that this term was very small, which would mean that g_2 is almost completely determined by g_1 (the Wandzura–Wilczek relation). This problem was traced in (Göckeler *et al.* 2000a), to a mixing of the original operator with a lower-dimensional operator. This additional operator mixes $\propto 1/a$ and so its renormalisation constant must be determined non-perturbatively. In (Göckeler *et al.* 2000a) this procedure was attempted, and led to results qualitatively consistent with the experimental values. Note that this is only a problem when using Wilson or clover like fermions, as we would expect the additional operator to appear like $\sim m_q \bar{q} \sigma \overset{\leftrightarrow}{D} q$ and hence vanish in the chiral limit. Thus there should be no mixing if one uses overlap fermions.

4.2 Twist four

Potential higher twist effects are present in the moment of a structure function, see Equation (3). These $O(1/Q^2)$ terms are composed of dimension 6, four quark matrix elements. A general problem is the non-perturbative mixing of these operators with the previous dimension 4 operators. At present results are restricted to finding combinations of these higher twist operators which do not mix from flavour symmetry. For the nucleon the $SU_F(3)$ flavour symmetry group must be considered, ie taking mass degenerate u, d and s quarks (Göckeler *et al.* 2002b) giving

$$\int_0^1 dx F_2(x, Q^2)\big|^{27, I=1}_{Nachtmann} = -0.0005(5)\frac{m_p^2 \alpha_s(Q^2)}{Q^2} + O(\alpha_s^2), \qquad (10)$$

for quenched Wilson fermions (ie part of data set 3). To access this moment experimentally needs very exotic combinations of moments from the measurement of the p, n, Λ, Σ and Ξ baryons and is not possible. Nevertheless this term is very small in comparison with the leading twist result, and might hint that higher twist contributions are small.

5   Miscellaneous pion and lambda results

Moments for the pion and rho structure functions were computed in (Best *et al.* 1997) for unimproved Wilson fermions. Using the Schrödinger Functional method, v_2 was recently calculated for the pion (Guagnelli *et al.* 2005) for both unimproved and $O(a)$-improved fermions. A higher twist matrix element for the pion has also been computed for quenched Wilson fermions (*ie* using part of data set 3)

$$\int_0^1 dx F_2(x, Q^2)\big|^{I=2}_{Nachtmann} = 1.67(64)\frac{f_\pi^2 \alpha_s(Q^2)}{Q^2} + O(\alpha_s^2), \qquad (11)$$

where the $SU_F(2)$ flavour symmetry group gives the combination $F_2^{I=2} = F_2^{\pi^+} + F_2^{\pi^-} - 2F_2^{\pi^0}$. This is again a rather small number, so although a rather exotic combination of pion matrix elements it might indicate that higher twist terms are small.

Finally there have been results for moments of Λ structure functions (Göckeler *et al.* 2002b) again using Wilson fermions (*ie* part of data set 3). These are potentially useful results as one can compare with nucleon spin structure and check violation of $SU_F(3)$ symmetry. First indications are that there is no evidence of flavour symmetry breaking in the matrix elements, *ie* that Λ and p are related by an $SU(3)_F$ flavour transformation.

6   Conclusions

Clearly the computation of many matrix elements giving low moments of structure functions is possible. We would like to emphasise that a successful computation is a fundamental test of QCD – this is not a model computation. There are however many problems to overcome: finite volume effects, renormalisation and mixing, continuum and chiral extrapolations and unquenching. At present although overall impressions are encouraging, it is still difficult to reproduce experimental/phenomenological results of (relatively) simple matrix elements, *eg* v_2. But progress is being made by the various groups working in the field. For example in comparison to our previous results (Göckeler *et al.* 1996) there are now non-perturbative Zs and considerations of both chiral and continuum extrapolations and some unquenched results are now available. While the continuum extrapolation seems to be 'just' a matter of more data at smaller lattice spacing, the chiral extrapolation does seem to present a problem, with no sign of any chiral logarithms being seen as predicted by χPT. Clearly everything depends on the data and the quest for better results should continue. To leave the region where constituent quark masses give a reasonable description of the data (*ie* linearity) unfortunately requires pion masses rather close to the physical pion mass. In this region fermions with better chiral properties will probably be needed, such as overlap, which in turn will need much faster machines.

Acknowledgments

I wish to thank my co-workers in the QCDSF and UKQCD Collaborations: A Ali Khan, T Bakeyev, D Galletly, M Göckeler, M Gürtler, P Hägler, T R Hemmert, A C Irving, B Joó, A D Kennedy, B Pendleton, H Perlt, D Pleiter, P E L Rakow, A Schäfer, G Schierholz, A Schiller, W Schroers, T Streuer, H Stüben, V Weinberg and J M Zanotti for a pleasant and profitable collaboration.

The numerical calculations have been performed on the Hitachi SR8000 at LRZ (Munich), on the Cray T3E at EPCC (Edinburgh) (Allton *et al.* 2002) on the Cray T3E at NIC (Jülich) and ZIB (Berlin), as well as on the APE1000 and Quadrics (QH2b) at DESY (Zeuthen). We thank all institutions. This work has been supported in part by the EU Integrated Infrastructure Initiative Hadron Physics, contract number RII3-CT-2004-506078 and by the DFG (Forschergruppe Gitter-Hadronen-Phänomenologie).

References

Allton C R *et al.* , 2002, *Phys Rev D* **65** 054502.
Bakeyev T *et al.* , 2004, *Nucl Phys Proc Suppl* **128** 82.
Best C *et al.* , 1997, *Phys Rev D* **56** 2743.
Chen J-W *et al.* , 1997, *Phys Lett B* **523** 107.
Cohen T D, 2001, *Phys Lett B* **529** 50.
Detmold W *et al.* , 2001, *Phys Rev Lett* **87** 172001.
Detmold W *et al.* , 2002, *Phys Rev D* **66** 054501.
Dolgov D *et al.* , 2002, *Phys Rev D* **66** 034506.
E155 Collaboration, 1999, *Phys Lett B* **458** 529.
Galletly D *et al.* , 2003, *Nucl Phys Proc Suppl* **129** 453.
Göckeler M *et al.* , 1996, *Phys Rev D* **53** 2317.
Göckeler M *et al.* , 2000a, *Phys Rev D* **63** 074506.
Göckeler M *et al.* , 2000b, *Nucl Phys B* **623** 287.
Göckeler M *et al.* , 2002a, *Nucl Phys Proc Suppl B* **119** 398.
Göckeler M *et al.* , 2002b, *Phys Lett B* **545** 112.
Göckeler M *et al.* , 2004, hep-ph/0410187.
Guagnelli M *et al.* , 2005, *Eur Phys J C* **40** 69.
Gürtler M *et al.* , 2004, hep-lat/0409164.
Hemmert T R *et al.* , 2003, *Phys Rev D* **68** 075009.
Martin A D *et al.* , 2002, *Eur Phys J C* **23** 73.
Martinelli G *et al.* , 1995, *Nucl Phys B* **445** 81.
Ohta S *et al.* , 2005, *Nucl Phys Proc Suppl* **140** 396.

Hadron tomography

Matthias Burkardt

Department of Physics, New Mexico State University,
Las Cruces, NM 88003, USA

1 Introduction

Generalised parton distributions (GPDs) have attracted significant interest since it has been recognised that they can not only be probed in deeply virtual Compton scattering experiments but can also be related to the orbital angular momentum carried by quarks in the nucleon (Ji 1997). However, remarkably little is still known about the physical interpretation of GPDs. In this talk, I want to address the question: *suppose, about 10 years from now, after a combined effort from experiment and theory, we know how these functions (ie GPDs) look like for the nucleon. What is it, in simple physical terms, that we will have learned about the structure of the nucleon?* (Above and beyond having learned something about the orbital angular momentum carried by the quarks.) In particular, I want to discuss another interesting piece of information that can be extracted from GPDs, namely *how partons are distributed in the transverse plane* and some possible connection with transverse single-spin asymmetries.

In nonrelativistic quantum mechanics, the physics of form factors is illucidated by transforming to the centre-of-mass frame and by interpreting the Fourier transform of form factors as charge distributions in that frame.

GPDs are the form factors of the same operators [light cone correlators $\hat{O}_q(x, \mathbf{0}_\perp)$] whose forward matrix elements also yield the usual (forward) parton distribution functions (PDFs). For example, the unpolarised PDF $q(x)$ can be expressed in the form

$$q(x) = \langle p, \lambda | \hat{O}_q(x, \mathbf{0}_\perp) | p, \lambda \rangle, \tag{1}$$

while the GPDs $H_q(x, \xi, t)$ and $E_q(x, \xi, t)$ are defined as (Ji 1997, Radyushkin 1997)

$$\langle p', \lambda' | \hat{O}_q(x, \mathbf{0}_\perp) | p, \lambda \rangle = \bar{u}(p', \lambda') \left(\gamma^+ H_q(x, \xi, t) + i \frac{\sigma^{+\nu} \Delta_\nu}{2M} E_q(x, \xi, t) \right) u(p, \lambda), \tag{2}$$

where $\Delta = p' - p$, $2\overline{p} = p + p'$, $t = \Delta^2$, $2\overline{p}^+\xi = \Delta^+$, and

$$\hat{O}_q(x, \mathbf{b}_\perp) = 2\overline{p}^+ \int \frac{dx^-}{4\pi} \overline{q}\left(-\frac{x^-}{2}, \mathbf{b}_\perp\right) \gamma^+ q\left(\frac{x^-}{2}, \mathbf{b}_\perp\right) e^{ix\overline{p}^+ x^-}. \tag{3}$$

Throughout this work we will suppress the Q^2-dependence of these matrix elements for notational convenience. In the end, the $\perp$ 'resolution' will be limited by $1/Q$.

In the case of form factors, non-forward matrix elements provide information about how the charge (*ie* the forward matrix element) is distributed in position space. By analogy with form factors, one would therefore expect that the additional information (compared to PDFs) contained in GPDs helps to understand how the usual PDFs are distributed in position space. Of course, since the operator $\hat{O}_q(x, \mathbf{0}_\perp)$ already 'filters out' quarks with a definite momentum fraction x, Heisenberg's uncertainty principle does not allow a simultaneous determination of the partons' longitudinal position, but determining the distributions of partons in impact parameter space is conceivable. Making this intuitive expectation more precise (*eg* what is the 'reference point', 'are there relativistic corrections', 'how does polarisation enter', 'is there a strict probability interpretation') will be the main purpose of these notes.

2   Impact parameter dependent parton distributions

In nonrelativistic quantum mechanics, the Fourier transform of the form factor yields the charge distribution in the centre-of-mass frame. In general, the concept of a centre of mass has no analogy in relativistic theories, and therefore the position space interpretation of form factors is frame dependent.

The infinite momentum frame (IMF) plays a distinguished role in the physical interpretation of regular PDFs as momentum distributions in the IMF. It is therefore natural to attempt to interpret GPDs in the IMF. This task is facilitated by the fact that there a is Galilean subgroup of transverse boosts in the IMF, which allows us to introduce a relativistic analogy to the nonrelativistic centre of mass in this frame (Soper 1977). For an eigenstate of P^+, one can define a (transverse) *centre of momentum* (CM)

$$\mathbf{R}_\perp \equiv -\frac{\mathbf{B}_\perp}{p^+} = \frac{1}{p^+} \int dx^- \int d^2\mathbf{x}_\perp T^{++}\mathbf{x}_\perp, \tag{4}$$

where $T^{\mu\nu}$ is the energy momentum tensor. Like its nonrelativistic counterpart, it satisfies $[J_3, R_k] = i\varepsilon_{kl}R_l$ and $[P_k, R_l] = -i\delta_{kl}$. These simple commutation relations enable us to form simultaneous eigenstates of $\mathbf{R}_\perp$ (with eigenvalue $\mathbf{0}_\perp$), P^+ and J_3

$$|p^+, \mathbf{R}_\perp = \mathbf{0}_\perp, \lambda\rangle \equiv \mathcal{N} \int d^2\mathbf{p}_\perp |p^+, \mathbf{p}_\perp, \lambda\rangle, \tag{5}$$

where $\mathcal{N}$ is some normalisation constant, and λ corresponds to the helicity when viewed from a frame with infinite momentum. For details on how these IMF helicity states are defined, as well as for their relation to usual rest frame states, see (Soper 1972).

In the following we will use the eigenstates of the $\perp$ centre-of-momentum operator (Equation 5) to define the concept of a parton distributions in impact parameter space

$$q(x, \mathbf{b}_\perp) \equiv \langle p^+, \mathbf{R}_\perp = \mathbf{0}_\perp, \lambda| \hat{O}_q(x, \mathbf{b}_\perp) |p^+, \mathbf{R}_\perp = \mathbf{0}_\perp, \lambda\rangle. \tag{6}$$

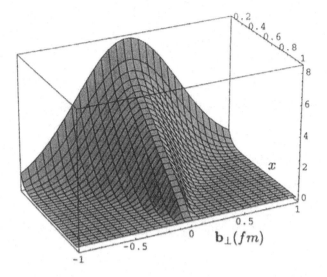

Figure 1. *Expected behaviour of parton distributions in impact parameter space* $q(x, \mathbf{b}_\perp)$.

It is straightforward to verify that the impact parameter dependent PDFs defined above (Equation 6) are the Fourier transform of H_q without any relativistic corrections (Burkardt 2000, Burkardt 2003, Diehl 2003)

$$q(x, \mathbf{b}_\perp) = |\mathcal{N}|^2 \int \frac{d^2\mathbf{p}_\perp}{2\pi} \int \frac{d^2\mathbf{p}'_\perp}{2\pi} \left\langle p^+, \mathbf{0}_\perp, \lambda \left| \hat{O}_q(x, \mathbf{b}_\perp) \right| p^+, \mathbf{0}_\perp, \lambda \right\rangle \quad (7)$$

$$= |\mathcal{N}|^2 \int \frac{d^2\mathbf{p}_\perp}{2\pi} \int \frac{d^2\mathbf{p}'_\perp}{2\pi} H_q(x, 0, -(\mathbf{p}'_\perp - \mathbf{p}_\perp)^2) e^{i\mathbf{b}_\perp \cdot (\mathbf{p}_\perp - \mathbf{p}'_\perp)}$$

$$= \int \frac{d^2\mathbf{\Delta}_\perp}{(2\pi)^2} H_q(x, 0, -\mathbf{\Delta}_\perp^2) e^{-i\mathbf{b}_\perp \cdot \mathbf{\Delta}_\perp}$$

and its normalisation is $\int d^2\mathbf{b}_\perp q(x, \mathbf{b}_\perp) = q(x)$. A similar interpretation exists for $\tilde{H}_q(x, 0, t)$ in terms of impact parameter dependent polarised quark distributions

$$\Delta q(x, \mathbf{b}_\perp) = \int \frac{d^2\mathbf{\Delta}_\perp}{(2\pi)^2} \tilde{H}_q(x, 0, -\mathbf{\Delta}_\perp^2) e^{-i\mathbf{\Delta}_\perp \cdot \mathbf{b}_\perp}. \quad (8)$$

Furthermore, $q(x, \mathbf{b}_\perp)$ has a probabilistic interpretation. With $\tilde{b}_s(k^+, \mathbf{b}_\perp)$ [$\tilde{d}_s(k^+, \mathbf{b}_\perp)$] as the canonical destruction operator for a quark [antiquark] with longitudinal momentum k^+ and $\perp$ position $\mathbf{b}_\perp$, one finds (Burkardt 2001)

$$q(x, \mathbf{b}_\perp) = \begin{cases} \sum_s \left| \tilde{b}_s(xp^+, \mathbf{b}_\perp) | p^+, \mathbf{0}_\perp, \lambda \rangle \right|^2 \geq 0 & \text{for } x > 0 \\ -\sum_s \left| \tilde{d}_s^\dagger(xp^+, \mathbf{b}_\perp) | p^+, \mathbf{0}_\perp, \lambda \rangle \right|^2 \leq 0 & \text{for } x < 0 \end{cases} \quad (9)$$

For large x, one expects $q(x, \mathbf{b}_\perp)$ to be not only small in magnitude (since $q(x)$ is small for large x) but also very narrow (localised valence core!). In particular, the $\perp$ width should vanish as $x \to 1$, since $q(x, \mathbf{b}_\perp)$ is defined with the $\perp$ CM as a reference point. A parton representation for $\mathbf{R}_\perp$ (4) is given by $\mathbf{R}_\perp = \sum_{i \in q,g} x_i \mathbf{r}_{i,\perp}$, where x_i $(\mathbf{r}_{i,\perp})$ is the momentum fraction ($\perp$ position) of the ith parton, and for $x = 1$ the position of active quark coincides with the $\perp$ CM.

3   Position space interpretation for $E(x, 0, -\mathbf{\Delta}_\perp^2)$

While both $H(x, 0, t)$ and $\tilde{H}(x, 0, t)$ are diagonal in helicity, $E(x, 0, t)$ contributes only to helicity flip matrix elements. In fact for $p^+ = p^{+\prime}$ (*ie* $\xi = 0$)

$$\int \frac{dx^-}{4\pi} e^{ip^+ x^- x} \left\langle P{+}\Delta,\uparrow \left| \bar{q}(0) \gamma^+ q(x^-) \right| P,\uparrow \right\rangle = H(x,0,-\mathbf{\Delta}_\perp^2) \,, \tag{10}$$

$$\int \frac{dx^-}{4\pi} e^{ip^+ x^- x} \left\langle P{+}\Delta,\uparrow \left| \bar{q}(0) \gamma^+ q(x^-) \right| P,\downarrow \right\rangle = -\frac{\Delta_x - i\Delta_y}{2M} E(x,0,-\mathbf{\Delta}_\perp^2) \,. $$

Since E is off-diagonal in helicity, it will only have a nonzero expectation value in states that are *not* eigenstates of helicity, *ie* if we search for a probabilistic interpretation for $E(x, 0, t)$ we need to look for it in states that are superpositions of helicity eigenstates. For this purpose, we consider the state

$$|X\rangle \equiv \left| p^+, \mathbf{R}_\perp = 0, X \right\rangle \equiv \left(\left| p^+, \mathbf{R}_\perp = 0, \uparrow \right\rangle + \left| p^+, \mathbf{R}_\perp = 0, \downarrow \right\rangle \right) / \sqrt{2}. \tag{11}$$

In this state, we find for the (unpolarised) impact parameter dependent PDF

$$q_X(x, \mathbf{b}_\perp) \equiv \langle X | O_q(x, \mathbf{b}_\perp) | X \rangle \tag{12}$$

$$= \int \frac{d^2 \mathbf{\Delta}_\perp}{(2\pi)^2} \left[H_q(x, 0, -\mathbf{\Delta}_\perp^2) + \frac{i\Delta_y}{2M} E_q(x, 0, -\mathbf{\Delta}_\perp^2) \right] e^{-i\mathbf{b}_\perp \cdot \mathbf{\Delta}_\perp} \,. $$

Upon introducing the Fourier transform of E_q

$$\mathcal{E}_q(x, \mathbf{b}_\perp) \equiv \int \frac{d^2 \mathbf{\Delta}_\perp}{(2\pi)^2} E_q(x, 0, -\mathbf{\Delta}_\perp^2) e^{-i\mathbf{b}_\perp \cdot \mathbf{\Delta}_\perp} \tag{13}$$

we thus conclude that E_q describes how the unpolarised PDF in the $\perp$ plane gets distorted when the nucleon target is polarised in a direction other than the z direction

$$q_X(x, \mathbf{b}_\perp) = q(x, \mathbf{b}_\perp) - \frac{1}{2M} \frac{\partial}{\partial b_y} \mathcal{E}_q(x, \mathbf{b}_\perp). \tag{14}$$

Due to the kinematics of DIS, it is the $j^+ = j^0 + j^z$ density of the quarks which couples to the electron: the electron in DIS couples more strongly to quarks which move towards the electron rather than away from it because if the quarks move towards (collision course) the electron the electric and magnetic forces add up, while if they move away the electric and magnetic forces act in opposite directions. For relativistic particles electric and magnetic forces are of the same magnitude. As a consequence, if the $\hat{z}$ axis is in the direction of the momentum of the virtual photon then the virtual photon couples only to the j^+ component of the quark current. Even though the j^0 component of the

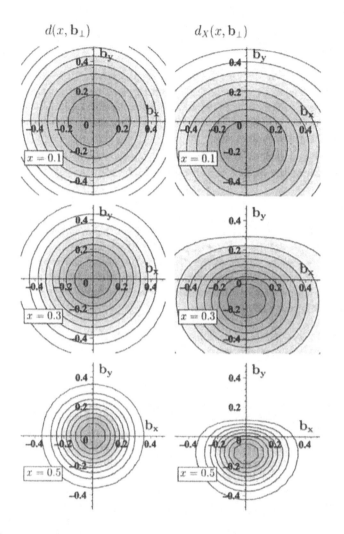

Figure 2. *Impact parameter dependent PDF for d quarks $d(x, \mathbf{b}_\perp)$ for $x = 0.1, 0.3,$ 0.5. Left column: unpolarised; right column: $d_X(x, \mathbf{b}_\perp)$ for proton polarised in $-x$ direction. The distributions are normalised to the central value $d(x, 0_\perp)$. For u quark the distortion is smaller and with opposite sign.*

current density is the same on the $+\hat{y}$ and $-\hat{y}$ sides of the nucleon, the j^z component has opposite signs on the $+\hat{y}$ and $-\hat{y}$ sides if the quarks have orbital angular momentum. Therefore the reason for the distortion is a combination of the fact that the electron 'sees' oncoming quarks better and the presence of orbital angular momentum.

Unfortunately, until more is known about $E(x, 0, -\Delta_\perp^2)$, one has to rely on model calculations in order to provide further details. However, the average $\perp$ flavour dipole moment can be related to the anomalous magnetic moment in a model independent way

$$d_q^y \equiv \int dx \int d^2\mathbf{b}_\perp b_y q_X(x,\mathbf{b}_\perp) = \frac{\int dx \int d^2\mathbf{b}_\perp \mathcal{E}_q(x,\mathbf{b}_\perp)}{2M} = \frac{\int dx E_q(x,0,0)}{2M} = \frac{\kappa_q}{2M} \quad (15)$$

where $e_q\kappa_q$ is the contribution from flavour q to the anomalous Dirac moment $F_2(0)$. In order to get some feeling for the order of magnitude, we consider a very simple model where only $q = u, d$ contribute to $F_2(0)$, one finds for example $\kappa_d \approx -2$ and therefore a mean displacement of d quarks of by about $0.2 fm$. For u quarks the effect is about half as large and in the opposite direction.

As a further illustration, we extend the model from the previous section to $E_q(x,0,t)$, by making the ansatz [the factor $\frac{1}{2}$ accounts for $\int dx H_u(x,0,0) = 2$]

$$E_u(x,0,t) = \frac{1}{2}\kappa_u H_u(x,0,t) \quad , \quad E_d(x,0,t) = \kappa_d H_d(x,0,t). \quad (16)$$

Results for d-quarks for this (oversimplified) model are shown in Figure 2. Of course, for a more realistic model (for a recent fit see Diehl 2004), the results will look different, and we will not know the details of the distortion until GPDs have been measured in experiments or in lattice QCD. However, even in the absence of detailed knowledge about GPDs, we have one model-independent constraint of the average magnitude of the distortion, which is provided by Equation 15.

4 Transverse single-spin asymmetries

Recently, the HERMES collaboration found a significant single-spin asymmetry (SSA), *ie* a left-right asymmetry in the transverse momentum distribution of produced mesons in the directions perpendicular to the nucleon spin, in semi-inclusive production of π and K mesons (HERMES 2004). Theoretically, two mechanisms have been proposed to explain such an asymmetry: the Sivers mechanism, where the final state interactions (FSI) give rise to a $\perp$ momentum asymmetry of the active quark already before it fragments into hadrons (Sivers 1991), and the Collins mechanism, where the asymmetry arises when a $\perp$ polarised quark fragments into mesons (Collins 2002 and references therein). The two mechanisms can be disentangled by also measuring the scattering plane of the electron and both have been found to contribute (HERMES 2004). Simple model calculations (Brodsky 2002) have revealed that, even at high energies, the FSI can indeed give rise to a non-vanishing transverse single-spin asymmetry for the active quark (Sivers mechanism). Those FSI can formally be incorporated into a definition of unintegrated parton densities by introducing appropriate Wilson line phase factors (Efremov 1981, Ji 2002, Collins 2003)

$$q(x,\mathbf{k}_\perp) = \int \frac{dy^- d^2\mathbf{y}_\perp}{16\pi^3} e^{-ixp^+y^- + i\mathbf{k}_\perp \cdot \mathbf{y}_\perp} \langle p,s|\bar{q}_U(y)U_{[\infty^-,\mathbf{y}_\perp,\infty^-,0_\perp]}\gamma^+ q_U(0)|p,s\rangle \quad (17)$$

with $q_U(0) \equiv U_{[\infty^-,0_\perp;0^-,0_\perp]}q(0)$ and $\bar{q}_U(y) \equiv \bar{q}(y)U_{[y^-,\mathbf{y}_\perp;\infty^-,\mathbf{y}_\perp]}$. The Us are path-ordered Wilson line gauge links, for example

$$U_{[0;\xi]} = P \exp\left(ig \int_0^1 ds\xi_\mu A^\mu(s\xi)\right) \quad (18)$$

connecting the points 0 and ξ. The choice of path for the gauge string in Equation 17 is not arbitrary but reflects the FSI along the path of the outgoing quark. Since the active quark in DIS is ultra-relativistic, this path goes from the position of the active quark to infinity along the light-cone. In addition, the two Wilson lines need to be connected at light-cone infinity. The choice of path for the segment at $x^- = \infty$ is arbitrary as long as the gauge field at $x^- = \infty$ is pure gauge. While in general all three gauge links in Figure 3 contribute, the gauge link segment at $x^- = \infty$ is important only in light-cone gauge (Belitsky 2003).

The Sivers mechanism is interesting for a variety of reasons: It vanishes under naive time-reversal and a non-trivial phase from the FSI is needed to give a nonzero effect. Therefore the Sivers mechanism provides information about the space-time structure of the target. Furthermore a nonzero Sivers mechanism implies a nonzero Compton amplitude involving nucleon helicity flip without quark helicity flip, ie it requires (nonperturbative) helicity nonconservation in the nucleon state (χSB!). Finally, in model calculation the mechanism requires interference between phases of wavefunction components that differ by one unit of orbital angular momentum and therefore the effect may provide novel insights about orbital angular momentum in the proton.

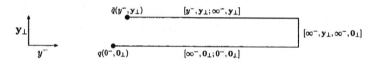

Figure 3. *Illustration of the gauge links in gauge invariant Sivers distributions (Equation 17).*

Although the introduction of Wilson line phase factors in gauge invariant parton densities has helped to understand why there can be a nonzero Sivers mechanism at high energies, it has at the same time obscured the underlying physics of the mechanism: The asymmetry arises from interference between phase factors from different partial waves and in addition requires a nontrivial phase contribution from the Wilson line. In Equation 17, time reversal invariance no longer implies a vanishing asymmetry. Indeed, under time reversal the direction of the gauge link changes and the only consequence of time reversal invariance for Equation 17 is opposite signs for the asymmetries in semi-inclusive pion production and Drell-Yan experiments (Collins 2002).

We will now illustrate how the Wilson line, together with the nucleon ground state wavefunction, conspires to provide a transverse asymmetry for the average transverse momentum for quarks of flavour q. Using integration by parts, it is straightforward to evaluate $\langle \mathbf{k}_{\perp q} \rangle \equiv \int dx \int d^2 \mathbf{k}_\perp q(x, \mathbf{k}_\perp) \mathbf{k}_\perp$ from Equation 17. Dropping all terms that vanish because of time-reversal invariance, this yields (Qiu 1991, Schäfer 1993, Boer 2003)

$$\langle \mathbf{k}_\perp \rangle \propto \left\langle p, s \left| \bar{q}(0)\gamma^+ \int_0^\infty d\eta^- G^{+\perp}(\eta)q(0) \right| p, s \right\rangle . \tag{19}$$

This result has a simple semi-classical interpretation: the gluon field strength tensor $G^{+\perp}(\eta)$ yields the $\perp$ component of the force from the spectators on the active quark. Integrating this force along the trajectory (for a ultrarelativistic particle, time=distance) of the outgoing quark then yields the $\perp$ impulse $\int_0^\infty d\eta^- G^{+\perp}(\eta)$ which the active quarks acquires from the FSI as it escapes the hadron. The average $\mathbf{k}_\perp$ is then obtained by correlating the quark density with the $\perp$ impulse. However, although this result nicely illustrates the physics of the contribution from the Wilson lines, it still does not tell us the sign/magnitude of the asymmetry. Indeed, early estimates for Equation 19 concluded that the resulting asymmetry should be very small (Schäfer 1993).

There are several reasons why one is interested to proceed with light-cone gauge $A^+ = 0$. First of all this is the most physical gauge for a light-cone description of hadrons, which is in turn the natural framework to describe DIS. Secondly, all light-like Wilson lines become trivial in this gauge. Furthermore, there exists already a rich phenomenology for light-cone wavefunctions of hadrons.

However, if one neglects the gauge link at $x^- = \infty$ in Equation 17 then the $\perp$ asymmetry vanishes in $A^+ = 0$ gauge: as we mentioned above the light-like gauge links are trivial in this gauge and without the gauge link at $x^- = \infty$, Equation 17 yields a vanishing asymmetry due to time-reversal invariance. In fact, as was revealed by explicit model calculations in (Brodsky 2002), very careful regularisation of the light-cone zero-modes (Burkardt 1996 and references therein) is required if one wants to calculate the asymmetry in $A^+ = 0$ gauge.

In Burkardt (2004a) it was demonstrated that in light-cone gauge the effect from the gauge link at $x^- = \infty$ is a nonzero Sivers effect which reads (to lowest order in g)

$$\langle k_q^i \rangle = -\frac{g}{4p^+} \int \frac{d^2 \mathbf{y}_\perp}{2\pi} \frac{y^i}{|\mathbf{y}_\perp|^2} \left\langle p, s \left| \bar{q}(0)\gamma^+ \frac{\lambda_a}{2} q(0)\rho_a(\mathbf{y}_\perp) \right| p, s \right\rangle, \qquad (20)$$

where $\rho_a(\mathbf{x}_\perp)$ is the total (quark+glue) colour charge density integrated along x^-

$$\rho_a(\mathbf{x}_\perp) = g \int dx^- \left[\bar{q}\gamma^+ \frac{\lambda_a}{2} q + f_{abc} A_b^i G_c^{i+} \right]. \qquad (21)$$

This result has again a very physical interpretation: the average $\mathbf{k}_\perp$ is obtained by summing over the $\perp$ impulse caused by the colour-Coulomb field (since we solved the constraint equations only to lowest order) of the spectators. Equation 20 is also very useful for practical evaluation of the Sivers effect from light-cone wavefunctions. Equation 17 involves the gauge field at $x^- = \infty$, which is very sensitive to the regularisation procedure, we have succeeded to express the asymmetry in terms of degrees of freedom at finite x^-, ie $\perp$ colour density-density correlations in the $\perp$ plane. Equation 20 can be directly applied to light-cone wavefunctions, without further regularisation.

If we want to proceed further, we need a model for the light-cone wavefunction. Here we do not want to consider a specific model, but rather the whole class of valence quark models, which may be useful for intermediate and larger values of x. In a valence quark model, since the colour part of the wavefunction factorises, the colour density-density correlations can be replaced by neutral density-density correlations

$$\left\langle \bar{q}(0)\gamma^+ \frac{\lambda_a}{2} q(0)\rho_a(\mathbf{y}_\perp) \right\rangle = -\frac{2}{3} \left\langle \bar{q}(0)\gamma^+ q(0)\rho(\mathbf{y}_\perp) \right\rangle \qquad (22)$$

and therefore

$$\langle k_q^i \rangle = \frac{g}{6p^+} \int \frac{d^2 \mathbf{y}_\perp}{2\pi} \frac{y^i}{|\mathbf{y}_\perp|^2} \langle p, s | \bar{q}(0)\gamma^+ q(0)\rho(\mathbf{y}_\perp) | p, s \rangle \qquad (23)$$

with $\rho(\mathbf{y}_\perp) = g \sum_{q'} \int dy^- \bar{q}'(y^-, \mathbf{y}_\perp) \gamma^+ q'(y^-, \mathbf{y}_\perp)$.

In the previous section we showed that the distribution of partons in the $\perp$ plane $q(x, \mathbf{b}_\perp)$ is significantly deformed for a transversely polarised target (Figure 2) and the mean displacement of flavour q ($\perp$ flavour dipole moment) is related to the anomalous magnetic moment contribution from that quark flavour (Equation 15). Qualitatively we expect that the sign and magnitude of the distortion is correlated with the sign of the density-density correlation.

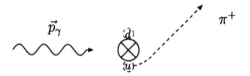

Figure 4. *The transverse distortion of the parton cloud for a proton that is polarised into the plane, in combination with attractive FSI, gives rise to a Sivers effect for u (d) quarks with a $\perp$ momentum that is on the average up (down).*

Using Equation 23 we therefore expect for the resulting Sivers effect $\langle k_u^y \rangle < 0$ and $\langle k_d^y \rangle > 0$ for a proton polarised in $+\hat{x}$ direction and we expect them to be roughly of the same magnitude. The FSI is attractive and thus it "translates" position space distortions (before the quark is knocked out) in the $+\hat{y}$-direction into momentum asymmetries that favour the $-\hat{y}$ direction and vice versa (Figure 4) (Burkardt 2004b).

In order to obtain a rough order of magnitude estimate, we approximate $|\mathbf{y}_\perp|^2 \approx \frac{2}{3} \langle r^2 \rangle$ in terms of the (electric) rms radius $\sqrt{\langle r^2 \rangle} \approx 0.86\,fm$. Furthermore, in the numerator we replace y^i by the average $\perp$ separation between u and d quarks, obtained from the average separation from the *centre of momentum* (15) $y^i \approx \frac{1}{2M}\left(\frac{\kappa^u}{2} - \kappa^d\right) \approx 0.29\,fm$ where the factor $\frac{1}{2}$ multiplying κ^u accounts for the number of up quarks. Taking into account the normalisation $\int \frac{d^2 \mathbf{y}_\perp}{2p^+} \langle p, s | \bar{d}(0)\gamma^+ d(0)\rho_u(\mathbf{y}_\perp) | p, s \rangle = 2g$, the average transverse momentum per down quark thus reads

$$\langle k_d^i \rangle = \alpha_s \frac{1}{M \langle r^2 \rangle} \left(\frac{\kappa_u}{2} - \kappa_d\right) \approx 50 MeV. \qquad (24)$$

Here we used $\alpha_s = 0.3$ since the typical $\perp$ momentum transfers are rather soft. For up quarks, the above estimate yields opposite sign and numerically the same result. However, taking into account that there are twice as many u quarks, the average $\perp$ momentum per u quark turns out to be half that for d quarks. Given the fact that made only a rough estimate, it is quite encouraging that we find a result that is roughly of the same order as the measured results (HERMES 2004).

5 Summary

Wilson line gauge links in gauge invariant Sivers distribution are a formal tool to include the final state interaction in semi-inclusive DIS experiments. The average transverse momentum due to these Wilson lines is obtained as the correlation between the quark density and the impulse from the spectators on the active quark is it escapes along its (almost) light-like trajectory.

In light-cone gauge $A^+ = 0$ only the gauge link at infinity contributes. After careful regularisation of the zero-modes, we succeeded in expressing the net asymmetry in terms of calor density-density correlation in the $\perp$ plane.

For a transversely polarised target the quark distribution in impact parameter space is transversely distorted due to the presence of quark orbital angular momentum: the j^+ current density is enhanced on the side where the quark orbital motion is head-on with the virtual photon. As the struck quark tries to escape the target, one expects on average an attractive force from the spectators on the active quark, *ie* the FSI convert a left-right asymmetry for the quark distribution in impact parameter space into a right-left asymmetry for the $\perp$ momentum of the active quark (Sivers effect).

The sign of the distortion in impact parameter space, and hence the sign of the Sivers effect, for each quark flavour is determined by the sign of the anomalous magnetic moment contribution (of course with the electric charge factored out) from that quark flavour to the anomalous magnetic moment of the nucleon.

References

Belitsky A, *et al.*, 2003, *Nucl Phys B* **656** 165.

Boer D *et al.*, 2003, *Nucl Phys B* **667** 201.

Brodksy S J *et al.*, 2002, *Phys Lett B* **530** 99.

Burkardt M, 1996, *Adv Nucl Phys* **23** 1.

Burkardt M, 2000, *Phys Rev D* **62** 071503.

Burkardt M, 2001, hep-ph/0105324.

Burkardt M, 2003, *Int J Mod Phys A* **18** 173.

Burkardt M, 2004a, *Nucl Phys B* **733** 185.

Burkardt M, 2004b, *Phys Rev D* **69** 057501.

Collins J C, 2002, *Phys Lett B* **536** 43.

Collins J C, 2003, *Acta Phys Polon B* **34** 3103.

Diehl M, 2003, *Phys Rept* **388** 41.

Diehl M *et al.*, 2004, hep-ph/0408173.

Efremov A V and Radyushkin A V, 1981, *Theor Math Phys* **44** 774.

HERMES collaboration, 2004, hep-ph/0408013.

Ji X, 1997, *Phys Rev Lett* **78** 610.

Ji X and Yuan F, 2002, *Phys Lett B*543.

Qiu J W and Sterman G, 1991, *Phys Rev Lett* **67** 2264.

Radyushkin A V, 1997, *Phys Rev D* **56** 5524.

Schäfer A *et al.*, 1993, *Phys Rev D* **47** 1.

Sivers D W, 1991, *Phys Rev D* **43** 261.

Soper D E, 1972, *Phys Rev D* **5** 1956.

Soper D E, 1977, *Phys Rev D* **15** 1141.

Tangerman R D and Mulders P J, 1995, *Phys Rev D* **51** 3357.

Dynamics of hard diffraction

Paul Hoyer

Department of Physical Sciences and Helsinki Institute of Physics
POB 64, FIN-00014 University of Helsinki, Finland

1 Rapidity gaps in Deep Inelastic Scattering

In the intuitive picture (Figure 1) of Deep Inelastic Scattering (DIS), $e + p \to e + X$, the incoming electron emits a photon of high virtuality Q^2, which scatters incoherently on single partons (quarks or gluons) in the target. In the target rest frame, the parton absorbs the large energy ν of the photon and is ejected from the target. The separation of colour charge between the struck parton and the target spectator system gives rise to a colour string. The breaking of the string fills the rapidity interval between the fastest and slowest particles with an approximately uniform distribution of hadrons.

Well before the experimental discoveries, Ingelman and Schlein (IS) proposed hard diffraction (Ingelman and Schlein 1985), and in particular diffractive DIS (DDIS), as a tool for studying the mechanisms underlying diffractive interactions. In Regge language, diffraction and the rapidity gaps which persist at high energy are associated with pomeron exchange; the structure of the 'pomeron' could then be elucidated.

In this approach the pomeron is a structure contained in hadron wave functions. The fraction of diffractive events is then expected to be similar in all hard processes. However, $p\bar{p}$ collider data (CDF 1997, CDF 2000, D0 2002) show that hard diffractive events with a pair of high-$p_\perp$ jets constitute only 1–2% of all jet events. Thus the diffractive fraction is considerably smaller in hadronically-induced events compared to the DDIS/DIS ratio of about 10%. This is consistent with the QCD factorisation theorem for diffractive processes (Collins 1998b), which requires universality only in hard diffraction induced by virtual photons. The pomeron apparently is generated by the reaction dynamics rather than being part of the initial hadron wavefunctions.

The energy dependence of the DDIS data also disagrees with the pomeron exchange model. In a Regge picture the ratio of diffractive to inclusive DIS cross sections would

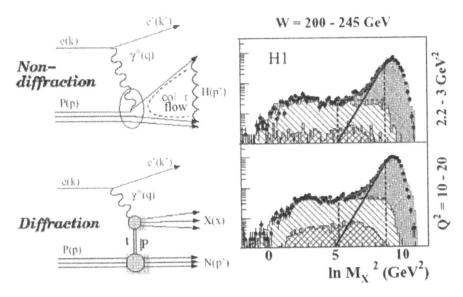

Figure 1. *In the intuitive picture (top left) of DIS the struck parton generates a colour string resulting in a uniform distribution of hadrons in rapidity. Data from (ZEUS 1993, ZEUS 1995, H1 1994, H1 1995) (right) shows that there is a large rapidity gap in about 10% of the DIS events. This is often phenomenologically described as due to "pomeron exchange" (bottom left).*

be expected to depend on the total hadronic mass W as

$$\frac{d\sigma^{\gamma^* p}_{diff}/dM_X}{\sigma^{\gamma^* p}_{tot}} \propto \frac{(W^2)^{2\bar{\alpha}_\mathbb{P}-2}}{(W^2)^{\alpha_\mathbb{P}(0)-1}} \simeq W^\rho \tag{1}$$

where M_X (the mass of the diffractive system) is fixed and $\rho \simeq 0.19$. The ZEUS data (Breitweg 1999) show $\rho = 0.00 \pm .03$, in clear contradiction to the Regge expectation. The observed W-independence of the ratio in Equation 1 suggests in fact a simple underlying relation between diffractive and inclusive DIS.

The phase of power-behaved high energy scattering amplitudes is fixed by analyticity and crossing symmetry: $A \propto \exp(-i\pi\alpha/2)s^\alpha$ for crossing-even amplitudes. Diffractive amplitudes have $\alpha \simeq 1$ and are thus dominantly imaginary. This shows that rescattering effects, involving on-shell intermediate states, must contribute in diffractive scattering. In the IS picture the pomeron is, on the other hand, part of the initial target wavefunction which has no dynamical phase (for stable targets such as the proton).

In the "Soft Colour Interaction" (SCI) model (Edin *et al.* 1996, Edin *et al.* 1997) of diffraction the partons emerging from the hard collision are assumed to exchange soft gluons with spectators. This redistributes colour and makes it possible for the target spectator system to emerge as a colour singlet. In the Monte Carlo implementation of the model no colour string then attaches to the target system, creating a rapidity gap in the hadronisation phase. With a single parameter (the likelihood of rescattering) the SCI model is able to quantitatively describe diffractive events both in DIS and in hadron induced diffraction.

Rescattering of the struck parton in the colour field of the target is an essential part of the standard leading twist dynamics of DIS, with observable effects such as shadowing in nuclear targets and single spin asymmetries (Brodsky *et al.* 2002a, Brodsky *et al.* 2002b). The rescattering is soft and transmitted by instantaneous 'Coulomb' gluon exchange. In this paper I shall discuss the qualitative features of diffractive scattering which emerge from such rescattering. This analysis provides a QCD basis for the scenario of the SCI model (Brodsky *et al.* 2004).

2 Parton distributions and rescattering

According to the QCD factorisation theorem (Collins and Soper 1982, Collins *et al.* 1985, 1988, 1998a, Bodwin 1985) the quark distribution of the nucleon is given by the matrix element

$$f_{q/N}(x_B, Q^2) = \frac{1}{8\pi} \int dx^- \exp(-ix_B p^+ x^- /2)$$
$$\times \langle N(p)|\overline{\psi}(x^-)\gamma^+ W[x^-; 0] \psi(0)|N(p)\rangle \qquad (2)$$

where all fields are evaluated at equal Light Front (LF) time $x^+ \equiv t + z = 0$ and small transverse separation $x_\perp \sim 1/Q$. The Wilson line

$$W[x^-; 0] = \text{P} \exp\left[ig \int_0^{x^-} dw^- A^+(w^-)\right] \qquad (3)$$

physically represents rescattering of the struck quark on target spectators. Equation 2 is derived in a 'parton model' frame where the virtual photon is moving in the *negative* z-direction. This ensures that the photon meets all constituents of the target at the same LF time, since it propagates with $z \simeq -t$ in the Bjorken limit. The LF matrix element (Equation 2) is invariant under boosts in the z-direction, and may thus be considered in the target rest frame.

It is important to remember that $x^+ = 0$ does not imply an instantaneous process in the sense of ordinary time. The 'Ioffe' coherence length of the virtual photon is $L_I \equiv \nu/Q^2 = 1/2m_N x_B$, where the momentum fraction x_B carried by the struck parton is typically small ($x_B \lesssim 0.01$) in diffractive scattering. Soft rescattering of the struck parton in the colour field of the target within a Ioffe length from the hard vertex contributes *coherently* to the DIS amplitude, and thus affects the cross section.

Only the longitudinal (A^+) component appears in the path-ordered exponential of Equation 3. This component has no x^+ derivative in the Lagrangian and is therefore "instantaneous" in x^+, as required for a matrix element evaluated at $x^+ = 0$. Transverse gluon exchange has a long formation length due to the Lorentz factor $E_q/m_q \propto \nu$ of the struck quark, and thus occurs too late to affect the inclusive cross section.

The path-ordered exponential (Equation 3) ensures the gauge invariance of the matrix element and appears superficially to have no physical effect, since it reduces to unity in LF gauge, $A^+ = 0$. On the other hand, based on general principles of quantum mechanics rescattering of the struck quark within the coherence length should occur and affect the DIS cross section at some level. Moreover, diffraction and shadowing effects are present in DIS data, and are normally attributed to rescattering.

The physical effects of rescattering were studied in some detail in the perturbative model of BHMPS (Brosky *et al.* 2002a). The LF gauge gluon propagator is

$$d_{LF}^{\mu\nu}(k) = \frac{i}{k^2 + i\varepsilon} \left[-g^{\mu\nu} + \frac{n^\mu k^\nu + k^\mu n^\nu}{k^+} \right] \tag{4}$$

where $n \cdot A = A^+$. The second term is absent in Feynman gauge and thus cannot contribute to gauge invariant amplitudes, but it does change the relative weights of individual Feynman diagrams. In particular, the poles at $k^+ = 0$ generated by the propagator in Equation 4 are absent from the full (gauge invariant) BHMPS amplitudes even though they occur in single diagrams. The contribution of a given diagram depends on the $i\epsilon$ prescription used at $k^+ = 0$, but the sum of all diagrams is prescription independent.

The BHMPS diagrams with struck quark rescattering vanish in the prescription $k^+ \to k^+ - i\epsilon$ due to a cancellation between the two terms in the square brackets of the LF propagator (Equation 4). This is consistent with the Wilson line reducing to unity. However, the remaining $k^+ = 0$ poles in diagrams involving interactions within the target spectator system (such as between p_2 and p' in Figure 2) then give a non-vanishing contribution to the scattering amplitude. In fact their contribution must be equal to that of the struck quark rescattering in Feynman gauge (the first term in Equation 4), as required by gauge invariance. LF gauge is thus subtle in that *interactions between spectators contribute to the DIS cross section* through the gauge dependent $k^+ = 0$ pole terms.

3 Diffraction in Deep Inelastic Scattering

Let us now consider more closely the diffractive DIS process $\gamma^*(q) + N(p) \to N(p') + X$. In addition to the standard DIS variables $Q^2 = -q^2$, $\nu = p \cdot q / m_N$ and $x_B = Q^2 / 2m_N \nu$ it is convenient to introduce the two additional Lorentz invariants

$$x_P = q \cdot (p - p') / q \cdot p \tag{5}$$

$$\beta = \frac{Q^2}{2q \cdot (p - p')} = \frac{x_B}{x_P} \approx \frac{Q^2}{Q^2 + M_X^2} \tag{6}$$

where M_X is the invariant mass of the diffractively produced system X. In the IS pomeron model (Ingelman and Schlein 1985) x_P is the momentum fraction carried by the pomeron and β plays the role of x_B in DIS on the pomeron. We take the invariant momentum transfer $t = (p - p')^2$ carried by the pomeron to be small.

3.1 Mechanism for diffraction

The perturbative model of DIS studied in (Brodsky *et al.* 2004) provides insights into the dynamics of diffractive DIS and shows why the *hard subprocess* is the same as in inclusive DIS, as implied by the diffractive factorisation theorem (Collins 1998b). Requiring a rapidity gap between the target and diffractive system imposes a condition only on the

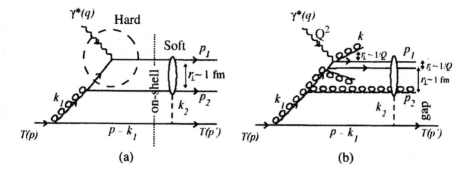

Figure 2. *Perturbative contributions to rescattering in DIS. The struck parton absorbs nearly all the photon momentum, $p_1^- \simeq 2\nu$, $p_1^+ \simeq 0$ (aligned jet configuration). In (a) the virtual photon strikes a quark and the diffractive system is formed by the $q\bar{q}$pair (p_1, p_2) which rescatters coherently from the target via 'instantaneous' longitudinal(A^+) gluon exchange with momentum k_2. In (b) the $Q\bar{Q}$ quark pair which is produced in the $\gamma^* g \rightarrow Q\bar{Q}$ subprocess has a small transverse size $r_\perp \sim 1/Q$ and rescatters like a gluon. The diffractive system is then formed by the $(Q\bar{Q})\,g$ system. The possibility of hard gluon emission close to the photon vertex is indicated. Such radiation (labelled k) emerges at a short transverse distance from the struck parton and is not resolved in the soft rescattering.*

soft rescattering of the struck quark, namely that the target system emerges as a colour singlet. As we shall see, this will not modify the Q^2-dependence of the diffractive parton distribution.

We refer to (Brodsky *et al.* 2004) for a detailed discussion of the properties of the DDIS model amplitudes shown in Figure 2. Here we only give the qualitative picture of the dynamics of $ep \rightarrow e'Xp'$ suggested by perturbation theory:

(i) A gluon (k_1) which carries a small fraction $k_1^+/p^+ \sim x_P$ of the proton momentum splits into (a) a $q\bar{q}$or (b) a gg pair. This is a soft process within the target dynamics, consequently the parton pair has a large transverse size ~ 1 fm.

(ii) The virtual photon is absorbed on (a) one of the quarks in the pair, or (b) scatters via $\gamma^* g \rightarrow Q\bar{Q}$ to a compact, $r_\perp \sim 1/Q$ quark pair. The struck quark (or $Q\bar{Q}$ pair) carries the asymptotically large photon momentum, $p_1^- \simeq 2\nu$. The parton (p_2) that did not interact with the photon also has large $p_2^- \simeq (m_q^2 + p_{2\perp}^2)/p_2^+$ owing to its small $p_2^+ \sim x_B p^+$ (but $p_2^-/p_1^- \propto 1/\nu$ vanishes in the Bjorken limit).

(iii) Multiple soft longitudinal gluon exchange (labelled k_2) turns the colour octet $q\bar{q}$ of Figure 2(a) or the $(Q\bar{Q})g$ of Figure 2(b) into a colour singlet diffractive system. The compact $Q\bar{Q}$ pair behaves as a high energy gluon since its internal structure is not resolved during the soft rescattering.

The rescattering which turns the diffractive system into a colour singlet occurs within the Ioffe length in the target. The colour currents of the gluon exchanges are thus shielded before a colour string can form between the target and the diffractive system. Consequently no hadrons are produced in the rapidity interval $\sim \log(1/x_B)$ between them.

The effective scattering energy of the diffractive system on the target spectator is given by $p_2^- \propto 1/x_B$. As required by analyticity, the crossing-even two-gluon exchange amplitude of Figure 2 is imaginary at low x_B, implying that the intermediate state between the two gluon exchanges is on-shell. Rescattering is necessary to generate the dominantly imaginary diffractive amplitudes.

3.2   Higher order effects at the hard vertex

In the above discussion we considered the hard virtual photon vertex only at lowest order. Just as in inclusive DIS, hard gluon emission and virtual loops give rise to a scale dependence in the parton distributions, and to corrections of higher order in α_s to the subprocess cross section. The perturbative gluons radiated at the hard vertex in Figure 2(b) have $k_\perp \gg p_{2\perp}$ and $k^+ \lesssim p_2^+$. Hence their rapidities $\sim \log(k^-/k_\perp) \sim \log(k_\perp/k^+) \gtrsim \log(p_{2\perp}/p_2^+)$ tend to be larger than the rapidity of the 'slow' parton p_2. The hadrons resulting from the hard gluon radiation therefore do not populate the rapidity gap. The gluons are radiated at a short transverse distance from the struck parton and their transverse velocity $v_\perp \sim k_\perp/k^- \sim k^+/k_\perp$ is small. The struck parton and its radiated gluons thus form a transversally compact system whose internal structure is not resolved in the soft rescattering.

According to the above discussion, the size of the rapidity gap and the soft rescattering are unaffected by higher order corrections at the virtual photon vertex. This is corroborated by the SCI Monte Carlo, where only small variations of the Δy_{max} distribution are observed when varying the parton shower cut-off and thereby the amount of perturbative radiation. Thus the Q^2-dependence of the diffractive parton distributions and the subprocess amplitudes are the same as in inclusive DIS, in accordance with the diffractive factorisation theorem (Collins 1998b).

3.3   Two gluon exchange models

Diffractive DIS has been modelled also using two-gluon exchange (Wüsthoff and Martin 1999, Bartels 1999, Martin *et al.* 2004). These models have been formulated in a frame where the photon is moving in the *positive* z-direction, so that $x^+ = t + z$ increases as the photon penetrates the target. In this 'dipole' frame the virtual photon splits into a $q\bar{q}$ pair (or, at higher orders, into $q\bar{q}G$, *etc.*) already an Ioffe distance *before* the target. The DIS cross section is then determined by the scattering of the $q\bar{q}$ colour dipole from the target. As this should be qualitatively similar to hadron-hadron scattering, previous experience can be used to model inclusive DIS as well as diffraction and shadowing. The simplest possibility is to describe diffraction via two-gluon exchange as in Figure 3.

The $q\bar{q}$ cross section in Figure 3 must be related to the $x^+ = 0$ matrix element (Equation 2) since the dipole frame is related to the parton frame (where the photon moves in the negative z-direction) by a rotation of 180° in the target rest frame. The matrix element (Equation 2) does not transform simply under this rotation, so the relation cannot be explicitly verified. Nevertheless, it is clear that the lower vertex in Figure 3 is

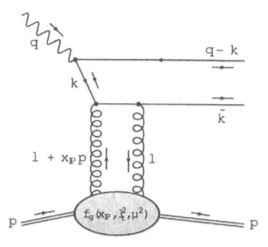

Figure 3. *Diagram contributing to a two-gluon exchange model for DDIS (Wüesthoff and Martin 1999, Bartels 1999, Martin et al. 2004). Since the upper vertex (diffractive system) has a large transverse size the lower vertex of this amplitude is not given by the generalised parton distribution of the target, contrary to what is indicated in the figure.*

not given by the Generalised Parton Distribution (GPD) of the target, since the $q\bar{q}$ pair in the upper vertex has a large transverse size. The situation is thus qualitatively different from *eg* deeply virtual meson production, $\gamma^* p \to \rho p$, where the $q\bar{q}$ pair forming the ρ has a transverse size of $\mathcal{O}(1/Q)$. Deeply virtual meson production is governed by the GPDs but suppressed by several powers of $1/Q^2$ compared to the leading twist DIS and DDIS processes.

There is thus an essential difference between the two-gluon exchange model and our rescattering picture (Brodsky *et al.* 2004). In Figure 2 the second (k_2) gluon represents rescattering via longitudinal (A^+) exchange – a process which contributes also in inclusive DIS. Diffractive events occur when the target remnant emerges as a colour singlet after the (multiple) soft rescattering.

4 Diffraction in hard hadron collisions

Our description of diffraction in deep inelastic lepton scattering can be extended to hard diffractive hadronic collisions. As required by dimensional scaling, only a single parton from the projectile and target participate in the hard subprocess. These leading twist subprocesses (including their higher order corrections) are the same for inclusive and diffractive scattering. The soft rescattering of the hard partons and their spectators is constrained by the requirement of a rapidity gap in the final state. The partonic systems on either side of the gap must be colour singlets in order to prevent the formation of a colour string in the later hadronisation phase.

The soft rescattering is quite different in hadron collisions as compared to DIS. In hadron collisions both the projectile and target spectator systems are coloured. The rescattering gluons (k_2 in Figure 2) can thus couple also to the projectile remnant. In Figure 4 the compact $q\bar{q}$ pair, which is created in the hard gluon–gluon collision, is not resolved by the soft rescattering and therefore retains its colour. Together with the projectile remnant it forms a transversally extended colour octet dipole which can rescatter softly from the target remnant. A rapidity gap is formed between the target remnant and the compact $q\bar{q}$ pair if the target remnant emerges as a colour singlet after the rescattering. The probability for this is, however, different from target neutralisation in DIS. This has been confirmed experimentally: The ratio of diffractive to inclusive cross sections is of the order of $\sim 1\%$ for a variety of hard processes observed at the Tevatron as compared to the $\sim 10\%$ ratio of DDIS/DIS. In the small-x region, these ratios are approximately independent of the momentum fractions in the proton. These observations are well accounted for by the SCI model (Enberg *et al.* 2001).

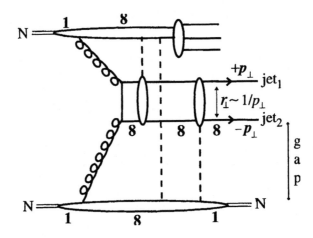

Figure 4. *Illustration of diffraction through rescattering in $NN \rightarrow 2$ jets $+ X$ in analogy with, and using the same notation as, the DIS case in Figure 2. The compact $q\bar{q}$ pair which forms the jets is assumed to be in a colour octet (**8**) configuration. This pair rescatters coherently and thus retains its colour.*

Events with two gaps and a central dijet system have also been observed at the Tevatron (CDF 2000). These are called double-pomeron exchange (DPE) corresponding to the conventional description where a pomeron is emitted from each of the colliding hadrons, followed by a hard scattering between one parton from each pomeron. In our framework these events have soft rescatterings involving both spectator systems, such that the colour of both interacting partons is screened and both spectator systems emerge as colour singlets.

Such events can be generated *eg* by diagrams like that of Figure 4 when the compact $q\bar{q}$ pair is created in a colour-singlet state and thus does not rescatter in either the projectile or target. In fact, the SCI model reproduces the empirical observations, both in absolute normalisation and in kinematical distributions such as the dijet invariant mass. Again, the underlying hard processes are the well-known perturbative QCD subprocesses of fully inclusive cross-sections. The appearance of one or more rapidity gaps depends on the soft rescatterings which affect the colour topology of the event.

5 Conclusions

Hard diffractive processes such as diffractive DIS provide new insight into the dynamics of QCD. We have emphasised that the subprocesses with large momentum transfer are universal in all inclusive reactions: they involve a *single constituent* from the projectile and target and are given by perturbative QCD. The parton distributions reflect the LF wave functions of the colliding particles *and* the soft rescattering of the partons emerging from the hard subprocess. The rescattering is mediated by longitudinal gluons and occurs 'instantaneously' in LF time as the partons pass the spectators. Hard partons which are radiated in the subprocess itself are not resolved by the soft longitudinal gluons which scatter coherently off the colour charge of the struck parton. Hence the Q^2 dependence of diffractive parton distributions is governed by the usual DGLAP equations, both for DDIS and in hadron-induced diffractive processes.

In a diffractive process the soft rescattering is constrained by the requirement that the diffractive systems on either side of the rapidity gap are colour singlets. Since the configurations of colour-charged spectators are different in virtual photon and the various hadron induced diffractive processes, this requirement means that diffractive parton distributions are process dependent. Comparisons of the parton distributions for different projectiles and rapidity gap configurations can thus give valuable information on the rescattering dynamics.

Our description of hard diffractive reactions provides predictions at several levels of accuracy:

1. The hard subprocesses are identical in inclusive and diffractive scattering. This applies also to hadron induced processes.

2. The Q^2 dependence of diffractive parton distributions is governed by the standard DGLAP equations. For DDIS this is a statement of the diffractive factorisation theorem (Collins 1998b).

3. The dependence on the fractional momentum x carried by the parton is similar for diffractive and inclusive distributions. This reflects our assumption that the overall momentum transferred in the rescattering is small, which is supported by data.

4. The dependence on the diffractive mass (or β) of the diffractive parton distributions arises from the underlying (non-perturbative) $g \to q\bar{q}$ and $g \to gg$ splittings in the case of quark and gluon distributions, respectively.

Acknowledgments

I am grateful to the organisers of HadronPhysics I3 and this workshop for their kind invitation. This paper is based on a collaboration with S. Brodsky, R. Enberg and G. Ingelman (Brodski *et al.* 2004). The research was supported by the Academy of Finland through grant 102046.

References

Bartels J, 1999, *Eur Phys J C* **7** 443.

Bodwin G T, 1985, *Phys Rev D* **31** 2616; Erratum 1986 *Phys Rev D* **34** 3932.

Brodsky S J *et al.* , 2002a, *Phys Rev D* **65** 114025.

Brodsky S J *et al.* , 2002b, *Phys Lett B* **530** 99.

Brodsky S J *et al.* , 2004, hep-ph/0409119.

CDF, Abe F *et al.* , 1997, *Phys Rev Lett* **79** 2636.

CDF, Affolder T *et al.* , 2000, *Phys Rev Lett* **85** 4215.

Collins J C and Soper D E, 1982, *Nucl Phys B* **194** 445.

Collins J C *et al.* , 1985, *Nucl Phys B* **261** 104.

Collins J C *et al.* , 1988, *Nucl Phys B* **308** 833.

Collins J C *et al.* , 1998a, *Phys Lett B* **438** 184.

Collins J C, 1998b, *Phys Rev D* **57** 3051; Erratum 1998 *Phys Rev D* **61** 019902.

D0, Abbott B *et al.* , 2002, *Phys Lett B* **531** 52.

Edin A *et al.* , 1996, *Phys Lett B* **366** 371.

Edin A *et al.* , 1997, *Z Phys C* **75** 57.

Enberg R *et al.* , 2001, *Phys Rev D* **64** 114015.

H1, Ahmed T *et al.* , 1994, *Nucl Phys B* **429** 477.

H1, Ahmed T *et al.* , 1995, *Nucl Phys B* **435** 3.

Ingelman G and Schlein P E, 1985, *Phys Lett B* **152** 256.

Martin A D *et al.* , 2004, hep-ph/0406224.

Wüesthoff M and Martin A D, 1999, *J Phys G: Nucl Part Phys* **25** R309.

ZEUS, Derrick M *et al.* , 1993, *Phys Lett B* **315** 481.

ZEUS, Derrick M *et al.* , 1995, *Phys Lett B* **346** 399.

ZEUS, Breitweg J *et al.* , 1999, *Eur Phys J C* **6** 43.

Deeply virtual Compton scattering: results & future

Wolf-Dieter Nowak

DESY,
D-15738 Zeuthen, Germany

1 Introduction

For more than two decades the momentum and spin composition of the nucleon and other hadrons has been investigated by now, preferentially using charged leptons as probes. A great variety of measurements was performed in order to study the underlying structure of quarks and gluons that constitute the fundamental degrees of freedom in Quantum Chromodynamics (QCD). Their momenta and angular momenta cannot yet be calculated from first principles, they are encoded in universal distribution functions whose determination is a central topic that embraces particle physics and hadron physics.

The longitudinal momenta and polarisations carried by quarks, antiquarks and gluons within a fast moving hadron are encoded in universal Parton Distribution Functions (PDFs). They are conveniently introduced in the description of the inclusive deep inelastic scattering (DIS) process, $e\,p \rightarrow e\,X$. The exchange of one virtual photon dominates this reaction at fixed-target kinematics with centre-of-mass energies $\sqrt{s} = \mathcal{O}(10 \text{ GeV})$, and it is still the major contribution at collider kinematics with $\sqrt{s} = \mathcal{O}(300 \text{ GeV})$. In the Bjorken limit of high virtuality Q^2 and large energy ν of the photon in the target rest frame, at a finite ratio $x_B = \frac{Q^2}{2m\nu}$ (m being the nucleon mass), the cross section factorises into that of a hard partonic subprocess and (a certain combination of) PDFs. For a parton of a given species, the unpolarised PDF represents the probability of finding it at a given fraction x_B of the nucleon momentum, while the polarised one describes the imbalance of probabilities between oppositely polarised partons. PDFs are called 'forward' distributions, as the inclusive $\gamma^* p$ cross section can be expressed through the optical theorem by the imaginary part of the forward Compton amplitude $\gamma^* p \rightarrow \gamma^* p$.

2   Generalised Parton Distributions

The theoretical framework of Generalised Parton Distributions (Dittes *et al.* 1988; Müller *et al.* 1994; Radyushkin 1996; Ji 1997; Blümlein *et al.* 1999) is capable of simultaneously treating several types of processes ranging from inclusive to hard exclusive lepton-nucleon scattering. Exclusive scattering is 'non-forward' in nature since the photon initiating the process is virtual and the final-state particle is usually real, forcing a small but finite t, the squared momentum transfer between initial and final nucleon states. GPDs depend on t and on Q^2, the hard scale of the process, and also on two longitudinal momentum variables. Through these dependences they carry information on two-parton correlations and on quark transverse spatial distributions (Burkardt 2000; Ralston and Pire 2002; Diehl 2002; Belitsky and Müller 2002, Burkardt 2003). A recent comprehensive theoretical review can be found in (Diehl 2003).

Presently the most intensely discussed GPDs are the chirally-even, or quark-helicity conserving GPDs F^q ($F = H, \tilde{H}, E, \tilde{E}$ and $q = u, d$). In order to constrain their non-forward behaviour, measurements can be performed of hard exclusive leptoproduction of a photon or meson, in processes leaving the target intact. The production of a real photon, *ie*, Deeply Virtual Compton Scattering (DVCS) $e\,p \rightarrow e\,p\,\gamma$, has two benefits:

i) it is considered to be the theoretically cleanest process that can be accessed experimentally in the foreseeable future and

ii) effects of next-to-leading order (Belitsky and Müller 1998; Ji and Osborne 1998; Mankiewicz *et al.* 1998) and sub-leading twist (see *eg* Anikin *et al.* 2000; Radyushkin and Weiss 2000; Belitsky *et al.* 2002) are under theoretical control.

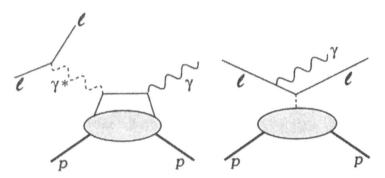

Figure 1. *Left: Deeply Virtual Compton Scattering. Right: Bethe-Heitler process (photon radiated by incoming or outgoing lepton).*

In the generalised Bjorken limit of large photon virtuality Q^2 at fixed x_B and t the dominant pQCD subprocess of DVCS is described by the 'handbag' diagram shown in the left panel of Fig 1. The internal variable x and the skewedness parameter ξ, with $\xi \simeq \frac{x_B}{2-x_B}$ in the Bjorken limit, describe the longitudinal momentum transfer between two partons: the parton (of flavour q) taken out of the proton carries the longitudinal momentum fraction $x + \xi$ and the one put back into the proton carries the fraction $x - \xi$. The GPD $F^q(x, \xi, t, Q^2)$ can then be considered as describing the correlation between these two partons at the given values of t and Q^2.

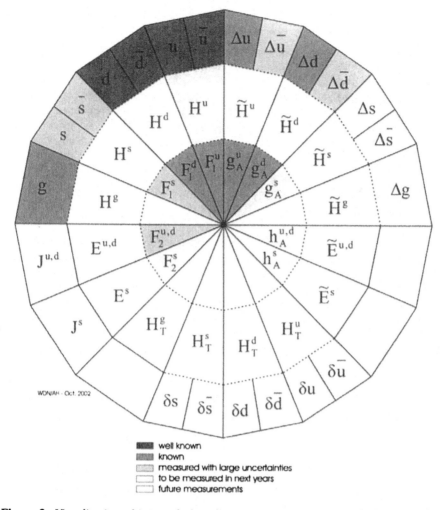

Figure 2. *Visualisation of (most of) the relevant Generalised Parton Distributions and their limiting cases, forward Parton Distributions and Nucleon Form Factors. Different colours illustrate the status of their experimental access (see legend). For explanations see text. The figure has been taken from (Nowak 2003).*

GPDs reduce to ordinary PDFs in the forward limit, *ie*, at vanishing momentum transfer. The first x-moments of GPDs are related to certain form factors measured in elastic lepton-nucleon scattering which describe the difference of the electromagnetic nucleon structure from that of a point-like spin-1/2 particle. A particular second moment of GPDs, for a given parton species $f = (u, d, g)$, is in the limit of vanishing t connected to the total angular momentum carried by these partons (see Equation 5). The latter finding (Ji 1997) stimulated strong interest in GPDs, as the total angular momenta carried by quarks and gluons in the nucleon constitute the hitherto missing pieces in the puzzle representing the momentum and spin structure of the nucleon.

Generalised Parton Distributions, as phenomenological functions, have to be parameterised. Two ansätze are most customary at present:

i) originally, the 'factorised ansatz' uses uncorrelated dependences on t and (x, ξ). The former is written in accordance with proton elastic form factors and the latter is based on double distributions (Radyushkin 1999) plus additional D-term (Polyakov and Weiss 1999). Double distributions are constructed from ordinary PDFs complemented with a profile function that characterises the strength of the ξ-dependence; in the limit $b \to \infty$ of the profile parameter b the GPD is independent on ξ. Note that b is a free parameter to be determined by experiment, separately for valence and sea quarks.

ii) measurements of elastic diffractive processes and, more recently, phenomenological considerations (Diehl *et al.* 2005; Guidal *et al.* 2004) suggest that the t-dependence of the $\gamma^* p$ cross section is entangled with its x_B-dependence. The 'Regge ansatz' for GPDs hence uses for the t-dependence of double distributions a soft Regge-type parameterisation $\sim |x|^{-\alpha' t}$ with $\alpha'_{soft} = 0.8...0.9 \, \text{GeV}^2$ for quarks.

The scheme presented in Fig 2 visualises the present experimental knowledge on the above mentioned functions. As main ingredients, GPDs are placed in the middle of three concentric rings. Their forward limits and moments are situated in the adjacent rings: PDFs in the outermost and nucleon form factors in the innermost one. Today's experimental knowledge of the different functions is illustrated in different colours from light (no data exist) to dark (well known). The emphasis in Fig 2 is placed on the physics message and not on completeness; some GPDs have been omitted. Empty sectors mean that the function does not exist, decouples from observables in the forward limit, or no strategy is known for its measurement. More details can be found in (Nowak 2003).

3   Deeply Virtual Compton Scattering

3.1   Compton form factors

The Bethe-Heitler (BH) process, or radiative elastic scattering, is illustrated in the right panel of Fig 1. Its final state is indistinguishable from that of the DVCS process, hence both mechanisms have to be added on the amplitude level. The differential real-photon leptoproduction cross section is given as

$$\frac{d\sigma}{dx_B dQ^2 d|t| d\phi} \propto |\tau_{BH}|^2 + |\tau_{DVCS}|^2 + \underbrace{\tau_{DVCS}\tau_{BH}^* + \tau_{DVCS}^*\tau_{BH}}_{I}. \tag{1}$$

Here ϕ is the azimuthal angle between the scattering plane, spanned by the incoming and outgoing leptons, and the production plane spanned by the virtual photon and the produced real photon (*cf* Fig 3. The BH amplitude τ_{BH} is exactly calculable using the knowledge of the elastic nucleon form factors. The DVCS contribution $|\tau_{DVCS}|^2$ can then be extracted by integrating over the azimuthal dependence of the cross section. In this case the interference term I vanishes to leading order in $1/Q$; its total contribution at collider kinematics was estimated to be at the percent level (Belitsky *et al.* 2002).

The twist-2 DVCS amplitudes can be represented in the convention of (Belitsky *et al.* 2002) as linear combinations of F_1 and F_2, the Dirac and Pauli elastic nucleon form

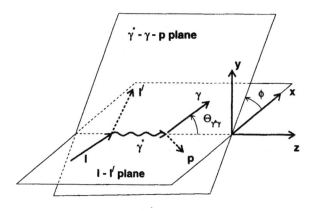

Figure 3. *Definition of the azimuthal angle ϕ in DVCS in the target rest frame.*

factors, with the Compton form factors (CFFs) $\mathcal{H}, \mathcal{E}, \widetilde{\mathcal{H}}, \widetilde{\mathcal{E}}$ (*cf* Equations 11, 14, 15, 16). These complex CFFs are flavour sums of convolutions of the corresponding leading-twist GPDs with the hard scattering kernels $C_q^{\mp}$ that are available up to NLO in pQCD (Belitsky and Müller 1998; Ji and Osborne 1998; Mankiewicz *et al.* 1998):

$$\mathcal{F}(\xi, t, Q^2) = \sum_q \int_{-1}^{1} dx \, C_q^{\mp}(\xi, x) \, F^q(x, \xi, t, Q^2). \tag{2}$$

Here the $-(+)$ sign in the superscript applies to the CFFs $\mathcal{F} = \mathcal{H}, \mathcal{E}$ ($\widetilde{\mathcal{H}}, \widetilde{\mathcal{E}}$), correspond-ing to the GPDs $F^q = H^q, E^q$ ($\widetilde{H}^q, \widetilde{E}^q$).

The real and imaginary parts of a CFF have different relationships to (the flavour sum over) the respective quark GPDs which are embodied. Taking Equation 2 at leading order in α_s (Belitsky *et al.* 2002), the imaginary part

$$\text{Im}\,\{\mathcal{F}\} = \pi \sum_q e_q^2 \left(F^q(\xi, \xi, t, Q^2) \mp F^q(-\xi, \xi, t, Q^2) \right) \tag{3}$$

directly probes the respective GPDs along the line $x = \pm \xi$. In contrast, through the real part of the CFF,

$$\text{Re}\,\{\mathcal{F}\} = -\sum_q e_q^2 \left[P \int_{-1}^{1} dx \, F^q(x, \xi, t, Q^2) \left(\frac{1}{x - \xi} \pm \frac{1}{x + \xi} \right) \right], \tag{4}$$

the integral over the respective GPDs is accessed, whereby the weighting by the propa-gators $1/(x \mp \xi)$ strongly enhances the contribution close to the line $x = \pm \xi$. The sign convention is the same as for Equation 2 and P denotes Cauchy's principal value.

Equations 3 and 4 show that in DVCS a GPD, at given values of t and Q^2, is essen-tially probed along the line $x = \pm \xi$, *ie* a complete mapping of a GPD in the (x, ξ)-plane is impossible and models of GPDs are to be constructed to calculate observables that have to be compared to corresponding experimental results in an iterative procedure.

Full (x, ξ)-mapping of GPDs is still possible, at least in principle:

i) once a large enough dynamic range in Q^2 is available in DVCS measurements, the known Q^2-evolution of GPDs can be used to constrain their x-dependence, similar as for the extraction of ordinary PDFs in DIS.

ii) in hard exclusive leptoproduction of a *virtual* photon (double DVCS or DDVCS) its virtuality, *ie* the effective mass of the produced lepton pair, is an additional variable that facilitates a complete mapping of GPDs. However, the DDVCS cross section is suppressed by an additional factor α_{em}^2, thereby making this reaction practically inaccessible in the foreseeable future (Guidal 2002).

The t-dependence of GPDs is directly accessible in DVCS although high experimental precision, *ie*, high statistical accuracy in conjunction with sufficient resolution is required to extrapolate to the limes $t \to 0$. The latter is of particular importance for the evaluation of the 2nd x-moment of the two 'unpolarised' GPDs $H^f + E^f$, which is related to the total angular momentum J^f of the parton species $f = (u, d, g)$, at a given value of Q^2 (Ji 1997):

$$J^f(Q^2) = \lim_{t \to 0} \frac{1}{2} \int_{-1}^{1} dx \, x \left[H^f(x, \xi, t, Q^2) + E^f(x, \xi, t, Q^2) \right]. \tag{5}$$

3.2   The interference term

The interference term I is of special interest, as the measurement of its azimuthal dependence opens experimental access to the *complex* DVCS amplitudes, *ie* to *both* their magnitude and phase (Diehl *et al.* 1997). This method of using the BH process as an 'amplifier' to study DVCS can be compared to holography (Belitsky and Müller 2002) in the sense that the phase of the Compton amplitude is measured against the known 'reference phase' of the BH process.

The full interference term can be filtered out by forming a cross section asymmetry, or difference, w.r.t. the charge of the lepton beam (Brodsky *et al.* 1972). The imaginary part of the interference term can be accessed by forming single-spin asymmetries, or differences, w.r.t. the spin of the lepton beam (Kroll *et al.* 1996) or of the target (Belitsky *et al.* 2002; Diehl 2003). Note that the measurement of cross section differences is favoured by theorists over that of asymmetries (Diehl 2003). Differences are free from azimuthal dependences of BH, DVCS and interference terms appearing in the denominator of an asymmetry and thereby complicating the separation of the relevant terms in the numerator. They allow easier separation of higher harmonics when compared to the evaluation of an asymmetry, while larger experimental systematic uncertainties may appear.

Each of the three terms in Equation 1 can be expressed as a Fourier series in ϕ (Diehl *et al.* 1997; Belitsky *et al.* 2002). For an unpolarised target, the interference term I can be written as

$$I = -\frac{K_I \, e_l}{\mathcal{P}_1(\cos\phi)\mathcal{P}_2(\cos\phi)} \times \tag{6}$$

$$\left\{ c_0^I + c_1^I \cos(\phi) + c_2^I \cos(2\phi) + c_3^I \cos(3\phi) + P_l \left[s_1^I \sin(\phi) + s_2^I \sin(2\phi) \right] \right\},$$

where K_I is a kinematic factor and $e_l = \pm 1$ is the charge of the lepton beam with longitudinal polarisation P_l. The virtual-lepton propagators $\mathcal{P}_{1,2}(\phi)$ of the BH process introduce an extra $\cos \phi$-dependence. The Fourier coefficient $c_1^I(s_1^I)$ is proportional to the real (imaginary) part of a certain linear combination of the four twist-2 CFFs $\mathcal{H}, \mathcal{E}, \widetilde{\mathcal{H}}, \widetilde{\mathcal{E}}$, the detailed expression depending on the target polarisation (*cf* Equations 11, 14, 15, 16). The coefficient c_0^I is related to approximately the same combination of CFFs as c_1^I, but it is kinematically suppressed by $1/Q$ (Belitsky *et al.* 2002; Diehl 2003). The coefficient c_1^I is sensitive to the D-term that was mentioned in Section 2 (Polyakov and Weiss 1999). The coefficients c_2^I and s_2^I describe twist-3 amplitudes and scale as $1/Q$, whereas c_3^I is α_S-suppressed at leading twist (Diehl 1997). In case of a polarised target, additional sums with analogous structure appear in Equation 6 (Diehl and Sapeta 2005). In particular, terms $s_3^I \sin(3\phi)$ appear where s_3^I is sensitive to contributions from gluon transversity, as in the case of c_3^I.

4 Azimuthal cross section asymmetries

4.1 Unpolarised target

The *beam-spin asymmetry* (BSA) for a longitudinally (L) polarised beam and an *unpolarised* (U) proton target is defined as

$$A_{LU}(\phi) = \frac{d\sigma^{\rightarrow}(\phi) - d\sigma^{\leftarrow}(\phi)}{d\sigma^{\rightarrow}(\phi) + d\sigma^{\leftarrow}(\phi)}, \tag{7}$$

where $\rightarrow (\leftarrow)$ denotes beam spin parallel (antiparallel) to the beam direction. Similarly, the *beam-charge asymmetry* (BCA) for an *unpolarised* beam of charge C scattering from an unpolarised proton target is defined as:

$$A_C(\phi) = \frac{d\sigma^{+}(\phi) - d\sigma^{-}(\phi)}{d\sigma^{+}(\phi) + d\sigma^{-}(\phi)}, \tag{8}$$

where the superscripts $+$ and $-$ denote the lepton beam charge.

Evaluating these asymmetries using Equations 1 and 6 to leading power in $1/Q$ in each contribution, and to leading order in α_S, only the $\sin \phi$ ($\cos \phi$) term remains in the numerator of the beam-spin (beam-charge) asymmetry. To the extent that the leading BH-term c_0^{BH} dominates the denominator, the products of the virtual-lepton propagators, $\mathcal{P}_1(\cos \phi)\mathcal{P}_2(\cos \phi)$, cancel. In this approximation, the azimuthal dependence of the beam-spin (beam-charge) asymmetry is reduced to $\sin \phi$ ($\cos \phi$):

$$A_{LU}(\phi) \propto \frac{1}{c_{0,U}^{BH}} s_{1,U}^I \sin \phi \propto Im \, \widetilde{M} \sin \phi, \tag{9}$$

$$A_C(\phi) \propto \frac{1}{c_{0,U}^{BH}} c_{1,U}^I \cos \phi \propto Re \, \widetilde{M} \cos \phi, \tag{10}$$

where the additional subscript (U) of the Fourier coefficients denotes the unpolarised target. It appears that both beam-charge and beam-spin asymmetries are sensitive to the *same* linear combination $\widetilde{M}$ of CFFs which describes an *unpolarised* proton target (Belitsky *et al.* 2002):

$$\widetilde{M} = \frac{\sqrt{t_0 - t}}{2m} \left[F_1 \, \mathcal{H} + \xi \, (F_1 + F_2) \, \widetilde{\mathcal{H}} - \frac{t}{4m^2} F_2 \, \mathcal{E} \right]. \tag{11}$$

Here $-t_0 = 4\xi^2 m^2 / (1 - \xi^2)$ is the minimum possible value of $-t$ at a given ξ.

Note that, almost independently of details of GPD models, the GPDs H^q are expected to dominate this expression, because i) the second term is suppressed by at least a factor of 10, as ξ is usually not larger than 0.2 even in fixed-target kinematics (*cf.* Fig 11) and the unpolarised contribution $\mathcal{H}$ is expected to dominate the polarised one $\widetilde{\mathcal{H}}$, in analogy to the forward case; ii) the third term is t-suppressed, by about a factor of 25 for typical t-values of about 0.15 GeV^2. For scattering on the proton, the GPD H^u will yield the major contribution to $\widetilde{M}$ because of u-quark dominance.

4.2   Polarised target

In case of a *polarised* proton target further sums appear in Equation 6, as mentioned above. They contain other linear combinations than $\widetilde{M}$, so that measurements of *target-spin asymmetries* deliver valuable additional experimental information.

The single-spin asymmetry w.r.t. to the polarisation of a *longitudinally* (L) polarised target (LTSA) is defined as

$$A_{UL}(\phi) = \frac{d\sigma^{\Leftarrow}(\phi) - d\sigma^{\Rightarrow}(\phi)}{d\sigma^{\Leftarrow}(\phi) + d\sigma^{\Rightarrow}(\phi)} \tag{12}$$

where $\Leftarrow$ ($\Rightarrow$) denotes target spin antiparallel (parallel) to the beam direction. In the above introduced approximation and for 'balanced' beam polarisation ($\langle P_l \rangle \approx 0$), its azimuthal dependence is also purely sinusoidal:

$$A_{UL}(\phi) \propto \frac{1}{c_{0,L}^{BH}} s_{1,L}^I \sin(\phi). \tag{13}$$

Neglecting terms of $\mathcal{O}(\xi^2)$ and higher, the Fourier coefficient $s_{1,L}^I$ is sensitive to a linear combination of CFFs different from Equation 11 (Belitsky *et al.* 2002; Diehl 2003):

$$s_{1,L}^I \quad \propto \quad \frac{\sqrt{t_0 - t}}{2m} \, \mathrm{Im} \left[F_1 \widetilde{\mathcal{H}} + \xi \, (F_1 + F_2) \left(\mathcal{H} + \frac{\xi}{1 + \xi} \mathcal{E} \right) \right.$$
$$\left. - \left(\frac{\xi}{1 + \xi} F_1 + \frac{t}{4m^2} F_2 \right) (\xi \widetilde{\mathcal{E}}) \right]. \tag{14}$$

A_{UL} is expected to be most sensitive to a combination of H^q and $\widetilde{H}^q$, because the kinematic suppression of the second term in Equation 14, as compared to the first one, may approximately compensate the expected dominance of the unpolarised GPDs H^q over their polarised counterparts $\widetilde{H}^q$. Hence both should become separable by combining this measurement with asymmetries measured on an unpolarised target. For not too small values of t there exists also some sensitivity to $(\xi \widetilde{\mathcal{E}})$, which is written in this way as $\widetilde{\mathcal{E}}$ itself is inversely proportional to ξ (Goeke *et al.* 2001; Diehl 2003).

For a *transversely* (T) polarised target, the definition of the single-spin asymmetry (TTSA) A_{UT} is more complicated. The additional dependence on the azimuthal angle ϕ_S of the spin vector creates "normal" (N) and "sideways" (S) components. In the approximation used for Equation 14, the corresponding Fourier coefficients contain yet further combinations of CFFs (Diehl and Sapeta 2005). The normal component reads:

$$s_{1,N}^I \propto -\frac{t}{4m^2} \operatorname{Im}\left[F_2\mathcal{H} - F_1\mathcal{E} + \xi(F_1 + F_2)(\xi\widetilde{\mathcal{E}})\right]. \tag{15}$$

This is known to be the *only* combination of CFFs where the GPDs E^q are not kinematically suppressed as compared to H^q. Hence DVCS measurements on a transversely polarised proton target, in particular of the normal contribution, appear to be indispensable for the evaluation of the total quark angular momentum through the Ji relation (5). An inherent complication for the measurement of this relation lies in the fact that both the GPDs H^q and E^q need to be measured towards lowest possible values of t, while the ϕ-dependence of the cross section disappears in the limit $t \to 0$; the relevant asymmetry is suppressed by a factor of $\sqrt{-t}/2m$ when extracting the GPDs H^q and even by a factor of $t/4m^2$ when extracting the GPDs E^q, as can be seen from a comparison of Equations 11 and 15.

The sideways component, written in the above used approximation, undergoes the same kinematic suppression as the normal component:

$$s_{1,S}^I \propto -\frac{t}{4m^2} \operatorname{Im}\left[F_2\widetilde{\mathcal{H}} + \xi(F_1 + F_2)\mathcal{E} - (F_1 + \xi F_2)(\xi\widetilde{\mathcal{E}})\right]. \tag{16}$$

It offers access to the imaginary part of a combination of *both polarised* CFFs, $\widetilde{\mathcal{H}}$ and $(\xi\widetilde{\mathcal{E}})$, although accompanied by the unpolarised CFF $\mathcal{E}$ whose ξ-suppression will presumably be compensated by its larger size.

4.3 Beyond DVCS

i) Similar as for differently polarised targets in DVCS, for hard exclusive processes other than DVCS, as *eg* meson production, the involved amplitudes embody other subsets of proton GPDs that are linearly combined by different kinematic suppression factors. In other words, a certain process is more sensitive to an individual GPD than another and hence a complete as possible determination of GPDs requires measurements of several hard exclusive reactions and/or final states.

ii) Every experiment covers a peculiar subspace of the (x_B, t, Q^2) phase space, with certain overlap regions between experiments as will be detailed at the beginning of section 6. Hence a certain GPD combination accessible through a certain cross section, cross section difference or cross section asymmetry will be surveyed by different experiments in partly overlapping subspaces only.

iii) In 'associated' DVCS, where the proton target does not stay intact, the formalism of angular analysis remains the same, while the accessible GPDs are different from those of the proton (Diehl 2003).

iv) Hard exclusive leptoproduction on nuclear targets proceeds either coherently, *ie* by scattering on the nucleus as a whole, or incoherently, *ie* on a single proton or neutron.

Coherent scattering proceeds preferentially at (very) small values of t. For small and medium values of t, the electromagnetic form factor of the neutron is small, leading to a small Bethe-Heitler cross section, as compared to DVCS. Hence in this t-region the interference term I is suppressed for scattering on the neutron and incoherent nuclear DVCS is expected to behave similarly to DVCS on the proton.

v) Gluon GPDs can be accessed in DVCS and vector meson production (Diehl 2003; Goloskokov and Kroll 2005), in particular at small ξ, *ie* at collider kinematics, but also in ϕ-production at fixed-target kinematics (Diehl and Vinnikov 2004).

5   Experimental results on DVCS

5.1   Collider experiments

The DVCS cross section has been measured in hard exclusive photon electroproduction at the HERA collider by the experiments H1 and ZEUS. It becomes accessible by integrating Equation 1 over its azimuthal dependence (see section 3.1). For the x_B-range accessible at collider kinematics ($10^{-2} \dots 10^{-4}$), two-gluon exchange plays a major role besides the above discussed quark-exchange handbag diagram, *ie*, *both* gluon and quark GPDs are probed simultaneously but only for very small skewedness values below 10^{-2}.

The analysis method used is similar in both experiments due to their similar geometry. As the outgoing proton remains undetected in the beam pipe, the event topology is defined by two electromagnetic clusters, the outgoing lepton and the produced real photon, and at most one associated track. Two events samples are selected:

i) in the 'DVCS-enriched' sample a hard, *ie* centrally produced real photon is required, while the outgoing lepton is measured under a small scattering angle w.r.t. the incoming one; still a high enough virtuality $Q^2 > 4\,\text{GeV}^2$ is ensured by requiring a large energy of the scattered lepton ($> 15\,\text{GeV}$).

ii) in the Bethe-Heitler dominated 'reference sample' the radiatively produced photon is emitted under a small angle w.r.t. the incoming lepton, and the outgoing lepton is measured in the central region.

A Monte Carlo simulation of the completely known BH process, which describes the reference sample, is used to subtract the BH contribution from the DVCS-enriched sample. The remainder of the spectrum is due to DVCS and possible additional background; no contribution from the interference term exists at leading twist, as the data are integrated over the azimuthal angle. The Q^2-dependence of the differential $\gamma^* p \rightarrow \gamma p$ cross section is illustrated in Fig 4.

The data shown are from H1, both earlier published (Adloff *et al.* 2001) and recent preliminary (Favart 2004), and from ZEUS (Chekanov *et al.* 2003), the latter based on substantially higher statistics. The solid curve shows a NLO pQCD calculation (Freund and McDermott 2002) using a GPD parameterisation based on MRST2001 PDFs and a Q^2-dependent t-slope $b(Q^2)$ describing the factorised t-dependence (Freund *et al.* 2003). Using instead CTEQ6 PDFs with the same t-slope (not shown) yields a very similar Q^2-dependence, but a different normalisation. For comparison a Colour Dipole model calculation (Donnachie and Dosch 2001) is also shown in the figure. Agreement between

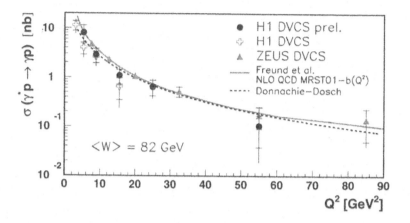

Figure 4. Q^2-dependence of the differential $\gamma^* p \to \gamma p$ cross section measured by H1 and ZEUS in comparison to a GPD-based NLO pQCD calculation. For comparison, a prediction of a Colour Dipole model is also shown. The figure is taken from Favart (2004).

all data sets and models can be seen in the log-scale representation of the Q^2-dependence, although there seem to be discrepancies at lower values of Q^2.

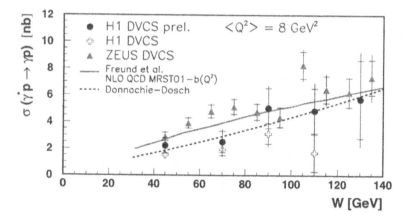

Figure 5. W-dependence of the differential $\gamma^* p \to \gamma p$ cross section measured by H1 and ZEUS, compared to a GPD-based NLO pQCD calculation. For comparison, a prediction of a Colour Dipole model is also shown . The figure is taken from Favart (2004).

The corresponding W-dependence is displayed in Fig. 5 for the same data sets and models, where W is the centre-of-mass energy of the system of virtual photon and proton.

The virtuality appears high enough to assign the observed steep rise with W to the nature of DVCS as a hard process, as increasing W implies decreasing x_B, where the parton densities in the proton show a fast rise. Most data reside at lower Q^2 ($\langle Q^2 \rangle = 8$ GeV^2), the region of possible discrepancies between data sets. Presently a 2σ difference exists between ZEUS and H1 data in the medium W-range. Differences of similar size exist between different model calculations. Note that, since the slope of the t-dependence of GPDs is still unknown, the normalisation of the GPD-based curves shown above remains arbitrary to some extent. Future measurements of the t-dependence of the DVCS cross section are therefore of high importance.

5.2   Fixed-target experiments

In sections 3 and 4 it was shown that in hard exclusive real-photon leptoproduction the interference of the Bethe-Heitler and Deeply Virtual Compton Scattering processes is a rich source for extracting a wealth of information on GPDs. In fact, the first published GPD-related experimental results were beam-spin asymmetries measured in DVCS on the proton by the fixed-target experiments HERMES at HERA (Airapetian *et al.* 2001) with a positron beam and by CLAS at Jefferson Laboratory (Stepanyan *et al.* 2001) with an electron beam.

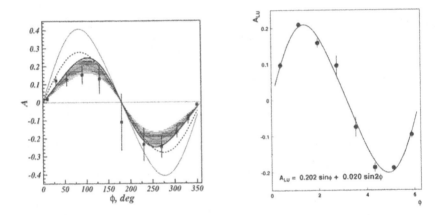

Figure 6. CLAS: *Azimuthal dependence of the beam-spin asymmetry. Left: earlier data at 4.25 GeV. Right: recent preliminary data at 5.75 GeV. Only statistical errors are shown.*

Note that opposite beam charges mean opposite signs of the measured BSAs, a fact that is not apparent when comparing Figures 6 and 7 due to different ϕ-ranges chosen. Meanwhile, more precise (preliminary) BSA measurements were presented by both experiments.

The average beam polarisation at CLAS (HERMES) was 70% (55%). Both experiments use the missing-mass technique to compensate for present incompletenesses of their detectors; at CLAS (HERMES) the real photon (recoiling proton) remains undetected. At CLAS, the missing-mass resolution cannot cleanly separate the $ep\pi^0$ from $ep\gamma$

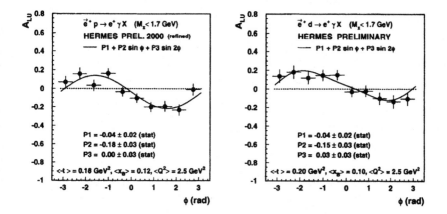

Figure 7. HERMES: *Azimuthal dependence of the beam-spin asymmetry on proton (left) and deuteron (right), measured at 27.6 GeV. Only statistical errors are shown.*

reactions. The electromagnetic calorimeter detects only photons above $8°$, ie, it misses most of the DVCS photons but detects usually one of the two π^0 decay photons. In the analysis of the 5.75 GeV data, background from π^0 decay is reduced by applying a corresponding veto. A cut $\theta_{\gamma\cdot\gamma} < 120$ mrad is used to select data at low t. The missing mass squared for the undetected real photon is restricted to $M_X^2 < 0.025$ GeV^2. The kinematics coverage for $W > 2$ GeV is $1.2 < Q^2 < 4$ GeV^2 and $0.1 < x_B < 0.5$, and the analysis is restricted to $-t < 0.5$ GeV^2. At HERMES, to account for the limited resolution of the spectrometer, an asymmetric missing-mass interval around the proton mass is chosen (called 'exclusive bin': $-1.5 < M_X < 1.7$ GeV), based on signal-to-background studies using a Monte Carlo simulation. Kinematic requirements to the outgoing lepton are $1 < Q^2 < 10$ GeV^2, $W^2 > 9$ GeV^2 and $\nu < 22$ GeV, implying $0.03 < x_B < 0.35$. For the results shown in Fig 7 (Ellinghaus *et al.* 2002b), the polar angle between virtual and produced real photon obeys $2 < \theta_{\gamma\cdot\gamma} < 70$ mrad. Monte Carlo studies show that the exclusive sample contains about 10% associated events, where the nucleon doesn't stay intact, and about 5% events from DIS fragmentation. Note that for the exclusive sample the variable t is calculated assuming the 3-particle final state $ep\gamma$, thereby considerably improving the t-resolution, and the analysis is restricted to $-t < 0.7$ GeV^2.

CLAS proton data for average kinematics ($\langle Q^2 \rangle = 1.25$ GeV^2, $\langle x_B \rangle = 0.19$, $\langle -t \rangle = 0.19$ GeV^2) are shown in Fig 6, the earlier (Stepanyan *et al.* 2001) BSA result in the left panel and the more recent preliminary result (Smith 2003) in the right one. The most recent (preliminary) BSA results from HERMES, on both proton and deuteron (Ellinghaus *et al.* 2002b), are shown in Fig 7. The average kinematics are $\langle Q^2 \rangle = 2.5$ GeV^2, $\langle x_B \rangle \simeq 0.10$ and $\langle -t \rangle \simeq 0.20$ GeV^2. All BSA data exhibit substantial sinusoidal asymmetries in accordance with the expectation given in Equation 9. The magnitude of the $\sin\phi$ component was fitted as $0.202 \pm 0.028_{stat} \pm 0.013_{sys}$ and $-0.23 \pm 0.04_{stat} \pm 0.03_{sys}$ for the published data from CLAS and HERMES, respectively, while 0.202 and $-0.18 \pm 0.03_{stat}$ were obtained from their recent preliminary data. The next-higher harmonic ($\sin(2\phi)$, see Equation 6) is found to be compatible with zero within the total experimental uncertainty in both experiments. No difference is seen when comparing the HERMES BSA results for proton and deuteron, which is not surprising when recalling the argument made in Note iv) of section 4.3. No published data exist yet for kinematic dependences of BSAs. Unpublished HERMES data (Ellinghaus 2004a) do not show any clear dependence on t, x_B or Q^2 within experimental uncertainties.

A beam-charge asymmetry measurement requires data for both beam charges. HERA is presently the only GeV-range accelerator that provides both electron and positron beams. It offers the additional flexibility of switching from time to time, for the same charge of the beam, the direction of its polarisation to reduce systematic effects.

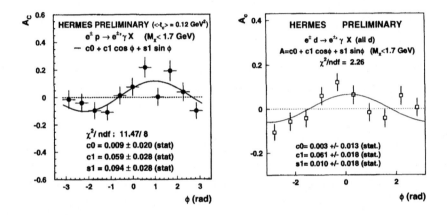

Figure 8. *HERMES: Azimuthal dependence of the beam-charge asymmetry on proton (left) and on unpolarised and vector-polarisation-balanced deuteron (right). Only statistical errors are shown.*

In Fig 8 preliminary BCA results from HERMES are shown (Ellinghaus 2004c), which were obtained from the same analysis as described above using $5 < \theta_{\gamma \cdot \gamma} < 45$ mrad. Both proton and deuteron data exhibit the expected $\cos \phi$-dependence with a similar sizeable magnitude, $0.059 \pm 0.028_{stat}$ and $0.061 \pm 0.018_{stat}$, again not a surprising agreement. The proton BCA also contains a significant $\sin \phi$-component that is caused by average beam polarisation values not vanishing individually for each beam charge. For clarity it has to be noted that an about twice as large deuteron BCA (not shown here) was reported earlier (Ellinghaus 2002a), obtained by discarding the low-t region with the requirement $15 < \theta_{\gamma \cdot \gamma} < 70$ which led to a larger average value of $\langle -t \rangle \simeq 0.27$ GeV^2. Both results are consistent because the BCA decreases with decreasing $-t$, as will be shown below.

Kinematic dependences are obtained by subdividing the data set into several bins, depending on the statistics for a given variable at a time. In each bin a fit of the azimuthal dependence is performed, including the same harmonics as indicated in the panels of Fig 8. The t-dependence of proton and deuteron BCA is shown in Fig 9, also obtained for $5 < \theta_{\gamma \cdot \gamma} < 45$ mrad (Ellinghaus 2004c). As expected, the signal becomes only sizeable from medium values of $-t$ on. Here proton and deuteron data agree, as discussed in Note iv) of section 4.3. Incoherent scattering on the neutron may become a

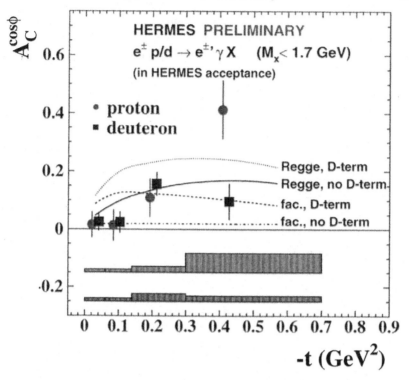

Figure 9. HERMES: *t-dependence of the* cos φ *component of the beam-charge asymmetry on proton and deuteron. Statistical (systematic) uncertainties are indicated by error bars (bands). The curves represent LO pQCD calculations using different GPD ansätze, see text.*

substantial contribution at larger $-t$-values, *ie* in the last t-bin, and compensate a further increase of the deuteron asymmetry. No effects are seen from coherent scattering on the deuteron bound state which would be present in the lowest $-t$-bin only. Superimposed to the experimental data are curves representing theoretical calculations (Vanderhaeghen *et al.* 2001) based on different GPD models (Vanderhaeghen *et al.* 1999). They are calculated at HERMES kinematics, separately for the average kinematics in each individual bin (Ellinghaus 2004a). On the basis of the available statistics, the data seem to favour the model with the Regge ansatz and no D-term contribution. From Fig 9 it can already be concluded, and it will be discussed in more detail in section 6.3, that BCA measurements possess a considerable discriminative power against different ansätze and parameterisations in GPD models. Note that no dependence on x_B or Q^2 is seen in unpublished BCA results (Ellinghaus 2004a).

By measuring DVCS on a longitudinally polarised *deuteron* target, HERMES obtained a preliminary result on the *longitudinal target-spin* asymmetry (LTSA, cf. Equations 12-14). Fits as explained above, including $\sin \phi$ and $\sin(2\phi)$ harmonics, yield asymmetries compatible with zero; both components are found to be smaller than 0.03 with the total experimental uncertainty being of the same order.

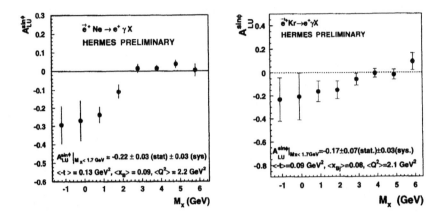

Figure 10. HERMES: *The* $\sin(\phi)$-*component of the beam-spin asymmetry on Neon and Krypton, shown in dependence on missing mass. Only statistical errors are given.*

DVCS on nuclear targets was briefly mentioned in Note iv) of section 4.3. Both experimental and theoretical information is scarce, especially for targets with an atomic number higher than that of deuterium. In Fig 10 preliminary BSA results of HERMES using Neon and Krypton targets are shown in dependence on the missing mass, using the proton mass to calculate the kinematic variables. The average kinematics are indicated in the panels, based on $2 < \theta_{\gamma^*\gamma} < 70$ mrad. As for all HERMES results discussed above, sizeable asymmetries appear only in the exclusive bin around the target (proton) mass, while they generally vanish at higher masses. It is clearly seen for both Neon and Krypton that already without separation of coherent and incoherent processes significant BSAs exist in the exclusive bin, while their interpretation can be attempted only after the separation. The fitted size of the $\sin \phi$-component is -0.22 ± 0.03 (-0.17 ± 0.07) for Neon (Krypton), without significant higher harmonics (Ellinghaus *et al.* 2002b). For the case of coherent hard exclusive processes on nuclei it was pointed out that information about the energy, pressure, and shear forces distributions inside nuclei will become accessible (Polyakov 2003).

6   Future DVCS measurements

In Fig 11 kinematics coverages are compared for DVCS measurements by existing or planned fixed-target experiments at CERN, HERA and JLAB. The kinematic limits are taken from (d'Hose 2002), (Ellinghaus 2004a and 2004b) and (Cardman *et al.* 2001).

As can be seen, the (x_B, Q^2)-regions of these fixed-target experiments do partly overlap, while in comparison to the collider experiments at HERA there is no overlap in x_B (fixed-target above 0.03, collider below 0.01) and only very little overlap in Q^2 (1...8 GeV^2 vs. 5...100 GeV^2). Higher x_B-values (> 0.3) can only be accessed at JLAB, an advantage of their relatively low beam energy. At moderate x_B, higher Q^2-values ($\simeq 8$ GeV^2) are reachable in the short-term at HERMES only. Later on, the upgraded JLAB

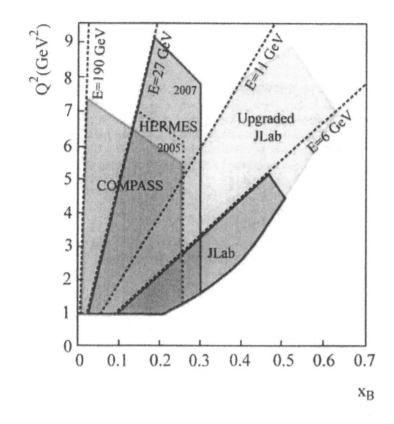

Figure 11. *Kinematics coverage for fixed-target experiments: i)* COMPASS *at 190 GeV; ii)* HERMES *at 27.6 GeV, dotted line for existing data ($\leq$ 2005), solid line for future (2005-2007) data with an integrated luminosity higher by about one order of magnitude; iii)* JLAB *experiments at 6 GeV (now), and at 11 GeV (after upgrade).*

will be able to also reach this region, by compensating their lower beam energy by a huge luminosity planned to be several orders of magnitude higher than that at other facilities.

6.1 HERA collider experiments

No published projections exist, to what extent the recent detector upgrades of the HERA collider experiments H1 and ZEUS will be beneficial to the ongoing and future measurements of DVCS, until the foreseen shutdown of the HERA accelerator in the middle of 2007. The newly installed spin rotators make the polarised beam also available to H1 and ZEUS. In both experiments microvertex detectors have been installed which will allow the precise measurement of the outgoing lepton track, and hence of the event vertex, so that the azimuthal angle of the photon can be determined with higher precision. Altogether, these upgrades will make it possible to also measure the azimuthal dependence of beam-spin and beam-charge asymmetry at collider kinematics. It remains to be shown, to what extent these future data sets will allow the determination of quark or gluon GPDs in the region of very small ξ.

6.2   Experiments at Jefferson Lab

Jefferson National Laboratory (JLAB) has approved two dedicated DVCS experiments to run at the 6 GeV longitudinally polarised electron beam with high luminosity. The first one, the high-resolution arm spectrometer E00-110 in Hall A (Bertin *et al.* 2000) is using both hydrogen and deuterium targets and finished data taking at the end of 2004.

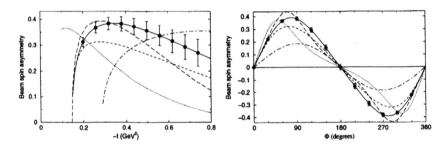

Figure 12. CLAS: *Projections for beam-spin asymmetries at 6 GeV: t-dependence at $\phi = 90°$ (left) and ϕ-dependence at $-t = 0.325\ GeV^2$ (right). Projected statistical errors are given at $Q^2 = 2 \pm 0.5\ GeV^2$ and $x_B = 0.35 \pm 0.05$, for which the solid (dashed) curve shows a calculation (Vanderhaeghen et al. 2001) with ξ-(in)dependent GPDs (Vanderhaeghen et al. 1999). The long-dashed curve shows a calculation including twist-3 effects. Other curves are for other kinematics. The figure is taken from Elouadrhiri (2002).*

It aims at a precise check of the Q^2-dependence of cross section differences in the reaction $ep \rightarrow ep\gamma$, for different beam helicities. The second experiment, E01-113 using the CLAS spectrometer in Hall B (Burkert *et al.* 2001), measures in early 2005 the kinematic dependence of the beam-spin asymmetry on t, ϕ, and x_B, for several fixed Q^2-bins. Also cross-section differences will be measured. As demonstrated in Fig 12, these dependences will be measured with a precision that will allow for a discrimination between certain parameter sets of GPD models.

Precision studies of hard exclusive scattering processes at fixed-target kinematics are among the main research programs driving the 12 GeV upgrade of the Continous Electron Beam Accelerator at JLab (Cardman *et al.* 2001). The high-duty-cycle and high-intensity beam (electrons only) will facilitate more accurate measurements of cross sections and single-spin asymmetries w.r.t. beam helicity and target spin. Running, *eg* 500 hours with the upgraded CLAS detector at a luminosity of 10^{35} cm^{-2}s^{-1} will yield a BSA with a precision about twice better than that for E01-113 (*cf* Fig 13), so that kinematic dependences can be studied in more detail. Projections (not shown) demonstrate (Cardman *et al.* 2001) that using, *eg* 8 bins in the range $0.2 < -t < 0.8$ in each of 3x3 cells in the $(2 < Q^2 < 5\ GeV^2,\ 0.2 < x_B < 0.6)$-plane, the beam-spin asymmetry may be measured with good statistical precision in most of the cells.

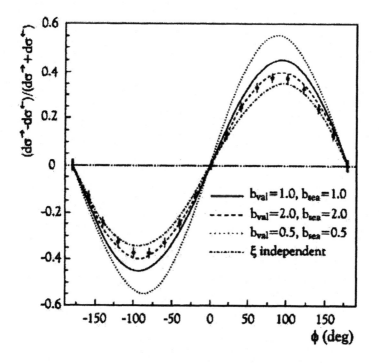

Figure 13. CLAS: *Projected statistical accuracy for a high-statistics BSA measurement at 11 GeV with an upgraded detector. Bins of* $Q^2 = (3 \pm 0.1)\ GeV^2$, $W = (2.8 \pm 0.15)$ *GeV, and* $-t = 0.3 \pm 0.1\ GeV^2$ *are used. GPD calculations (Vanderhaeghen et al. 2001) are shown for different combinations of profile parameters (Vanderhaeghen et al. 1999). The figure is taken from Mecking (2002).*

No plans are published to also install a positron beam at JLAB, so that no high-precision measurements of beam-charge asymmetries can be expected.

6.3 New results expected from HERMES

Between 2002 and the middle of 2005, HERMES data are taken with a transversely po-larised hydrogen target, allowing the evaluation of *transverse target-spin asymmetries* (*cf* Equations 15, 16). Based on an anticipated data sample of about $0.15\ \text{fb}^{-1}$, a first attempt was made to evaluate the sensitivity to GPDs, in DVCS and hard exclusive ρ^0-production on the proton (Ellinghaus *et al.* 2005). Assuming u-quark dominance, the sensitivity to the GPD E_u was studied and, through a model for it, also to the total angular momentum J^u (cf. Equation 5). For both reactions, the projected total experimental 1σ-uncertainty is equivalent to a range of about 0.12 in J^u, so that a significant result can be expected.

The newly built HERMES recoil detector (HERMES Collaboration 2001) will sur-

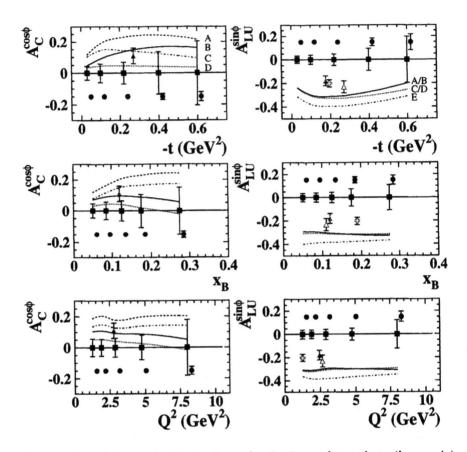

Figure 14. *Left (right) panel: Existing data and projections on beam-charge (beam-spin) asymmetry on the proton, shown as* cos φ *(sin φ) component. For explanations see text. Notes: i) the more recent t-dependence shown in Fig 9 is not included here; ii)* HERMES *average kinematics are used for the displayed model calculations, the average* CLAS *kinematics are lower (higher) in* Q^2 *(x_B) by about a factor of two, as it can also be seen in the figure. The figure is taken from Ellinghaus (2004b).*

round the (unpolarised) internal gas target. By measuring the hitherto undetected recoil proton and/or other low-momentum particles, it will serve several purposes:

i) the slow recoil proton can be identified measuring its large energy deposition in the (diamond-shaped) double-layer double-sided Silicon strip detector.

ii) in conjunction with a double (stereo)-layer scintillating fiber detector possible additional tracks will be identified, so that the exclusivity of the reaction can be established and contaminations from DIS fragmentation and associated (resonance) production will both be reduced to ≤1%.

For 2005-2007, HERMES recoil detector operation is planned with an unpolarised hydrogen target, sharing about equally the running time between both beam charges. In Fig 14 projected accuracies are shown for beam-charge and beam-spin asymmetries, confronted to different GPD model predictions that are explained in Table 1.

Model	D–Term	b_{val}	b_{sea}	Ansatz t–dependence
A	Yes	1	∞	Regge
B	No	1	∞	Regge
C	Yes	1	∞	factorised
D	No	1	∞	factorised
E	Yes/No	1	1	factorised

Table 1. *Parameter sets for GPD model predictions calculated (Vanderhaeghen et al. 2001) on the basis of (Vanderhaeghen et al. 1999) in (Ellinghaus 2004a).*

Error bars shown are total experimental uncertainties, *ie* statistical and systematic uncertainty added in quadrature. The statistical accuracy at a given point in one of the variables $(-t, x_B, Q^2)$ includes integration over the other two variables. Existing data already discussed above, included for completeness, are shown at average kinematics: Preliminary HERMES data are represented by closed triangles for BCA (Ellinghaus 2002a) and 2000 BSA (Ellinghaus 2002b), but open triangles for 96/97 BSA (Airapetian *et al.* 2001); open crosses show BSA data from CLAS (Stepanyan *et al.* 2001). Projected total experimental uncertainties for future BCA and BSA results from HERMES are shown in dependence on $-t$, x_B, and Q^2 (Ellinghaus 2004b): Black squares show the precision of the soon expected final results from 1996-2000 proton data and red circles show the projected precision for 1 fb^{-1} of data from recoil detector running in 2005-2007. Clearly, in the last case a finer binning will be possible for lower values of the respective variables.

Somewhat earlier (Korotkov and Nowak 2002a) the azimuthal dependence of the beam-charge asymmetry was studied for a certain class of GPD models (Vanderhaeghen *et al.* 1999; Goeke *et al.* 2001); projections for larger values of x_B are shown in Fig 15. For the models chosen, there seems to be a clear sensitivity to the existence of the D-term, although it was also shown that the D-term contribution can be replaced by equivalent tuning of other model parameters (Belitsky *et al.* 2002). The right panel of the same figure indicates that from the anticipated data set even 2-dimensional dependences can be mapped to some extent, here showing the t-dependence of the beam-spin asymmetry for two distinct regions in x_B (Korotkov 2001).

When measuring a beam-spin asymmetry at HERMES kinematics, the imaginary part of $\widetilde{M}$ (cf. Equation 11 and text thereafter) will be dominated by $Im\,\mathcal{H}$, *ie* by the GPDs H^q. This suggests a possible way for a first measurement of the quantity $\sum_q e_q^2 \left(H^q(\xi, \xi, t) - H^q(-\xi, \xi, t) \right)$. Its dependence on the skewedness variable ξ is shown in Fig 16 (Korotkov and Nowak 2002b) for two different GPD parameterisations (see caption). Measuring on a proton target, u-quark dominance can be used to obtain a coarse mapping of the function $(H^u(\xi, \xi, t) - H^u(-\xi, \xi, t))$ as a function of t and ξ, *ie* x_B. This function is sometimes referred to as 'singlet' combination, as in the forward limit of vanishing t (and ξ) it reduces to the unpolarised singlet quark PDF $u(x_B) + \bar{u}(x_B)$.

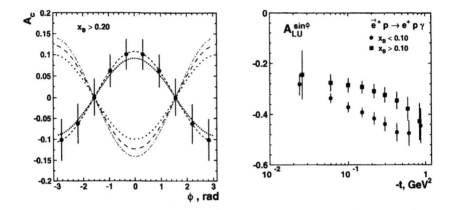

Figure 15. HERMES: *Projections for 2005-07 running with a recoil detector, based on 2 fb⁻¹. Left: φ-dependence of beam-charge asymmetry in the region $x_B > 0.20$. GPDs calculated without D-term, where dashed-dotted means ξ-independent and long-dotted (long-dashed) means ξ-dependent with profile parameter $b = 1$ ($b = 3$), are confronted to those including a D-term, denoted by dotted (dashed) instead. Right: t-dependence of the $\sin \phi$-component of the beam-spin asymmetry, for two distinct regions in x_B. The figures are taken from (Korotkov and Nowak 2002a) and (Korotkov 2001).*

Altogether, the new data set expected from HERMES running in 2005-2007 with a recoil detector will have greatly improved capabilities to discriminate between different GPD models.

6.4   Future DVCS results from COMPASS

When considering leptoproduction by muons instead of electrons, the strength of radiative elastic scattering is reduced by the squared ratio of the beam particle masses, $(m_e/m_\mu)^2$ (Mo and Tsai 1969). The relative contributions of Bethe-Heitler and DVCS processes to real photon leptoproduction vary strongly with beam energy, the former dominates the latter at electron beam energies of 27.5 GeV (Korotkov and Nowak 2002a) and below. Hence at HERMES and CLAS the DVCS cross section contribution is very hard to access experimentally. Instead, for a muon beam the DVCS process is already dominant over the

BH one at an energy of 200 GeV, making COMPASS at CERN the only set-up that is able to measure $|\tau_{DVCS}|^2$ at moderate values of x_B. At 100 GeV, DVCS and BH contribution are of comparable size, suggesting this lower energy for a measurement of the beam-charge asymmetry. Like in the case of the HERA electron or positron beam, the CERN SPS muon beam can be produced with either charge. Unlike the former case, its helicity is fixed and hence always non-zero for individual beam charges. The resulting non-zero $\sin \phi$-component in the beam-charge asymmetry drops out when symmetrising the BCA, *ie*, when calculating it only over a range of π.

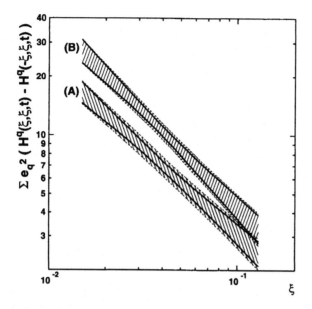

Figure 16. HERMES: *Projected extraction of* $Im\,\mathcal{H}$, *measuring DVCS at* HERMES *with a Recoil Detector in 2005-07, based on* $2\,fb^{-1}$. *The projection B (A) is calculated using a* ξ-*(in)dependent GPD, corresponding to the dash-dotted (long-dotted) line in the left panel of the previous figure. Solid lines enclose the projected fully correlated* 1σ *error band in the region* $-t \leq 0.15\,GeV^2$. *The shaded area outside of a band indicates a possible systematic uncertainty of the extraction method used. The figure is taken from (Korotkov and Nowak 2002b).*

A possible GPD experiment at COMPASS (Burtin *et al.* 2003) would use a 2.5 meter long liquid hydrogen target to achieve a luminosity comparable to that of HERMES, *ie*, approximately 10^{32} s^{-1}cm^{-2}. Several detector upgrades are necessary to reduce photon background from π^0 and to detect the recoil proton. The anticipated accuracy of a DVCS cross section measurement at $E_\mu = 190$ GeV amounts to a few % (Burtin *et al.* 2003). Running at 100 GeV will allow to measure the azimuthal dependence of the beam-charge asymmetry in bins of x_B and Q^2 with good statistical accuracy, as can be inferred from Fig 17 (d'Hose *et al.* 2002), with the two curves representing two different GPD ansätze, see caption.

7 Conclusions

Deeply Virtual Compton Scattering appears to be the presently best tool to pursue the in-depth study of the angular momentum structure of the nucleon. Interpreting the rich body of present and future data within the theoretical framework of generalised parton distributions, it can be expected that severe constraints to different ansätze and parameterisations will emerge.

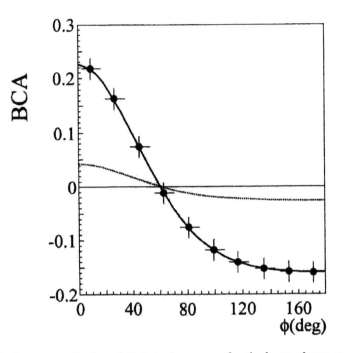

Figure 17. COMPASS: *Projected statistical accuracy for the beam-charge asymmetry from 100 GeV running, 3 months each per beam charge, with an upgraded apparatus. The statistical uncertainty shown is for the bin $0.03 < x_B < 0.07$, $1.5 < Q^2 < 2.5 GeV^2$ while integrating over $0.06 < -t < 0.3 \ GeV^2$. Solid and dotted curve show a nonfactorised and a Regge-type GPD ansatz. The figure is taken from d'Hose et al. (2002).*

Final analysis results from existing JLAB and HERMES data sets are expected soon to give a first glimpse on kinematic dependences of beam-spin and beam-charge (HERMES only) asymmetries on proton and deuteron (HERMES only). Final data on proton and deuteron longitudinal target-spin asymmetries can be expected from HERMES, as well as on the A-dependence of BSAs measured on several nuclei.

HERMES data taking with a transversely polarised hydrogen target in 2003-2005 is expected to yield information on transverse target-spin asymmetries, which may be capable of giving first experimental hints on the total angular momentum of the u-quark.

JLAB experiments running in 2004 and 2005 and HERMES running with a recoil detector in 2005-2007 are expected to deliver quite accurate kinematic dependences of asymmetries and cross section differences. This data will presumably allow a first look to the ξ-dependence of the unpolarised 'valence' u-quark GPD.

Independent information is expected from possible COMPASS running in the last quarter of the decade, in particular beam-charge asymmetry and DVCS cross section will be measured with good precision at moderate values of x_B.

The collider experiments H1 and ZEUS are measuring DVCS at very small values of x_B. They have obtained the Q^2 and W-dependence of the DVCS cross section and will attempt to measure its t-dependence, as well. Based on recent upgrades they will attempt to obtain results on beam-spin and beam-charge asymmetries from data taking in 2005-2007.

The extraction of firm information on GPDs from experimental data will continue to constitute a complicated task. As can be judged from today, a major step in precision towards multi-dimensional mapping of generalised parton distributions can be made once the 12 GeV upgrade of the JLab electron beam facility will have become reality. This data will then allow to perform a global fit based on high-precision beam-spin (and target-spin) asymmetries, aiming at a simultaneous determination of several accessible GPDs in dependence on t, x_B, and Q^2. Note that a considerable model-dependence will remain for the (internal) x-dependence of GPDs, as it can be accessed only via beam-charge asymmetry measurements that are not possible at JLAB. Nevertheless, eventually completely new knowledge may become available on the 3-dimensional angular momentum structure of the nucleon.

Acknowledgments

I am deeply indebted to M Diehl and F Ellinghaus for many beneficial discussions and most valuable comments on the manuscript. My thanks go to the University of Glasgow, namely to G Rosner and R Kaiser, for the invitation to present this exciting subject at the I3HP European Topical Workshop, held in early September 2004 in front of the most beautiful golf scenery of St. Andrews.

References

Adloff C *et al.* [H1 collaboration], 2001, *Eur Phys J C* **24** 517.
Airapetian *et al.* , [HERMES collaboration], 2001, *Phys Rev Lett* **87** 182001.
Anikin I V *et al.* , 2000, *Phys Rev D* **62** 071501.
Belitsky A V and Müller, 1998, *Phys Lett B* **417** 129.
Belitsky A V *et al.* , 2002, *Nucl Phys B* **629** 323.
Belitsky A V and Müller, 2002, *Nucl Phys A* **711** 118c.
Berger E *et al.* , 2001, *Phys Rev Lett* **87** 142302
Bertin P *et al.* , 2000, CEBAF experiment E00-110.
Blümlein J *et al.* , 1999, *Nucl Phys B* **560** 283; 2000, *Nucl Phys B* **581** 449.
Brodsky S J *et al.* , 1972, *Phys Rev D* **6** 172.
Burkardt M, 2000, *Phys Rev D* **62** 071503; Erratum ibid. **66** 119903.
Burkardt M, 2003, *Int J Mod Phys A* **18** 173.
Burkert V *et al.* , 2001, CEBAF experiment E01-113
Burtin E *et al.* , 2003, *Nucl Phys A* **721** 368c.
Cano F and Pire B, 2004, textit *Eur Phys J A* **19** 423
Cardman L S *et al.* , 2001, *White Paper: The Science driving the 12 GeV Upgrade of CEBAF*,
    http://www.jlab.org/12GeV/collaboration.html.
Chekanov S *et al.* [ZEUS collaboration], 2003, *Phys Lett B* **573** 46.
Diehl M *et al.* , 1997, *Phys Lett B* **411** 193.
Diehl M, 2002, *Eur Phys J C* **25** 223; Erratum-ibid, 2003, **31** 277.
Diehl M, 2003, *Physics Reports* **388** 41.
Diehl M and Vinnikov A V, 2004, arXiv:hep-ph/0412162.

Diehl M *et al.* , 2005, *Eur Phys J C* **39** 1.

Diehl M and Sapeta S, 2005, *in preparation.*

Dittes F M, 1988, *Phys Lett B* **209** 325.

Donnachie A and Dosch H G, 2001, *Phys Lett B* **502** 74.

Ellinghaus F [HERMES collaboration], 2002a, *Nucl Phys A* **711** 171c.

Ellinghaus F *et al.* [HERMES collaboration], 2002b, arXiv:hep–ex/0212019, *AIP Conf Proc* **675** 303, Ed Makdisi I Y *et al.*

Ellinghaus F, 2004a, PhD thesis, Humboldt University Berlin/D, *DESY-THESIS-2004-005.*

Ellinghaus F [HERMES collaboration], 2004b, to be published in *Proc. DIS'04, Strbske Pleso/SK.*

Ellinghaus F [H1, ZEUS and HERMES collaborations], 2004c, arXiv:hep–ex/0410094, to be published in *Proc. ICHEP'04, Beijing, China.*

Ellinghaus F *et al.* , 2005, *in preparation.*

Elouadrhiri L, 2002, *Nucl Phys A* **711** 154c.

Favart L [H1 collaboration], 2004, *Eur Phys J C* **33** S509.

Freund A, 2000, *Phys Lett B* **472** 412.

Freund A and McDermott M F, 2002, *Phys Rev D* **65** 091901; *Eur Phys J C* **23** 651.

Freund A *et al.* , 2003, *Phys Rev D* **67** 036001.

Goeke K *et al.* , 2001, *Prog Part Nucl Phys* **47** 401.

Goloskokov S V and Kroll P, 2005, arXiv:hep-ph/0501242.

Guidal M, 2002, *Nucl Phys A* **711** 139c.

Guidal M *et al.* , 2004, arXiv:hep/ph-0410251.

HERMES collaboration, 2001, *A Large Acceptance Recoil Detector for HERMES*, DESY PRC 01-01.

d'Hose N *et al.* , 2002, *Nucl Phys A* **711** 160c.

Ji X, 1997, *Phys Rev Lett* **78** 610; *Phys Rev D* **55** 7114.

Ji X and Osborne J, 1998, *Phys Rev D* **58** 094018.

Korotkov V A, 2001, *talk at '267. WE-HERAEUS-Seminar on Generalised Parton Distributions'*, Bad Honnef (Germany), Nov 2001; *also in:* HERMES collaboration 2001.

Korotkov V A and Nowak W-D, 2002a, *Eur Phys J C* **23** 455.

Korotkov V A and Nowak W-D, 2002b, *Nucl Phys A* **711** 175c.

Kroll P *et al.* , 1996, *Nucl Phys A* **598** 435.

Mankiewicz L *et al.* , 1998, *Phys Lett B* **425** 186.

Mecking B A, 2002, *Nucl Phys A* **711** 330c.

Mo L W and Tsai Y S, 1969, *Rev Mod Phys* **41** 205.

Müller D *et al.* , 1994, *For Phys* **42** 101.

Nowak W-D, 2003, 'The Spin Structure of the Nucleon' 37, Ed Steffens E and Shanidze R, Kluwer Academic Publishers.

Polyakov M V and Weiss C, 1999, *Phys Rev D* **60** 114017.

Polyakov M V, 2003, *Phys Lett B* **555** 57.

Radyushkin A V, 1996, *Phys Lett B* **380** 417; *Phys Rev D* **56** 5524.

Radyushkin A V, 1999, *Phys Rev D* **59** 014030.

Radyushkin A V and Weiss C, 2001, *Phys Rev D* **63** 114012.

Ralston J P and Pire B, 2002, *Phys Rev D* **66** 111501.

Smith E S [CLAS collaboration], 2003, *AIP Conf Proc* **698** 129, Ed Parsa Z.

CLAS collaboration, Stepanyan S *et al.* , 2001, *Phys Rev Lett* **87** 182002.

Vanderhaeghen M *et al.* , 2001, *Computer Codes for DVCS and BH Calculations*, priv com.

Vanderhaeghen M *et al.* , 1999, *Phys Rev D* **60** 094017.

The HERMES Recoil Detector

Björn Seitz

II. Physikalisches Institut, University of Giessen
35392 Giessen, Germany

1 Introduction

The angular momentum structure of the nucleon is one of the central questions of present day hadron physics. The spin of the nucleon is composed of the spins and orbital angular momenta of its constituents, the quarks and gluons. The HERMES experiment at DESY was designed to study the spin structure of the nucleon using inclusive and semi-inclusive Deep Inelastic Scattering processes (Rith 2002). Impressive results on the contribution of different quark flavours to the nucleon spin were obtained recently (Airapetian *et al.* 2004).

Inclusive and semi-inclusive measurements, however, cannot access the orbital angular momentum of quarks and gluons. The only viable although indirect access to this quantity known presently is through so-called Generalised Parton Distributions (Ji 1997, Goecke *et al.* 2001, Nowak 2004). The cleanest process to measure them is Deeply Virtual Compton Scattering, the electroproduction of a real photon (Belitsky and Müller 2002). This process leads to the same final state as the Bethe-Heitler process. Hence these two processes interfere quantum-mechanically. At HERMES energies, the comparatively large BH cross-section can serve as a tool to access the DVCS process. The interference allows direct access to both the real and imaginary parts of the scattering amplitude which are linked to certain linear combinations of Generalised Parton Distributions. Two observables making use of this interference were studied at HERMES so far: the beam helicity asymmetry A_{LU} and the beam charge asymmetry A_C on an unpolarised target as a function of the azimuthal angle ϕ between the lepton scattering plane and the production plane. Current HERMES data on these observables are shown as open symbols in Figure 1 together with predictions for an extended running period using the recoil detector described here.

The present HERMES set-up allows only the detection of the scattered lepton and the produced photon. The DVCS events are then identified by requiring the missing mass M_X^2 to be close to the mass of the proton. The resolution in M_X^2 due to the HERMES

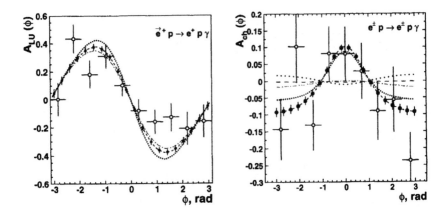

Figure 1. *Projections on the two main observables in DVCS at HERMES energies, the beam charge asymmetry A_C (left) and the beam spin asymmetry A_{LU} (right). The open symbols indicate the current HERMES data, the closed symbols the projected results for an integrated luminosity of $2\,fb^{-1}$. The curves show predictions from various GPD parametrisations (Kaiser et al. 2002).*

spectrometer is $\Delta M_X^2 \approx 1.5$ GeV^2, so that background from the excitation of nucleon-resonances or non-resonant pion production can contaminate the DVCS sample. Also kinematic variables of the process can only be calculated from these two particles detected. This limits in particular the resolution and range in the Mandelstam variable t, describing the momentum transfer to the target nucleon.

In order to improve the detection capabilities for hard exclusive reactions like DVCS, the HERMES collaboration will install a new recoil detector for the envisaged final two years of the HERA-II running period. The study of hard exclusive processes requires a recoil detector surrounding the target. The detector focuses on positive identification of recoiling protons, improving the transverse momentum reconstruction of these particles and rejecting non-exclusive background events (Kaiser *et al.* 2002).

The detector design is driven by the angular and momentum distributions of protons recoiling in DVCS events. The kinematic distribution for these protons as a function of momentum p and polar angle θ is shown in Figure 2. A complete coverage in the azimuthal angle ϕ is desired as well as a polar angle coverage of $0.1 < \theta < 1.35$ rad. Expected momenta reach from below 0.1 GeV/c up to 1.5 GeV/c.

Two complementary detectors are needed to cover the whole kinematic range. Low momentum particles will be detected by a silicon strip detector (SSD) while higher momenta will be covered by a scintillating fibre tracker (SFT). The SSD will operate inside the beam vacuum to measure very low momenta.

The HERMES Recoil Detector shown in Figure 3 will thus consist of three main components: a two-layer SSD surrounding the target cell inside the beam vacuum, a SFT in a longitudinal magnetic field of 1 T and a photon detector consisting of three layers of tungsten radiators and scintillators (Seitz 2004). The fibre tracker and the silicon

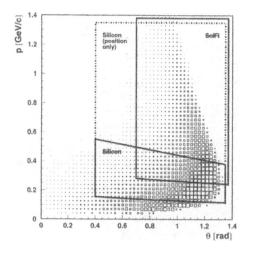

Figure 2. *Distribution of recoil protons from DVCS events as a function of the polar angle θ and their momentum p. The areas where the silicon and scintillating fibre detectors provide space point and particle identification information are indicated as well as the area where only space points are delivered from the silicon detector.*

detectors were optimised to detect recoil protons in a momentum range from 0.1 to 1.5 GeV/c. Their particle identification properties allow the discrimination of protons from pions over the whole momentum range. Momentum determination for low momentum particles stopping inside the silicon detector is performed using the $\Delta E - E$ technique. The momentum of fast particles will be determined by their bending in the magnetic field. The photon detector is used for particle identification of high-momentum particles, for π° reconstruction and for the rejection of background from neutral particles.

In the following, the main detector components will be detailed starting from the inside out describing the silicon detector, the scintillating fibre tracker and the photon detector. Another section will describe the tracking and particle identification algorithms envisaged for the recoil detector at HERMES. The projected performance for DVCS measurements using the recoil detector and a summary conclude the paper.

2 Silicon detector

Detecting very low energetic recoil protons requires a segmented detector as close as possible to the target cell. Inside the beam vacuum a two-layer silicon detector will be installed (Hristova 2004). The innermost silicon layers should detect protons starting from the lowest momenta possible while covering a polar angular acceptance $0.1 < \theta < 1.35$ rad. The lower momentum cut-off is determined by the material between the target and the detector. By using silicon sensors and placing them inside the scattering chamber vacuum this cut-off can be as low as 135 MeV/c. In this momentum range the energy

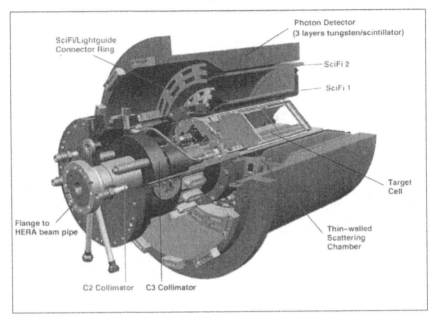

Figure 3. *View of the HERMES Recoil Detector showing its different subdetector components as well as the target cell and beam pipe. For a better view of the detectors the superconducting magnet was omitted from this view.*

deposited in silicon by a proton is a steep function of the momentum. Therefore the proton momentum can be computed from the energy deposition in the two silicon layers. If the proton stops in the second silicon layer the system acts as a $\Delta E - E$ detector.

The SSD consists of 16 double-sided TIGRE detectors (MICRON Semiconductors Ltd) of 300 μm thickness covering an active area of 99×99 mm^2 each. They feature 128 strips, 702 μm wide, on each side with a strip pitch of 758 μm. Strips on either side are perpendicular to each other allowing space point reconstruction. The SSD is sensitive to energy depositions from minimum ionising particles (MIP) up to the large energy deposition expected from protons stopping in either silicon layer. To cover the large dynamic range expected from these signals the TIGRE sensors are connected to the read-out electronics using a fan-out which splits the signal by capacitive charge division (Reinecke 2004). Every sensor strip is connected to a high and a low gain channel allowing signals from 4 fC up to 270 fC to be digitised. The readout itself is based on HELIX128-3.0 chips.

The response and efficiency of the SSD was tested in beams at DESY using electrons as MIP as well as at proton beams at Erlangen University and GSI (Murray 2004). The test beam at DESY features a 6 GeV/c electron beam which was used to test the position dependence of the SSD performance with MIPs. These measurements were performed using a Si–telescope with very precise spatial resolution as a reference detector. Figure 4 shows the response of the test module to electrons of 6 GeV/c as a function of the relative position along the strip. The efficiency of the module is measured to be 98.73%. The energy response was found to be homogenous along the strip.

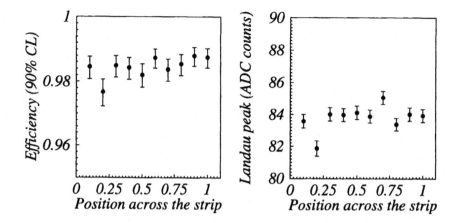

Figure 4. *Efficiency (left) and energy response (right) of the SSD measured with a 6 GeV electron beam at DESY as a function of the relative position across the strip.*

3 Scintillating fibre detector

Particles of higher momenta will be measured in two barrels consisting of 4 layers of 1 mm Kuraray SCSF–78 M scintillating fibres each. A stereo angle of 10° between each two adjacent layers of each barrel allows space point reconstruction. The diameters of the barrels are 220 mm and 370 mm, respectively. The barrels are built as a self-supporting structure to minimise the material traversed by the particles.

Scintillating fibres are an ideal tool to combine energy and position measurements of charged particles in an intermediate momentum range. They offer a high granularity while keeping the mechanical construction and material density at a minimum. A fibre thickness of 1 mm ensures sufficient light output even for minimum ionising particles while for application with the HERMES Recoil Detector the position resolution is still smaller than the effect of multiple scattering. The chosen fibres combine superior radiation hardness, higher light yield and better mechanical durability compared to other products available.

The readout of the fibres is done via 3.5 m long light guides of Kuraray clear fibres connected to Hamamatsu H-7548 64 channel PMTs. The readout is based on front-end cards using GASSIPLEX chips storing the energy deposition in each fibre. They were developed for the HADES RICH detector and adapted for use with fast PMTs. In addition, the Dynode 12 signals of the PMTS are used to extract fast timing information. They will be stored in multi–event, multi–hit TDCs in order to discriminate events in coincidence with the HERMES spectrometer trigger from random background.

A mixed proton/pion beam was used at GSI to test prototypes of the SFT together with prototypes for the other subdetectors. The set up is shown in Figure 5. External Scintillators were used for triggering (S1 and S2), particle identification via Time-of-Flight (S2 and S0) and efficiency determination (S3 and S4). The incidence position was measured in front of the scintillating fibre modules using a MWPC.

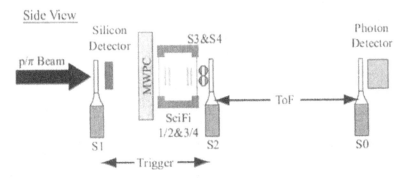

Figure 5. *Setup of the recoil detector test measurements performed at GSI. Prototypes from all subdetectors were exposed to a mixed proton/pion beam in the relevant momentum range. A ToF system and a MWPC served as independent reference detectors.*

The scintillating fibre detector prototype consisted of four 64-channel modules arranged in parallel and stereo layer pairs resembling the final setup as closely as possible. For read-out and analysis chain the final components were used.

Beam momenta from 300 MeV/c to 900 MeV/c were used to test the response, efficiency and particle identification properties of the SFT detector. Test beam results from protons and pions taken with these prototype modules are shown in Figure 8 (left). The detailed analysis and resulting response to protons and pions for various momenta can be found in Hoek (2004).

The PID response (see Section 5) of a combination of 4 staggered SFT modules is compared to the particle discrimination reached by the time-of-flight technique. The particle discrimination properties reach the design values. Figure 6 shows the detection efficiency of one module for protons and pions as a function of the incident momentum. The analysis of these data uses an elaborate clustering algorithm to take various sources of cross talk into account. For use as a tracker a high efficiency per layer is mandatory for a good tracking efficiency.

4 Photon detector

The Recoil Detector is complemented by a detector array optimised to detect photons. This photon detector consists of three segmented scintillator barrels where each segment is read out by wave length shifting fibres. Starting from the inside out the scintillator barrels are preceded by 2 radiation lengths of tungsten converter for the first layer and 1 radiation length of tungsten in front of the other two layers, respectively. The innermost barrel is segmented parallel to the beam line into 60 parts. The 2 × 44 segments of the two outer layers form an angle of ±45° with respect to the beam line.

The read-out of the photon detector will be done by using multi anode photomultipliers of the same type as will be used for the scintillating fibre counter. Due to the low total channel number commercial VME electronics will be used to digitise the deposited energy signal.

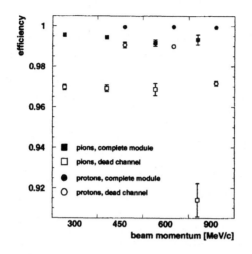

Figure 6. *Efficiency of a single scintillating fibre module for proton and pion detection measured for four different momenta at the GSI test beam. The closed symbols show a completely working module, the open symbols the efficiency for a module with one dead channel. Data points corresponding to proton and pion signals are slightly shifted for better visibility.*

The purpose of the photon detector is threefold. It should detect and thus reject background from neutral particle production. Simulations predict an efficiency of 70 % to detect a single photon from a π° decay. For charged particles the information from the photon detector can be used to enhance the PID properties of the complete setup at high momenta. Also, it can be used to reconstruct π° with an efficiency of 12%.

Prototype modules of this detector were also tested at GSI. Its energy response and PID capabilities were studied and found to be in agreement with expectations. Last but not least, the photon detector will deliver a trigger signal on cosmic ray particles for the global alignment of the complete detector system.

5 Tracking and particle identification

One of the central issues in building the Recoil Detector is to not only detect the recoil proton from hard exclusive reactions, but to also measure its momentum as precisely as possible in order to enhance the definition of the reaction kinematics.

Tracking with the setup chosen for the HERMES Recoil Detector provides several challenges.

- *Momentum range*: The tracking algorithm has to deal with a large momentum range from slow protons to MIP. Two methods have thus to be applied: measuring the momentum using the energy deposition inside the SSD for the low momentum range and using the bending in a 1 T solenoidal field using space points from the SSD and SFT.

- *Energy loss*: In the momentum range considered for DVCS protons the error in neglecting energy losses in active and passive detector materials can be $\mathcal{O}(10\%)$ of the particle's total energy.

- *Multiple scattering*: Effects from multiple scattering will be comparable to or larger than the detector resolution.

- *Low number of space points*: Including the interaction vertex reconstructed from the existing HERMES spectrometer, an ideal track will only have 5 space points.

- *Stopping particles*: Low momentum particles might stop in various places inside the HERMES Recoil Detector. These cases will have to be treated separately.

The tracking process can be divided into two separate steps, namely track finding and track fitting. Track finding is basically a classification problem where individual space points need to be grouped into a track candidate. Track fitting will then determine the parameters of the track described by these space points.

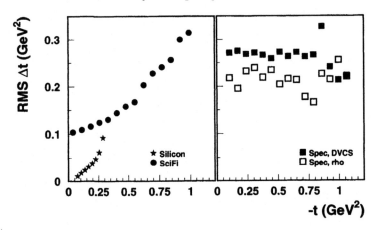

Figure 7. *Expected resolution in t (left) for the SSD ($\star$) and SFT ($\bullet$) compared to the resolution obtained by the existing HERMES setup (right). A significant improvement especially at low values of t is obvious.*

For track finding a search cone starting from a track seed will be defined. The information on space points found will be used to update the search cone until enough space points to form a track are found. Short tracks will be treated as low momentum particles. Their momentum will be calculated from the deposited energy. For full tracks the bending inside the magnetic field will be used. A track following method takes the inhomogeneities in the magnetic field into account (Osborne 2004). In a next step, energy loss and multiple scattering will be incorporated using a Kalman filter approach.

The treatment of energy loss in the tracking algorithm alone justifies the effort of particle identification. One of the underlying philosophies of the HERMES Recoil Detector is that all detector components should not only deliver spatial coordinates for tracking, but also provide ΔE information to be used for particle identification.

The particle identification scheme employed for the HERMES Recoil Detector follows the PID scheme developed for the existing spectrometer as close as possible. It is based on conditional probabilities for the individual detector responses to the particle types in question.

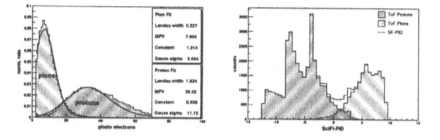

Figure 8. *Parent distributions for proton and pion at $P = 600 MeV/c$ determined for a single SFT module (left). The fit is shown together with the experimental data sample where particles were identified via ToF. The resulting PID distributions for 3 staggered detectors are shown in the right hand figure. A clear separation into protons and pions using the SFT alone is visible.*

Each individual subcomponent of the HERMES recoil detector acts as a position sensitive detector and a detector measuring the energy loss. Thus, for each charged particle trajectory a measurement on $\vec{p}$ and ΔE_{det} will be available. From this a purely detector dependent probability function (the so called *parent distribution*) for a particle of type i and momentum p leaving a signal ΔE inside the detector can be defined. These parent distributions are either determined in a test experiment or from data requiring stringent cuts on the reaction. From these parent distributions the decadic logarithm of the probabilities for two particle species can be formed, the so called PID quantity. It is purely detector dependent and additive allowing an easy combination of different subdetectors. Figure 8 show the results obtained at the GSI test beam. Other subdetectors show similar performance and will be added subsequently giving a clear discrimination for various particle species over the whole momentum range.

6 Projections and summary

Based on the expected performance of the HERMES Recoil Detector projections of its physics potential were evaluated. They are calculated for Deeply Virtual Compton Scattering based on various model approaches for Generalised Parton Distributions (Kaiser *et al.* 2002). The enhanced statistical precision as well as the improved systematic accuracy and resolution in t will allow a binning of DVCS beam charge and beam spin asymmetries as shown in Figure 1 in x and t. The projected results for these quantities as a function of t corresponding to an integrated luminosity of $2fb^{-1}$ are shown in Figure 9.

The Recoil Detector will run with similar luminosity of electrons and positrons of

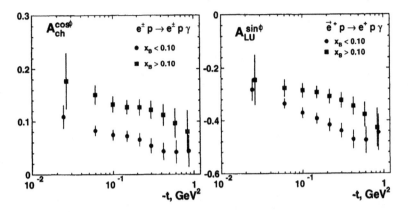

Figure 9. *Projected result on A_C (left) and A_{LU} (right) as a function of t for two different bins in x for an integrated luminosity of $2fb^{-1}$. The systematic accuracy will be comparable to the statistical precision.*

both polarisation states. Possible target nuclei range from proton and deuteron to Helium, Neon, Krypton and Xenon allowing the study not only of proton and neutron, but of nuclear effects as well. Running the Recoil Detector will produce a large data sample on nucleon and nuclear unpolarised targets allowing a wealth of reactions to be studied.

It's main emphasis, however, is the detection of recoiling nucleons from hard exclusive processes. Prototype results and simulations prove that it is well suited for this task and will provide unprecedented data on these reactions.

The physics implications in terms of a possible extraction of nucleon GPDs are further discussed in Nowak (2004).

References

Airapetian A *et al.*, 2004, submitted to *Phys Rev D.*
Belitsky A and Müller D, 2002, *Nucl Phys B* **629** 323.
Goecke K *et al.* , 2001, *Prog Part Nucl Phys* **47** 401.
Hoek M, 2004, Contribution to these proceedings.
Hristova I, 2004, Contribution to these proceedings.
Ji X, 1997, *Phys Rev Lett* **78** 610.
Kaiser R *et al.* , 2002, The HERMES Recoil Detector, Technical Design Report HERMES 02–003.
Korotkov V A and Nowak W D, 2002, *Eur Phys J C* **23** 455.
Murray M, 2004, Contribution to these proceedings.
Nowak W D, 2004, Contribution to these proceedings.
Osborne A, 2004, Contribution to these proceedings.
Reinecke M, 2004, *IEEE Trans Nucl Sci* **51** 1111.
Rith K, 2002, *Prog Part Nucl Phys* **49** 245.
Seitz B, 2004, *Nucl Instrum Methods A* (in press).

Prospects for GPD measurements with COMPASS at CERN

Nicole d'Hose

CEA-Saclay, SPhN-DAPNIA-DSM
F91191 Gif-sur-Yvette Cedex, France

1 Introduction

1.1 GPDs: 3-dimensional picture of the nucleon

One of the main open questions in the theory of strong interactions is to understand how the nucleon is constructed from quarks and gluons, the fundamental degrees of freedom in QCD. An essential tool to investigate nucleon structure is the study of deep inelastic scattering processes where individual quarks and gluons are resolved. The parton densities that can be extracted from *inclusive* deep inelastic scattering (DIS) describe the distribution of longitudinal momentum ("longitudinal" refers to the direction of the fast moving nucleon in the centre of mass of the virtual photon - nucleon collision). Nevertheless they do not carry any information about the distribution of partons in the transverse plane. In this sense inclusive deep inelastic scattering provides us with a 1-dimensional image of the nucleon. In recent years it has become clear that much more detailed information, encoded in Generalised Parton Distributions (GPDs) (Ji 1997, Radyuskin 1997) can be obtained from hard *exclusive* processes such as deeply virtual Compton scattering (DVCS) and hard exclusive meson production (HEMP). The transverse component of the non-zero momentum transfer between the initial and final nucleon gives access to information about the spatial distribution of partons in the transverse plane. This "mixed" longitudinal parton momentum and transverse coordinate representation provides a 3-dimensional image of the partonic structure of the nucleon (Burkardt 2000, Belitsky and Müller 2002a, Ralston and Pire 2002).

Not only the momentum but also the polarisation of the target can be changed by hard exclusive scattering, which leads to a rich spin structure of GPDs. The GPDs will help to unravel the nucleon spin puzzle because they provide, thanks to a sum rule (Ji 1997), a measurement of the total angular momentum carried by partons, comprising the spin and the orbital angular momentum. Moreover, it has also been shown that GPDs make a connection between ordinary parton distributions and elastic form factors and hence between the principal quantities which so far have provided information on nucleon structure.

Physically we can expect that the dynamics of partons will give rise to a separation between fast and slow partons, quarks and gluons, in their transverse spatial distribution. For example, lattice calculations (Negele *et al.* 2004) have shown that slow partons tend to stay at a larger distance from the nucleon centre than the fast partons. The chiral dynamics approach (Strikman and Weiss 2004) has also demonstrated that at larger transverse distance the gluon density is generated by the "pion cloud" of the nucleon which implies that the transverse size of the nucleon increases if x_{Bj} drops significantly below m_π/m_N. This is the kinematical domain of study at COMPASS. Knowledge of the transverse size of parton distribution, which is reachable by hard exclusive lepto-production of photons or mesons, is indubitably interesting in itself, and will provide also important parameters in modelling hadron-hadron collisions at LHC or RHIC.

1.2 DVCS and HEMP: tools to study GPDs

Deeply Virtual Compton Scattering (DVCS) as well as Hard Exclusive Meson Production (HEMP) are the most suitable tools to learn about GPDs. Ji (1997), Radyuskin (1997) and Collins *et al.* (1997) have shown that deeply virtual Compton scattering can be factorised into a hard-scattering part (exactly calculable in pQCD) and a non-perturbative nucleon structure part (see illustration in Figure 1).

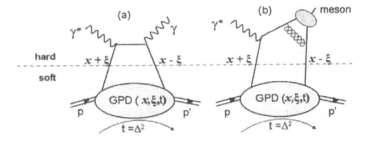

Figure 1. *Handbag diagrams for the (a) DVCS and (b) HEMP amplitudes described at leading order.*

The factorisation is valid when the finite momentum transfer $t = \Delta^2 = (p - p')^2$ to the target remains small compared to the photon virtuality Q^2. In the so-called "handbag" diagram, the lower blob represents the soft structure of the nucleon and can be described, at leading order in $1/Q$, in terms of four GPDs which conserve quark helicity: $H, \tilde{H}, E$ and $\tilde{E}$. The GPDs reflect the structure of the nucleon independently of the reaction which probes the nucleon. They can also be accessed through the hard exclusive electroproduction of mesons $\pi, \rho^0, \omega, \phi...$ for which the factorisation (Collins *et al.* 1997) implies the extra condition that the virtual photon must be longitudinally polarised.

The GPDs depend upon three kinematic variables: x, ξ and t. The variable x is the average longitudinal momentum fraction of the active quarks in the quark loop (see Figure 1), while ξ is the longitudinal momentum fraction of the transfer Δ necessary to transform a virtual photon into a real photon or a meson, and is related to x_{Bj} as $\xi = x_{Bj}/(2 - x_{Bj})$ in the Bjorken limit. The third variable, $t = \Delta^2$, is the momentum transfer between the initial and final nucleons, which contains in addition to the longitudinal component a transverse one. This leads to information about the spatial transverse distribution of partons in addition to the longitudinal momentum distribution.

At present GPDs are still mostly unknown, apart from constraints from their generic properties linked to parton densities and form factors. The different models are based on the dynamics of partons. The amplitudes of the hard exclusive processes is an integral over x of the GPDs as the variable x flows along the loop (see Figure 1). To extract GPDs from this convolution, the strategy will be to parameterise the GPDs and to determine the parameters by a fit to the data.

2 GPD measurements with a high energy μ beam

A complete set of measurements of both Hard Exclusive Meson Production (HEMP) with a large set of mesons ($\rho, \omega, \phi, \pi, \eta$...) and Deeply Virtual Compton Scattering (DVCS) can be performed with the 100 GeV muon beam and the high resolution COMPASS spectrometer completed by a recoil detector for which a conceptual design will be described in the last section.

2.1 Complementarity of the kinematic domains investigated at JLab, DESY and CERN

Experiments have already been undertaken at very high energy at the HERA collider (H1 collaboration 2001 and ZEUS collaboration 2003) to study mainly the gluon GPDs at very small x_{Bj} ($\leq 10^{-2}$). Larger values of x_{Bj} have been investigated in fixed target experiments at JLab (CLAS collaboration 2001) (at 6 GeV, with an upgrade at 11 GeV planned around 2010) and HERMES (HERMES collaboration 2001) (at 27 GeV). The goal of an experiment is to study DVCS and HEMP at fixed x_{Bj} in a large range in Q^2 in order to control the factorisation and the dominance of the handbag diagram for the studied reaction (see Figure 1). Figure 2 shows that the maximum energy of the lepton beam will limit the domain in Q^2 at fixed x_{Bj}. Hence the high energy of the muon beam provides a large advantage to COMPASS. The experimental program using COMPASS at CERN with a muon beam of 100 GeV will give access mainly to three bins in x_{Bj} (presented in Figure 2): $[0.05 \pm 0.02]$; $[0.1 \pm 0.03]$; $[0.2 \pm 0.07]$ over a large range of Q^2 ($1.5 \leq Q^2 \leq 7.5$ GeV^2). The range in Q^2 for COMPASS is at present limited to a maximum of 7.5 GeV^2 not due to the energy of the muons but due to a reasonable time limit of 6 months for data taking for a DVCS experiment, assuming a muon flux of $2 \cdot 10^8$ μ per SPS spill. It has to be noted that an increase of the number of muons per spill by a factor 2 would increase the range in Q^2 up to about 11 GeV^2.

At COMPASS a domain of intermediate x_{Bj} can be explored which is sensitive to both valence and sea quarks as well as gluons. The chiral dynamics approach, developed

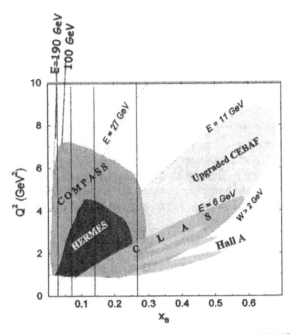

Figure 2. *Kinematical coverage for various planned or proposed DVCS experiments. The limit $s \geq 6$ GeV^2 assures to be above the resonance domain, and $Q^2 > 1.5$ GeV^2 allows to reach the Deep Inelastic regime. The range in Q^2 at COMPASS is at present limited up to 7.5 GeV^2 due to the luminosity determined by a maximum of $2 \cdot 10^8$ μ per SPS spill.*

by Strikman and Weiss (2004), shows that at larger transverse distance from the nucleon centre the gluon density is generated by the "pion cloud" and so the transverse size of the nucleon increases if x_{Bj} drops significantly below m_π/m_N. This can be studied in the kinematical domain of COMPASS.

2.2 DVCS with the two charged states of polarised muon beam

Deeply virtual Compton scattering is accessed by photon lepto-production: $lp \rightarrow l'p'\gamma$. In this reaction, the final photon can be emitted either by the leptons (Bethe-Heitler process) or by the proton (genuine DVCS process). Which mechanism dominates at given Q^2 and x_{Bj} depends mainly on the lepton beam energy E_l. Large values of $1/y = 2m_p E_l x_{Bj}/Q^2$ favour DVCS and small values of $1/y$ favour Bethe-Heitler. The Bethe-Heitler process is completely calculable in QED, with our knowledge of the elastic form factor at small t. The high energy muon beam available at CERN makes it possible to vary the kinematics to favour one of these two processes.

When the DVCS contribution dominates over the BH contribution, the cross section is essentially the square of the DVCS amplitude which, at leading order, has the form:

$$\mathcal{A}(\gamma_T^*) \sim \int_{-1}^{+1} \frac{f(x,\xi,t)}{x - \xi + i\epsilon}dx \sim \mathcal{P}\int_{-1}^{+1} \frac{f(x,\xi,t)}{x - \xi}dx - i\pi f(\xi,\xi,t) \qquad (1)$$

where $\mathcal{A}$ represents the dominant $\gamma^*p \to \gamma p$ amplitudes for the transverse γ^* polarisation and f stands for a generic GPD and $\mathcal{P}$ for Cauchy's principal value integral. The kinematics fixes t and $\xi \sim x_{Bj}/2$.

Since GPDs are real valued due to time reversal invariance, the real and imaginary parts of the DVCS amplitude contain very distinct information on GPDs. The imaginary part depends on the GPDs at the specific values $x = \xi$. The real part is a convolution of the GPDs with the kernel $1/(x - \xi)$ (see Equation 1). To extract the GPDs from this convolution the strategy will be similar to the one used in DIS. The GPDs will be adequately parametrised and the parameters will be determined by a fit to the data.

The real and imaginary parts can be accessed separately through the interference between BH and DVCS. To see how it works (Diehl *et al.* 1997, Belitsky *et al.* 2002b), let us consider an unpolarised target and discuss the dependence of the cross section on the angle φ between leptonic and hadronic planes, and on the charge e_l and longitudinal polarisation P_l of the muon beam:

$$
\begin{aligned}
\frac{d\sigma(\ell p \to \ell p\gamma)}{d\varphi} \\
= \; &\mathrm{A_{BH}}(\cos\varphi, \cos2\varphi, \cos3\varphi, \cos4\varphi) \\
&+ \mathrm{A_{INT}}(\cos\varphi, \cos2\varphi)[\; e_\ell\, [c_1 \cos\varphi \,\mathrm{Re}\,e\mathcal{A}(\gamma_T^*) \; + \; c_2 \cos2\varphi \,\mathrm{Re}\,e\mathcal{A}(\gamma_L^*) \; + \; ...] \\
&\qquad\qquad + e_\ell P_\ell \,[s_1 \sin\varphi \,\mathrm{Im}\,m\mathcal{A}(\gamma_T^*) \; + \; s_2 \sin2\varphi \,\mathrm{Im}\,m\mathcal{A}(\gamma_L^*) \;]] \\
&+ \mathrm{A_{VCS}}(\cos\varphi, \cos2\varphi, P_\ell \sin\varphi)
\end{aligned}
\qquad (2)
$$

where $\mathrm{A_{BH}}$, $\mathrm{A_{VCS}}$, $\mathrm{A_{INT}}$, c_i, s_i are known expressions and $\mathcal{A}$ represents $\gamma^*p \to \gamma p$ amplitudes for different γ^* polarisation. The scaling predictions give leading twist-2 and twist-3 contributions for $\mathcal{A}(\gamma_T^*)$ and $\mathcal{A}(\gamma_L^*)$ respectively. With muon beams one naturally reverses both charge and helicity at once, but we see how all four expressions in the interference can be separated: in the cross section difference $\sigma(\mu^{+\downarrow}) - \sigma(\mu^{-\uparrow})$ the Bethe-Heitler contribution A_{BH} drops out and the real parts of $\mathcal{A}(\gamma_{T,L}^*)$ can be accessed. Using angular analysis $\mathcal{A}(\gamma_T^*)$ and $\mathcal{A}(\gamma_L^*)$ can be separated which allows a test of the scaling predictions $\mathcal{A}(\gamma_T^*) \sim Q^0$ and $\mathcal{A}(\gamma_L^*) \sim Q^{-1}$ of the factorisation theorem. In the sum of the cross section $\sigma(\mu^{+\downarrow}) + \sigma(\mu^{-\uparrow})$ the imaginary parts of $\mathcal{A}(\gamma_{T,L}^*)$ can be separated from the Bethe-Heitler and VCS contributions by their angular dependence, since their coefficients change sign under $\varphi \to -\varphi$ whereas the other contributions do not.

Figure 3 shows the azimuthal distribution of the beam charge asymmetry (BCA) which could be measured at COMPASS with 100 GeV muon beams for different (x_{Bj}, Q^2) domains. Statistical errors are evaluated for 150 days of data taking with a 25% global efficiency. The data allow a good discrimination between different models. Model 1 (Vanderhaeghen *et al.* 1999) uses a simple ansatz to parametrise GPDs based on nucleon form factors and parton distributions and fulfils the GPD sum rules. Model 2 (Goeke *et al.* 2001) is more realistic because it correlates the x and t dependence. This takes into account the fact that the slow partons tend to stand at a larger distance

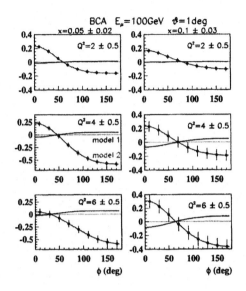

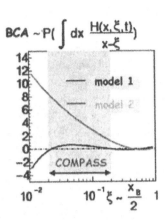

Figure 3. *Projected error bars for a measurement of the azimuthal angular distribution of the beam charge asymmetry measurable at COMPASS at $E_\mu = 100$ GeV and $|t| \leq 0.6$ GeV^2 for 2 domains of x_{Bj} ($x_{Bj} = 0.05 \pm 0.02$ and $x_{Bj} = 0.10 \pm 0.03$) and 3 domains of Q^2 ($Q^2 = 2 \pm 0.5$ GeV^2, $Q^2 = 4 \pm 0.5$ GeV^2 and $Q^2 = 6 \pm 0.5$ GeV^2) obtained in 150 days of data taking with a global efficiency of 25% and with $2 \cdot 10^8$ μ per SPS spill ($P_{\mu+} = -0.8$ and $P_{\mu-} = +0.8$) and a 2.5m long liquid hydrogen target*

Figure 4. *The beam charge asymmetry (BCA) is mainly controlled by the real part of DVCS amplitude and this figure presents the evolution of this amplitude for the dominant GPD H as a function of x_{Bj}. The domain which can be investigated at COMPASS gives large deviation between the models 1 and 2, where only the last model takes care about the different spatial distributions for fast and slow partons.*

from the nucleon centre than the fast partons. A gradual increase of the t-dependence of $H(x, 0, t)$ is considered from larger to smaller values of x. The parametrisation: $H(x, 0, t) = q(x)e^{t<b_\perp^2>} = q(x)/x^{\alpha t}$ is used where $<b_\perp^2> = \alpha \cdot ln1/x$ represents the increase of the nucleon transverse size with energy. The domain of intermediate x_{Bj} reachable at COMPASS is related to the observation of sea quarks or meson "cloud" or also gluons and it provides a large sensitivity to this three-dimensional picture of partons inside a hadron as can be seen in Figure 4.

2.3 HEMP

The GPDs reflect the structure of the nucleon independently of the reaction which probes the nucleon. In this sense they are universal quantities and can not only be accessed through DVCS but also through the hard exclusive leptoproduction of mesons. The vector meson channels $\rho^{0,\pm}, \omega, \phi, \ldots$ are sensitive only to the GPDs H and E, while the pseudo-scalar channels $\pi^{0,\pm}, \eta, \ldots$ are sensitive only to $\tilde{H}$ and $\tilde{E}$. In comparison DVCS depends on the four GPDs: H, E, $\tilde{H}$ and $\tilde{E}$. This property makes the hard meson production reactions complementary to the DVCS process as it provides an additional tool or filter to disentangle the different GPDs. Quark and gluon GPDs contribute both to the meson production as the GPDs for gluons enter at the same order in α_s as those for quarks. Decomposition with respect to quark flavour and gluon contributions can be realised through the different combinations obtained with a set of mesons. For example:

$$H_{\rho^0} = \frac{1}{\sqrt{2}}(\frac{2}{3}H^u + \frac{1}{3}H^d + \frac{3}{8}H^g)$$

$$H_\omega = \frac{1}{\sqrt{2}}(\frac{2}{3}H^u - \frac{1}{3}H^d + \frac{1}{8}H^g)$$

$$H_\phi = -\frac{1}{3}H^s - \frac{1}{8}H^g \qquad (3)$$

Nevertheless HEMP processes involve a second non-perturbative quantity which is the meson distribution amplitude, describing the coupling of the meson to the $q\bar{q}$ (or gluon) pair produced in the hard scattering (Figure 1). This complexity leads to more constraints in the applicability of the GPDs formalism which relies on factorisation requiring the the virtual photon to be longitudinally polarised.

The different scaling predictions (Vanderhaeghen *et al.*, 1997) for photon and meson production are shown in Figure 5. In leading twist the DVCS cross section $d\sigma/dt$ is predicted to behave as $1/Q^4$ whereas the meson longitudinal cross sections will obey a $1/Q^6$ scaling (due to the "extra" gluon exchange for the mesons). It is clear that the production of ρ^0 vector meson provides the largest counting rates compared to other mesons. With its decays into two charged particles whose invariant mass gives a clear resonance signal, this channel can be easily selected with the present COMPASS spectrometer.

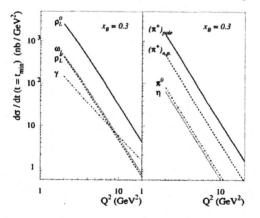

Figure 5. *Scaling behaviour of the leading order predictions for the forward differential leptoproduction cross section $d\sigma_L/dt$ on the proton for vector mesons (left panel) and pseudo scalar mesons (right panel). Also shown is the scaling behaviour of the forward DVCS cross section (dashed-dotted line in left panel).*

In the case of the production of a pseudo-scalar meson (with spin zero) where its polarisation is unobserved, the longitudinal cross section must be extracted from $\sigma_T +$

ϵ σ_L by a Rosenbluth separation requiring at least two different beam energies. For the production of a ρ meson (or for any vector meson) the angular distribution of decay $\rho \to \pi^+\pi^-$ contains information on the ρ helicity. Experimental data obtained in the E665 (1997), ZEUS (2000), H1 (2000) collaborations and presently in the COMPASS (2002) collaboration have indicated that the helicity of the photon in the γ^*p centre of mass system is approximately retained by the vector meson, a phenomenon known as s-channel helicity conservation (SCHC). This can be used to translate a measurement of the cross sections for transverse and longitudinal ρ mesons into a measurement of σ_T and σ_L for transverse and longitudinal photons, without resorting to a Rosenbluth separation.

The validity of SCHC can be tested in detail with the complete angular distribution of the meson production obtained with a longitudinal polarised beam which provides the full set of ρ density matrix elements. The analyses of the present COMPASS 2002 data demonstrate a copious production of ρ^0 mesons, in a range of Q^2 significantly larger than those of the previous experiments. Figure 6 presents the ratio $R = \sigma_L/\sigma_T$ determined by the decay angular distribution. The 2002 COMPASS data are limited to Q^2 up to about 5 GeV^2 while the new 2003 data will provide results up to 27 GeV^2 due to the enlarged Q^2 trigger. The ratio R increases with Q^2 and reaches a value larger than 1 at Q^2 around 2 or 3 GeV^2, providing a favourable region for a GPDs study where the longitudinal contributions become dominant.

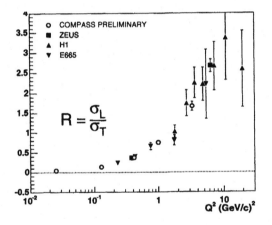

Figure 6. $R = \sigma_L/\sigma_T$ *obtained with the spin density matrix element* r^{04}_{00} *determined from the ρ decay angular distribution. The COMPASS 2002 preliminary results are compared to the published data of E665 (1997), ZEUS (2000) and H1 (2000).*

Precise simulations of exclusive ρ^0 and π^0 production have been performed (Pochodzalla *et al.* 1999) and it is undoubtedly that HEMP and DVCS, which can be realised at the same time in the same setup, provide different and complementary facets for the GPDs study.

3 Beam and target requirements

The highest luminosity that can be reached at COMPASS will be required for these exclusive measurements. These experiments will use 100-190 GeV/c muons from the M2 beam line. Presently limits on radiation protection in the experimental hall imply that the maximum flux of muons expected is $2 \cdot 10^8$ muons per SPS spill (5.2s spill duration, repetition each 16.8s). Under these circumstances, we can reach a luminosity of $\mathcal{L} = 4 \cdot 10^{32}$ $cm^{-2}s^{-1}$ with the present polarised ^{6}LiD or NH_3 target of 1.2 meter length, and only $\mathcal{L} = 1.3 \cdot 10^{32}$ $cm^{-2}s^{-1}$ with a new liquid hydrogen target of 2.5 meter length which has to be designed for this proposal.

It has to be noted that these experiments will benefit directly from every improvement of the muon flux in order to increase statistics and enlarge the kinematical domain in Q^2.

In order to get useful cross sections, it is necessary to perform a precise absolute luminosity measurement. This has already been achieved by the NMC Collaboration within a 1% accuracy. The integrated muon flux was measured continuously by two methods: either by sampling the beam with a random trigger (provided by the α emitter Am^{241}) or by sampling the counts recorded in 2 scintillators hodoscope planes used to determine incident beam tracks. The beam tracks were recorded off-line, in the same way as the scattered muon tracks to exactly determine the integrated usable muon flux.

μ^+ and μ^- beams of 100 GeV energy, with the largest intensity as well as exactly opposite polarisation (to a few %) are required. The muons are provided by pion and kaon decay and are naturally polarised. The pions and kaons come from the collision of the 400 GeV proton beam (of $1.2 \cdot 10^{13}$ protons per SPS cycle) on a Be primary target. A solution is under study and consists of:

1) selecting 110 GeV pion beams from the collision and 100 GeV muon beams after the decay section in order to maximise the muon flux;

2) keeping the collimator settings constant which define the pion and muon momentum spreads (both the collimator settings in the hadron decay section and the scrapper settings in the muon cleaning section) in order to fix the μ^+ and μ^- polarisations at exactly the opposite value ($P_{\mu^+} = -0.8$ and $P_{\mu^-} = +0.8$);

3) fixing N_{μ^-} close to $2 \cdot 10^8$ μ per SPS spill with the longest 500mm Be primary target.

4) the number of N_{μ^+} will be about a factor 2 larger than that of N_{μ^-}.

4 Detectors necessary to complement the high resolution COMPASS forward spectrometer

4.1 Necessity to complement the present setup

The outgoing muon scattered in the very forward direction (below $1°$) will be measured in the present high resolution COMPASS spectrometer. The outgoing meson or photon are emitted at larger angle, but must be below $10°$ in order to be detected by the RICH detector or the two COMPASS calorimeters (ECAL1 + ECAL2), mainly constituted of lead-glass blocks of excellent energy and position resolution and high rate capability. The recoil proton scatters at large angle and small momentum ($\leq$ 750 MeV/c) which cannot be detected by the present COMPASS apparatus. Furthermore, the missing mass energy technique using the energy balance of the scattered muon and photon or meson is not accurate enough at the given high beam energy (the resolution in missing mass which is required is $(m_p + m_\pi)^2 - m_p^2 = 0.25$ GeV^2 and the experimental resolution which can be achieved is larger than 1 GeV^2). Thus, a recoil detector has to be designed to measure the proton momentum precisely and insure the exclusivity of these processes.

The DVCS reaction is surely the most delicate reaction to perform because a final state must be selected that includes one muon, one photon and one low energy proton among many competing reactions listed below:

1) Hard Exclusive π^0 Production $\mu p \rightarrow \mu p \pi^0$ where π^0 decays in two photons, for which the photon with higher energy imitates a DVCS photon, and the photon with smaller energy is emitted at large angle outside of the acceptance or its energy is below the photon detection threshold.

2) Diffractive dissociation of the proton $\mu p \rightarrow \mu \gamma N^*$ with the subsequent decay of the excited state N^* in $N + k\pi$. (The low energy pions are emitted rather isotropically).

3) Inclusive Deep Inelastic Scattering with, in addition to the reconstructed photon, other particles produced outside the acceptance or for which tracks are not reconstructed due to inefficiency.

Moreover, the background must be taken into account. It includes beam halo tracks with hadronic contamination, beam pile-up, particles from the secondary interactions and external Bremsstrahlung.

4.2   Optimisation of the detector design

A simulation has been carried out in order to define the proper geometry of the detector complementing the present COMPASS setup. The goal was to maximise the ratio of DVCS events over DIS events for a sample of events with one muon and one photon in the COMPASS spectrometer acceptance plus only one proton of momentum smaller than 750 MeV/c and angle larger than 40° (typical DVCS kinematics). Using the event generator code PYTHIA 6.1 which generates all Deep Inelastic Scattering (DIS) processes with many γ and π^0 production possibilities, the experimental parameters such as maximum angle and energy threshold for photon detection and maximum angle for charged particle detection have been tuned. With photon detection extended up to 24 degrees and above an energy threshold of 50 MeV and with charged particle detection up to 40 degrees, the number of DVCS events estimated with models is more than an order of magnitude larger than the number of DIS events over the whole useful Q^2 range (see Figure 7).

4.3   Quality of the calorimetry and extension

The two calorimeters ECAL1 and ECAL2 are necessary for DVCS and π^0 production. They mainly consist of lead-glass blocks called GAMS. They are cells of 38.4 × 38.4 × 450 mm^3. The typical characteristics of such a calorimeter are:

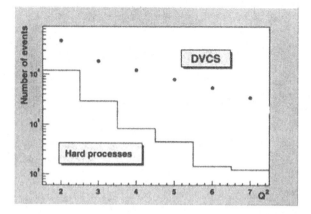

Figure 7. *Number of events for DVCS (dots) and DIS (histogram) processes as a function of Q^2 for a selection of events with only one muon, one photon and one recoiling proton.*

- energy resolution: $\sigma P_\gamma / P_\gamma = 0.055/\sqrt{P_\gamma} + 0.015$

- position resolution: $\sigma_x = 6.0/\sqrt{P_\gamma} + 0.5$ in mm

- high rate capability: 90% of signal within 50ns gate with no dead time

- effective light yield: about 1 photoelectron per MeV; hence low energy photons down to 20 MeV can be reconstructed.

At 10 GeV a separation efficiency of 100% can typically be achieved for a minimum distance between two photon tracks at the face of the calorimeter of D = 4 cm. This excellent performance of the calorimeters will play a key role in the perfect separation between DVCS events and hard π^0 events.

Moreover it has been shown with the simulation that it would be useful to extend the calorimetry from 10° up to an angle of 24° in order to separate contributions with only one photon, two photons and more.

4.4   Recoil detection

One possible solution to complement the present COMPASS setup is presented in Figure 8. It consists of one recoil detector described below, an extended calorimeter from 10 to 24 degrees, and a veto for charged forward particles up to 40 degrees. This calorimeter has to work in a crowded environment and in a magnetic fringe field of SM1.

The recoil detector is based on a time of flight (ToF) measurement between two barrels of scintillating slabs read at both ends. The inner barrel (noted A) (2.8m length) surrounding the target should be made of slats as thin as possible (4mm) to allow low momentum proton detection. Thicker (5cm) and longer (4m) slabs should be used for the outer barrel (noted B). An accurate t measurement implies a timing resolution of 200ps. External interleaved layers of scintillator and lead should be added to detect extra neutral particles and give an estimate of the background. The combination of time of flight measurement and the energy loss in the various sensitive detectors would provide

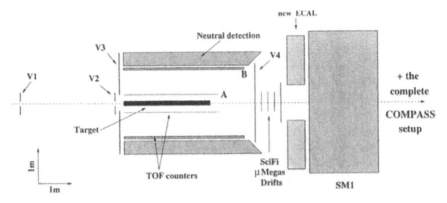

Figure 8. *Proposal for a detector complementing the COMPASS setup. A recoil detector, an extended calorimeter from 10 to 24 degrees, and a veto (V4) for charged forward particles until 40 degrees have been added.*

discrimination of events in this fully hermetic detector. We have tested the concept of this detector using the present muon beam at nominal flux and a simplified setup (one sector of reduced length). A resolution of 300ps has been achieved. The performance of this ToF system is limited by the number of photo-electrons that are collected, dispersion due to counter length and high rates (see the contribution by Michael Seimetz to these proceedings). The extension to long (about 4m) scintillators has to be studied carefully and the technology has to be improved to achieve a still better resolution. The construction of a fully functional prototype segment of the appropriate length is part of the Joint Research Activity GPDs within the European Integrated Infrastructure Initiative I3HP.

References

Belitsky A V and Müller D, 2002a, *Nucl Phys A* **711** 118.
Belitsky A V, Müller D, Kirchner A, 2002b, *Nucl Phys B* **629** 323.
Burkardt M, 2000, *Phys Rev D* **62** 071503.
CLAS collaboration, Stepanyan S *et al.*, 2001, *Phys Rev Lett* **87** 182002.
Collins J C, Frankfurt L and Strikman M, 1997, *Phys Rev D* **56** 2982.
Diehl M *et al.* , 1997, *Phys Lett B* **411** 193.
Goeke K, Polyakov M V, Vanderhaeghen M, 2001, *Prog Part in Nucl Phys* **47** 401, implementation by Mossé L and Vanderhaeghen M.
H1 collaboration, Adloff C *et al.* , 2001, *Phys Lett B* **517** 47.
HERMES collaboration, Airapetian A *et al.* , 2001, *Phys Rev Lett* **87** 182001.
Ji X, 1997, *Phys Rev Lett* **78** 610; 1997, *Phys Rev D* **55** 7114.
Negele J W *et al.* , 2004, *Nucl PhysProc Suppl* **128** 170.
NMC collaboration, 1992, Amaudruz P *et al.* , *Phys Lett B* **295** 159.
Pochodzalla J *et al.* , hep-ex/9909534.
Radyushkin A W, 1997, *Phys Rev D* **56** 5524.
Ralston J P and Pire B, 2002, *Phys Rev D* **66** 111501.
Strikman M and Weiss C, 2004, *Phys Rev D* **69** 054012.
Vanderhaeghen M, Guichon P A M, Guidal M, 1999, *Phys Rev D* **60** 094017.
ZEUS collaboration, Chekanov S *et al.* , 2003, *Phys Lett B* **573** 46.

The magnetic moment of the Δ^+

Evangeline Downie

Department of Physics & Astronomy, University of Glasgow,
Glasgow, G12 8QQ, UK

1 Introduction

The magnetic dipole moment of the $\Delta^+(1232)$ (μ_{Δ^+}) has never been experimentally measured with sufficient precision for hadron model evaluation. The only previous measurement is that of Kotulla *et al* (2002) which gave $\mu_{\Delta^+} = 2.7^{+1.0}_{-1.3}(stat) \pm 1.5(syst) \pm 3(theor)\ \mu_n$. However μ_{Δ^+} has been predicted frequently using a variety of theoretical approaches which have yielded a broad range of values. If we can measure this fundamental property experimentally, we can then assess the validity of the various models. We are currently performing the experiment, described herein, using the Glasgow tagged photon spectrometer at the MAMI-B 855 MeV electron accelerator in Mainz, Germany.

2 Background of the experiment

Most members of the baryon decuplet decay via the strong force and so have exceptionally short mean lifetimes, precluding the use of spin precession in a magnetic field to determine their magnetic dipole moments. However, a pioneering measurement of the magnetic dipole moment of the Δ^{++} by Nefkens *et al* (1978) provided a method which has been adapted for our measurement as calculated by Drechsel *et al* (2000).

As the Δ^+ has such a short lifetime (10^{-24} s), its Breit-Wigner width (mass uncertainty) is quite large (around 120 MeV), and it is therefore possible to create a high energy Δ^+ which then decays radiatively to a lower energy Δ^+. We intend to form a Δ^+ at the upper end of its mass range by using tagged photons to excite the protons in a liquid hydrogen target. This high energy Δ^+ will then decay to a Δ^+ lower down in the mass distribution. This $\Delta\Delta\gamma$ vertex has an amplitude that is dependant on μ_{Δ^+} and thus μ_{Δ^+} will have an affect on both the differential cross-section and photon asymmetry of any reaction in which this vertex is involved.

Due to the short lifetime of the Δ^+, we will actually detect the production photon in combination with the products of the low energy decay: a proton and a π^0, which itself decays into two back-to-back photons. This gives an overall reaction of $\gamma p \to \Delta^+ \to \Delta^+\gamma' \to p\pi^0\gamma' \to p\gamma\gamma\gamma'$. The three-photon, single-proton final state is predicted to have a very small total cross-section, of the order of 60 nb.

As a test of the reaction model and also to provide a consistency check on our experimental analysis, we will also analyse a parallel channel where the Δ decays to an $n\pi^+$ final state resulting in an overall reaction of $\gamma p \to \Delta^+ \to \Delta^+\gamma' \to n\pi^+\gamma'$.

3 Experimental setup

We are using the Crystal Ball and TAPS to provide photon spectroscopy, two MWPCs from the old DAPHNE detector for vertex tracking in the Crystal Ball and a new plastic ΔE Particle Identification Detector designed specifically for this experiment. This setup provides an excellent 4π detector system of great sensitivity which, combined with the high quality beam from MAMI-B and the resulting tagged photon beam from the high resolution Glasgow Tagged Photon Spectrometer, should enable us to evaluate this small cross section with greatly improved experimental precision.

4 Current status

The experiment has been running since August 2004 and we have currently collected several hundred hours of data. We have performed test analyses and find that we can clearly identify and measure the energy of charged and neutral pions and protons and also that we can cleanly identify neutrons. We have also performed a basic analysis of the photon asymmetry of $\gamma p \to \pi^0 p$ and find that it agrees with, and has achieved better precision than, previous measurements indicating that we fully understand our detector system and polarised photon production facilities.

Thus we are confident that we can provide an invaluable experimental input to the theoretical discussion of QCD in the non-perturbative region by a sound evaluation of the five-fold differential cross section and photon asymmetry of the reaction $\gamma N \to N\pi\gamma'$.

Acknowledgments

The work reported in this talk was carried out by the CB@MAMI collaboration at the Institut für Kernphysik, Universität Mainz, Germany. It was supported by the UK EPSRC.

References

Drechsel D *et al*, 2000, *Phys Lett B* **484** 236.
Kotulla M *et al*, 2002, *Phys Rev Lett* **89** 272001 .
Nefkens B M K *et al*, 1978, *Phys Rev D* **18** 3911.

Measurement of the neutron electric form factor at MAMI

Derek I Glazier

Department of Physics & Astronomy, University of Glasgow,
Glasgow, G12 8QQ, UK

Electron scattering can be aproximated by the exchange of a virtual photon with four-vector q. The response of a nucleon to such an interaction is parameterised by the nucleon electromagnetic form factors, $G_E^N(Q^2)$ and $G_M^N(Q^2)$, which are related to the internal electric and magnetic structure respectively. They are functions of $Q^2 = -q^2$ only, and this governs the distance scales probed by the photon. Determining the neutron electric form factor G_E^n, has been complicated by the lack of free neutron targets and its small magnitude compared to G_M^n. Recently these problems have been overcome by exploiting the sensitivity of the electromagnetic form factors to polarisation observables. Quasielastic scattering of longitudinally polarised electrons off neutrons bound in light nuclei, (D, ^{3}He) allow measurements of asymmetries, dependent on the interference of G_E^n and G_M^n, that have small systematic uncertainties from nuclear binding effects.

Recent experiments at the 3-spectrometer facility at MAMI measured G_E^n using the double polarised $D(\vec{e}, e'\vec{n})p$ reaction. Here the ratio of the non-zero neutron polarisation components is proportional to the ratio of electric to magnetic form factors:

$$R_P = \frac{P_x}{P_z} = -\frac{1}{\sqrt{\tau + \tau(1+\tau)\tan^2 \vartheta_e/2}} \cdot \frac{G_{E,N}}{G_{M,N}} \tag{1}$$

where ϑ_e is the electron scattering angle and $\tau = Q^2/4M_n^2$. Our measurement of this ratio for $Q^2 = 0.3$, 0.6 and 0.8 $(\text{GeV}/c)^2$ will be described below.

A longitudinally polarised electron beam with energies up to 883 MeV, currents of $10-15$ μA, and typical polarisation around 80% was incident on a 5 cm long liquid deuterium target. The quasi-elastically scattered electrons were detected in a high resolution magnetic spectrometer, defining the reaction Q^2 and tagging the recoiling neutron. The neutrons were detected, and their polarisation analysed, in a neutron polarimeter consisting of two segmented walls of long plastic scintillator bars and a spin-precession magnet. The magnet was approximately 3 m from the target, the first wall 6 m from the target and the second wall a further 3 m out. The scintillation light was collected at both ends of a bar allowing the hit position and thus momentum direction of the neutrons to be determined. Thin plastic "veto" detectors, situated in front of the neutron detectors, were used to identify neutral particles. In addition cuts on time-of-flight and the reconstructed electron were used to identify neutrons from quasi-elastic scattering.

Determination of the neutron polarisation was carried out using the first wall as a nucleon target: polarised nucleon-nucleon scattering results in an "up-down" asymmetry in the azimuthal scattering angle Φ_n, that is proportional to the transverse polarisation P_T and the analysing power averaged over the detector acceptances A_{eff}:

$$A = \frac{\sqrt{N^+(\Phi_n)N^-(\Phi_n + \pi)} - \sqrt{N^+(\Phi_n + \pi)N^-(\Phi_n)}}{\sqrt{N^+(\Phi_n)N^-(\Phi_n + \pi)} + \sqrt{N^+(\Phi_n + \pi)N^-(\Phi_n)}} = A_{\text{eff}} P_T \sin \Phi_n \quad (2)$$

where $N^\pm(\Phi_n)$ represents the Φ_n distribution between 0 rad and π rad, with beam helicity = ± 1. This exploits the spin flip of the ejected neutron with flipping of the beam helicity to give an asymmetry independent of neutron detection efficiency. The asymmetry was measured through detection of the scattered neutron in the second scintillator wall allowing reconstruction of Φ_n from the hit positions. Without spin-precession P_T is given by P_x, the x-component of the polarisation transfered to the neutron from the electron. Spin-precession, by an angle χ, using a vertically aligned magnetic field, allows P_T and thus the measured asymmetry to also become sensitive to P_z : $A(\chi) = A_{max} \sin(\chi - \chi_0)$, where $A_{max} = A_{eff} \sqrt{P_x^2 + P_z^2}$ and $\tan(\chi_0) = P_x/P_z$. Hence measuring $A(\chi)$ for 7 values of χ, allowed χ_0 and thus P_x/P_z to be determined.

This ratio then required two corrections before G_E^n was extracted. First a correction due to the Fermi motion of the neutron, which requires the polarimeter to be 180° symmetric around q to give results consistent with the case of a free nucleon, small deviations from this symmetry lead to corrections of around 1%. The second was due to nuclear effects on the polarisation transfered to a neutron bound in deuterium. The size of this effect was calculated using a model by Arenhövel to calculate the polarisation transfer event-by-event with, and without, nuclear effects: It ranged from 5% for 0.8 $(\text{GeV}/c)^2$ to 16% for 0.3 $(\text{GeV}/c)^2$. G_E^n was finally extracted from the polarisation ratio and the known magnetic form factor G_M^n, using eqn.(1).

The full results are available in Glazier *et al.* (2005).

Acknowledgments

I would like to thank Michael Seimetz for his contributions to this talk as well as the rest of the MAMI-A1 collaboration for their work on the experiment. Thanks are due to the MAMI accelerator and polarised beam groups. This work was supported by the Deutsche Forschungsgemeinschaft (SFB 443) and the UK Engineering and Physical Sciences Research Council.

References

Glazier D I *et al.* , 2005, *Eur Phys J A* **24** 101.

Single π^0 and η photoproduction off protons at CB-ELSA

Olivia Bartholomy

Helmholtz-Institut für Strahlen- und Kernphysik,
Rheinische Friedrich-Wilhelms-Universität Bonn, Germany

1 Introduction

Photoproduction experiments are a sensitive tool for the investigation of baryon resonances providing information complementary to πN-scattering experiments. A variety of resonances are predicted to contribute to different photoproduction channels although they are not seen in πN scattering (Capstick 1992, Capstick and Roberts 1994). Previously published data sets on the channels $\gamma p \rightarrow p\pi^0$ and $\gamma p \rightarrow p\eta$ do not cover the resonance region over the full solid angle, which is now achieved with the CB-ELSA (Crystal Barrel at the ELectron Stretcher Accelerator at Bonn) experiment to a greater degree of completeness. Using a partial-wave analysis, masses, widths, and photocouplings were extracted from our data along with several other data sets. For further information see Anisovich *et al.* (2005), Bartholomy *et al.* (2005) and Credé *et al.* (2005).

2 Differential cross sections

The differential cross sections of the reactions $\gamma p \rightarrow p\pi^0$ and $\gamma p \rightarrow p\eta$ are shown in Figure 3 of Bartholomy *et al.* (2005) and Figure 2 of Credé *et al.* (2005). From these data, total cross sections were obtained by summing over the measured data points and extrapolating to the non-measured region using the results of the partial-wave analysis (PWA) (Figure 4 of Bartholomy *et al.* (2005) and Figure 3 of Credé *et al.* (2005)).

The strongly varying angular distributions of $\gamma p \rightarrow p\pi^0$ already indicate that many resonances contribute. The measurements start in the mass region of the $P_{33}(1232)$, which is visible through the shape of the cross section, proportional to $\cos^2 \theta_{cm}$. The data at higher energies cannot easily be interpreted without the use of a PWA. Complicated structures appear and remain up to highest energies, while an increasingly large fraction of the cross section is due to t-channel exchange as indicated by the rise of $d\sigma/d\Omega$ for mesons in the forward direction.

The situation for $\gamma p \to p\eta$ is much simpler. The η in the final state only allows N* resonances in the intermediate state (isospin filter). This leaves fewer unknown parameters for the analysis of these distributions. Again, resonances contribute up to high energies, and the importance of t-channel exchange increases at higher photon energies.

Our results agree with previously measured data in regions where a comparison is possible. The CB-ELSA experiment adds new data points and extends the measured region in energy and angle which is essential to determine resonance contributions.

3   Results of a partial-wave analysis

Results of the PWA (see Anisovich *et al.* 2005) include confirmation of known resonances in both channels. In the fits, which include data sets from Bartholomy *et al.* (2005) and Credé *et al.* (2005) in addition to the present data, evidence for a new resonance $D_{15}(2070)$ and indications for a $P_{13}(2200)$ resonance are found. The first is especially important for the description of the CB-ELSA data on η photoproduction.

Along with the $S_{11}(1535)$ and the $P_{13}(1720)$, the $D_{15}(2070)$ fits in well with the pattern of strongly contributing resonances in $\gamma p \to N^* \to p\eta$. From a harmonic-oscillator model, spin $S = 1/2$ and orbital angular momenta $L = 1, 2$ and 3 can be assigned to these states, coupling to the measured quantum numbers $J = 1/2^-, 3/2^+$ and $5/2^-$ ($J = L - S$). These resonances decay into Nη without spin flip.

4   Conclusions

The CB-ELSA experiment has measured high-statistics, high-quality data on photoproduction of single mesons. Analyses of other single- and double-meson channels are underway. We expect even more insight into the spectrum of baryon resonances from polarisation data already taken with the CB-ELSA experiment in combination with the TAPS (Two-Arm Photon Spectrometer) detector and from double-polarisation experiments planned for the near future.

The work reported in this talk was carried out as part of the CB-ELSA collaboration.

References

Anisovich A *et al*, 2005, *Eur Phys J A* **24** 111.
Bartholomy O *et al* [CB-ELSA collaboration], 2005, *Phys Rev Lett* **94** 012003.
Capstick S, 1992, *Phys Rev D* **46** 2864.
Capstick S and Roberts W, 1994, *Phys Rev D* **49** 4570.
Credé V *et al* [CB-ELSA collaboration], 2005, *Phys Rev Lett* **94** 012004.

Inclusive photoproduction of light mesons at HERA

Anna Kropivnitskaya

Institute for Theoretical and Experimental Physics,
Bolshaya Cheremushkinskaya 25, 117 218 Moscow, Russia

Measurements are presented of the inclusive photoproduction of the η, ρ^0, $f_0(980)$ and $f_2(1270)$ neutral mesons in ep interactions at HERA at an average γp collision energy of 210 GeV. Inclusive cross-sections, differential in rapidity and transverse momentum, are shown and a comparison is made with measurements of the photoproduction of other particle species at HERA.

The process by which quarks and gluons convert to colourless hadrons is one of the outstanding problems in particle physics. The theory of perturbative quantum chromodynamics (QCD) is not applicable and phenomenological models based on the laws of thermodynamics are often used (Hagedorn 1965, Tsallis 1988, Beck 2000). High energy particle collisions which give rise to large multiplicities of particles, produced with low values of transverse momentum, provide an opportunity to study hadronisation. This paper presents precision measurements of the inclusive photoproduction of η, ρ^0, $f_0(980)$ and $f_2(1270)$ mesons made by the H1 experiment, in ep collisions, at HERA. The measurements are from the 2000 running period with an integrated luminosity of 38.7 pb^{-1}.

The η mesons are reconstructed from their two-photon decay mode. The ρ^0, $f_0(980)$ and $f_2(1270)$ mesons are reconstructed through their $\pi^\pm\pi^\mp$ decay mode. The kinematic interval accepted for the reconstructed neutral mesons, in rapidity space, was $|y_{lab}| < 1$.

In Figure 1 one of the universal features observed in the behaviour of long-lived hadrons (Rostovtsev 2001) is investigated for the light resonances measured here. The double differential cross sections for η, ρ^0, $f_0(980)$ and $f_2(1270)$ production are presented as a function of $m + p_T$, where m is the meson's nominal mass. The cross sections follow closely the same power law function as observed for pions at the same γp collision energy, once allowance has been made for the different isospin and spin of the various hadron species. Within the measured rapidity interval the resonance production rates are flat in rapidity.

The significant ρ^0 mass shift observed in this experiment and by the OPAL and DEL-

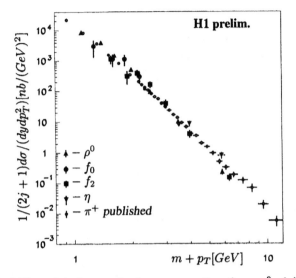

Figure 1. *Differential photoproduction cross sections for η, ρ^0, $f_0(980)$ and $f_2(1270)$ mesons plotted as a function of $(m + p_T)$, where m is the nominal meson mass. The open symbols show the π^+ production cross section, calculated from measurements of the charged particle photoproduction spectrum.*

PHI collaborations (Abreu *et al* 1995, Acton *et al* 1992) requires further investigation. It is important to understand the origin of this effect in the context of the expected distortion of the nominal mass and width of hadronic resonances in heavy ion collisions at RHIC, where a similar effect could be attributed to the formation of a quark-gluon plasma.

Acknowledgments

I am grateful to the HERA machine group whose outstanding efforts have made this experiment possible. I thank the engineers and technicians for constructing and maintaining the H1 detector, funding agencies for financial support, the DESY technical staff for continual assistance, and the DESY directorate for the hospitality which they extend to the non-DESY members of the collaboration. This work is partially supported by CRDF.

References

Abreu P *et al* [DELPHI collaboration], 1995, *Z Phys C* **65** 587.
Acton P D *et al* [OPAL collaboration], 1992, *Z Phys C* **56** 521.
Beck C, 2000, *Physica A* **286** 164.
Hagedorn R, 1965, *Suppl Nuovo Cimento* **3** 147.
Rostovtsev A, 2001, in *proc 31st Int Symp on Multiparticle Dynamics*, Datong, China.
Tsallis C, 1988, *J Stat Phys* **52** 479.

Hadron physics at the CELSIUS/WASA facility

Karin Schönning and Samson Keleta Negasi

Department of Radiation Sciences, Uppsala University,
Uppsala, Sweden

1 Introduction

The CELSIUS/WASA facility in Uppsala, Sweden, is a 4π-multidetector system for studies of meson production at the kinematic threshold in nucleon-nucleon collisions and for studies of η-meson decays. The objective of the meson production program is to investigate the role of baryon resonances in the production mechanism and to study final state interactions of the nucleon-nucleon system, the meson-nucleon system and the meson-meson system. Studies of the η-meson decays provide a sensitive test of Chiral Perturbation Theory and its prediction of the quark mass difference. Symmetries of the Standard Model, such as C, CP and CPT, can be tested by studying rare η-decays that violate these symmetries.

2 The CELSIUS/WASA facility

The WASA detector system is integrated into the 82 metres long CELSIUS (Cooling with Electrons and Storing of Ions from the Uppsala Synchrocyclotron) storage ring at The Svedberg Laboratory (TSL). The maximum kinetic energy for protons is 1450 MeV with electron cooling provided up to a proton beam energy of 550 MeV. A pellet target is developed to enable a 4π detector geometry and provides the necessary high luminosity. Its working principle is that small droplets are formed by pressing liquid hydrogen or deuterium through a narrow, vibrating nozzle. The droplets are then frozen into pellets which fall through a narrow pipe into the scattering chamber where they cross the CELSIUS beam line (Ekström 2002). This gives a well defined and, in principle, background free interaction point. The WASA detector (Zabierowski 2002) consists of a Forward

Detector (FD), a Central Detector (CD) and a Zero Degree Spectrometer or Tagging Spectrometer (TS). The FD measures the angle and kinetic energy of charged particles, in particular scattered beam and recoil particles such as protons, deuterons and ^{3}He. The CD measures the angle and kinetic energy of charged particles and photons, mainly from meson decays. It consists of a Mini Drift Chamber (MDC) for measuring charged particles, a Plastic Scintillating Barrel (PSB) and a Scintillator Electromagnetic Calorimeter (SEC). The TS uses the CELSIUS dipole magnets to detect recoil particles emitted at very small angles; particles that would otherwise escape detection in the beam pipe.

3 Results and future prospects

Exclusive measurements of double pion and triple pion production in proton-proton collisions have already been performed with the present WASA setup. The cross section for $pp \rightarrow pp\pi^0\pi^0$ at a beam energy of 1360 MeV was measured by Koch (2004) to be $\sigma_{2\pi^0} = 197.5\mu b$ ($\pm18.4\% \pm 8.6\%$). This is three times lower than the value predicted by Alvarez-Ruso (1998). Several kinematic distributions of the final state particles in $pp \rightarrow pp\pi^0\pi^0$ show a significant deviation from phase space. The formation of a $\Delta\Delta$ resonance pair seems to be the main reaction mechanism at this energy (Koch 2004).

The $pp \rightarrow pp\pi^0\pi^0\pi^0$ and the $pp \rightarrow pp\pi^+\pi^-\pi^0$ reactions will give important contributions to the background in η-decay experiments. The cross section of the former has never been measured previously and no theoretical model exists yet. The upper limit, at a beam energy of 1360 MeV, has now been measured to be $\sigma_{3\pi^0} < 1.5$ (±0.5)μb (Koch 2004). The cross section of the latter reaction was also measured by Jacewicz (2004), at the same energy, to be $\sigma_{\pi^+\pi^-\pi^0} = 4.6$ ($\pm1.2^{+0.7}_{-0.9}$)μb.

The TSL will be closed down after the summer of 2005. In the autumn of 2005 the WASA detector will be moved to the COSY facility in Jülich, Germany. Before then, a large amount of data will be taken for studies of reactions such as $pd \rightarrow \pi^0\eta \, ^3He$; $pd \rightarrow \omega \, ^3He$; $pd \rightarrow \eta \, ^3He$; $pd \rightarrow pd\eta$; $pn \rightarrow pn2\pi$, $pn3\pi$, $pn\eta$ and $pd \rightarrow p\Lambda\Theta^+$.

Acknowledgments

We would like to thank our helpful colleague Henrik Pettersson and the rest of the CELSIUS/WASA collaboration members.

References

Alvarez-Ruso L *et al*, 1998, *Nucl Phys A* **633** 519.
Ekström C *et al*, 2002, *Physica Scripta T* **99** 169–173.
Jacewicz M, 2004, *Measurement of the reaction pp* $\rightarrow$ *pp$\pi^+\pi^-\pi^0$ with CELSIUS/WASA*, PhD thesis, Uppsala University.
Koch I, 2004, *Multi pion production in proton-proton collisions*, PhD thesis, Uppsala University.
Zabierowski J *et al*, 2002, *Physica Scripta T* **99** 159–169.

The generalised GDH sum rule

Vincent A Sulkosky

The College of William and Mary,
Williamsburg, VA, USA

We have made a precise measurement of the spin-dependent $^3\vec{\text{He}}(\vec{e}, e')$ inclusive cross sections and asymmetries to evaluate the generalised Gerasimov–Drell–Hearn (GDH) integral at Q^2 below 0.3 GeV^2 (Chen et al 1997). The GDH sum rule (Gerasimov 1966, Drell and Hearn 1966) relates the helicity dependent photoproduction cross sections for scattering circularly polarised photons off a longitudinally polarised target. For spin 1/2 targets,

$$\int_{\nu_{th}}^{\infty} [\sigma_{\frac{1}{2}}(\nu) - \sigma_{\frac{3}{2}}(\nu)]\frac{d\nu}{\nu} = -\frac{2\pi^2\alpha\kappa^2}{M^2} \tag{1}$$

where κ, M are the anomalous magnetic moment and mass of the target, ν_{th} is the inelastic threshold, ν is the photon energy, and $\frac{1}{2}$ ($\frac{3}{2}$) corresponds to the case of the photon helicity being parallel (anti-parallel) to the target spin.

The GDH integral can be generalised to finite Q^2. One method is to replace the photoproduction cross sections with the electroproduction cross sections:

$$I(Q^2) = \int_{\nu_0}^{\infty} [\sigma_{\frac{1}{2}}(\nu, Q^2) - \sigma_{\frac{3}{2}}(\nu, Q^2)]\frac{d\nu}{\nu} \tag{2}$$

Other generalisations exist that include a flux factor or can be extended in terms of the g_1 spin dependent structure function. These all reduce to the original GDH sum rule at $Q^2 = 0$. Due to its connection with the Bjorken sum rule, the extension to finite Q^2 provides a bridge from non-perturbative QCD to perturbative QCD. This bridge allows a comparison of the measured quantity to theoretical predictions over the entire Q^2 range.

Experiment E97-110 was conducted in Hall A at the Thomas Jefferson National Accelerator Facility (JLab) with longitudinally polarised beam currents up to 10 μA, at several incident energies from 1.15 to 4.4 GeV. The beam polarisation was measured with both Møller and Compton polarimeters. The average polarisation was $\sim 75\%$.

One of the Hall A high resolution spectrometers was used along with a septum magnet to detect the electrons at scattering angles of 6° and 9°. The septum magnet was required for small angles as the minimum spectrometer angle is 12.5°. The spectrometer detectors include drift chambers for particle tracking, scintillators for triggering the data acquisition, and pre-shower, shower and Cherenkov detectors for particle identification. Using these detectors, the pion contamination was reduced by a factor of 10^4.

A polarised ^{3}He target with 40% average polarisation in beam and 10^{36} $(cm^2 \cdot s)^{-1}$ luminosity was used as an effective polarised neutron target. Data were acquired with both longitudinal and transverse target polarisation configurations. The target polarisation was measured by two methods of polarimetry: NMR and EPR.

The new low Q^2 measurements of the extended GDH integral allow us to test the dynamics of Chiral Perturbation Theory (CPT), to learn about the spin structure of ^{3}He and the neutron, and to determine if the slope of the integral changes sign, which would allow an extrapolation to the real photon point. Our data complements the data from the previous Hall A experiment E94-010 below $Q^2 = 0.1$ GeV^2. Figure 1 shows the expected quality of the results for the neutron extended GDH integral. Currently the data analysis is underway: The false asymmetries are consistent with zero, preliminary elastic asymmetries are in good agreement with the world data, and understanding of the spectrometer optics is close to completion.

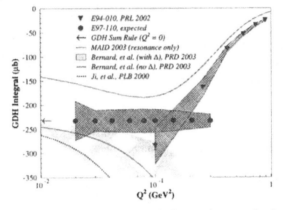

Figure 1. *E97-110 expected results along with the previous results from E94-010. The MAID phenomenological model and two CPT calculations are shown.*

Acknowledgments

This work was supported in part by the US Department of Energy (DOE) and was contributed for the JLab Hall A collaboration. The Southeasten Universities Research Association operates JLab for the DOE (contract DE-AC05-84ER40150, modification M175).

References

Chen J-P *et al.* , 1997, Jefferson Lab proposal E97-110.
Drell S D and Hearn A C, 1966, *Phys Rev Lett* **16** 908.
Gerasimov S B, 1966, *Sov J Nucl Phys* **2** 430.

ΔG/G at COMPASS

Sebastien Procureur

DAPNIA/SPhN,
Orme des Merisiers, 91191 Gif/Yvette, France

1 The nucleon spin

The nucleon is a composite object, made up with quarks and gluons, so we can decompose its spin as the contributions from spin and orbital momenta of these particles:

$$\frac{1}{2} = \frac{1}{2}\Delta\Sigma + \Delta G + L_q^z + L_g^z. \tag{1}$$

$\frac{1}{2}\Delta\Sigma$ is the contribution of spin-1/2 quarks, ΔG is the contribution of spin-1 gluons, and L is the orbital momentum of quarks and gluons along the flight direction. All these terms are almost unconstrained in QCD, except $\Delta\Sigma$ which can be estimated via the so called Ellis-Jaffe sum rule: $\Delta\Sigma = 0.6$, with the assumption that $\Delta s = 0$. But in the late '80s, this value was ruled out by experiments (EMC, SMC, SLAC) which measured $\Delta\Sigma$ to be ≈ 0.2. However, it was then realized that what they measured was not directly the quark contribution, but $\Delta\Sigma - \frac{3\pi}{2}\alpha_s\Delta G$. Thus, the Ellis-Jaffe prediction and the experimental value were found not to be incompatible, provided ΔG is large and positive. It also showed that we need ΔG to access $\Delta\Sigma$, and that's why several experiments were dedicated to its measurement, including RHIC and COMPASS.

2 The COMPASS experiment

COMPASS is a polarised Deep Inelastic Scattering experiment using a 160 GeV/c polarised muon beam on a fixed deuterium target. The beam delivers a high flux of 2.10^8 muons in 4.8s, every 16s. The liquid deuterium target is split into 2 cells of 60cm, polarised by Dynamic Nuclear Polarisation. Because we want to get rid of instrumental asymmetries, the cells have opposite polarisations; for the same reason, we also use a dipole which allows us to periodically reverse this polarisation. The outgoing particles are detected in a two-stage spectrometer, one part being dedicated to small angle particles, and the other part to large angle ones. The spectrometer contains around 200 tracking planes (MicroMegas, DC, GEM,...). Particle identification is performed by using calorimeters and a Ring Imaging CHerenkov detector (RICH).

3 Measuring $\Delta G/G$

The lowest order process in QCD in which $\frac{\Delta G}{G}$ is involved is the so-called photon-gluon fusion, where a virtual photon from the muon interacts with a gluon from the nucleon to give a quark anti-quark pair. But this process is α_s-suppressed compared to the Leading Order (LO) process, so tagging corresponding events is necessary. In COMPASS, this can be done by two different channels; namely open charm production and high P_T hadron pair production.

3.1 Open charm analysis

In this channel, PGF events are tagged using the detection of a D meson, decaying into a kaon-pion pair. Indeed, because of the charm mass, it is very unlikely the charm comes from the nucleon. So this channel is theoretically very clean. Unfortunately, statistics are rather low: in a preliminary analysis of the 2002 data, a few hundred events were found (using a D* tagging). But large hardware and software improvements are underway, and a much larger number of events is expected. All in all, the projected error on $\frac{\Delta G}{G}$ after the 2004 run is expected to be:

$$\delta(\frac{\Delta G}{G}) = 0.24.$$

3.2 High P_T channel

Here PGF events are tagged with the detection of two hadrons with large transverse momentum with respect to the virtual photon direction. Statistics are much larger in this case, but the interpretation of the measured asymmetry is more difficult: If LO processes are indeed suppressed by this selection, other processes with non-zero asymmetry can produce such a hadron pair, thus adding terms into the expression of the asymmetry. These processes are essentially QCDC processes, where the outgoing quark radiates a gluon, and the so-called resolved photon processes, where the virtual photon fluctuates in hadronic components. For the 2002 data, the following value for this asymmetry was obtained: $\frac{A_{LL}}{D} = -0.065 \pm 0.036$ (stat) ± 0.010 (syst).

Now, to extract $\frac{\Delta G}{G}$, we need Monte Carlo studies not only to estimate the analysing power and the ratio of PGF processes, but also to estimate the contribution of background processes. It is also foreseen that a cut of $Q^2 > 1$ will be applied in order to get rid of background processes. The projected error on $\frac{\Delta G}{G}$ in this analysis, after the 2004 run, is expected to be:

$$\delta(\frac{\Delta G}{G}) = 0.05, \text{ for all } Q^2; \quad \delta(\frac{\Delta G}{G}) = 0.16, \text{ for } Q^2 > 1.$$

Relativistic Glauber theory for $A(e, e'p)$ reactions

Bart Van Overmeire

Department of Subatomic and Radiation Physics,
Ghent University, Belgium

1 Introduction

Exclusive $A(e, e'p)$ reactions at high Q^2 ($Q^2 \geq 0.5$ (GeV/c)^2) are used to study several aspects of nuclear and nucleon structure in a region where one expects that both hadronic and partonic degrees-of-freedom may play a role. Amongst the physics issues which can be investigated with electromagnetically induced proton knockout reactions are the mean-field properties of nuclei, their short-range structure and the possible medium modifications of nucleons when they are embedded in a dense hadronic medium like the nucleus. Other subjects of current investigation include nuclear transparency and relativistic effects in nuclei.

The extraction of physical information from the $A(e, e'p)$ measurements depends strongly on the availability of a reliable model for the description of the final-state interactions (FSI) which the struck proton undergoes in its way out of the target nucleus. Popular frameworks to treat the FSI effects in modelling $A(e, e'p)$ reactions can roughly be divided in two major classes. At "low" energies (ie proton lab momenta $p_p \leq$ 1 GeV/c), most models use optical potentials to determine the scattering wavefunction of the knocked-out proton. The parameters in these optical potentials are obtained from global fits to elastic proton-nucleus $p + A$ scattering data.

2 Glauber theory

At 'higher' energies ($p_p \geq 1$ GeV/c), the use of optical potentials can no longer be justified in view of the highly inelastic character and diffractive nature of the proton-nucleon pN scattering cross sections and a valid alternative is provided in terms of

Glauber theory (Glauber and Matthiae 1970), which is a multiple-scattering extension of the eikonal approximation. In this framework, the effects of FSI are calculated directly from the elementary proton-nucleon $p + N$ scattering data and one assumes that the ejected proton only undergoes elastic or mildly inelastic collisions with the 'frozen' spectator nucleons. Whereas most Glauber-inspired $A(e, e'p)$ models are factorised and use non-relativistic wavefunctions and electron-proton couplings, we have developed a formulation of Glauber theory for $A(e, e'p)$ reactions that is both relativistic and unfactorised, and dubbed it the relativistic multiple-scattering Glauber approximation (RMSGA) (Ryckebusch *et al.* 2003).

3   Numerical results

The relative contributions from single- and multiple-scatterings to the Dirac-Glauber phase have been investigated. The Dirac-Glauber phase is a function which reflects the overall FSI effect of the consecutive cumulative scattering of the fast proton on the 'frozen' point scatterers of the residual nucleus. For the target nucleus ^{4}He, it turned out that single-scatterings give a reasonable account of the complete scattering processes and the average number of scatterers which the proton encounters in its way out of the nucleus is found to be of the order of one. For ^{12}C and heavier target nuclei, on the other hand, Glauber calculations restricted to single-scatterings substantially overestimate the FSI mechanisms. For a ^{208}Pb target nucleus convergence of the multiple-scattering series which implements all FSI effects is only reached when quadruple-scattering terms are included.

Relativistic effects are observed to have a negligible impact on the distorting and absorptive effect of the FSI which the ejected proton undergoes. For some $A(e, e'p)$ observables, however, inclusion of relativity is essential. Because relativistic effects in the description of the FSI are found not to exceed the few percent level, it can be excluded that this could be attributed to the FSI. At low missing momenta, the effects of relativity are mostly confined to the interference structure functions $\mathcal{R}_{TL}$ and $\mathcal{R}_{TT}$. For the dominant $\mathcal{R}_L$ and $\mathcal{R}_T$ response functions the effect is marginal.

Acknowledgments

The author would like to thank J Ryckebusch and P Lava. This work has been supported by the Fund for Scientific Research-Flanders (FWO).

References

Glauber R J, and Matthiae G, 1970, *Nucl Phys B* **21** 135.
Ryckebusch J *et al.* , 2003, *Nucl Phys B* **728** 226.

Hybrid mesons in the flux-tube model

Jozef Dudek

Rudolf Peierls Centre for Theoretical Physics,
1 Keble Road, Oxford, OX1 3NP, UK

Virtually all experimentally observed mesons have J^{PC} quantum numbers in the set allowed to states made from a quark and an antiquark moving with some relative orbital angular momentum. This set excludes possibilities like $0^{+-}, 1^{-+}$ and 2^{+-}.

Mesons are states within QCD, so we might sensibly ask what the gluonic field is doing in mesons? One interpretation of the quantum number systematics described above is that the gluonic field is in its ground state with $J^{PC} = 0^{++}$ so that the J^{PC} of the meson is totally determined by the state of the quarks - we label these mesons "conventional". Since QCD at the hadronic scale is strongly coupled we might expect local excitations of the gluonic field to be possible, probably with quantum numbers other than 0^{++}, with the inclusion of a valence quark pair such an excited state is known as a hybrid meson.

While a spectrum of such states seems to be inevitable in QCD, specific predictions of their masses and decay properties require non-perturbative techniques or, where these are not available, models. One scheme suited to this task is the flux-tube model (Isgur and Paton 1985, Kokoski and Isgur 1987).

Flux-tube model

This model has its origin in the strong coupling limit of Hamiltonian QCD on a spatial lattice. In this limit the lowest energy $q\bar{q}$ states have quark and anti-quark linked by the minimal possible number of occupied gauge links, *ie* we have string-like configurations of gluonic flux between the quark and antiquark. This structure has some further support from Euclidean space-time lattice studies, and has the nice property of reducing to a linear potential model in the non-relativistic limit for conventional mesons.

In this picture, the excited gluonic field in hybrids corresponds to transverse oscillations of the string joining the quarks. In the adiabatic limit where the string oscillates much faster than the typical quark speeds, we obtain a modified potential with respect to conventional mesons and hybrid mesons of larger mass.

This model has been extended to include decays via the flux-tube breaking mechanism, which bears a close resemblance to the phenomenologically successful $^3P_0$ model for conventional decays. Hybrid decays to pairs of conventional mesons are found to obey certain selection rules - whose origin is the symmetry of the tube excitation. In particular there are no decays to pairs of identical states, and some considerable suppression of *eg* $\pi\rho$.

This selection rule is something of a problem when compared with experiment - the $1^{-+}\pi\rho$ partial wave in $\pi p \to \pi\pi\pi p$ (Adams *et al.* 1998) has an apparently resonant peak at 1600 MeV, which cannot, by dint of its J^{PC} be a conventional meson.

Quark currents: electromagnetic, weak, strong

Isgur (1999) proposed another way in which hybrids can interact. A current acting on the quark will cause it to oscillate and as it does so it will drag the flux-tube along with it. This oscillation can hence excite a mode of the string-like tube and thus excite a conventional meson to a hybrid meson. For the particular case of a photon current such an interaction is of interest as there is a proposal to produce hybrids in photoproduction in the GlueX experiment at Jefferson Lab.

This mechanism was examined for electromagnetic, weak and strong currents (Close and Dudek 2003, 2004a, 2004b). It was found that excitation of certain exotic hybrid mesons in photoproduction should occur at a rate comparable to that of conventional excited mesons.

The missing mass spectrum in the inclusive decay $B \to J/\psi X$ is reported to have an excess above model expectations in the $m_X \sim 2$ GeV region. One possible explanation for this is the presence of a kaonic hybrid meson being produced in the decay $B \to J/\psi K_{\mathcal{H}}$, where the weak quark-level current excites the flux-tube. This was investigated in the flux-tube model and it was found that the branching ratio would be sufficient to explain the experimental "excess".

The possibility of bypassing the selection rule suppressing $\pi\rho$ decays of hybrids was investigated. By considering the pion to be a pointlike current, which by virtue of its isospin can only couple to the quarks and not to the tube, the symmetry which gave rise to the selection rule is maximally broken and considerable $\pi\rho$ branching ratios of hybrid mesons were found.

References

Adams G S *et al.* [E852 Collaboration], 1998, *Phys Rev Lett* **81** 5760.
Close F E and Dudek J J, 2003, *Phys Rev Lett* **91** 142001.
Close F E and Dudek J J, 2004a, *Phys Rev D* **69** 034010.
Close F E and Dudek J J, 2004b, *Phys Rev D* **70** 094015.
Isgur N and Paton J, 1985, *Phys Rev D* **31** 2910.
Isgur N, 1999, *Phys Rev D* **60** 114016.
Kokoski R and Isgur N, 1987, *Phys Rev D* **35** 907.

What can lattice QCD say about pentaquarks?

Abdullah Shams Bin Tariq

School of Physics and Astronomy, University of Southampton, SO17 1BJ, UK
and Physics Dept., Rajshahi University, Rajshahi 6205, Bangladesh

The recent excitement surrounding the (non-)observation of pentaquark states has placed some expectation on the lattice community. However there are some, not always well-appreciated, practical limits to what can be done, as well as a few areas of promise. So far few calculations are available and most conclusions have to be taken with a good deal of caution. This contribution reviews the present status of lattice pentaquark calculations, spells out what can and cannot be done and discusses plans for future work.

Against the couple of hundred non-lattice papers, only a few lattice papers/review articles (Sasaki 2004, Csikor *et al.* 2003, 2004, Chiu and Hsieh 2004) have appeared on the archive, and only one (Csikor *et al.* 2003) has been published at the time of writing. Burkert *et al.* (2004) write: *"Lattice QCD is currently not providing fully satisfactory predictions for the Θ^+. One group finds no signal, three groups find a signal at about the right mass, two at negative parity, one at positive parity."*

In lattice spectroscopy one typically looks at zero momentum 2-pt correlators. These have contributions from all states with the operator quantum numbers and going to large time pulls out the lowest energy state. The problem is that the Θ^+ is above the KN threshold and any operator creating the Θ^+ will also create a lower energy KN state. There are suggestions to use operators $\langle \Theta^+|\mathcal{O}_\Theta|0\rangle \gg \langle KN|\mathcal{O}_{KN}|0\rangle$, but with $\langle KN|\Theta^+\rangle \neq 0$ whatever operator is used, the lowest state will be the KN, and other states will decay into it. However, a scattering state has a characteristic volume dependence, and a check of this dependence is crucial to ascertain whether any observed state is a scattering or bound state. As stated by Davies (2004), *"It is the theory of QCD that we are simulating, not a model. However, we are limited in the things we can calculate."*

Lattice calculations may also allow us to learn about the parity. Note that most quark models and the chiral soliton model predict a positive parity for the Θ^+. Of the calculations available so far, Sasaki (2004) uses Wilson fermions and an operator of type $[ud][u\gamma_5 d]\bar{s}$ to obtain $J^P = \frac{1}{2}^-$. Two plateaux are seen in the effective mass plot with a kink, which is not well understood. Another one by Csikor *et al.* (2003) uses Wilson fermions and operators of type $[u\gamma_5 d][u\bar{s}\gamma_5 d]$. They analyse using a cross-correlator method with two operators, one operator having KN-type colour contractions and obtain $J^P = \frac{1}{2}^-$. They claim to observe one state to be the KN s-wave and another above to be the Θ^+, arguing that for this volume the next KN state is expected to be higher than this one extracted using this cross-correlator technique. The work by Chiu and Hsieh (2004), on the Θ^+ and Θ_c^+, use Domain-Wall fermions with operator $[ud][u\gamma_5 d]\bar{s}$ and, for the Θ^+, obtain $J^P = \frac{1}{2}^+$. They claim that others get it wrong because of using non-chiral fermions. However, none of these works checks the volume dependence to establish whether the observed state is a scattering or bound state. As summarised by Csikor *et al.* (2004), *"... it cannot be ruled out that the pentaquark states observed so far on the lattice turn out to be mixtures of nucleon-kaon scattering states."*

The only work studying volume effects (Mathur *et al.* 2004) observes a dependence characteristic of KN scattering states, *ie* they do not see the Θ^+. They use overlap fermions and operators of type $[u\gamma_5 d][\bar{s}u\gamma_5 d]]$. They do have a small volume, but do a very careful analysis addressing most issues rigorously.

With these significant obstacles for the Θ^+, the question is: Is there something that can be done well on the lattice, that may be relevant to pentaquarks? Evidently, states above threshold are horrendously difficult to simulate. But could there be any below threshold? Indeed, the Jaffe-Wilczek model predicts stable Θ_c^+ and Θ_b^+. Though the H1 experiment may have already seen Θ_c^+ in the DN-channel (Atkas *et al.* 2004), there is a reasonable consensus that as the heavy quark gets heavier, less is gained in terms of mass from binding the heavy quark into the baryon. There indeed is a reasonable chance that Θ_b^+ will be stable and it may be possible to predict or reject a stable Θ_b^+ using lattice QCD, which could be a non-trivial contribution. This is one idea being explored by the UKQCD collaboration, where one idea is to use a static b-quark (Allton *et al.* 2004). It should also be possible, though slightly more difficult to look for a possible stable Θ_c^+.

Another approach under consideration for states above threshold is to attempt spectral function decomposition using Maximum Entropy techniques. This, however, will be an exploratory attempt.

This work was carried out in the UKQCD collaboration. Discussions with Chris Allton, Jonathan Flynn, Gregorio Herdoiza and Federico Mescia are gratefully acknowledged.

References

Aktas A *et al.* [H1 collaboration], 2004, *Phys Lett B* **588** 17 .
Allton C, Flynn J and Herdoiza G, 2004, *Work under consideration.*
Burkert V D *et al.* , 2004, nucl-ex/0408019.
Chiu T W and Hsieh T H, 2004, hep-ph/0403020.
Csikor F *et al.* , 2003, *J High Energy Phys* **0311** 070.
Csikor F *et al.* , 2004, hep-lat/0407033.
Davies C T H, 2004, *Lectures given to SUSSP58 Hadron Physics Summer School, St Andrews.*
Mathur N *et al.* [Kentucky collaboration], 2004, *Phys Rev D* **70** 074508.
Sasaki S, 2004, *Phys Rev Lett* **93** 152001.

Quantum weights of quasi-particles in QCD at finite temperature

Nikolay Gromov

Faculty of Physics, St Petersburg State University,
Ulianovskaya 1, 198 904, St Petersburg, Russia

The functional determinant for quantum oscillations about periodic instantons with non-trivial holonomy has been computed exactly by Diakonov *et al.* (2004). The gauge field and ghost determinants are functions of the Polyakov line, the temperature, Λ_{QCD}, and the separation between the Bogomol'nyi–Prasad–Sommerfeld (BPS) monopoles (or dyons) which constitute the periodic instanton.

The finite-temperature behaviour of any theory is specified by the partition function. The partition function of the Yang-Mills (YM) theory, at temperature T, can be represented as a Euclidean path-integral with the usual YM action over all periodic configurations with period $1/T$. Configurations with finite action are classified by the behaviour of the Wilson loop that goes along the time direction or simply holonomy at spatial infinity. In the semi-classical approximation one has to find all the classical configurations. A particularly important class of such configurations are self-dual ones.

There are two known generalisations of self-dual instantons to non-zero temperatures. One is the periodic instanton of Harrington and Shepard studied in detail by Gross, Pisarski and Yaffe (1981). These periodic instantons, also called *calorons*, have trivial holonomy *ie* its value belongs to the group centre.

The other generalisation (KvBLL) has been constructed a few years ago by Kraan and van Baal (1998) and by Lee and Lu (1998); it has been named *caloron with non-trivial holonomy*. It is also a self-dual solution of the Yang–Mills equations of motion with an integer topological charge. The significant feature of the caloron with a unit topological charge is that it can be viewed as "made of" two BPS monopoles or dyons in the case of the SU(2) gauge group.

Dyons are self-dual solutions of the Yang–Mills equations of motion with *static* (*ie* time-independent) action density, which have both the magnetic and electric field at infinity decaying as $1/r^2$.

There is a range of evidence that dyons may play an essential role in the confinement-deconfinement phase transition. Precisely these objects determine the physics of the *supersymmetric* YM theory where in addition to gluons there are gluinos, *ie* Majorana (or Weyl) fermions in the adjoint representation. It should be noted that the KvBLL calorons and dyons seem to be observed in lattice simulations below the phase transition temperature.

To study this possible scenario quantitatively, one first needs to find out the 1-loop quantum weight of dyons, or the probability with which they appear in the Yang–Mills partition function. The caloron quantum weight depends on the separation of the constituent dyons r_{12}. This dependence gives rise to the quantum interaction between dyons. At large separations it has a linear plus Coulomb form. When the holonomy is not too far from trivial dyons inside the caloron attract each other, but if the holonomy is far enough from the trivial values dyons experience a strong linearly increasing repulsion.

We give here a simplified expression, obtained in the limit when the separation between dyons situated at $z_{1,2}$ is much larger than their core sizes:

$$
\begin{aligned}
\mathcal{Z}_{\mathrm{KvBLL}} \;=\; & \int d^3 z_1\, d^3 z_2\, T^6\, (2\pi)^{\frac{8}{3}}\, C \\
\times\; & \left(\frac{8\pi^2}{g^2}\right)^4 \left(\frac{\Lambda e^{\gamma_E}}{4\pi T}\right)^{\frac{22}{3}} \left(\frac{\mathrm{v}}{2\pi T}\right)^{\frac{4\mathrm{v}}{3\pi T}} \left(\frac{\bar{\mathrm{v}}}{2\pi T}\right)^{\frac{4\bar{\mathrm{v}}}{3\pi T}} \\
\times\; & \exp\left[-2\pi\, r_{12}\, P''(\mathrm{v})\right]\, \exp\left[-V^{(3)} P(\mathrm{v})\right],
\end{aligned} \tag{1}
$$

where the dependence on holonomy is encoded in the asymptotic value of $\mathrm{v} \equiv \sqrt{A_4^a A_4^a}$ at spatial infinity, $\bar{\mathrm{v}} \equiv 2\pi T - \mathrm{v}$, and the overall factor C is a combination of universal constants: numerically $C = 1.031419972084$. Λ is the QCD scale parameter in the Pauli–Villars scheme.

I am overwhelmingly grateful to V Petrov, D Diakonov and S Slizovskiy for their collaboration on the work reported here and to the "Dinastija" foundation for financial support.

References

Diakonov D *et al.* , 2004, *Phys Rev D* **70** 036003.
Gross D J, Pisarski R D and Yaffe L G, 1981, *Rev Mod Phys* **53** 43.
Kraan T C and van Baal P, 1998, *Phys Lett B* **428** 268.
Lee K and Lu C, 1998, *Phys Rev D* **58** 025011.

Vacuum condensates of dimension two in pure gluodynamics

Angelo Raffaele Fazio

Bogoliubov Laboratory of Theoretical Physics, Joint Institute for Nuclear Research, Dubna 141980, Russian Federation

1 $SU(2)$ Yang-Mills in the maximal Abelian gauge

Vacuum condensates of dimension two in Yang-Mills theories have witnessed increasing interest in recent years, both from the theoretical point of view as well as from lattice simulations, which have provided evidence for an effective dynamically generated gluon mass. My aim is to define a renormalisable effective potential for these condensates and evaluate it in an analytic form. The condensates arise as a non-trivial solution of the gap equation corresponding to the minimisation of the effective potential. In this contribution I will derive analytically the complete effective potential, at two ghost loops, in the framework of the formalism of the composite operator. I shall consider the maximal Abelian gauge $SU(2)$ Yang-Mills action in the four dimensional continuum Minkowski space:

$$
S = \int d^4x \left[-\frac{1}{4g^2} F^a_{\mu\nu} F^{a\mu\nu} - \frac{1}{4g^2} F_{\mu\nu} F^{\mu\nu} - \frac{1}{2\alpha} \left(D^{ab}_\mu A^{b\mu} \right)^2 + \right.
$$

$$
\left. + \bar{c}^a D^{ab}_\mu D^{\mu bc} c^c - \varepsilon^{ab} \varepsilon^{cd} \bar{c}^a c^d A^b_\mu A^{c\mu} - \frac{\alpha}{4} \epsilon^{ab} \epsilon^{cd} \bar{c}^a \bar{c}^b c^c c^d \right]. \tag{1}
$$

I have chosen the diagonal generator of the gauge group $SU(2)$ as Abelian charge and have made the following decomposition for the gluon, ghost and anti-ghost fields respectively: $(A^{\mu a}, A^\mu), (c^a, c), (\bar{c}^a, \bar{c})$, where $a = 1, 2$ denotes off-diagonal components of the Lie-algebra valued fields. D^{ab}_μ is the covariant derivative defined with respect to the diagonal component A_μ of the Lie Algebra valued connection. $F^a_{\mu\nu}$ and $F_{\mu\nu}$ are the field strength components. In (1) A_μ is gauge fixed by the Landau condition.

2 Two-loop ghost-antighost condensation

The generalised effective potential for this problem will depend only on the complete propagators of the theory $G(x,y)$ for the off-diagonal ghosts; $\Delta_a(x,y)$ and $\Delta(x,y)$ respectively for off-diagonal gluons and diagonal gluons:

$$
\begin{aligned}
V(G,\Delta_a,\Delta) &= -\imath \int \frac{d^4 p}{(2\pi)^4} \text{tr}\left[\log(S^{-1}(p)G(p)) - S^{-1}(p)G(p) + 1\right] \\
&+ \frac{\imath}{2} \int \frac{d^4 p}{(2\pi)^4} \text{tr}\left[\log(D_a^{-1}(p)\Delta_a(p)) - D_a^{-1}(p)\Delta_a(p) + 1\right] \quad (2)\\
&+ \frac{\imath}{2} \int \frac{d^4 p}{(2\pi)^4} \text{tr}\left[\log(D^{-1}(p)\Delta(p)) - D^{-1}(p)\Delta(p) + 1\right] \\
&+ V_2(G,\Delta_a,\Delta)
\end{aligned}
$$

In the previous formula all space-time and gauge indices have been suppressed. $S(p)$, $D(p)$ and $D_a(p)$ are the free propagators. In order to focus on ghost-antighost condensation consider the approximation in which $\Delta_a(p) = D_a(p)$ and $\Delta(p) = D(p)$. In this case V_2 includes the contribution of diagrams which are two-particle irreducible with respect to the ghost-antighost lines only, in the approximation of bare vertices. To compute the effective potential (2) I make the following *ansatz* for the ghost propagator:

$$
G^{ab}(p) = -\imath \frac{p^2 \delta^{ab} + \varphi(p^2)\epsilon^{ab}}{p^4 + \varphi^2(p^2)} \tag{3}
$$

From the mass gap equation of the potential (2), we find the following asymptotyc behaviour for $\varphi(p^2)$, disregarding the tadpole terms:

$$
\varphi(p^2) = \begin{cases} \varphi & |-p^2| \le \Lambda^2 \\ \varphi \times (-\frac{p^2}{\Lambda^2})^{-\varepsilon} & |-p^2| \gg \Lambda^2 \end{cases} \tag{4}
$$

with Λ a fixed scale of mass and

$$
\varepsilon = \frac{g^2}{4\pi^2} + O(g^2). \tag{5}
$$

The two-loop effective potential is

$$
\begin{aligned}
V(\varphi) &= \varphi^2 \left(\frac{1}{32\pi^2} - \frac{1}{\varepsilon} + \frac{3g^2}{512\pi^4} - \frac{g^2}{512\pi^2}\right) + \frac{3g^2}{512\pi^4\varepsilon}\varphi^2 \\
&+ \left(\frac{1}{32\pi^2} - \frac{3g^2}{512\pi^4}\right)\varphi^2 \log\left(\frac{\varphi^2}{\Lambda^4}\right) + \frac{\alpha\varphi^2}{256\pi^4}\left(\frac{1}{2}\log\left(\frac{\varphi^2}{\Lambda^4}\right) - \frac{1}{\varepsilon}\right)^2. \tag{6}
\end{aligned}
$$

I conclude that in the weak-coupling regime the ghost-antighost condensation is not destroyed at two-loop order for $\alpha > 0$. The value of the condensate is:

$$
\log\left(\frac{\varphi^2}{\Lambda^4}\right) = \frac{8\pi^2}{g^2} + \frac{32\pi^2}{g\sqrt{\alpha}} - \frac{1}{2} - \frac{8\pi^2}{\alpha}. \tag{7}
$$

Matching Regge theory to the operator product expansion (OPE)

Sergey S Afonin

V A Fock Department of Theoretical Physics, St Petersburg State University, 1 ul Ulyanovskaya, 198504, St Petersburg, Russia

It is well known in hadron phenomenology that the masses squared of mesons with given quantum numbers are approximately linear functions of their number of radial excitation n. In the present work we propose a systematic method to take into account possible corrections to the linear trajectories for vector (V), axial-vector (A), scalar (S), and pseudoscalar (P) cases in a consistent way with the bosonic string model. The string picture gives an equal slope of trajectories for the all V, A, S and P cases since this quantity is proportional to the string tension depending on glue-dynamics only. Our method is based on the consideration of two-point correlators of quark currents in the large-N_c limit of QCD ('t Hooft 1974, Witten 1979). On the one hand, by virtue of confinement they are saturated by an infinite set of narrow meson resonances in that limit, on the other hand their high-energy asymptotic behaviour is provided by perturbation theory and the Operator Product Expansion (OPE) (Shifman et al 1979) due to the asymptotic freedom of QCD.

Let us consider the following *ansatz* for the meson mass spectrum:

$$m_J^2(n) = M_J^2 + a n + \delta_J(n), \tag{1}$$

where $J \equiv V, A, S, P$ and the correction to linear trajectory, $\delta_J(n)$, models a possible deviation from the string picture in QCD. It turns out (Afonin et al 2004) that for consistency with the OPE one needs the following conditions on the spectra of masses $m_J^2(n)$ and residues in the V, A and S, P channels ($F_{VA}^2(n)$ and $G_{SP}^2(n)$ respectively):

$$m_{V,A}^2(n) = M^2 + an + A_m^{V,A} e^{-B_m n}, \tag{2}$$

$$m_{S,P}^2(n) = \overline{M}^2 + an + A_m^{S,P} e^{-B_m n}, \tag{3}$$

$$F_{V,A}^2(n) = a\left(\frac{1}{8\pi^2}\left(1 + \frac{\alpha_s}{\pi}\right) + A_F^{V,A} e^{-B_F n}\right), \tag{4}$$

$$G^2_{S,P}(n) = a\left(\frac{3}{16\pi^2}\left(1 + \frac{11\alpha_s}{3\pi}\right) + A^{S,P}_G e^{-B_G n}\right), \qquad (5)$$

where $B_{m,F,G} > 0$ and the constants $A^{V,A}_{m,F}$, $A^{S,P}_{m,G}$, $B_{m,F,G}$ have to be fitted.

Our numerical calculations (Afonin *et al* 2004) have shown that non-linear corrections to the mass spectrum result in a slight change of the linear trajectories. The exponentially decreasing contributions to the meson residues are small. However, they control substantially the values of condensates.

The conclusions that we gained from our analysis are as follows:

- The matching to the OPE cannot be achieved by a simple linear parameterisation of the mass spectrum, the linear trajectory *ansatz*.

- The convergence of the generalised Weinberg sum rules requires the universality of slopes and intercepts for parity conjugated trajectories.

- There must exist deviations from the linear trajectory *ansatz* triggered by chiral symmetry breaking. These deviations must decrease at least exponentially with n.

- There are also deviations from constant residues (decay constants) $F^2(n)$ (or for the quantities $G^2(n)$ in the scalar case). The analytic structure of OPE again imposes an exponential (or faster) decrease on these deviations.

- For heavy states, the D-wave vector mesons have to decouple from asymptotic sum rules. This fact implies the exponential (or faster) decrease in the corresponding decay constants $F^2_D(n)$.

- Our results seem to exclude a light $\sigma(600)$ particle as a quarkonium state and rather favour the non-linear realisation of chiral symmetry with the lightest scalar of mass approximately 1 GeV, its chiral partner being the $\pi'(1300)$.

- As a consequence of our approach the quantities L_8, L_{10} and Δm_π are obtained, in satisfactory agreement with the phenomenology.

- The dimension two gluon condensate λ does not affect the results and can be neglected.

Acknowledgments

The work reported in this talk was carried out jointly with A A Andrianov, V A Andrianov and D Espriu. It was supported by The Program "Universities of Russia: Basic Research" (Grant 02.01.016).

References

Afonin S S, Andrianov A A, Andrianov V A, and Espriu D, 2004, *JHEP04* **039** 1.
Shifman M A, Vainstein A I, and Zakharov V I, 1979, *Nucl Phys B* **147** 385, 448.
't Hooft G, 1974, *Nucl Phys B* **72** 461.
Witten E, 1979, *Nucl Phys B* **160** 57.

Strangeness production in a coupled channels framework

Olaf Scholten

Kernfysisch Versneller Instituut, University of Gronningen,
9747 AA Groningen, The Netherlands

The description of Kaon and Φ-meson production in an effective field theory offers several interesting challenges such as channel coupling effects, gauge invariance and analyticity of the amplitude.

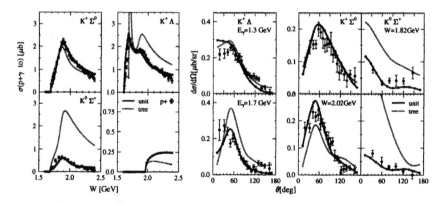

Figure 1. *Energy dependence of total cross sections (left) and angular distributions (right) for photo-induced strangeness production on the proton.*

The simplest approach to investigate channel-coupling (CC) effects is the K-matrix formalism. In (Kondratyuk 2000) this approach is discussed extensively for Compton and pion scattering on the nucleon. In Figure 1 results of a K-matrix calculation for the strangeness production sector are compared with recent data (Glander 2004, Lawall 2004). The full calculation (thick solid curve) shows good agreement with the data. The model includes all established resonances below 1.7 GeV is crossing symmetric and gauge invariant. The procedure of Davidson-Workman 2001 (DW) has been used for gauge restoration. The results of a calculation at the tree level calculation (thin dashed curve) shows the importance of CC effects. For certain kinematics the differences are huge. For others where they seem small, such as in the total cross section for the $K - \Lambda$ channel, large differences do show in the angular distributions. For the extraction of resonance parameters coupled channels effects thus have to be considered. In the calculation phenomenological form factors were introduced which forces the use of a procedure for gauge restoration. Figure 2 (left) shows that the results depend strongly on the procedure followed for gauge restoration. Minimal substitution (Ohta 1998, Kondratyuk 2000) fails to describe the data. In the fit procedure DW has been used which differs significantly from that of Janssen 2002 (SJ). The large difference between the Ohta and DW procedures resides in the convection current, the co-called A_2-amplitude.

To investigate the dependence on the gauge-restoration scheme, a schematic microscopic model for photo-induced $K - \Lambda$ production has been developed (Korchin 2003) in which the loop-dressing of the vertex involves a scalar particle. Effective form factors are generated at the tree level while the model allows for a self-consistent gauge-invariant calculation. The results for the A_2-amplitude are compared with the DW and the Ohta procedure in Figure 2. A full microscopic approach for strangeness production thus needs to be developed where form factors at the tree-level can be interpreted as real parts of loop corrections. This is crucial for interpreting the data (Kondratyuk 2000).

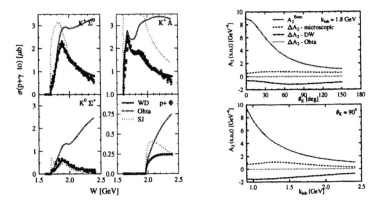

Figure 2. *Left: Dependence on the gauge-restoration scheme. Right: Microscopic and phenomenological descriptions for the gauge restoration term are compared to a microscopic calculation, see (Korchin 2003).*

References

Davidson R M and Workman R, 2001, *Phys Rev C*63 058201.
Glander K-H *et al.* , 2004, *Eur Phys J A* **19**, 251.
Janssen S *et al.* , 2002, *Phys Rev C* **66,** 035202.
Korchin A Yu and Scholten O, 2003, *Phys Rev C* **68** 045206.
Kondratyuk S and Scholten O, 2002, *Phys Rev C* **62** 025203.
Lawall R, 2004, private communication.
Ohta K, 1989, *Phys Rev C* **40** 1335.
Scholten O, 2004, *Phys Rev C* **64** 24004.

Nucleon properties in the perturbative chiral quark model

Jan Kuckei

Institut für Theoretische Physik, Universität Tübingen
Auf der Morgenstelle 14, D-72076 Tübingen, Germany

The basis of the perturbative chiral quark model is an effective chiral Lagrangian describing the valence quarks of baryons as relativistic fermions moving in an external field (static potential) $V_{\text{eff}}(r) = S(r) + \gamma^0 V(r)$ with $r = |\vec{x}|$, which in the SU(3)-flavour version are supplemented by a cloud of Goldstone bosons (π, K, η). By treating Goldstone fields as small fluctuations around the three-quark core we have the linearised effective Lagrangian

$$\mathcal{L}_{\text{eff}}(x) = \overline{\psi}(x)[i\ \not{\partial} - S(r) - \gamma^0 V(r)]\psi(x) + \frac{1}{2}[\partial_\mu \hat{\Phi}(x)]^2 - \overline{\psi}(x)S(r)i\gamma^5 \frac{\hat{\Phi}(x)}{F}\psi(x) + \mathcal{L}_{\chi SB}(x)$$

(1)

The additional term $\mathcal{L}_{\chi SB}$ contains the chiral-symmetry-breaking mass contributions for quarks and mesons: $\mathcal{L}_{\chi SB}(x) = -\overline{\psi}(x)\mathcal{M}\psi(x) - \frac{B}{2}Tr[\hat{\Phi}^2(x)\mathcal{M}]$ Here, $\hat{\Phi}$ is the octet matrix of pseudoscalar mesons, $F = 88$ MeV is the pion decay constant in the chiral limit, $\mathcal{M} = \text{diag}\{\hat{m}, \hat{m}, m_s\}$ is the mass matrix of current quarks (we restrict to the isospin symmetry limit $m_u = m_d = \hat{m}$) and $B = - < 0|\overline{u}u|0 > /F^2$ is the quark condensate constant. We rely on the standard picture of chiral symmetry breaking and for the masses of pseudoscalar mesons we use the leading term in their chiral expansion:

$$M_\pi^2 = 2\hat{m}B, \qquad M_K^2 = (\hat{m} + m_s)B, \qquad M_\eta^2 = \frac{2}{3}(\hat{m} + 2m_s)B$$

(2)

The following set of parameters is chosen in our evaluation: $\hat{m} = 7$ MeV, $m_s = 25\hat{m}$, $B = M_{\pi+}^2/2\hat{m} = 1.4$ GeV The meson masses satisfy the Gell-Mann-Oakes-Renner and the Gell-Mann-Okubo relation. In addition, the linearised effective Lagrangian fulfils the PCAC requirement. To derive the properties of baryons, which are modelled as bound states of valence quarks surrounded by a meson cloud, we formulate a perturbation theory. At zeroth order, the unperturbed Lagrangian simply describes the nucleon by three relativistic valence quarks which are confined by the effective one-body potential in the Dirac equation. For the unperturbed three-quark ground-state we introduce the notation $|\phi_0\rangle$ with the normalisation $\langle \phi_0|\phi_0 \rangle = 1$. We expand the quark field

ψ in the basis of potential eigenstates as $\psi(x) = \sum_\alpha b_\alpha u_\alpha(\vec{x}) \exp(-i\mathcal{E}_\alpha t)$ where the quark wavefunctions $\{u_\alpha\}$ in orbits α are solutions of the Dirac equation with the static potential $V_{\text{eff}}(r)$. The expansion coefficients b_α are the corresponding single quark annihilation operators. For the description of baryon properties we use the effective potential $V_{\text{eff}}(r)$ with a quadratic radial dependence $S(r) = M_1 + c_1 r^2$ and $V(r) = M_2 + c_2 r^2$ with the particular choice $M_1 = \frac{1-3\rho^2}{2\rho R}$, $M_2 = \mathcal{E}_0 - \frac{1+3\rho^2}{2\rho R}$, $c_1 \equiv c_2 = \frac{\rho}{2R^3}$. Here, $\mathcal{E}_0$ is the single-quark ground-state energy; R are ρ are the two parameters of our model. They are related to the ground-state quark wavefunction by

$$u_0(\vec{x}) = N \exp\left[-\frac{\vec{x}^2}{2R^2}\right] \begin{pmatrix} 1 \\ i\rho\,\vec{\sigma}\vec{x}/R \end{pmatrix} \chi_s \chi_f \chi_c, \qquad (3)$$

where $N = [\pi^{3/2} R^3 (1 + 3\rho^2/2)]^{-1/2}$ is a normalisation constant; χ_s, χ_f, χ_c are the spin, flavour and colour quark wavefunction, respectively. The parameter ρ is related to the axial charge g_A of the nucleon calculated in zeroth-order (or 3-quark-core) approximation by $g_A = \frac{5}{3}\left(1 - \frac{2\rho^2}{1+\frac{3}{2}\rho^2}\right)$. Therefore, ρ can be replaced by g_A using this matching condition. In our calculations we use the value g_A=1.25, so that we have only one free parameter, R. In the numerical studies R is varied in the region from 0.55 fm to 0.65 fm. The expectation value of an operator $\hat{A}$ is set up as

$$\langle \hat{A} \rangle = \langle \phi_0 | \sum_{n=1}^\infty \frac{i^n}{n!} \int d^4 x_1 \dots \int d^4 x_n T[\mathcal{L}_I(x_1) \dots \mathcal{L}_I(x_n) \hat{A}] | \phi_0 \rangle \qquad (4)$$

The interaction Lagrangian $\mathcal{L}_I$ may contain photon fields which are introduced by minimal substitution in the derivatives of the quark- or meson-fields, respectively. For the evaluation we apply Wick's theorem with the appropriate propagators for quarks and mesons. For the quark field we use a Feynman propagator for a fermion in a binding potential with

$$iG_\psi(x,y) = \langle \phi_0 | T\{\psi(x)\overline{\psi}(y)\} | \phi_0 \rangle = \theta(x_0 - y_0) \sum_\alpha u_\alpha(\vec{x}) \overline{u}_\alpha(\vec{y}) e^{-i\mathcal{E}_\alpha(x_0-y_0)} \qquad (5)$$

and for the meson fields we adopt the free boson Feynman propagator with

$$i\Delta_{ij}(x - y) = \langle 0 | T\{\Phi_i(x)\Phi_j(y)\} | 0 \rangle = \delta_{ij} \int \frac{d^4 k}{(2\pi)^4 i} \frac{\exp[-ik(x-y)]}{M_\Phi^2 - k^2 - i\epsilon} \qquad (6)$$

For further details and applications of our model to various nucleon properties, such as electric and magnetic form factors, sigma terms or $N - \Delta$ transition, we refer to the literature given in the references.

References

Lyubovitskij V E et al. , 2000, Phys Rev D 63 054026.
Lyubovitskij V E et al. , 2001a, Phys Rev C 64 065203.
Lyubovitskij V E et al. , 2001b, Phys Rev C 65 025202.
Cheedket S et al. , 2002, Eur Phys J A 20 317.
Pumsa-ard K et al. , 2003, Phys Rev C 68 015205.
Inoue T et al. , 2003, Phys Rev C 69 035207.

Gauge-invariance in hadronic scattering processes

Cedran J Bomhof

Vrije Universiteit Amsterdam,
NL-1081 HV Amsterdam, The Netherlands

Schematically, the scattering cross section of deep inelastic scattering (DIS) in the parton model can be written as

$$\mathrm{d}\sigma^{h\ell} \propto \sum_q \int \mathrm{d}x \, f^q(x) \, \mathrm{d}\hat{\sigma}^{q\ell}(x) \tag{1}$$

Here $\mathrm{d}\hat{\sigma}^{q\ell}$ is the partonic scattering cross section that can be calculated perturbatively and the $f^q(x)$ are the distribution functions of parton species q with momentum fraction x. These distribution functions contain all the non-perturbative information of the hadrons. Hence they cannot (at present) be calculated from first principles. However, they can be measured in certain types of experiments and then be used to make predictions about other hadronic scattering processes. At leading twist there are three different quark-distribution functions: The *unpolarised distribution* $f_1^q(x) = q(x)$ of unpolarised quarks of flavour q, the *longitudinal polarisation distribution* or *helicity distribution* $g_1^q(x) = \Delta q(x)$ of longitudinally polarised quarks in a longitudinally polarised hadron, and the *transverse polarisation distribution* or *transversity distribution* $h_1^q(x) = \delta q(x)$ corresponding to transversely polarised quarks in a transversely polarised hadron.

The functions f_1^q and g_1^q can be measured relatively easily in DIS and measurements of g_1^q revealed that the quark helicities only account for about 20% of the helicity of the hadron. A more complete discussion on these issues can be found in (Ryckbosch 2004). For a theoretical description of the distribution functions they must be properly defined. An appropriate framework is provided by the quark-correlator (Soper 1977):

$$\Phi_{ij}(k) = (2\pi)^{-4} \int \mathrm{d}^4\xi \, e^{ik\cdot\xi} \, \langle P,S| \overline{\psi}_j(0) \, \mathcal{U}(0,\xi) \, \psi_i(\xi) \, |P,S\rangle \tag{2}$$

All the distribution functions are projections of the quark-correlator:

$$f_1(x) = \int \mathrm{d}k^- \mathrm{d}^2k_T \, \mathrm{Tr}\left[\gamma^+ \Phi(k)\right] \tag{3}$$

$$S_L\, g_1(x) \;=\; \int dk^- d^2k_T \,\mathrm{Tr}\left[\gamma^+\gamma_5\Phi(k)\right] \tag{4}$$

$$S_T^i\, h_1(x) \;=\; \int dk^- d^2k_T \,\mathrm{Tr}\left[i\sigma^{i+}\Phi(k)\right] \tag{5}$$

The operator $\mathcal{U}(0,\xi)$ is the gauge link. The gauge link is necessary to obtain a colour gauge-invariant definition of the quark-correlator. Colour gauge-invariance of the quark-correlator is a prerequisite for the distribution functions that are defined via the quark-correlator to be gauge-invariant and have any physical relevance.

As stated above, the gauge link $\mathcal{U}(0,\xi) \equiv \mathcal{P}\exp ig\int_C dz \cdot A(z)$ is a path-ordered exponential connecting the fields at 0 and ξ along a certain path C. This has important implications for processes where the intrinsic transverse momenta of the partons play a role. In this case the points 0 and ξ in the definition of the quark-correlator will also be separated in the transverse direction ξ_T. If $\xi_T=0$ the gauge link runs along the light-cone, if $\xi_T\neq 0$ it must include a transverse piece (Brodsky *et al.* 2002, Belitsky *et al.* 2003), leading to different possible integration paths. The integration path C is determined by the hard scattering part of that process. In fact, it can explicitly be calculated by taking all interactions of the current quark with the spectator partons into account (Boer and Mulders 2000). For example, in semi-inclusive deep inelastic scattering (SIDIS: $\ell + h_1 \to \ell' + h_2 + X$) the points 0 and ξ are connected to each other via light-cone future-infinity. In Drell-Yan scattering (DY: $h_1 + \bar{h}_2 \to \ell + \bar{\ell} + X$), however, they are connected via light-cone past-infinity. Due to the different gauge links in SIDIS and DY, the difference of the average transverse momentum of the quarks in the incoming proton in these two processes does not necessarily have to vanish. Instead, it is proportional to an object that is called a *gluonic pole* (Boer *et al.* 2003). The gluonic poles play a role in single spin asymmetries measured in experiment.

Recent studies have shown that scattering processes with more complicated partonic subprocesses have more complicated link structures (Bomhof *et al.* 2004). One new effect is an extra winding of the integration path C around light-cone future and past-infinity before connecting the points 0 and ξ. Another is the occurrence of traced Wilson loops in the quark-correlator. All these new structures occur in hadronic pion production processes, *eg* $p^\uparrow p \to \pi X$ and $\bar{p}^\uparrow p \to \pi X$. Such processes have been studied *eg* at FERMILAB and RHIC. The exact form of the gauge links in these processes is now under investigation. Other topics for future research are the implications that the new link structures have on universality and factorisation (Collins and Metz 2004).

References

Belitsky A V, Ji X, Yuan F, 2003, *Nucl Phys B* **656** 165.
Boer D, Mulders P J, 2000, *Nucl Phys B* **569** 505.
Boer D, Mulders P J, Pijlman F, 2003, *Nucl Phys B* **667** 201.
Bomhof C J, Mulders P J, Pijlman F, 2004, *Phys Lett B* **596** 277.
Brodsky S J, Hwang D S, Schmidt I, 2002, *Phys Lett B* **530** 99.
Collins J C, Metz A, 2004, hep-ph/0408249.
Feynman R P, 1969, *Phys Rev Lett* **23** 1415.
Ryckbosch D, 2004, Contribution to these proceedings.
Soper D E, 1977, *Phys Rev D* **15** 1141.

Baryon observables in a covariant Faddeev approach

Markus Kloker

Institute for Theoretical Physics, Tübingen University,
Auf der Morgenstelle 14, D-72076 Tübingen, Germany

Recent experimental results emphasise the complicated nature of baryons. An example is provided by the ratio of the electric to the magnetic form factor of the proton measured at Jefferson Lab (Gayou *et al* 2002, Jones *et al* 2000). This ratio surprisingly decreases with increasing photon virtuality. Such findings encourage the study of relativistic, explicitly covariant models of baryons (Alkofer and von Smekal 2001, Maris and Roberts 2003).

The starting point of our Poincaré covariant model is the relativistic 3-quark Faddeev equation, which can be derived from the Dyson equation of the 6-point function by neglecting three-particle interactions. Assuming separability, we expand the two-quark correlation function in terms of non-pointlike scalar and axial-vector diquarks, introducing diquark-quark vertex functions χ^a and diquark propagators. One thereby arrives at a set of Bethe-Salpeter equations for the fully Poincaré covariant baryon wavefunctions (Figure 1). Binding of quarks and diquarks takes place via a quark exchange interaction and is therefore related to the Pauli principle for three-quark states.

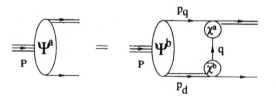

Figure 1. *The Bethe-Salpeter equation for the spinorial baryon-quark-diquark wavefunction Ψ.*

The calculation of the nucleon form factors (Oettel and Alkofer 2003) is done in a current conserving fashion, with two-loop diagrams included. Using a quark propagator fitted to meson observables and modeling the diquark propagator as a series of quark loops, one has a model with one free parameter: namely, the width of the axial-vector diquark. The results of the calculation are in reasonable agreement with experiments. As an example we provide the ratio $\mu G_E/G_M$ from Oettel and Alkofer (2003), which is compared with the experimental data of Jones *et al* (2000).

In an exploratory study, the radiative decay $\Sigma^0 \rightarrow \Lambda\gamma$ was calculated within our model. To preserve the Ward-Takahashi identity for the photon coupling two-loop contributions, like the coupling to the exchange quark and seagull diagrams, were taken into account. Using free propagators slightly overestimates the experimental value.

It is important that all elements of the model can in principle be determined from QCD calulations. Following this path we will try to determine the t-matrix and include effects of the meson cloud in future investigations of baryon form factors.

Ratio of Sachs form factors (Proton)

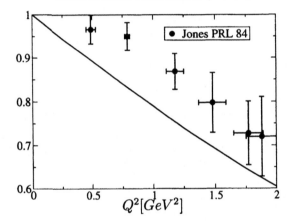

Figure 2. *Ratio of electric and magnetic form factors (Oettel and Alkofer 2003). The data are taken from Jones et al (2000).*

Acknowledgments

The work reported in this talk was carried out jointly with R Alkofer, A Höll and C D Roberts. It was supported by COSY (contract 41445395), by the German DFG (contract GRK683) and by the US DOE (contract W-31-109-ENG-38).

References

Alkofer R and von Smekal L, 2001, *Phys Rep* **353** 281.
Gayou O *et al.*, 2002, *Phys Rev Lett* **88** 092301.
Jones M K *et al.*, 2000, *Phys Rev Lett* **84** 1398.
Maris P and Roberts C D, 2003, *Int J Mod Phys E* **12** 297.
Oettel M and Alkofer R, 2003, *Eur Phys J A* **16** 95, and references therein.

Direct observation of ππ-scattering in πXe-interactions at 2.3 and 3.5 GeV/c

Bogdan Slowinski

Faculty of Physics, Warsaw University of Technology, Poland
Institute of Atomic Energy, Otwock-Swierk, Poland

The problem of mesonic degrees of freedom of nucleons and nuclei has received much consideration from various points of view (Ericson and Weise 1988, Neudatchin *et al.* 2004). Approaches to the problem include the use of pions as a probe, as well as the consideration of the collective pionic modes in nuclei and the role of pions in hadronic structures. We present the results of a comparative analysis of the experimental works, in which the emission of secondary pions produced in different channels of interactions of pions with nuclear targets at several GeV has been investigated. In particular, we examine our data on neutral pions created in peripheral πXe interactions obtained with a Xenon bubble chamber, which allows the detection of pions within a 4π geometry of their emission angles and with no limitation on their energy up to about 5 GeV (Slowinski 1977).

This gives us a possibility to analyse the correlation between the directly measured quantities: emission angle Θ_x and total energies E_x of π^0-mesons. Of particular interest are the so-called quasi-two-body channels which are predominantly one-step intranuclear collisions only without appreciable admixture of secondary interactions (Slowinski 1999). At the same time neutral pions do not experience intranuclear electromagnetic interactions and so they carry out information about the target on which they have been produced. Therefore a kinematic correlation between Θ_x and E_x as a simple consequence of momentum and energy conservation may give information about the intranuclear target mass involved in these interactions. Our results show a clear concentration of experimental points around the kinematic curves corresponding to an intranuclear target of the rest mass close to the pion mass (Slowinski 1977). Our experimental data are compared with the results of an intranuclear cascade simulation (Bielakov 1960), which do not reveal the observed correlation. Figure 1 shows typical scatter plots for neutral pions produced in πXe interactions at 2.3 and 3.5 GeV/c where only one neutral pion and one fast charged particle (proton and negative pion, correspondingly) are observed in the final state.

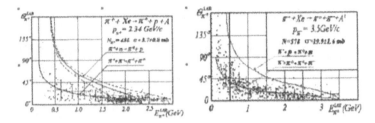

Figure 1. *Scatter plot of experimental points for π^0 from the reaction $\pi^+Xe \rightarrow \pi^0pA$ at 2.3GeV/c (left) and $\pi^{--}Xe \rightarrow \pi^0\pi^{--}A$ at 3.5GeV/c (right). Solid curves correspond to the kinematics of two-particle reactions. Dashed curves in the left figure mark the kinematical region due to Fermi oscillations of intranuclear nucleons. Crosses mark the regression curves for experimental points. (Slowinski 1977).*

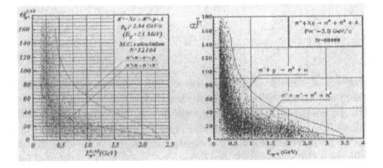

Figure 2. *Similar to Figure 1, but simulated by using the FRITIOF cascade code (Slowinski 1974) (performed by A. Galoyan).*

The charged pions produced in analogous reactions do not exhibit this property so well (Slowinski 1977, Uzhinskii 1996). Moreover, the correlation of neutral pions gradually vanishes when the number of secondary particles increases, as can be seen in Figure 2 for the case of the reaction π^+Xe at 2.3 GeV/c in which several pions are produced (Slowinski 1977).

References

Bielakov V A *et al.*, 1960, *ZhETF* **39** 937.
Ericson T and Weise W, 1988, *Pions and Nuclei*, Clarendon Press, Oxford.
Slowinski B, 1974, *Sov J Nucl Phys* **19** 301.
Slowinski B, 1977, *JINR Communication* 1-10932.
Slowinski B, 1999, *JINR Communication* E1-99-93.
Uzhinskii V V, 1996, *JINR Communication* E2-96-192.

Capability computing for computational hadron physics

Norbert Attig

John von Neumann Institute for Computing (NIC), Research Centre Jülich
52425 Jülich, Germany

1 Introduction

The EU project *Integrated Infrastructure Initiative on Hadron Physics (I3HP)* mainly focuses on experimental research in hadron physics. Consequently, a lot of European accelerator institutions are involved in this project and offer transnational access (TA) to their facilities. However, a number of theoretical groups are integrated in I3HP and their investigations, mainly based on lattice quantum chromodynamics (QCD) simulations, need access to high-end supercomputer facilities. The Research Centre Jülich is operating one of Europe's most powerful supercomputers and is able to offer TA via the national supercomputing centre *John von Neumann Institute for Computing (NIC)* to researchers within the framework of I3HP.

2 John von Neumann Institute for Computing (NIC)

This institute is a joint foundation of the Helmholtz centres Research Centre Jülich and DESY (Deutsches Elektronensynchrotron). It is acting as a German national supercomputing centre. NIC is managed by a board of directors. A scientific council gives recommendations with respect to the scientific programme of NIC and the allocation of supercomputing resources to the NIC projects. Within NIC the Central Institute for Applied Mathematics (ZAM) of the Research Centre Jülich provides the major production systems to the scientific community of NIC. Also within NIC, the Centre for Parallel Computing at the DESY laboratory in Zeuthen makes special-purpose computers, like APEmille and apeNEXT, available to research projects in elementary particle physics. Furthermore, NIC maintains research groups, which are dedicated to supercomputer-oriented investigation in selected fields of physics and other natural sciences. Presently a research group in Jülich works in the field of complex systems; another group at the DESY laboratory in Zeuthen works in the field of elementary particle physics.

NIC currently serves 150 projects in computational science and engineering with more than 400 scientists at 45 sites in Germany, in particular at universities. For a few years now, research groups from outside of Germany also have a chance to access the supercomputer facilities at NIC. There will be even more European links in the framework of European infrastructure projects like DEISA (Distributed European Infrastructure for Supercomputing Applications) and I3HP. A further push towards European users, especially from new eastern member states, is in progress.

The current supercomputer in Jülich used by NIC applications, called *Jump*, is an IBM system based on the p690 architecture. Each p690 frame consists of 32 processors (POWER4+, 1.7 GHz), 128 GByte main memory, and a peak performance of 218 GFlops. The supercomputer is a cluster of 41 p690 frames connected by a high performance switch. Its characteristics are: 1,312 processors, 5.2 TByte main memory, 56 TByte disk space, an aggregate peak performance of 8.9 TFlops, and a LINPACK performance of 5.6 TFlops. The bandwidth is currently 1,400 MByte per link and the latency is 6.5 μs. With these characteristics *Jump* is top-ranked in Europe and very well-suited for capability computing, *eg* lattice QCD simulations, which are performed by theoretical hadron researchers.

3   I3HP and NIC-TA

Within I3HP there are only two institutions offering TA to supercomputer resources: NIC, which provides CPU time, and ZIB (Zuse Institute Berlin), which provides access to storage capacity. Up to now, only the I3HP networking activity N2 *Computational Hadron Physics (ComHP)* is requesting these resources. This should increase significantly within the next few months!

The EU is funding 500,000 GFlops hours on *Jump* at Jülich for four years for non-German researchers. This number corresponds to 73,400 processor hours on the system which means roughly 2,000 processor hours per month, counting from now on. This is of course not sufficient for large-scale lattice QCD simulations, however, it should be used in an appropriate manner to demonstrate a valuable usage of supercomputing resources in the framework of the I3HP project. Additionally, the EU is funding the participation of I3HP users in NIC seminars and training courses at Jülich.

Researchers who are interested in requesting part of these resources should visit the web page http://www.fz-juelich.de/nic/i3hp-nic-ta, where a lot of information is given about the application procedure and the usage of the supercomputer facility. It is also advisable to coordinate any access plan with the management of the ComHP project. Last but not least, a scientific proposal has to be written and forwarded to the chairman of the NIC scientific council. It will be reviewed by the NIC resource allocation committee under the constraints of already granted resources. Finally, the decision of this committee will be communicated and, in case of successful evaluation, the investigations can start.

Finite-size effects in Lattice QCD with dynamical Wilson fermions

Boris Orth

Fachbereich Mathematik und Naturwissenschaften, Bergische Universität Wuppertal, Gaussstrasse 20, D-42119 Wuppertal, Germany

1 Introduction

Due to limited computing resources the choice of parameter values for a full Lattice QCD simulation always amounts to a compromise between having as small a lattice spacing, as light quark masses, and as large a volume as possible. Aiming at pushing unquenched simulations with Wilson's action towards the computationally expensive regime of small quark masses, our GRAL project addresses the question whether some computing time can be saved by sticking to lattices with rather modest numbers of grid sites, and extrapolating the finite-volume results to infinite volume (prior to the usual chiral and continuum extrapolations). In this context the systematic volume dependence of simulated pion and nucleon masses has been investigated and compared with an analytic prediction by Lüscher and with results from Chiral Perturbation Theory (ChPT) (Orth 2004). We have analysed data from Hybrid Monte Carlo simulations with the standard two-flavour Wilson action at two different lattice spacings of $a \approx 0.08$ fm and 0.13 fm. The considered quark masses amount to approximately 85 and 50% (at the smaller a) and 36% (at the larger a) of the strange quark mass. At each quark mass we have examined at least three different lattices with $L/a = 10$ to 24 sites in the spatial directions ($L = 0.85$–2.04 fm).

2 Finite-size effects in pion and nucleon masses

If enclosed in a box whose linear size L is much larger than the Compton wave length of the pion, a single hadron is practically unaffected by the finite volume. When L is decreased, the virtual pion cloud surrounding the particle is slightly distorted, and if periodic boundary conditions have been imposed, pions may be exchanged "around the world". As a consequence the mass of the hadron is shifted relative to its asymptotic value by terms of order $e^{-m_\pi L}$ (Lüscher 1986). At still smaller values of L the quark wave functions of the enclosed hadron are "squeezed" and one observes rapidly increasing finite-volume effects approximately proportional to some negative power of L (Fukugita *et al.* 1992).

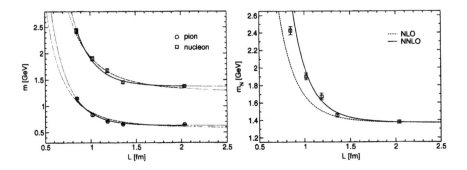

As the left figure illustrates for the example of our largest quark mass, such a behaviour is indeed borne out by the data. Empirically, both the pion and the nucleon data can be well described by exponentials (solid curves); for comparison, the graph also shows (inferior) fits to a power law (dashed curves). Comparing our data with available analytic predictions it turns out that in the case of the pion neither Lüscher's asymptotic finite-size formula with input from infinite-volume ChPT up to NNLO, nor the full LO ChPT result in finite volume (Gasser and Leutwyler 1987) apply below lattice sizes of about 1.5 fm: the predicted effects (Colangelo and Dürr 2004) are still too small. By contrast, the volume-dependence of our nucleon mass data is reproduced remarkably well even down to lattice sizes of only about 1 fm by a recent NNLO formula from relativistic Baryon ChPT (Ali Khan 2004), as displayed in the right graph for the largest quark mass ($m_\pi \approx 640\,\text{MeV}$). This formula is a promising example of how parametrisations suitable for controllable infinite-volume extrapolations may be found quite generally within the very Effective Field Theories that also describe the quark mass dependence.

Acknowledgments

The work sketched above has been carried out in collaboration with Klaus Schilling and Thomas Lippert. I thank the NIC in Jülich and Zeuthen for computing time and support.

References

Ali Khan A *et al.* [QCDSF-UKQCD Collaboration], 2004, *Nucl Phys B* **689** 175.
Colangelo G, Dürr S, 2004, *Eur Phys J C* **33** 543.
Fukugita M, *et al.* , 1992, *Phys Lett B* **294** 380.
Gasser J, Leutwyler H, 1987, *Phys Lett B* **184** 83.
Lüscher M, 1986, *Commun Math Phys* **104** 177.
Orth B, 2004, WUB-DIS 2004-03.

N to Δ transition in lattice QCD

Constantia Alexandrou

Department of Physics, University of Cyprus
CY-1678 Nicosia, Cyprus

The N to Δ transition form factors encode important information on hadron deformation and have been studied carefully in recent experiments (Mertz *et al.* 2001, Joo *et al.* 2002). This work presents the first lattice evaluation of the momentum dependence of the magnetic dipole, M1, the electric quadrupole, E2, and the Coulomb quadrupole, C2, transition amplitudes. They are calculated in the quenched approximation on a lattice of size $32^3 \times 64$ at $\beta = 6.0$ with Wilson fermions with sufficient accuracy to exclude a zero value of E2 and C2 at low Q^2. This accuracy is achieved by applying two novel methods:

- An interpolating field for the Δ that allows a maximum number of statistically distinct lattice measurements contributing to a given Q^2.

- Extraction of the transition form factors by performing an over-constrained analysis of the lattice measurements using all lattice momentum vectors contributing to a given Q^2 value (Hägler *et al.* 2003).

The evaluation of the three-point function $G_\sigma^{\Delta j^\mu N}(t_2, t_1; \mathbf{p}\,', \mathbf{p}; \Gamma)$ is described in detail in a series of papers (Alexandrou *et al.* 2004). The used kinematics correspond to the Δ being produced at rest so that $\mathbf{q} \equiv \mathbf{p}' - \mathbf{p} = -\mathbf{p}$ and $Q^2 = -q^2$. We fix $t_2/a = 12$ and search for a plateau as a function of t_1, the time slice where the current couples to a quark. The present analysis includes 200 configurations at $\kappa = 0.1554, 0.1558$ and 0.1562 corresponding to $m_\pi/m_\rho = 0.64, 0.59$ and 0.50 respectively. The nucleon mass at the chiral limit was used to set the lattice spacing a, obtaining $a^{-1} = 2.04(2)$ GeV. Using the optimal sink that was constructed for the Δ the results for $\mathcal{G}_{M1}^*$ are shown in Figure 1 as a function of Q^2, where

$$\mathcal{G}_{M1}^* \equiv \frac{1}{3} \frac{1}{\sqrt{1 + \frac{Q^2}{(m_N + m_\Delta)^2}}} \mathcal{G}_{M1} \quad . \tag{1}$$

Results in the chiral limit are obtained from a linear extrapolation in m_π^2. The same figure also shows the experimental values as extracted from the measured cross sections using the phenomenological model MAID (Drechsel *et al.* 1999). Although the lattice data in the chiral limit lie higher than the MAID data both data sets are well described by the phenomenological ansatz $\mathcal{G}_a(Q^2) = \mathcal{G}_a(0)\left(1 + \alpha\,e^{-\gamma Q^2}\right)G_E^p(Q^2)$ for $a = M1, E2$ and $C2$, where $G_E^p(Q^2) = 1/(1 + Q^2/0.71)^2$ is the proton electric form factor.

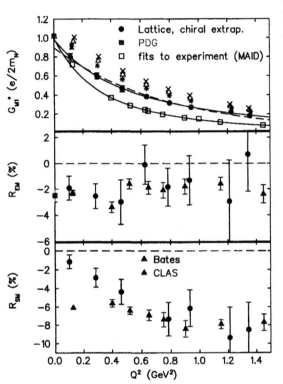

Figure 1 also shows the ratios EMR $\equiv R_{EM} = -\frac{\mathcal{G}_{E2}(q^2)}{\mathcal{G}_{M1}(q^2)}$ and CMR $\equiv R_{SM} = -\frac{|q|}{2m_\Delta}\frac{\mathcal{G}_{C2}(q^2)}{\mathcal{G}_{M1}(q^2)}$ in the chiral limit obtained by performing a linear extrapolation in m_π^2. As expected, both EMR and CMR become more negative as they approach the chiral limit. The quenched results for EMR are accurate enough to exclude a zero value at low Q^2. With our current statistics they are in agreement with the experimental measurements. Whether the apparent discrepancy for CMR at low Q^2 is a significant deficiency of quenched QCD or a problem with a single data point remains to be resolved by new experimental measurements that are currently being analysed.

Figure 1. *Top: $\mathcal{G}_{M1}^*$ as function of Q^2 at $\kappa = 0.1554$ (crosses), $\kappa = 0.1558$ (open triangles), $\kappa = 0.1562$ (asterisks) and in the chiral limit (filled circles). Open squares show $\mathcal{G}_{M1}^*$ extracted from measurements using MAID (Drechsel et al. 1999). The solid lines show fits to the ansatz $\mathcal{G}_a(Q^2) = \mathcal{G}_a(0)\left(1 + \alpha\exp(-\gamma Q^2)\right)G_E^p(Q^2)$. The dashed line is a fit to the lattice data using $a\exp(-bQ^2)$. Middle: EMR, Bottom: CMR, both in the chiral limit (filled circles) compared to experiment (filled triangles).*

References

Alexandrou C et al. , 2004, *Phys Rev D* **69** 114506.
Alexandrou C et al. , 2004, hep-lat/0409122; hep-lat/0408017.
Drechsel D et al. , 1999, *Nucl Phys B* **645** 145.
Hägler Ph et al. , LHPC and SESAM collaborations, 2003, *Phys Rev D* **68** 034505.
Joo K et al. , 2002, *Phys Rev Lett* **88** 122001.
Mertz C et al. , 2001, *Phys Rev Lett* **86** 2963.

Dynamical overlap

Nigel Cundy

Department of Physics, Universität Wuppertal
Gaussstrasse 19, Wuppertal, Germany

1 HMC and overlap fermions

We are developing a hybrid Monte-Carlo (HMC) algorithm (Cundy *et al.* 2004b, Fodor *et al.* 2003, Duane *et al.* 1987) to generate dynamical gauge field ensembles with overlap fermions, which satisfy a lattice chiral symmetry (unlike Wilson fermions) and have no doublers (unlike staggered fermions). However, adapting HMC to overlap fermions presents difficulties not found in the Wilson or staggered formulations.

HMC generates gauge field ensembles using a molecular dynamics (MD) update followed by a metropolis accept/reject step. For the MD update we introduce a momentum field, $\Pi_\mu(x)$, conjugate to the gauge fields $U_\mu(x)$, and a random spinor field $\Phi(x)$. This allows us to define a MD "energy"

$$E = \frac{1}{2} \sum_{x,\mu} \Pi_\mu^2(x) + S_g(U) + \Phi^\dagger \frac{1}{D^\dagger D} \Phi, \tag{1}$$

where S_g is some suitable gauge action, and D is a suitable Dirac operator. We then perform a numerical integration of the classical equations of motion of the system, $dU/d\tau = i\Pi U$, and $dE/d\tau = 0$, over a fictitious computer time τ. The integration procedure must be area conserving, reversible, and ergodic. The leapfrog integration scheme, which uses $\tau/\Delta\tau$ time steps, has these properties, and conserves energy up to $O(\Delta\tau^2)$. The overlap Dirac operator is (Neuberger 1998)

$$D = 1 + \mu + (1 - \mu)\gamma_5 \text{sign}(Q)(1 - \sum_l |\psi^l\rangle\langle\psi^l|)$$

$$+(1 - \mu)\sum_l |\psi^l\rangle\langle\psi^l| \text{sign}(\lambda_l),$$

where μ is a mass parameter, Q is the hermitian Wilson operator with a negative mass, and ψ^l are the lowest eigenvectors of Q with eigenvalues λ_l. Solving the equation of motion $\dot{E} = 0$ involves differentiating $D^\dagger D$ to generate the fermionic force F. The fermionic force is given in (Cundy *et al.* 2004b).

2 HMC with overlap fermions

Adapting HMC to overlap fermions presents a number of unique problems:

1. The calculation of the matrix sign function in the overlap operator requires $O(100)$ calls to the Wilson operator. To calculate the fermionic force, we need to invert the overlap operator, which is computationally costly.

2. The δ-function in the fermionic force has to be treated exactly to prevent large energy conservation violations (Fodor *et al.* 2003, Cundy *et al.* 2004b).

3. We need the energy conservation up to $O(\Delta\tau^2)$ otherwise the trajectory length will be too large at low masses.

4. At low masses, the conjugate gradient (CG) routine used to invert $D^\dagger D$ will not converge if D has a zero mode (an eigenvector which has zero eigenvalue when $\mu = 0$), because the condition number of $D^\dagger D$ is too high.

5. At low masses, the fermionic force can become unacceptably large in the presence of the zero mode, leading to a very low acceptance rate for non-trivial topological sectors. The method of chiral projection (Bode *et al.* 1999) will overcome this.

3 Conclusions

Dynamical overlap simulations are possible, and will soon be practical.

Acknowledgments

I would like to thank Thomas Lippert and Stefan Krieg for many useful discussions. This work is supported by EU grant MC-EIF-CT-2003-501467.

References

Arnold G *et al.* , 2003, hep-lat/0311025.
Bode A *et al.* , 1999, hep-lat/9912043.
Cundy N *et al.* , 2004a, hep-lat/0405003.
Cundy N *et al.* , 2004b, hep-lat/0409029.
Duane S *et al.* , 1987, *Phys Lett B* **195** 216.
Fodor Z *et al.* , 2003, hep-lat/0311010.
Fodor Z *et al.* , 2004, hep-lat/0409070.
Krieg S *et al.* , 2004, hep-lat/0409030.
Neuberber H, 1998, *Phys Rev D* **57** 5417.

Recent overlap results

Volker Weinberg

Institut für Theoretische Physik, Freie Universität Berlin,
D-14196 Berlin, Germany

1 Introduction

As initially suggested by (Ginsparg and Wilson 1982) and rediscovered by Hasenfratz (1998) more than fifteen years later, one can implement chiral symmetry on the lattice by requiring the condition $\gamma_5 D + D\gamma_5 = aD\gamma_5 D$ for a lattice implementation of the Dirac operator D. One possible solution of this equation is the overlap operator D_N (Neuberger 1998), which is defined by $D_N = \frac{\rho}{a}(1 + \mathrm{sgn}(X))$, $X = D_W - \frac{\rho}{a}$, where we use the Wilson operator D_W as the 'kernel' in the overlap construction. The tunable parameter ρ is set to 1.4, as this was found to be the optimal choice concerning the condition number of D_W and the locality of D_N. To compute the numerically very expensive 'sign function' $\mathrm{sgn}(X) = \frac{X}{\sqrt{X^\dagger X}}$, we use a polynomial approach using minmax polynomials (Giusti et al. 2003) and project out $\mathcal{O}(10)$ eigenvalues of the Wilson Kernel to treat the sign function on the corresponding subspace exactly.

2 Spectral properties at T=0

In contrast to Wilson fermions, the index theorem is valid exactly on the lattice for overlap fermions and allows an unambiguous computation of the topological charge and susceptibility, χ_{top}, using a spectral approach. The spontaneous breaking of chiral symmetry by the dynamical creation of a non-vanishing chiral condensate, $\langle \overline{\Psi}\Psi \rangle$, is related to the spectral density of the Dirac operator near zero by the famous relation $\langle \overline{\Psi}\Psi \rangle = -\pi/V\rho(0)$ (Banks and Casher 1979). Computing the lowest 140 eigenvalues on $\mathcal{O}(250)$ configurations on a $16^3\,32$ lattice at $\beta = 8.45$ using the Arnoldi-algorithm, we obtained $\chi_{top} = (187(5)\mathrm{MeV})^4$ and $\langle \overline{\psi}\psi \rangle^{\overline{MS}}(2\,\mathrm{GeV}) = (263(9)\,\mathrm{MeV})^3$ (Galletly et al. 2003). For small eigenvalues, the computed spectral density averaged over topological sectors (Figure 1 (left)) nicely matches the prediction of chiral random matrix theory, which only incorporates the symmetries and universal properties of the real theory.

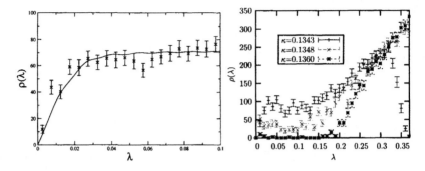

Figure 1. *The spectral density $\rho(\lambda)$ at T=0 (left) and at 3 values of κ_{sea} in the vicinity of the chiral phase transition (right).*

3 Chiral phase transition

To investigate the spectral properties of the overlap Dirac operator in the vicinity of the chiral phase transition, we concentrate on 3 values of $\kappa_{sea} = 0.1343, 0.13480$ and 0.1360 at fixed coupling $\beta = 5.2$ on $\mathcal{O}(250)$ $16^3 8$ finite temperature configurations. To smooth the configurations we perform one step of APE-smearing.

The transition is determined (Bornyakov *et al.* 2004) to occur at a critical coupling $\kappa_t = 0.1344(1)$, which equals $T_c = 210(3)$ MeV in physical units, while the 3 chosen κ values correspond to $T/T_c = 0.98, 1.06$ and 1.28 respectively. One can clearly see (Figure 1 (right)) a non-vanishing chiral condensate in the chiral broken phase at $\kappa = 0.1343$ below the phase transition and a large gap in the spectrum at $\kappa = 0.1360$ in the chiral restored phase. However, at $\kappa = 0.1348$, a value which was determined by the Polyakov-loop method to be above the transition, we find a non-vanishing tail of small eigenvalues extending down to zero, which might be an artefact of our hybrid fermionic approach.

Acknowledgments

The work was done within the QCDSF and DIK (DESY-ITEP-Kanazawa) collaborations together with M. Gürtler, E.-M. Ilgenfritz, G. Schierholz and T. Streuer.

References

Banks T, and Casher A, 1979, *Nucl Phys B* **169** 103.
Bornyakov V G *et al.* , 2004, *arXiv:hep-lat/0401014*.
Galletly D *et al.* , 2004, *Nucl Phys Proc Suppl* **129** 453.
Ginsparg P H, and Wilson K G, 1982, *Phys Rev D* **25** 2649.
Giusti L *et al.* , 2003, *Comput Phys Commun* **153** 31.
Hasenfratz P, 1998, *Nucl Phys Proc Suppl* **63** 53.
Lüscher M, and Weisz P, 1985, *Commun Math Phys* **97** 59.
Neuberger H, 1998, *Phys Lett B* **417** 141; *Phys Lett B* **427** 353.

Generalised parton distributions from lattice QCD

James M Zanotti

John von Neumann-Institut für Computing NIC / DESY
15738 Zeuthen, Germany

Generalised Parton Distributions (GPDs) are able to provide insights into the complex interplay of longitudinal momentum and transverse coordinate space as well as spin and orbital angular momentum degrees of freedom in the nucleon. The $(n-1)^{th}$ moments of the (generalised) parton distributions are revealed through the extraction of the generalised form factors (GFFs), $A_{n,2i}^q(\Delta^2)$, $B_{n,2i}^q(\Delta^2)$ and $C_n^q(\Delta^2)$ from the (non-)forward matrix elements of the appropriate local twist-2 operators.

Burkardt (2003) has shown that the spin-independent and spin-dependent generalised parton distributions $H(x,0,\Delta^2)$ and $\tilde{H}(x,0,\Delta^2)$ arising from matrix elements with purely transverse momentum transfer $(\Delta_\perp)$ gain a physical interpretation when Fourier transformed to impact parameter space

$$q(x,\vec{b}_\perp) = \int \frac{d^2\Delta_\perp}{(2\pi)^2}\, e^{-i\vec{b}_\perp \cdot \vec{\Delta}_\perp} H(x,0,-\Delta_\perp^2)\,, \tag{1}$$

(and similar for the polarised $\Delta q(x,\vec{b}_\perp)$) where $q(x,\vec{b}_\perp)$ is the probability of finding a quark with longitudinal momentum fraction x and at transverse position (or impact parameter) $\vec{b}_\perp$. Burkardt also argued that $H(x,0,-\Delta_\perp^2)$ becomes $\Delta_\perp^2$-independent as $x \to 1$ since, physically, we expect the transverse size of the nucleon to decrease as x increases, ie $\lim_{x\to 1} q(x,\vec{b}_\perp) \propto \delta^2(\vec{b}_\perp)$. As a result, we expect the slopes of the moments of $H(x,0,-\Delta_\perp^2)$ in $\Delta_\perp^2$ to decrease as we proceed to higher moments. This is also true for the polarised moments of $\tilde{H}(x,0,-\Delta_\perp^2)$.

Moments of GPDs are amenable to lattice calculations. Thus, they offer a promising way to link phenomenological observations to first principle theoretical considerations. In this talk we report on recent unquenched results obtained by the QCDSF collaboration. For full details of our simulations, we refer the reader to Göckeler et al. (2004).

In Figure 1 (left) we show the $\Delta^2 = t$-dependence of $A_{n0}(t)$ for $n = 1, 2, 3$. The form factors have been normalised to unity to make a comparison of the slopes easier and we fit the form factors with a dipole form. We note here that the form factors

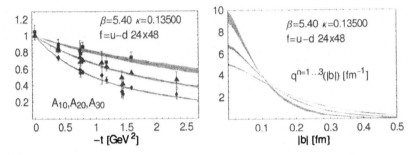

Figure 1. *Change of slope of the GFFs (left) and narrowing of the corresponding $b_\perp$ dependent GFFs (right); $n = 1$ corresponds to the lowest (widest) curve in the left (right)-hand plot.*

for the unpolarised moments are well separated and that their slopes decrease with increasing n, which is consistent with the observation that $\lim_{x \to 1} H(x, 0, t)/q(x) = 1$ (Burkardt 2003). For the polarised moments, we observe a similar scenario, however here the change in slope between the form factors is not as large. This is to be compared with the results from (Hägler 2003) which reveal no change in slope between the $n = 2$ and $n = 3$ polarised moments. The normalised impact parameter dependent moments $q^{n=1\cdots3}(b_\perp)$ with $\int db_\perp q^{n=1\cdots3}(b_\perp) = 1$ are shown in the right plot of Figure 1. Here we clearly observe the anticipated narrowing of the distributions in impact parameter space, going from $n = 1$ to $n = 3$.

Acknowledgments

Additional authors involved in this work: M Göckeler, P Hägler, R Horsley, D Pleiter, P E L Rakow, A Schäfer and G Schierholz.

References

Burkardt M, 2003, *Int J Mod Phys A* **18** 173.
Diehl M, 2002, *Eur Phys J C* **25** 223.
Diehl M *et al.* , 1997, *Phys Lett B* **411** 193.
Göckeler M *et al.* , 1997, hep-ph/9711245.
Göckeler M *et al.* , 2004, hep-lat/0409162.
Göckeler M *et al.* , 2004, hep-lat/0410023.
Hägler P *et al.* , 2003, *Phys Rev D* **68** 034505.
Hägler P *et al.* , 2003, hep-lat/0312014.
Ji X, 1997, *Phys Rev Lett* **78** 610.
Ji X, 1998, *J Phys G: Nucl Part Phys* **24** 1181.
Müller D *et al.* , 1994, *For Phys* **42** 101.
Radyushkin A V, 1997, *Phys Rev D* **56** 5524.

Mass storage capacity for computational hadron physics

Hinnerk Stüben

Konrad-Zuse-Zentrum für Informationstechnik Berlin (ZIB)
Takustrasse 7, 14195 Berlin, Germany

1 Introduction

In the framework of the *HadronPhysics Integrated Infrastructure Initiative (I3HP)* which is funded by the European Community in the Sixth Framework Programme ZIB offers Transnational Access to its state-of-the-art mass storage system. Access is offered primarily but not limited to the Computational Hadron Physics community. In this note the role of mass storage in computational hadron physics is explained, and the configuration of the storage system used is outlined.

2 The role of mass storage in the computational pipeline

The underlying theory which is studied is quantum chromodynamics (QCD). Simulations of QCD are typically carried out in three steps. The first step is a large scale Hybrid Monte Carlo calculation. Such a calculation requires the fastest parallel computers available. The sustained performance needed is at the order of 10^{12} floating point operations per second. Even at this speed the whole Hybrid Monte Carlo computation takes months. Outcome of the Hybrid Monte Carlo run is a set of so-called configurations. This set is considered as a representative sample of states of the gluon field.

From the sample of configurations all physical quantities can be calculated. This is done in step two where typically correlation functions are calculated for each configuration. In the third step the final results are obtained from fits to the calculated correlation functions. Due to the stochastic nature of the Monte Carlo method the results have a statistical error. The error decreases with the square of the computing time. The error analysis is part of the third step.

In contrast to step one step two requires about two orders of magnitude less of computing power. Hence step two can by performed on local PC clusters. Step three is executed on a desktop PC.

In principle steps two and three have to be carried out for every physical quantity to be calculated. Different quantities can and are in practice calculated by different members of a collaboration using their local computers.

The configurations generated in step one are very precious. Hence they are carefully stored, typically on a mass storage system. The size of one sample of configurations is in the 10^{12} Byte range. The advantage of a mass storage system is that the data is available on-line. However, there are limitations in accessing such systems especially for international collaborations.

Today, copying data to the places where step two is run is done manually by one or two people. Therefore data is preferably copied in large chunks and local disk capacity becomes a bottle-neck. The goal of this Access Activity is to increase productivity by reducing manual work and removing the disk space bottle-neck. This can be achieved by employing suitable software which allows the structuring of the data in a transparent way and makes it feasible that all members of a collaboration can use the data storage system. In addition collaborations can decide to make their configurations public which will broaden the possibilities of research.

3   Hard- and software configuration

The main components of the mass storage system at ZIB are two StorageTek Powderhorn 9310 data silos. Attached to the silos are 14 T9940B tape drives. The total capacity is 12,000 cassettes. Currently each cassette can store 200 GByte of data. The base data management server is a Silicon Graphics O300 running SGI's Data Migration Facility (DMF). The system is connected to the internet via a 155 Mbit/s line.

For the Transnational Access a higher level data server was added. On this server the *dCache* software is running which provides the features mentioned above. *dCache* is a joint venture between the Deutsches Elektronen-Synchrotron (DESY) and the Fermi National Accelerator Laboratory (Fermilab).

Acknowledgments

The activity presented is funded by the European Community under contract number RII3-CT-2004-506078.

References

http://www.infn.it/eu/i3hp.
http://www.zib.de.
http://www.dcache.org.

Non-perturbative versus perturbative effects in generalised parton distributions

Sigfrido Boffi

Dipartimento di Fisica Nucleare e Teorica, Università di Pavia
and Istituto Nazionale di Fisica Nucleare, Sezione di Pavia, Italy

Different inputs at the hadronic scale have been considered in the Quantum Chromo-dynamic (QCD) evolution of Generalised Parton Distributions (GPDs) to study the sensitivity of the results to the non-perturbative nature of the low-scale hadronic structure. In particular, the meson-cloud model was assumed to include sea quarks in the partonic content at the hadronic scale. As an example, in Figure 1 the singlet and non-singlet quark GPDs as well as the gluon GPD are plotted assuming GPDs in the form of double distributions, with input parton distributions derived in the relativistic hypercentral Constituent Quark Model (CQM) for valence quarks implemented with the meson-cloud model.

As a general result of the present investigation, evolution in the Dokshitzer-Gribov-Lipatov-Altarelli-Parisi (DGLAP) region is not very sensitive to the input. The reason is twofold. At large x the valence contribution dominates at the input hadronic scale and, as the scale increases, the distributions are swept from the DGLAP domain to lie entirely in the Efremov-Radyushkin-Brodsky-Lepage (ERBL) region. In addition, evolution never pushes the input distributions from the ERBL to the DGLAP region which then evolves independently of the ERBL input. In contrast, the ERBL region is rather sensitive to the input including, or not including, the sea.

Due to the mixing between singlet and gluon distributions in the evolution equations, even with a vanishing gluon distribution at the hadronic scale, one obtains a significant gluon GPD, peaked around $x = 0$, in the ERBL region after evolution. The peak height depends on the input and is higher when the sea is included.

Acknowledgments

The work was carried out in collaboration with Barbara Pasquini (University and INFN, Pavia) and Marco Traini (Univeristy of Trento and ECT*).

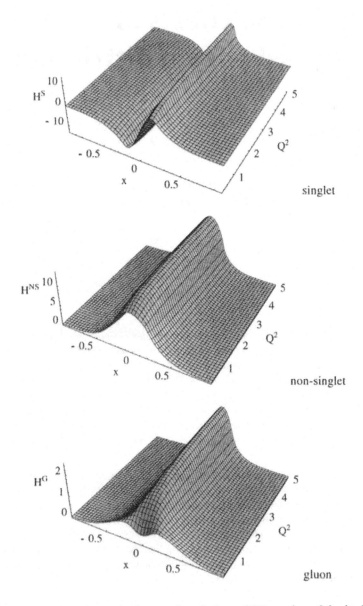

Figure 1. *Singlet and non-singlet quark and gluon GPDs at $\xi = 0.2$ obtained from NLO evolution within the double-distribution model using parton distributions of the hypercentral CQM with the sea contribution at the initial scale $Q_0^2 = 0.27 \, GeV^2$.*

Neutron spin structure measurements in JLab Hall A

Kees de Jager

Jefferson Lab,
12000 Jefferson Avenue, Newport News, VA 23606, USA

Recently, the high polarised luminosity available at Jefferson Lab (JLab) has allowed the study of the nucleon spin structure at an unprecedented precision, enabling us to access the hard-to-reach valence quark (high-x) region and also to accurately map the intermediate to low Q^2 region. The high-x region is of special interest, because this is where the valence quark contributions are expected to dominate. With sea quarks and explicit gluon contributions expected not to be important, it is a clean region to test our understanding of nucleon structure.

In 2001, JLab experiment E99-117 (Zheng *et al.* 2004) was carried out in Hall A to measure A_1^n with high precision in the x region from 0.33 to 0.61 (Q^2 from 2.7 to 4.8 GeV^2). Asymmetries from inclusive scattering of a highly polarised 5.7 GeV electron beam on a high pressure (> 10 atm) (both longitudinal and transversely) polarised ^{3}He target were measured. Beam polarisation was measured with a Møller polarimeter and a Compton polarimeter. The average beam polarisation was 78% $\times$ (1 $\pm$ 0.03). The ^{3}He target was polarised by spin exchange with optically pumped Rubidium. The average in-beam polarisation was 40% $\times$ (1 $\pm$ 0.04). The scattered electrons were detected with two high-precision spectrometers with their standard detector packages (scintillators for trigger, vertical drift chambers for tracking, gas Cherenkov counters and shower counters for particle identification).

Parallel and perpendicular asymmetries were extracted for ^{3}He. After taking into account the beam and target polarisation and the dilution factor, they were combined to form $A_1^{^3He}$. Using the most recent model (Bissey *et al.* 2002), nuclear corrections were applied to extract A_1^n. Final results on A_1^n are shown in the left panel of Figure 1. For clarity, not all theoretical predictions are shown. A more complete list is given in (Zheng *et al.* 2004).

The experiment greatly improved the precision of data in the high-x region. This is the first evidence that A_1^n becomes positive at large x, showing clear SU(6) symmetry breaking. The results are in good agreement with the LSS 2001 pQCD fit to previous world neutron data (Leader *et al.* 2002) (long-dashed curve) and the statistical model (Bourrelly *et al.* 2002) (dotted curve). The trend of the data is consistent with the RCQM predictions (the shaded band). The data disagree with the predictions from the leading-order pQCD models (short-dashed and dash-dotted curves).

Assuming the strange sea quark contributions are negligible in the region $x > 0.3$, the polarised quark distribution functions $\Delta u/u$ and $\Delta d/d$ were extracted from our neutron data combined with the world proton data. The results are shown in the right panel of Figure 1, along with predictions from the RCQM (full curves), leading-order pQCD (short-dashed curves), the LSS 2001 fits (long-dashed curves) and the statistical model (dotted curves). The results agree well with RCQM predictions as well as the LSS 2001 fits and statistical models but are in significant disagreement with the predictions from the leading-order pQCD models assuming hadron helicity conservation. This suggests that effects beyond leading-order pQCD, such as the quark orbital angular momentum, may play an important role in this kinematic region.

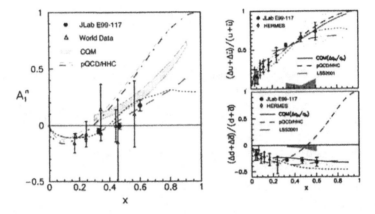

Figure 1. *The new results for A_1^n (left) and $\Delta u/u$ and $\Delta d/d$ (right), compared with existing data and theoretical predictions (see text for further details).*

Acknowledgments

The work presented was supported by the U. S. Department of Energy (DOE) contract DE-AC05-84ER40150 Modification NO. M175, under which the Southeastern Universities Research Association operates the Thomas Jefferson National Accelerator Facility.

References

Bissey F *et al.* , 2002, *Phys Rev C* **65** 064317.
Bourrelly C *et al.* , 2002, *Eur Phys J C* **23** 487.
Leader E *et al.* , 2002, *Eur Phys J C* **23** 479.
Zheng X *et al.* , 2004, *Phys Rev Lett* **92** 012004.

Simulation of the DVCS-TOF detector for COMPASS

Michael Seimetz

DSM/DAPNIA/SPhN
CEA Saclay, 91191 Gif sur Yvette, France

2   Aim of the simulation

Deeply Virtual Compton Scattering (DVCS) offers experimental access to Generalised Parton Distributions (GPD), and thereby to a detailed understanding of the partonic phase space of the nucleon. A planned DVCS measurement at the COMPASS facility at CERN (d'Hose 2004) in the exclusive channel $\mu p \to \mu' \gamma p$ requires the construction of an additional proton detector in the target region.

The experimental concept foresees the use of two concentric barrels of plastic scintillators allowing the determination of the proton momentum with the time-of-flight technique. Apart from high rate capability, these large-area scintillators must have an excellent timing resolution of less than 200 ps. The feasibility of these ambitious goals will be demonstrated with a prototype of this DVCS-TOF detector until 2006. As a first step, we have performed a simulation of the production of fluorescence light, and its propagation inside the scintillators and light guides using the program LITRANI (Gentit) which we have extended by several new methods.

3   Physical input of the simulation

The energy deposition of protons inside the scintillator material BC 408 is described using Landau's universal distribution function (Wilkinson 1996). The scintillation efficiency is a function of the proton kinetic energy (Cecil et al. 1979). A realistic description of light production is achieved by parameterising the timing properties of BC 408 in terms of rise and decay times (Nam 2002). Photons are produced according to the spectral emittance of the scintillator material, with a peak at 425 nm. Similarly, the spectral quantum efficiency of the photocathodes is taken into account to model photon detection.

Furthermore, new light guide geometries have been defined. Twisted strips of plexiglass are used as adapters between the end planes of flat and broad scintillators and the entrance faces of photomultipliers. In our simulation these strips are made of a large number of short segments, where each of the latter is twisted by a small angle around its longitudinal axis. The side planes are subdivided into flat triangles. In addition it is possible to require an overall 90° bend of the entire guide.

We have tested our program via comparison to published experimental data (Smith *et al.* 1999) and other simulations (Massam 1977). The numbers of detected photoelectrons and attenuation lengths are in reasonable agreement. Transmission properties of curved light guides are also well reproduced if their shape is approximated with a sufficiently large number of segments.

4 Applications to the DVCS-TOF detector

As first application to our experimental setup we have calculated the transmission properties of various possible light guides for the scintillators of layers A and B. Optimum lengths for both twisted-strip and troncoidal configurations have been determined where the number of transmitted photons, originating from a δ-shaped pulse at the centre of the scintillator, shows a maximum, and the arrival time of the photons at the photocathode peaks within the first 2 ns.

Using these geometries, photomultiplier signals produced by protons crossing the scintillators at different hit positions, angles, and with different momenta in the range 0.3-0.8 GeV/c have been simulated. They differ considerably both in shape as well as in the integrated number of photoelectrons. Therefore an analysis of the pulse form will be necessary to discriminate protons from background events, and to achieve the desired timing resolution. An Analogue Ring Sampler developed at Saclay for the ANTARES experiment (Feinstein 2003) is well suited for this purpose.

Acknowledgments

This work has been supported by the European Union's Sixth Framework Program.

References

Cecil R A *et al.* , 1979, *Nucl Instrum Methods* **161** 439.
d'Hose N, 2004, Contribution to these proceedings.
Feinstein F, 2003, *Nucl Instrum Methods A* **504** 258.
Gentit F-X, LITRANI, http://gentit.home.cern.ch/gentit/litrani/.
Massam Th, 1977, *Nucl Instrum Methods* **141** 251.
Nam J W *et al.* , 2002, *Nucl Instrum Methods A* **491** 54.
Smith E S *et al.* , 1999, *Nucl Instrum Methods A* **432** 265.
Wilkinson D H, 1996, *Nucl Instrum Methods A* **383** 513.

The ALICE transition radiation detector readout electronics

Jorge Mercado

Physikalisches Institut der Universität Heidelberg
Philosophenweg 12, D-69120 Heidelberg, Germany

1 Introduction

ALICE (A Large Ion Collider Experiment) is the only dedicated heavy ion experiment at the CERN Large Hadron Collider (LHC) where Pb nuclei will be accelerated up to $\sqrt{s} = 5.5$ TeV per nucleon pair. Within ALICE, a transition radiation detector (TRD) has been designed in order to provide electron identification in the central barrel at momenta in excess of 2 GeV/c as well as fast (6 μs) triggering capability for high transverse momentum ($p_t \geq 3$ GeV/c) processes.

2 The ALICE TRD

The ALICE TRD will cover the pseudorapidity range $|\eta| \leq 0.9$ with a total of 540 individual detectors arranged in 6 radial layers covering a total area of 736 m^2. The layers are subdivided into 18 azimuthal and 5 longitudinal sections (CERN/LHCC 2001-021). Each detector consists of a sandwich radiator, a combination of Rohacell$^©$ and polypropylene fibre mats of 48 mm thickness; it is followed by a drift chamber with a 30 mm drift gap and a 7 mm amplification gap read out via a segmented cathode pad plane glued to a multi-layer carbon fibre honeycomb backing. The chambers are operated with a Xe/CO_2 (85:15) mixture (total volume about 28 m^3) in order to achieve a high conversion probability for transition radiation. The readout electronics of the 1.2 million individual pads is mounted directly on the back of the detectors. The anticipated total radiation thickness of all six layers of the detectors is approximately 15%.

3   The ALICE TRD readout electronics

Beyond the 1.2 million analog channels which are digitised during the 2 μs drift time, the TRD also implements an on-line trigger which is capable of tracking all of the up to 16,000 charged particles within the six detector layers with a very tight time budget of 6 μs for all digitisation and processing.

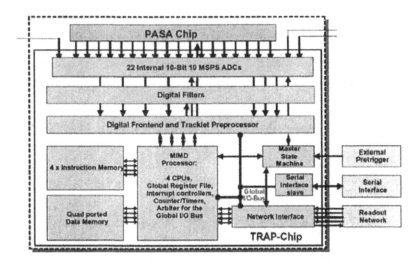

Figure 1. *Architecture of the various TRD on-detector electronics building blocks.*

The on-detector pads feed a charge-sensitive preamplifier (PASA) whose noise is determined by its input capacity, therefore requiring its proximity to the pad planes. The preamplifier also implements first-level shaping and tail cancellation functionality. Here, the greatest design effort is directed towards noise reduction. The differential PASA outputs are digitised with a custom 10-bit ADC at 10 MHz. The remainder of the TRD electronics chain implements a short 64-word single event buffer plus a tracklet processor (TRAP), which identifies potential high-p_t track candidates for further processing.

The read-out is performed in two stages: first, during the trigger processing, where all tracklet candidates are shipped within 600 ns from the 65664 TRAP chips (Figure 1) to the global tracking unit (GTU) for merging of the six detector layers; second, the event buffer is read out in case of an accept of the event (Lindenstruth and Musa 2004). The high speed TRD read-out is performed with 1080 2.5 GBit optical links. The appropriate serialiser chips are designed in full custom for on-detector operation and are attached to the TRAP chips. They form the interface between the on-detector electronics and the off-detector GTU.

References

CERN/LHCC, 2001, ALICE TRD Technical Design Report, CERN 2001-021.
Lindenstruth V and Musa L, 2004, *Nucl Instrum Methods A* **522** 33.

A fibre tracker with particle identification for the HERMES Recoil Detector

Matthias Hoek

II. Physikalisches Institut
University of Giessen, D-35392 Giessen, Germany

1 Introduction

The HERMES Recoil Detector aims at detecting recoiling nucleons of exclusive processes, *eg* Deeply Virtual Compton Scattering or exclusive ρ production. The scintillating fibre tracker (SFT), which is part of the Recoil Detector, allows the momentum reconstruction of these particles in the momentum range from 250 MeV/c up to 1400 MeV/c (Seitz 2004). In addition, particle identification capabilities are required to reject positively charged pions from intermediate Δ-decays.

2 Design and construction

The HERMES Recoil Detector consists of a silicon detector, the SFT and a photon detector, surrounded by a superconducting solenoid. The magnetic field enables momentum reconstruction by measuring the bending of the tracks. The SFT consists of two concentric barrels of radii of 109 and 183 mm. Each barrel consists of two times two fibre layers with stereo angles of 0° and 10° each, allowing the reconstruction of the space point of the particle track. Kuraray SCSF-78 fibres were chosen, due to their superior light yield and radiation hardness. A diameter of 1 mm was chosen to match the tracking resolution to contributions from multiple scattering. Each barrel layer is assembled using pre-shaped modules, which consist of 64 fibres arranged in two layers. The downstream fibre end-faces are machined to optical quality and aluminised. The reflecting end-face enhances the light yield by 20–30%. The upstream fibre end-faces are glued into customised connectors and machined to optical quality as well. The barrels are read out with Hamamatsu H-7548 64 channel PMTs via 4 m long light guides made of clear fibres.

3   Performance

The SFT module response was investigated in a test beam experiment at GSI (Murray 2004). Taking cross talk on the PMT cathode into account, a clustering algorithm was developed, combining adjacent PMT channels above a threshold of 1.0 photo electron (p.e.) to clusters. The energy response was extracted for pions and protons at various momenta, shown in Figure 1. The mean signal for minimum ionising particles (MIP) corresponds to 7.2 p.e., in agreement with previous simulations.

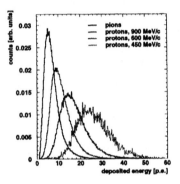

Figure 1. *Energy response of a scintillating fibre module*

A PID algorithm allows proton/pion separation up to 600 MeV/c with a pion contamination of less than 10% (Sommer 2003). The detection efficiency per layer was measured to exceed 98% for MIPs. First tracking studies indicate a resolution of a two-layer module of $<300\,\mu m$.

4   Outlook

The SFT is currently being assembled and will be tested and aligned in a dedicated test beam experiment at DESY. Prototypes tested at GSI fulfilled the design criteria. The complete Recoil Detector will be installed in the HERMES experiment in 2005 for a data taking period of two years.

References

Murray M, 2004, Contribution to these proceedings.
Seitz B, 2004, Contribution to these proceedings.
Sommer W, 2003, Diplom Thesis, University of Giessen, Germany.

Silicon Detector in HERMES

Ivana Hristova

DESY,
15738 Zeuthen, Germany

1 Introduction

In 2005, after ten years of running with a polarised gas target, the HERMES experiment (Ackerstaff *et al.* 1998) will replace the target cell with a new one surrounded by a Recoil Detector for large angle tracking, to explore the physics of Generalised Parton Distributions (GPDs). The Recoil Detector and GPDs are introduced elsewhere in these proceedings.

Recoiling protons with low momenta from Deeply Virtual Compton Scattering will be measured with a silicon strip detector (SSD) placed inside the HERA ring vacuum. Two sensor layers separated by 1.5 cm will surround the target cell in a diamond shape. With a total surface area of 0.16 m^2 and 8192 readout channels, the SSD will determine the momenta of protons between 0.1–0.5 GeV/c from the ionisation energy loss in the sensors. For higher energy particles two space points for tracking will be provided.

Due to the small scale of the project and the relatively small amount of time and manpower, an existing tested design for the silicon sensor (TIGRE) and the readout chip (HELIX) are used. To increase the dynamic range from 40 fC to the required 280 fC each sensor strip is read out by two HELIX channels, one of which is connected via a charge division capacitor (Reinecke *et al.* 2004). The control electronics and the ADC are adopted from the Lambda wheel detector (Steijger 2000). All silicon detector modules are produced and currently tested with laser, slow protons and 1 MIP electron beams.

2 The silicon detector–module design and assembly

The silicon sensors, supplied by Micron Semiconductor Ltd, are made of 6-inch n-type silicon with $\langle 1\text{-}0\text{-}0 \rangle$ crystal orientation and resistivity of a few kΩcm. The wafers are double-sided single-metal, with perpendicular strips, 300 ± 15 μm thick and the strip implant is achieved by Boron/Phosphorus-doping on the p/n-side.

Strips are 702 μm wide with 1 μm implant depth and 56 μm separation. The n-side strips are separated by p-stop implants. The 97.3 × 97.3 mm^2 active area (128 strips on each side) is surrounded by guard and bias rings. As significant background-induced bias currents are expected, the polysilicon bias resistance for a single strip is only a few MΩ. The coupling capacitance is realised by a 0.8 μm SiO$_2$ layer. Strips are covered by a 1.75 μm aluminium grid-shaped layer which yields the design capacitance of 1 nF. Each strip has 2 bond pads for AC-coupling and is supported by 0.8 μm of polysilicon in order to avoid penetrating the SiO$_2$ dielectric. An oxide passivation layer of 0.5 μm (up to 1.3 μm between the strips) covers almost all of the surface to protect the sensor. For each sensor I/V–C/V scans, strip and stability tests were performed. The typical measured values are 50 V full depletion voltage, 6.5/5.9 MΩ bias resistance, 9/7 pF interstrip and 34/54 pF total strip capacitance for p/n-side.

To place the readout electronics at one end of a detector module, external 50 μm thick polyimide foils (flexleads) are used to bring the p- and n-side readout lines parallel to each other. The flexleads have one wiring layer with a 5 μm copper metallisation, onto which 5 μm nickel and 0.1 μm gold is grown chemically. The traces have widths and minimum distances of 70 μm. For one module two silicon sensors are glued into a 2 mm thick support frame and the flexleads are then glued to the frame. The frame is made of aluminium-nitride ceramic (Shapal M). The connection between sensors and flexleads is realised by ultrasonic bonding with 17.5 μm thick aluminium wires. The hybrid, which hosts the readout and control electronics of a module, is a four-layer design with polyimide (kapton) cores. The total thickness is 200 μm, of which 65 μm are from the four wiring layers and the rest are the sum of the kapton cores. Two hybrids (for p- and n-side strips) are glued to both sides of a 500 μm thick aluminium heatsink. The upper two layers of the hybrid are lead out as a 25 cm long flexible part for the interface to the outer electronics. HELIX128-3.0 was chosen as the analog readout chip. To extend the dynamic range an array of 256 charge division capacitors and 4 chips are required per hybrid. Thin-film pitch adapters (Siegert TFT) are used to widen the chip pitch from 41.4 μm to 182 μm. In addition the hybrid contains a radiation monitor (NMRC, Ireland), a temperature sensor (P1000), LVDS receivers (DS90lv019) for digital control, and an analog line driver (MAX435) for the output signals. The passive components are assembled in an automatic reflow solder process. The active components are glued to the hybrid and wire-bonded resulting in about 1500 bonds per hybrid. The silicon sensors and the hybrids are tested separately and afterwards both parts are connected together.

Acknowledgments

I gratefully acknowledge the support of I-M Gregor, R Kaiser, M Kopytin, B Krauss, W Lange, W-D Nowak, M Reinecke, J Stewart, A Vandenbroucke and C Vogel.

References

Ackerstaff K *et al.* , 1998, *Nucl Instrum Methods A* **417** 230.
Reinecke M *et al.* , 2004, *IEEE Trans Nucl Sci* **59** 1111.
Steijger J J M, 2000, *Nucl Instrum Methods A* **447** 55.

Track reconstruction for the HERMES Recoil Detector

Andrew Osborne

Department of Physics and Astronomy, University of Glasgow,
Glasgow, G12 8QQ, United Kingdom

The Recoil Detector is the latest planned upgrade to the HERMES experiment at DESY, Hamburg, and will enable event-level exclusivity in the study of Deeply Virtual Compton Scattering (DVCS) processes in which a proton and a virtual photon with high energy and virtuality interact to produce a real photon and the intact target proton. Currently the missing mass resolution of the HERMES spectrometer is insufficient to identify these exclusive events individually and to separate them from events in which the real photon is accompanied by a Δ resonance instead of a proton. As a result exclusivity can only be established at the level of a data sample. The aim of the Recoil Detector is to establish exclusivity on a per-event-basis by detecting all reaction products. Additionally, the Recoil Detector will directly measure the proton momentum, achieving an improvement in t-resolution by about an order of magnitude for small values of t.

Low energy proton and pion momentum (110-500MeV/c) reconstruction is provided in the Recoil Detector by a silicon detector used as a dE/dx telescope. Higher energy proton momenta (300-1200MeV/c) are measured by a scintillating fibre (SciFi) detector through bending of the track in a 1 Tesla magnetic field. Overall, these detector systems provide a total of only four detector planes in which to measure the particle path. This low multiplicity presents a challenging environment for tracking. In addition, the inhomogeneous magnetic field distorts the shape of the track from the usual helical form.

Tracking in the Recoil Detector is split into a track finding step and a track fitting step. Despite $\vec{B}$ field inhomogeneity, current Monte Carlo studies assert that a homogeneous approximation suffices for both tasks. At present, however, field map resolution in the Monte Carlo is rough at 2cm per bin, around 10% of the total tracking radius. The track finder makes the assumption of a helical track shape, and searches the hit database for those hits which in their proper combination satisfy the equation of a helix. Constraints placed on parameters such as the helix radius can be varied (at run-time if necessary) depending on the quality and noise level of the environment.

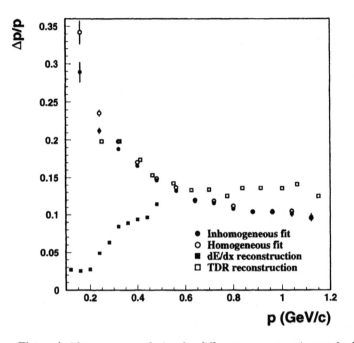

Figure 1. *Momentum resolution for different reconstruction methods.*

Fitting performance (Figure 1) nonetheless improves with an inhomogeneous fit. For fitting the original 2cm field is interpolated using a spline fit to obtain a field in 1cm bins. Further investigations have shown that as interpolation improves field resolution, fitting performance worsens; a finer (and consequently more realistic) field map is planned for use in the Monte Carlo in order to carry out a fair comparison. A generalised method of performing field map look-ups is currently under development; these libraries will provide a simple way of obtaining consistent field map measurements between the Monte Carlo and the tracking.

Further Monte Carlo studies will determine how much the track is affected by the inhomogeneity and if a homogeneous assumption is sufficient. Although tracking algorithms are in a state where reconstruction can be performed given either outcome, the inhomogeneous fit has the added advantage of a natural accommodation of energy loss and scattering, the effects of which are still being investigated.

References

Bugge L and Myrheim J, 1981, *Nucl Instrum Methods* **179** 365.
Hart J C and Saxon D H, 1984, *Nucl Instrum Methods* **220** 309.
HERMES collaboration, Kaiser R *et al.*, 2002, DESY-PRC 02-01.
Saxon D H, 1985, *Nucl Instrum Methods A* **234** 258.
Seitz B, Contribution to these proceedings.

Test beams used in the HERMES Recoil Detector project

Morgan Murray

Department of Physics and Astronomy, University of Glasgow,
Glasgow, G12 8QQ, United Kingdom

1 Introduction

The HERMES Recoil Detector project uses three testing facilities: GSI, near Frankfurt; Erlangen Tandem at Erlangen University and the T22 test beam at DESY, Hamburg.

2 GSI test facility

The purpose of the GSI test facility was three-fold: firstly to check the response of the silicon sub-detector to particles of a non-normal incidence; secondly to test the efficiency of the scintillating fibre (SciFi) sub-detector and thirdly to test the particle identification (PID) capabilities of the SciFi detector (Hoek 2004). The GSI beam provided a mixed proton/pion beam in the momentum range 300 to 900 MeV/c. A separate trigger system was set up alongside the tested sub-detectors and this was also used as a time-of-flight system. A silicon sub-detector module was placed on an X-Yϕ table and rotated between $-45°$ and $+45°$ with respect to the beam. A calibration was made between angle and signal strength for the silicon detector and the efficiency of the SciFi sub-detector was >95% for both types of particles in the momentum range at which this sub-detector will be responsible for tracking in the final setup. The PID was also within design parameters.

3 Erlangen test facility

The Erlangen Tandem accelerator supplies a constant flow of slow protons in the energy range of 4-9 MeV to a silicon module held in vacuum with a water cooling system.

The setup is used to calibrate the silicon sub-detector's response to a range of energies, corresponding to protons with a momentum of 80-130 MeV/c. Data output is analysed in ROOT, but calibration data is similar in format to that of the T22 facility.

4 Test beam T22 at DESY, Hamburg

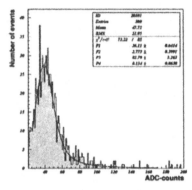

Figure 1. *A typical analysed output from the TESA program. In this example, the peak of the Landau-Gauss convolution is at 52 ADC channels, corresponding to 0.128V.*

The T22 test beam at DESY, Hamburg, provides a bunched current of e^+/e^- in a user-definable energy range of 1-6 GeV. This range is important for the testing of the silicon sub-detector of the Recoil Detector, as it corresponds to a minimum ionising particle (MIP) for silicon. The module under test sits inside of a three-detector reference telescope built to test the ZEUS Micro Vertex Detector (MVD) (Gregor *et al.* 2004). The readout is performed by a Labview program, also adapted from the MVD test setup. Data is read into an n-tuple data format structure and analysed by a set of PAW macros.

The analysed data is presented as a set of graphs, plotting the number of observed events versus ADC channel. The distribution is a Landau-Gauss convolution, the peak of which has a direct correlation with the energy deposited in the silicon (PDG 2004). For our detector, using MIPs, this energy deposition is about 116 keV.

Acknowledgments

I gratefully acknowledge my "Recoil" colleagues who contributed information for this talk: M Hartig, M Hoek, I Hristova, M Reinecke, B Seitz, J Stewart and C Vogel.

References

Gregor I M *et al.* , 2004, HERMES 04-019.
Hoek M, 2004, Contribution to these proceedings.
Particle Data Group, 2004, *Phys Lett B* **592** 1.

Hadron physics summer school and workshop

List of participants

Sergey S Afonin,
V A Fock Department of Theoretical Physics, St Petersburg State University,
1 ul Ulyanovskya, 198 504, St Petersburg, Russia,
Sergey.Afonin@pobox.spbu.ru

Conrado Albertus,
Departamento de Física Moderna, Universidad de Granada,
E-18071 Granada, Spain,
albertus@ugr.es

Constantia Alexandrou,
Department of Physics, University of Cyprus,
CY-1678 Nicosia, Cyprus,
alexand@ucy.ac.cy

Norbert Attig,
John von Neumann Institut für Computing, Forschungszentrum Jülich,
52425 Jülich, Germany,
n.attig@fz-juelich.de

Gunnar Bali,
Department of Physics and Astronomy, University of Glasgow,
Glasgow G12 8QQ, UK,
g.bali@physics.gla.ac.uk

Olivia Bartholomy,
Helmholtz - Institut für Strahlen- und Kernphysik, Universität Bonn,
Nussallee 14-16, D-53115 Bonn, Germany,
bartholomy@hiskp.uni-bonn.de

Stanislav Belostotski,
Petersburg Nuclear Physics Institute,
St Petersburg, Gatchina, 188350 Russia,
belostot@hermes.desy.de

Mike Birse,
Department of Physics and Astronomy, University of Manchester,
Manchester M13 9PL, UK,
mike.birse@man.ac.uk

Sigfrido Boffi,
Dipartimento di Fisica Nucleare e Teorica, Università degli Studi di Pavia and INFN,
Sezione di Pavia, Pavia, Italy,
Sigfrido.Boffi@pv.infn.it

Cedran J Bomhof,
Department of Physics and Astronomy, Vrije Universiteit,
Amsterdam, De Boelelaan, NL-1081 HV Amsterdam,The Netherlands,
cbomhof@nat.vu.nl

Derek Branford,
Department of Physics and Astronomy, University of Edinburgh,
Edinburgh EH9 3JZ, UK,
db@ph.ed.ac.uk

Stanley J Brodsky,
Stanford Linear Accelerator Center, Stanford University,
Stanford, California 94309, USA,
sjbth@slac.stanford.edu

Grzegorz Brona,
Faculty of Physics, Warsaw University,
Hoża 69, PL-00-681 Warsaw, Poland
grzegorz.brona@puw.edu.pl

Matthias Burkardt,
Department of Physics, New Mexico State University,
Las Cruces, NM 88003, USA
burkardt@nmsu.edu

Volker D Burkert,
Thomas Jefferson National Accelerator Facility,
12000 Jefferson Ave, Newport News, VA 23606, USA,
burkert@jlab.org

Frank E Close,
Rudolf Peierls Centre for Theoretical Physics, University of Oxford,
1 Keble Road, Oxford OX1 3NP, UK,
f.close1@physics.ox.ac.uk

Richard Codling,
Department of Physics and Astronomy, University of Glasgow,
Glasgow G12 8QQ, UK,
r.codling@physics.gla.ac.uk

Philip L Coltharp,
Department of Physics, Florida State University,
315 Keen Building, Talahassee, FL 32306-4350, USA
coltharp@hadron.physics.fsu.edu

Nigel Cundy,
Department of Physics, Bergische Universität Wuppertal,
Gaussstrasse 20, D-42119 Wuppertal, Germany,
cundy@theorie.physik.uni-wuppertal

Christine T H Davies,
Department of Physics and Astronomy, University of Glasgow,
Glasgow G12 8QQ, UK,
c.davies@physics.gla.ac.uk

Kees de Jager,
Thomas Jefferson National Accelerator Facility,
12000 Jefferson Ave, Newport News, VA 23606, USA,
kees@jlab.org

Sara Della Monaca,
Dipartimento Fisica, Università degli studi di Trento,
Via Sommarive 14, I-38050 POVO, Italy,
dmonaca@science.unitn.it

Nicole d'Hose,
DSM/DAPNIA/SPhN,
F-91191 Gif-sur-Yvette Cedex, France,
ndhose@cea.fr

Evangeline Downie,
Department of Physics and Astronomy, University of Glasgow,
Glasgow G12 8QQ, UK,
e.downie@physics.gla.ac.uk

Jozef Dudek,
Rudolf Peierls Centre for Theoretical Physics,
1 Keble Road, Oxford, OX1 3NP, UK
dudek@thphys.ox.ac.uk

Andrey Elagin,
Joint Institute for Nuclear Research (JINR),
Joliot-Curie 6, 141980, Dubna, Moscow Region, Russia,
elagin@nusun.jinr.ru

Kathrin Essig,
Helmholtz - Institut für Strahlen- und Kernphysik, Universität Bonn,
Nussallee 14-16, D-53115 Bonn, Germany,
essig@hiskp.uni-bonn.de

Laura Fabbietti,
Physik-Department E12, Technische Universität München,
James-Franck-Strasse, D-85748 Garching, Germany,
Laura.Fabbietti@physik.tu-muenchen.de

Angelo Raffaele Fazio,
Joint Institute for Nuclear Research (JINR),
Joliot-Curie 6, 141980, Dubna, Moscow Region, Russia,
raffaele@thsun1.jinr.ru

Teresa Fernández-Caramés,
Departament de Física Tèorica, Facultat de Física, Universitat de València,
C Dr Moliner 50, E-46100, Burjassot, València, Spain
teresa.fernandez@uv.es

Fabrizio Ferro,
Dipartimento di Fisica, Politecnico di Torino,
10129 Torino, Corso Duca degli Abruzzi 24, Italy,
fabrizio.ferro@polito.it

Klaus Foehl,
Department of Physics and Astronomy, University of Edinburgh,
Edinburgh EH9 3JZ, UK,
k.foehl@ph.ed.ac.uk

Léon Gaillard,
School of Physics and Astronomy, University of Birmingham,
Birmingham B15 2TT, UK,
lxg866@bham.ac.uk

Yury Gavrikov,
Petersburg Nuclear Physics Institute,
St Petersburg, Gatchina, 188350 Russia,
gury@pnpi.spb.ru

Linda Gerén,
Department of Physics, University of Stockholm,
SE-106 91 Stockholm, Sweden,
geren@physto.se

Derek I Glazier,
Department of Physics and Astronomy, University of Glasgow,
Glasgow G12 8QQ, UK,
d.glazier@physics.gla.ac.uk

Meinulf Göckeler,
Institut für Theoretische Physik, Universität Regensburg,
Universitätsstrasse 31, 93053 Regensburg, Germany,
meinulf.goeckeler@physik.uni-regensburg.de

Nikolay Gromov,
Faculty of Physics, St Petersburg State University,
Ulianovskaya 1, 198 904, St Petersburg, Russia,
nik.gromov@pobox.spbu.ru

Eric Gutz,
Helmholtz - Institut für Strahlen- und Kernphysik, Universität Bonn,
Nussallee 14-16, D-53115 Bonn, Germany,
gutz@hiskp.uni-bonn.de

Brian Hannafious,
Department of Physics and Astronomy, University of New Mexico,
Albequerque, NM 87131, USA,
Physics@nmsu.edu

Thomas Hemmert,
Physik Department, Technische Universität München,
D-85747 Garching, Germany,
themmert@physik.tu-muenchen.de

Matthias Hoek,
Physikalisches Institut, Universität Giessen,
35392 Giessen, Germany,
Matthias.Hoek@physik.uni-giessen.de

Roger Horsley,
Department of Physics and Astronomy, University of Edinburgh,
Edinburgh EH9 3JZ, UK,
rhorsley@ph.ed.ac.uk

Paul Hoyer,
Department of Physical Sciences & Helsinki Institute of Physics, University of Helsinki,
POB 64, FIN-00014 Helsinki, Finland,
paul.hoyer@helsinki.fi

Ivana Hristova,
DESY,
15738 Zeuthen, Germany,
ivana.hristova@desy.de

Boris L Ioffe,
Institute for Theoretical and Experimental Physics,
Bolshaya Cheremushkinskaya 25, 117 218 Moscow, Russia,
ioffe@itep.ru

David G Ireland,
Department of Physics and Astronomy, University of Glasgow,
Glasgow G12 8QQ, UK,
d.ireland@physics.gla.ac.uk

Matti Järvinen,
Department of Physical Sciences & Helsinki Institute of Physics, University of Helsinki,
POB 64, FIN-00014 Helsinki, Finland,
matti.o.jarvinen@helsinki.fi

Ralf Kaiser,
Department of Physics and Astronomy, University of Glasgow,
Glasgow G12 8QQ, UK,
r.kaiser@physics.gla.ac.uk

James D Kellie,
Department of Physics and Astronomy, University of Glasgow,
Glasgow G12 8QQ, UK,
j.kellie@physics.gla.ac.uk

Markus Kloker,
Institut für Theoretische Physik, Universität Tübingen,
Auf der Morgenstelle 14, D-72076 Tübingen, Germany,
kloker@tphys.physik.uni-tuebingen.de

Tamas Kovacs,
Department of Physics, Bergische Universität Wuppertal,
Gaussstrasse 20, D-42119 Wuppertal, Germany,
kovacs@theorie.physik.uni-wuppertal.de

Stefan Krieg,
Department of Physics, Bergische Universität Wuppertal,
Gaussstrasse 20, D-42119 Wuppertal, Germany,
krieg@theorie.physik.uni-wuppertal.de

Anna Kropivnitskaya,
Institute for Theoretical and Experimental Physics,
Bolshaya Cheremushkinskaya 25, 117 218 Moscow, Russia,
kropiv@mail.desy.de

Jan Kuckei,
Institut für Theoretische Physik, Universität Tübingen,
Auf der Morgenstelle 14, D-72076 Tübingen, Germany,
kuckei@tphys.physik.uni-tuebingen.de

Jeffrey Lachniet,
Department of Physics, Carnegie Mellon University,
Pittsburgh, PA 15213, USA,
lachniet@ernest.phys.cmu.edu

Tim Ledwig,
Institut für Theoretisch Physik II, Ruhr-Universität Bochum,
D-44780 Bochum, Germany,
tim.ledwig@tp2.ruhr-uni-bochum.de

Thomas Lippert,
John von Neumann Institut für Computing, Forschungszentrum Jülich,
52425 Jülich, Germany,
th.lippert@fz-juelich.de

Agnes Lundborg,
Department of Radiation Sciences, University of Uppsala,
Box 535, 751 21 Uppsala, Sweden
agnes.lundborg@tsl.uu.se

I J Douglas MacGregor,
Department of Physics and Astronomy, University of Glasgow,
Glasgow G12 8QQ, UK,
i.macgregor@physics.gla.ac.uk

Nicolas Matagne,
Institute of Physics B5, Université de Liège,
Sart Tilman, B-4000 Liège, Belgium,
nmatagne@ulg.ac.be

Vincent Mateu,
Departamento de Física Téorica and IFIC, Centro Mixto Universidad de Valencia-CSIC,
Institutos de Investigación de Paterna, Aptd 22085, 46071 Valencia, Spain,
mateu@ific.uv.es

Judith McGovern,
Department of Physics and Astronomy, University of Manchester,
Manchester M13 9PL, UK,
Judith.McGovern@man.ac.uk

Bryan McKinnon,
Department of Physics and Astronomy, University of Glasgow,
Glasgow G12 8QQ, UK,
b.mckinnon@physics.gla.ac.uk

Jorge Mercado,
Physickalisches Institut, Ruprecht-Karls-Universität Heidelberg,
Philosophenweg 12, D-69120 Heidelberg, Germany,
mercado@uni-heidelberg.de

Volker Metag,
II Physikalisches Institut, Universität Giessen,
D-35392, Giessen, Germany,
Volker.Metag@exp2.physik.uni-giessen.de

Bernard Metsch,
Helmholtz - Institut für Strahlen- und Kernphysik, Universität Bonn,
Nussallee 14-16, D-53115 Bonn, Germany,
metsch@itkp.uni-bonn.de

Morgan Murray,
Department of Physics and Astronomy, University of Glasgow,
Glasgow G12 8QQ, UK,
m.murray@physics.gla.ac.uk

Samson Keleta Negasi,
Department of Radiation Sciences, University of Uppsala,
Box 535, 751 21 Uppsala, Sweden
samson@tsl.uu.se

Wolf-Dieter Nowak,
DESY,
22603 Hamburg, Germany,
nowakw@ifh.de

Boris Orth,
Department of Physics, Bergische Universität Wuppertal,
Gaussstrasse 20, D-42119 Wuppertal, Germany,
orth@theorie.physik.uni-wuppertal.de

Andrew Osborne,
Department of Physics and Astronomy, University of Glasgow,
Glasgow G12 8QQ, UK,
a.osborne@physics.gla.ac.uk

Henrik Pettersson,
Department of Radiation Sciences, University of Uppsala,
Box 535, 751 21 Uppsala, Sweden
Henrik.Pettersson@tsl.uu.se

Martin Poghosyan,
Faculty of Physics, Yerevan State University,
1 Alex Manoogian Street, 375049 Yerevan, Republic of Armenia
mpogos@server.physdep.r.am

Sebastien Procureur,
DSM/DAPNIA/SPRN,
F-91191 Gif-sur-Yvette Cedex, France,
sebastien.procureur@cern.ch

Dan Protopopescu,
Department of Physics and Astronomy, University of Glasgow,
Glasgow G12 8QQ, UK,
d.protopopescu@physics.gla.ac.uk

Paul E L Rakow,
Department of Mathematical Sciences, University of Liverpool,
Liverpool L69 3BX, UK,
rakow@amtp.liv.ac.uk

Angels Ramos,
Departament d'Estructura i Constituents de la Matèria, Universitat de Barcelona,
Diagonal 647, 08028 Barcelona, Spain,
ramos@ecm.ub.es

Amir H Rezaeian,
Department of Physics, UMIST,
PO Box 88, Manchester M60 1QD, UK,
REZAEIAN@diroc.phy.umist.ac.uk

Günther Rosner,
Department of Physics and Astronomy, University of Glasgow,
Glasgow G12 8QQ, UK,
g.rosner@physics.gla.ac.uk

Giuseppe Russo,
Dipartimento di Fisica e Astronomia dell'Università di Catania,
Corso Italia 57, I-95129 Catania, Italy,
Giuseppe.Russo@ct.infn.it

Dirk Ryckbosch,
Department of Subatomic and Radiation Physics, Universiteit Gent,
Proeftuinstraat 86, 9000 Gent, Belgium,
dirk@inwfsun1.rug.ac.be

Gerrit Schierholz,
DESY,
15738 Zeuthen, Germany,
Gerrit.Schierholz@desy.de

Olaf Scholten,
Kernfysisch Versneller Instituut, University of Gronningen,
9747 AA Groningen, The Netherlands,
Scholten@kvi.nl

Karin Schönning,
Department of Radiation Sciences, University of Uppsala,
Box 535, 751 21 Uppsala, Sweden
karin.schonning@tsl.uu.se

Michael Seimetz,
DAPNIA/SPhN, CEA Saclay,
F-91191 Gif-sur-Yvette Cedex, France,
mseimetz@cea.fr

Björn Seitz,
II Physikalisches Institut, Universität Giessen,
D-35392, Giessen, Germany,
Bjoern.Seitz@exp2.physik.uni-giessen.de

Adrian Sevcenco,
Institute for Space Sciences,
PO Box MG-6, Ro 76900 Bucharest, Romania,
hasegan@venus.nipne.ro

Sergey Slizovskiy,
Faculty of Physics, St Petersburg State University,
Ulianovskaya 1, 198 904, St Petersburg, Russia,
S.Slizovskiy@pobox.spbu.ru

Bogdan Slowinski,
Faculty of Physics, Warsaw University of Technology,
Pl Politechniki 1, 00-661 Warsaw, Poland,
slowb@if.pw.edu.pl

Emil Stan,
Institute for Space Sciences,
PO Box MG-6, Ro 76900 Bucharest, Romania,
hasegan@venus.nipne.ro

Marco Statera,
INFN, Dipartimento di Fisica, Università di Ferrara,
Sezione de Ferrara, 44100 Ferrara, Italy,
Marco.Statera@desy.de

Hinnerk Stueben,
Konrad-Zuse-Zentrum für Informationstechnik Berlin,
D-14195 Berlin, Germany,
stueben@zib.de

Vincent A Sulkosky,
Physics Department, The College of William and Mary,
PO Box 8795, Williamsburg, VA 23187, USA,
nova@camelot.physics.wm.edu

Claire Tarbert,
Department of Physics and Astronomy, University of Edinburgh,
Edinburgh EH9 3JZ, UK,
c.tarbert@ed.ac.uk

Abdullah Shams Bin Tariq,
Department of Physics, Rajshahi University,
Rajshahi 6205, Bangladesh,
asbt@hep.phys.soton.ac.uk

Neil Thompson,
Department of Physics and Astronomy, University of Glasgow,
Glasgow G12 8QQ, UK,
n.thompson@physics.gla.ac.uk

Dmitry Utkin,
Institute for Theoretical and Experimental Physics,
Bolshaya Cheremushkinskaya 25, 117 218 Moscow, Russia,
utkin@itep.ru

Martin G van Beuzekom,
NIKHEF,
PO Box 41882, 1009 DB, Amsterdam, The Netherlands,
martinb@nikhef.nl

Bart Van Overmeire,
Department of Subatomic and Radiation Physics, Universiteit Gent,
Proeftuinstraat 86, 9000 Gent, Belgium,
Bart.VanOvermeire@ugent.be

Denis Veretennikov,
Petersburg Nuclear Physics Institute,
St Petersburg, Gatchina, 188350 Russia,
denis@pnpi.spb.ru

Timciuc Vladlen,
DGAP, Moscow Institute of Physics and Technology,
Institutskii per 9, 141700 Dolgoprudny, Russia,
vladlen-t@mail.ru

Daniel P Watts,
Department of Physics and Astronomy, University of Edinburgh,
Edinburgh EH9 3JZ, UK,
dwatts1@ph.ed.ac.uk

Volker Weinberg,
Institut für Theoretische Physik, Freie Universität Berlin,
D-14196 Berlin, Germany,
volker.weinberg@physik.fu-berlin.de

Wolfram Weise,
Physik Department, Technische Universität München,
D-85747 Garching, Germany,
weise@physik.tu-muenchen.de

Urs Wenger,
NIC/DESY,
15738 Zeuthen, Germany,
urs.wenger@desy.de

Lars Westerberg,
Department of Radiation Sciences, University of Uppsala,
Box 535, 751 21 Uppsala, Sweden
lars.westerberg@tsl.uu.se

Christine Wilson,
Department of Physics and Astronomy, University of Glasgow,
Glasgow G12 8QQ, UK,
c.wilson@physics.gla.ac.uk

Azusa Yamaguchi,
Department of Physics and Astronomy, University of Edinburgh,
Edinburgh EH9 3JZ, UK,
v1ayamag@ph.ed.ac.uk

Zhenyu Ye,
DESY,
22603 Hamburg, Germany,
yezhenyu@desy.de

James M Zanotti,
NIC/DESY,
15738 Zeuthen, Germany,
jzanotti@ifh.de

Hadron physics

Index